P. CAVERS

WEED CONTROL HANDBOOK:
PRINCIPLES

Weed Control Handbook: Principles

ISSUED BY THE

BRITISH CROP PROTECTION COUNCIL

AND EDITED BY

R.J. HANCE
BSc PhD CChem FRSC
Brook Hill Woodstock Oxford

AND

K. HOLLY
BSc PhD
36 Sunderland Avenue,
Oxford

EIGHTH EDITION

BLACKWELL SCIENTIFIC PUBLICATIONS

OXFORD LONDON EDINBURGH

BOSTON MELBOURNE

© 1963, 1965, 1968, 1977, 1982, 1990
Blackwell Scientific Publications
Editorial offices:
Osney Mead, Oxford OX2 0EL
8 John Street, London WC1N 2ES
23 Ainslie Place, Edinburgh EH3 6AJ
3 Cambridge Center, Suite 208
  Cambridge, Massachusetts 02142, USA
107 Barry Street, Carlton
  Victoria 3053, Australia

First published 1958
Second edition 1960
Third edition 1963
Fourth edition 1965
Fifth edition 1968 reprinted 1970
Sixth edition 1977
Seventh edition 1982
Eighth edition 1990

Set by Best-Set Typesetters Ltd, Hong Kong
Printed and bound in Great Britain
at The University Press, Cambridge

DISTRIBUTORS

UK
  Marston Book Services Ltd
  PO Box 87
  Oxford OX2 0DT
  (*Orders:* Tel: 0865 791155
        Fax: 0865 791927
        Telex: 837515)

USA
  Publishers' Business Services
  PO Box 447
  Brookline Village
  Massachusetts 02147
  (*Orders*: Tel: (617) 524-7678)

Canada
  Oxford University Press
  70 Wynford Drive
  Don Mills
  Ontario M3C 1J9
  (*Orders*: Tel: (416) 441-2941)

Australia
  Blackwell Scientific Publications
  (Australia) Pty Ltd
  107 Barry Street
  Carlton, Victoria 3053
  (*Orders*: Tel: (03) 347-0300)

British Library
Cataloguing in Publication Data

Weed control handbook: principles.
  1. Great Britain. Weeds. Control
  measures
  I. Hance, R.J. II. Holly, K.
  III. British Crop Protection Council
  632'.58'0941
  ISBN 0-632-02459-3

Library of Congress
Cataloging-in-Publication Data

Weed control handbook: principles / issued
  by the British Crop Protection Council.
  —8th ed. / edited by R.J. Hance, K.
  Holly.
    p.   cm.
  ISBN 0-632-02459-3
  1. Weeds—Control—Handbooks,
  manuals, etc. 2. Weeds—Control—
  Great Britain—Handbooks, manuals, etc.
  3. Herbicides—Handbooks, manuals,
  etc. I. Hance, R.J. II. Holly, K.
  III. British Crop Protection Council.
  SB611.W354 1990 632'.58—dc20
  89–37187 CIP

Contents

Introduction

This eighth edition of the *Weed Control Handbook* has been considerably revised in response to the substantial technical and conceptual changes that have taken place since the seventh edition was published in 1982. The underlying theme continues, which is to provide basic information about weeds, herbicides and the principles governing weed control and vegetation management which should be of interest not only in the UK and other temperate climatic regions but also elsewhere.

The system of having chapters written by a team of specialists headed by a 'manager' has been changed, each chapter now having an identified author or authors. However, as the Acknowledgements at the end of many of the chapters show, the total number of experts who have contributed remains at the same level as in previous editions. As editors, we have attempted to minimize overlap between chapters but at the same time have tried to ensure that each chapter can be used alone, without reference to the others. We have, of course, not attempted to influence the fundamental viewpoints of authors so the reader will notice differences in approach and emphasis, for example with respect to the use of herbicides. Similarly, some authors have felt that a Further reading list would benefit their chapter while others have not.

In this edition, the sequence of chapters has been changed so that *The properties of herbicides* precedes those that deal with the various aspects of herbicide use and behaviour. Formulation and application of herbicides is now considered separately from the methodology of application. Similarly, the subject of herbicides in plants is now dealt with in two chapters, one concerned with entry and transport and the other with biochemical aspects, while there is a new chapter on plant resistance to herbicides. Weed control in amenity areas is now combined with that in non-agricultural land and a new chapter on the ecological consequences of weed control systems has been introduced. Many other chapters have undergone major revision, particularly those dealing with weed biology, evaluation of new herbicides and with legislative aspects of herbicide use which have changed substantially as a result of the Food and Environment Protection Act 1985. Because of current uncertainties about the organization of state funded research and development in the UK this section has been omitted from the relevant chapter.

As before, chemicals are, where possible, identified by their British Standards Institution (BSI) common names and crops are referred to by their common names, but botanical names are included in the index. There is a departure from previous practice in that, in the interests of ease of reading, weeds are also referred to by their common names except where botanical names are necessary to avoid ambiguity. The botanical names of all weeds mentioned are listed in the Appendix.

On behalf of the British Crop Protection Council we should like to express our gratitude to the time and effort so freely given by the contributors. We also thank Brenda Barnard, Maurice Bone, Andy Smith and Charles Worthing for the figures of chemical structures and the publishers, Blackwell Scientific Publications Ltd, for their co-operation.

Disclaimer

Every effort has been made to ensure that all information in this *Handbook* is correct at the time or writing but the Authors, the Editors, the British Crop Protection Council and the Publishers do not accept liability for any error or omission in the content, or for any loss, damage or any other accident arising from the use of information mentioned herein.

It is essential to follow the instructions on the Approved label before handling, storing or using any crop protection product. Approved 'off-label' uses are made entirely at the risk of the user.

Oxford 1989   R.J. Hance
                                                        K. Holly

1 / The biology of weeds

A.M. MORTIMER

*Department of Environmental and Evolutionary Biology, The University of Liverpool,
PO Box 147, Liverpool L69 3BX*

Introduction

Weeds are easy to define, in one sense. They are *natural hazards* to the activities of man. In this vein the seventh edition of this *Handbook* defined a weed as 'a plant growing where it is not wanted' and the constitution of the European Weed Research Society uses 'any plant or vegetation interfering with the objectives of people'. A shorter expression of the same principle is the widely used phrase 'a plant out of place'. Therefore, whether or not a plant is a weed depends on the circumstances in which it is growing. For example *Cynodon dactylon* is quoted as one of the world's worst weeds but in some places it is better known as a valuable lawn grass (Bermuda grass) or as a hay and fodder crop. Similarly a crop plant which carries over into the next crop in a rotation is a weed. The status of a plant as a weed may vary historically; couch grass is currently a serious weed in the UK but in the 19th century it was regarded sometimes as 'a valuable source of fodder'.

Like insects and disease organisms, weeds may interfere with systems of land management which may have objectives as disparate as crop production or running a railway. Most management systems use monocultures (as in arable crops or forestry) or simple mixtures of species (as in leys) in order to maximize crop yield or profitability and in industrial situations zero vegetation may be desirable. However the 'natural' vegetation of an area would in most cases contain a mixture of species. Therefore ingress of unwanted plants into an area used for crop production or an industrial purpose can be seen as an intrinsic adjustment towards a more 'natural' plant community.

Weed species however are not easy to define in general on the basis of a narrow set of taxonomic, morphological or phenological properties. In consequence, floral guides in the popular and scientific press that group weed species, almost invariably employ a contemporary habitat-based classification that strongly reflects land usage and rarely does one species occupy every classified habitat. Since the world relies on just 12–15 major crop species it is hardly surprising that only some 250 plant species are regarded as important weeds, of which only about 75 are thought to be responsible for 90% of crop losses attributable to weeds. Weed species do have one common characteristic, notably the ability to persist in the face of repeated habitat disturbance and periodic and near total destruction of above ground biomass. This characteristic may be achieved by a variety of means.

Definition of weeds through crop damage

The potential threat that weeds pose to crop production is considerable. However it remains almost impossible to place an exact financial figure on the actual cost incurred by this threat or the actual realized damage because of the difficulties in obtaining appropriate records. It has been suggested that British agriculture has spent in excess of £100 million per annum on herbicides this decade to alleviate damage that would be otherwise caused by weeds and worldwide expenditure on herbicides outstrips that on either insecticides or fungicides.

The financial damage resulting from weed infestations in a crop production system occurs through both yield reduction, and in lowered value (market price) of the crop. There is therefore a quantity and quality component to the damage and simple economics dictate that for cost effective land use, crop yield and price must be maximized whilst the cost of weed control needs to be minimized. Damage that occurs as a commodity unit price loss, which may often be large, rarely bears a relation to the abundance of weeds above a critical threshold. Thus the presence of poisonous seed of black nightshade in peas to be processed for canning, or wild oat seed in cereal grain above legal purity standards or the presence of common ragwort in hay, dramatically reduces the relative worth of the crop.* Quantitative damage on the other hand is related to the abundance of weeds and the proportional damage occurs either directly as yield loss or indirectly in effecting the efficiency of the overall production process. Table 1.1 provides a classification of four land use systems in relation to components of damage. This is simplistic and the distinction between quality and quantity is somewhat arbitrary and dependent upon the land use system. Neither does it pretend to be complete but rather to illustrate the range of types of losses that may occur. Specific examples can be found in later chapters which deal with weed control in particular situations. This form of classification is easiest to envisage and construct for arable and horticultural crops. Here yield loss is proportional (in a specific way) to the abundance (density) of weeds (see later), through either competition for resources for growth or parasitism. Some weeds living in close association with crops may act as alternate hosts for pathogens or pests. In some instances species may be regarded as weeds because they are alternate hosts in the life cycles of plant pathogens for example between barberry and blackstem rust (*Puccinia graminis*) but they are often non-specific. Various species in the family Cruciferae act as host for *Plasmodiophora brassicae* (clubroot) that infects brassicas and many weed species suffer parasitism from nematodes and thus act as secondary hosts in addition to crops. In this way they enable the pest or pathogen to bridge cropping seasons. Similarly legumes and members of the Chenopodiaceae act as hosts for aphids that attack field and broad beans.

Similar counterparts in commodity yield reduction may be envisaged for grassland and for forestry. In forest plantations in particular a very few tree species are cropped and all others are considered as weeds. The overall aim of weed control is to remove competition from weeds until the tree canopy is dense and hence prohibits weed invasion. Weed control practices are therefore particularly focused: (a) upon the control of weeds in nurseries where the aims, and many of the weed species, are very similar to those in market gardening; as well as (b) on the control of shrub and woody weeds during the establishment and early years of plantation growth. Woody and scrub weeds such as gorse, rhododendron, blackberry, broom and bracken have proved major weeds in this latter respect.

In amenity and conservation areas, yield reduction may be seen in different terms as the loss of species of aesthetic or management interest as

* The term 'seed' is used in this chapter and elsewhere to describe the natural dispersal unit and not in a strict botanical sense.

Chapter 1

Table 1.1 The influence of weeds on loss in a range of land use systems

Category of financial loss	Cause of loss to			
	Arable and horticultural crops	Grassland	Forestry	Amenity / Nature Reserves
Quantity:				
Commodity yield reduction	Components of crop yield through competition, weed parasitism Alternate hosts to plant pathogens and viruses and insect pests	Herbage loss through competition Loss of grazing land through invasion Alternate hosts to plant pathogens and viruses and insect pests	Yield loss in nursery and mature plantations Alternate hosts to plant pathogens and viruses and insect pests	Loss of desirable species through long-term competitive exclusion / succession
Production inefficiency	Clean seedbed preparation Increased time and labour in the harvesting process Crop damage in the application of weed control agents	Lowered efficiency of conversion to grazer biomass. Weeds change stock prehension, palatability and digestibility of herbage Physical damage to animal (e.g. by thorns)	Scrub clearance for new plantation Interference with logging operations	Interference with commercial activities Prevention of spread of 'injurious' weeds under legal statute
Quality:				
Commodity price loss	Smaller size, poorer appearance and crop contamination	Poisoning of stock	Reduced appearance and vigour	Amenity / conservation 'value'

weeds invade and as such damage may be more difficult to evaluate financially, although the relative value and earning power of the land may be assessed. In these situations the effect of weeds may be more readily viewed as a loss in quality. Examples of woody weeds in nature reserves include birch species, Scots pine and gorse invading lowland heaths, and hawthorn invading chalk grasslands. In nature reserves in contrast to forestry, there are usually few weed species in comparison to the mixed 'crop' community assemblage and the major objective of weed management is to prohibit the invasion of the community, often at its periphery.

Herbage yield reduction in grassland may arise from competition of weeds with more favoured herbage grasses or by species replacement as grassland regresses to natural pasture. In the worst case weeds, for instance rushes or bracken, may predominate and lead to a loss in absolute area of grazing land.

Weeds also affect the efficiency of the land use system causing opportunity costs as finance is allocated to weed control that could be utilized alternatively. For example the knowledge that weed seeds are present in arable soil or as thorny or poisonous plants in pasture necessitates weed control measures that are prophylactic to the coming crop, seed sown or flock of grazing animals. Whilst an essential part of a good husbandry programme, such control might not be implemented if weeds were absent. In Californian forage and vegetable crops it has been estimated that the cost of tillage for weed control alone far exceeds the losses which are due to animal pests or plant diseases. Reduced efficiency in the production system may also occur if weeds interfere with both mechanical and manual harvesting operations (see for examples Chapters 11, 12 & 13). In the same vein, weeds may interfere with the conversion of biomass by stock, quantitatively by influencing the palatability and digestibility of herbage and qualitatively by reducing the value of the herbage if poisonous plants are present.

Aims of weed control

Weed control may be practised to maximize profit or to minimize risk of damage to the crop. This translates into the deployment of management practices, that either *contain* weeds to levels that do not incur unacceptable economic injury or that aim to *eradicate* weed species from the cropped locality. It would be gratifying to be able to argue that the former strategy is commonplace but there is little evidence at present to suggest that decision making is made on these criteria. This is in part due to risk averseness in farmers and growers but also due to a continuing lack of knowledge of the ecology of weeds. Certainly in the 1970s the relatively low cost of pesticides in overall UK farm budgets may have promoted insurance spraying or spraying aimed at the goal of eradication. These points are developed in later sections of this chapter.

Development of a weed flora

Pre-adaptation, evolution and alien immigration are the three modes by which a weed species may become incorporated into the weed flora of managed plant communities or agro-ecosystems.

Pre-adaptation

Pre-adapted weeds may be defined as those species that are resident within a plant community within dispersal distance of the crop and come to predominate within the crop through a change in management practice. In this circumstance there is no prerequisite for genetic change in the species; rather potential weed species 'wait in the wings for the call to the stage of a crop production system'. Whether or not this call is answered will depend upon the ecology of the non-crop species and the exact nature of crop management practices. It is the combined effects of management practice(s) and the pattern of crop development through time that acts as a template of inter-specific selection resulting in particular species becoming weeds. It is for this very reason that it is not easy to predict which species, from for instance a hedgerow or wasteland, may expand into a newly created, managed habitat. Only by considering the suite of life history characters possessed by a species can any (tentative) prediction be made. One obviously important character is the ability to germinate rapidly after soil disturbance. Many common arable and horticultural weeds (fat-hen, common chickweed and knotgrass) are colonists of bare areas and Neolithic and early agriculture selected such species since their germination is promoted by the simple act of soil disturbance and vegetation destruction in land clearance.

The development of modern agriculture in the 20th century has provided further examples of the ways and rapidity with which species may be incorporated into the weed flora. Before the advent, post-1950, of the extensive use of herbicides against broad-leaved weed species in cereals, grass weeds were reported as much less common than now. The inference from this observation is that removal of dicotyledenous species has in some way, possibly through reduced intra-weed competition, allowed grass weeds to increase. Careful experimentation also has shown that the abundance of sheep's sorrel in hill grassland is certainly regulated by other dicoty-ledenous and grass species. As agriculture has become more mechanized and the size and structure of land holdings increased, particularly in the USA, many weeds have extended their ranges spectacularly whereas others that were formerly widespread have all but disappeared. In Europe and UK improved threshing and seed cleaning procedures, changes in soil fertility and the introduction of new crops have resulted in the decline of species such as pheasant's eye, corncockle, mugwort, rye brome and corn marigold from arable fields. The trend towards reduced cultivations in arable production provides a more recent example of inter-specific selec-tion of pre-adapted species. Barren brome is a grass weed with no per-sistent seed bank whose seedlings germinate close to the soil surface. When deep ploughing occurs, seeds dispersed from hedgerow populations into the edge of the crop are buried and seedlings fail to establish. The use of minimum tillage techniques coupled with the increased tendency of farmers to sow winter cereals early has enabled this species to escape the regulation that would have normally occurred with deep ploughing or during seedbed preparations for spring sown cereals. These changes, coupled with the fact that sterile brome is a strong competitor with winter cereals and disperses seed before harvest, has led to its rapid emergence as an arable weed from its historical place as a hedgerow and wasteland species.

There is an increasing tendency for crop species themselves to be weeds to succeeding crops in a rotation sequence. Cereals arising from seed shed at harvest are one example and the persistence and multiplica-tion of volunteer potatoes can cause serious difficulties in vining peas, sugar-beet and other vegetable crops. In a narrow sense then crops too are pre-adapted. However they are much less likely to persist naturally outside the immediate cropped environment.

Evolution

The evolution of weeds may be seen at three levels: (a) speciation – genetic change in a taxon and the evolution of a species with characters favouring growth in disturbed and unpredictable habitats; (b) race formation in which locally adapted races of a weed species occur; and (c) crop mimicry where the association of crop and weed is so intimately associated with man that it has led to the evolution of mimetic weeds.

Table 1.2 Comparison of two closely related species of *Ageratum* from the tropics. After Baker and Stebbins (1965)

Ageratum microcarpum	*Ageratum conyzoides*
Non weed	Weed
Little plasticity	Plastic growth form
Perennial	Annual
Slow flowering	Quick flowering
Low night temperatures for flowering	No night temperature flowering requirement
Self-incompatible	Self-compatible
Diploid	Tetraploid

The Compositae as a family contains several examples of both weedy and non-weedy taxa. One particularly well researched example concerns *Ageratum* spp. *A. conyzoides* is an annual species that has escaped from the New World tropics and become one of the commonest weeds in warmer parts of the world. Comparison with a closely related species, *A. microcarpum* reveals a strikingly different set of life history characters (Table 1.2). *A. conyzoides* can complete its life cycle in less than 2 months, to flower over a wide range of temperatures and display extraordinary plasticity. 'Very tiny plants with only a single flower head will mature seed in extreme conditions of crowding, waterlogging or drought; on the other hand they can grow to two feet in height with hundreds of flower heads' (Baker, 1965). By contrast *A. microcarpum* is not very plastic, is a perennial and is slow to flower. Studies indicate that *A. conyzoides* has evolved as a polyploid relative of *A. microcarpum*.

Where agricultural practices are continued for a sufficient length of time and sufficient genetic variation occurs within a species, locally adapted races of weeds are likely to arise. There is now a wealth of evidence that agro-ecotypes may arise in response to a host of selective forces. One of the earliest observations of race formation was of the selection of dwarf forms of fool's parsley and upright hedge parsley in cereal crops following the introduction of the reaper. The regular use of fertilizers has also resulted in the rapid evolution of edaphic races of sweet vernal-grass in the Park Grass plots at Rothamsted Experimental Station, UK and genetic changes in pasture grasses in response to grazing are also well known. The regular mowing of lawns and bowling greens may also result in the evolution of dwarf and prostrate races of common weeds, such as annual meadow-grass.

Crop mimicry by weeds results from the co-evolution of the crop and the weed under the hand of man. In considering cases of mimicry the critical distinction is that the mimic (the weed) functionally imitates the model (the crop) so that the operator (man) is unable to discriminate effectively between model and mimic. Two major forms of crop mimicry may be distinguished that correspond to periods in the life cycle of the crop where resemblance of model and mimic are greatest — 'vegetative' and 'seed' mimics. In the former close similarity occurs during seedling and vegetative growth periods to the extent that the agency of weed control — for instance hand removal of weeds — is deceived. Most cases of vegetative mimicry involve the Gramineae, reflecting the superficial similarity among seedlings of members of the family. For mimicry to occur however there must be demonstrable evolution in the weed species towards increased similarity with the crop. This process is often enhanced by gene exchange from crop to weed (introgression) as occurs in *Zea mexicana* (teosinte), a common weed of *Zea mais* (maize) in Central America. As a result during the vegetative phase of growth, teosinte and maize are virtually indistinguishable having for instance similar leaf size and plant colour and escape hand weeding. Although species are identifiable, and therefore discriminated against, at flowering and harvesting, seed of teosinte is still returned to the seedbed by the cultural practices employed, thus ensuring persistence of the weed–crop mixture.

A classic and historical example of seed mimicry involves gold of pleasure, an annual weed of flax in field crops in Eurasia. A specialized race of *C. sativa* has evolved and populations of this race flower at the same time as the crop and, because fruits do not dehisce prematurely, are harvested with the flax. The seed of other weed species that are also harvested are removed by the age-old technique of winnowing but those races of *C. sativa* that possess similar winnowing characteristics to flax escape removal from the grain. The seed characteristics that determine the winnowing properties of *C. sativa* have been shown to be genetically controlled and the repeated sowing of flax and *C. sativa* selects for an intimate association of crop and weed.

The examples discussed above briefly illustrate the types of changes that may occur over evolutionary time given sufficient genetic variation and intensity and duration of selection. From the many investigations conducted it is evident that weed species do retain considerable genetic variation and may evolve. Tracing this evolutionary path is often difficult as both ancestral origins and past selection pressures may be unknown and at best inferred. There is the suggestion that cleavers may either have undergone race formation in cereal crops recently in the UK or that current

Table 1.3 Sample of alien introductions that have become established in the British flora

Date of probable introduction	Species	Probable source of introduction
1763	Rhododendron	Iberian peninsula
1808	Slender speedwell	Caucasus
1809	Hoary cress	Central and southern Europe
1825	Common field speedwell	Western Asia
1842	Canadian pondweed	North America
1860	Gallant soldier	South America
1871	Pineappleweed	North East Asia
1902	Tarweed	USA
1917	Winter wildoat	Central Asia

arable agricultural practices are promoting a display of the considerable phenotypic variation that the species is known to possess.

Modern examples of evolution in weeds in the UK involve herbicide resistance (Chapter 8). Where races of weeds evolve a similar spectrum of reactions to herbicides as the crop in which they grow an additional form of mimicry might be construed. This however is not strictly the case since changes in the weed population may arise in the absence of the crop model and cases of herbicide resistance are best considered as instances of chemical selection.

Alien immigration

Import of seed and commodities can be a potent source of propagules of alien plants so that in 1947, 91 alien species were listed in the neighbourhood of Southampton docks. Earlier a publication in 1919 recorded 348 aliens by the River Tweed which had long been a centre of wool importation from many parts of the world. Whilst additional species are continually being introduced into the UK, both inadvertently by industry and consciously by seed firms, botanic gardens and private collectors, few alien species succeed in establishing themselves as damaging weeds. Table 1.3 lists some of those that have and their approximate date of introduction and probable geographical origin.

Divining the reasons why one species succeeds where another does not, raises some of the same issues that were considered when discussing pre-adaptation. An important difference however is that alien species at introduction may often be small in population size and in addition ill-represent the range of genotypes that occur in parental populations. Even so, some species particularly Canadian pondweed and Gallant soldier have spread geometrically within Britain in recorded history. Neither is the spread of aliens limited to grasses or herbs. Sycamore, possibly a Roman introduction to Britain, but more likely an arrival in the 15th century is a forest weed. Rhododendron is another woody adventive that was deliberately introduced from the Iberian peninsula via Kew and subsequently cultivated by man. In the last 15–20 years wild rhododendron has undergone noticeable range expansion both in North Wales, South West Ireland and Eastern England. The explanations for this are unknown but it may be that: (a) after a period of establishment rhododendron populations are simply expanding geometrically; (b) that changes in land management have encouraged the spread of species; and (c) that through natural selection or hybridization the species has evolved and become adapted to British habitats. A combination of characters would appear to contribute to its success: prolific seed production; longevity of adult plants; freedom from pests and diseases; vigorous vegetative regrowth when cut; unpalatability to grazing animals; and shade tolerance.

A commonly advanced reason for the rapid expansion of an alien introduction is escape from natural regulation from 'above' in the food chain. Although 600 species (animals, fungi, viruses) have been recorded on rhododendron worldwide, in Britain there would appear to be no obvious pests or diseases. Changes in land management practice may have also contributed to the species' expansion — for instance stocking rates of sheep, and withdrawal of lime subsidies. Thus conditions for seedling establishment may have been improved — seedlings are killed by direct application of lime to acidic soils on which rhododendron thrives. High stock rates and consequent overgrazing of pasture species coupled with the unpalatability of the weed may also place rhododendron at an advantage. Furthermore, there remains the possibility that the free flowering characteristic of the species has allowed introgression with other *Rhododendron* relatives.

This example encapsulates the difficulties posed when assessing the reasons for the 'weediness' of a species. Regulation of the abundance of a plant species may be above from predation, from below through restricted habitat availability and in the case where neither are constraining factors, changing abundance may be solely determined by the intrinsic demographic properties of the species.

Very often agro-ecosystems along with other landscapes are mosaics of very different habitats (in arable agro-ecosystems components being hedgerow, headland, cropped land and wasteland) and the ability of a species to disperse effectively between similar habitat components is crucial to its success. Habitat contiguity and habitat size may severely restrict the spread of a species and are important components in the consideration of rarity. Firewood (or rosebay willowherb) provides a striking example of this relationship. Though native to Britain since Glacial times, this species was scarce at the turn of the century. Its association with the habitat conditions arising out of heath fires is well known. A plausible explanation for its present day widespread abundance is, firstly, that there has been an increase in available habitat resulting from the frequency with which fires have occurred especially since 1914 with the extensive felling of woodlands and other industrial activities; and secondly because new habitats have occurred at distances within the dispersal range of the wind borne achenes of the species. Typically habitat contiguity is maintained in a linear manner along roads and railways along which fires may occur.

Features that may confer 'weediness'

The foregoing discussion illustrates that the development of a weed flora and the incorporation of a weed species in it depends upon a range of factors operating both on ecological and evolutionary timescales. A species may become a weed simply through the fortuitous possession of the appropriate characters that enable it to exploit a niche created by a particular land usage practice. Weed abundance is therefore proportional to habitat size given that the species may disperse effectively amongst non-contiguous areas of habitat. Such species have been considered 'minor' because their 'weediness' is a function of phenotypic characters that enable population growth in a particular habitat. The damage that they may cause may well not be minor!

'Major' weeds on the other hand have been characterized as those that possess 'a general all-purpose genotype' that ensures a strategy of species persistence under high levels of unpredictability in habitat conditions. Such species may be weeds which are common to a large number of cropping situations where either land management or husbandry practices create a diversity of habitats. Ecological facets of such a 'general purpose' strategy that immediately suggest themselves include high reproductive capacity, well developed powers of dispersal, and short life cycles. Moreover the maintenance and acquisition of genetic diversity, is crucial to the long term

Table 1.4 Life history characteristics of a plant species that if combined would result in an 'ideal' weed. After Baker and Stebbins (1965)

1. Seed germination requirements fulfilled in many environments
2. Discontinuous germination (through internal dormancy mechanisms) and considerable longevity of seed
3. Rapid growth through the vegetative phase to flowering
4. 'Seed' production in a wide range of environmental conditions; tolerant and plastic
5. Continuous seed production for as long as conditions for growth permit
6. Very high 'seed' output in favourable environmental circumstances
7. Self-compatible but not completely self-pollinating or apomictic
8. Possession of traits for short and long distance dispersal
9. When cross-pollinated, unspecialized pollinator visitors or wind pollinated
10. If a clonal species, has vigorous vegetative growth and regeneration from fragments
11. If a clonal species, has brittleness of leafy parts ensuring survival of main plant
12. Shows strong inter-specific competition by special mechanisms (e.g. allelopathic chemicals)

persistence of a weed species in a wide habitat range. Combining all of these considerations it is possible to construct a tentative blueprint for the 'ideal' weed species (Table 1.4). As its author H.G. Baker drily pointed out there is luckily no claimant at the present time and neither should we expect one to arise, *a priori*. Whilst providing a framework for comparative assessment of morphological, physiological and genetical characters which *may* enable a species to increase in abundance and become 'weedy', it is only by consideration of the population ecology of species in a range of habitats that the likelihood of species persistence at levels damaging to man's interest may be critically considered. With this strong reservation, some generalizations may be made. The characteristics are listed in chronology of the development of an individual plant and it is convenient to follow them.

Seedling emergence and seedbanks

Episodic seedling emergence from a persistent seedbank results in the occurrence of pulses of growing plants which, in an unpredictable habitat, may ensure some flowering and seeding adults. Species that display germination and dormancy characteristics that enable a response to a wide range of environmental cues therefore express a 'generalist strategy' and might be expected to occur widely. Fig. 1.1 illustrates the periodicity of seedling emergence of some of the more common arable weeds. Practical knowledge of periodicity of germination is of significance since it is a major factor determining the association of weeds with cropping systems and

(a)

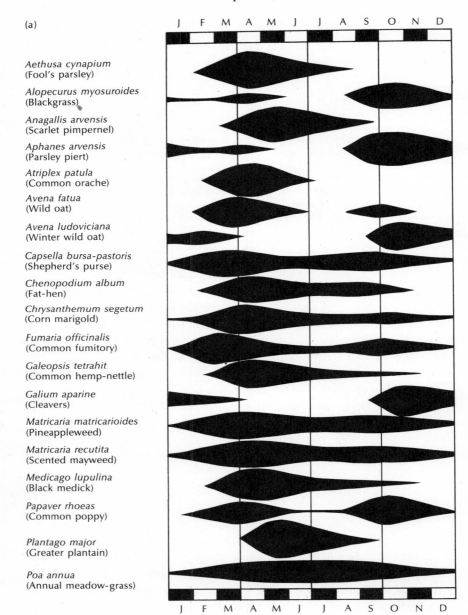

Fig. 1.1 (a) and (b)   The main germination periods for some common annual weeds of arable land. Dark areas indicate relative germination to show seasonal periodicity.

(b)

J F M A M J J A S O N D

Polygonum avicutare
(Knotgrass)

Polygonum convolvulus
(Black bindweed)

Polygonum persicaria
(Redshank)

Ranunculus arvensis
(Corn buttercup)

Raphanus raphanistrum
(Wild radish)

Senecio vulgaris
(Groundsel)

Sinapis arvensis
(Charlock)

Solanum nigrum
(Black nightshade)

Sonchus asper
(Prickly sow-thistle)

Spergula arvensis
(Corn spurrey)

Stellaria media
(Common chickweed)

Thlaspi arvense
(Field penny-cress)

Tripleurospermum maritimum
spp. *inodorum*
(Scentless mayweed)

Urtica urens
(Small nettle)

Veronica hederifolia
(Ivy-leaved speedwell)

Veronica persica
(Common field speedwell)

Vicia hirsuta
(Hairy tare)

Viola arvensis
(Field pansy)

J F M A M J J A S O N D

enables a degree of forecasting as to which weed species may occur in the seedbed. As Fig. 1.1 illustrates, no species behaves in exactly the same way although three main patterns can be distinguished: predominantly autumn germinators, predominantly spring germinators and year-round germination. These patterns are observed regardless of soil cultivation and the physiological mechanisms underlying them are numerous, extremely varied and beyond the scope of this chapter to explore in depth. Seedling emergence patterns reflect the seasonal variation in edaphic and climatic factors and the extent to which weed species respond to aspects of these changes as stimuli or cues.

Ecologically, seed dormancy has two important aspects. The first is that there is a stage in the life history of the species that is resistant to extreme adverse conditions. The second is the synchronization of resistant and non-resistant stages with the appropriate adverse/benign periods of time. Dormancy strategies can be either predictive or consequential. A predictive dormancy strategy reflects adaptation to predictable seasonal environments, seeds entering dormancy in advance of adverse conditions. Contrastingly, a consequential strategy reflects a direct response to adverse conditions. In weeds, predictive seed dormancy is generally referred to as innate dormancy and consequential seed dormancy as 'enforced' or 'induced'.

Innate dormancy may be brought about by several different mechanisms but is genetically determined and if present is the condition of the seed (or fruit) at the time of dissemination from the parent. 'After-ripening' requirements may involve defined qualitative and quantitative temperature changes and enable germination to be cued to seasonal environmental changes. Thus germination of fresh seed of knotgrass is delayed until soil temperature begins to rise in spring. In contrast seeds of ivy-leaved speedwell begin to germinate in autumn only after exposure to high summer temperatures. Other mechanisms that confer innate dormancy are the presence of hard seed coverings or chemical germination inhibitors that must be leached before germination can proceed. Germination inhibitors may be predictably eluted with autumnal rains in Britain but the scarification of hard seed coats may be less so although passage through the gut of animals feeding intensively prior to winter hibernation or winter food shortages may constitute an appropriate cue. Many weed species display adaptive variability in germination requirements amongst the seeds produced by an individual plant. In the Chenopodiaceae seed lots from single plants can be differentiated into those that will germinate readily in autumn whilst others require exposure to low winter temperatures and germinate in spring. This feature is a somatic seed polymorphism since parental plants give back progeny of both germination types regardless of the seed type from which they themselves arose. In some species seed with differing germination requirements may be recognized morphologically (Chenopodiaceae) whilst in others (dock and wild oats) the dormancy status differs according to the location of the seed on the infructescence.

Consequential dormancy is an adaptive response to seasonal environments with an unpredictable level of adversity. In these circumstances there clearly is an advantage to only entering a dormant state if adverse conditions do appear and in responding to favourable conditions when they reappear. Seeds are *enforced* in dormancy therefore by the absence of the appropriate environmental conditions for growth and once these restrictions are removed germination proceeds. In many species, seed burial within the soil profile or under litter enforces dormancy and it is only after exposure to light that germination occurs. Many weeds exhibit a second form of consequential dormancy known as *induced* dormancy. Here dormancy is induced in seed held in enforced or innate dormancy such that a specific cue must be received before germination or return to enforced dormancy can occur.

Flux between enforced and induced states of dormancy occurs in many species and may be dependent upon seasonal variation in temperature. Seeds of knotgrass and redshank in the soil are induced into dormancy as soil temperature rises in late spring and only after exposure to a period of low (chilling) temperatures can germination take place. This mechanism ensures that germination is restricted to the early part of the year and both species are spring germinators. Experimental studies have exemplified the importance of interactions among abiotic factors of the soil environment on germination and dormancy. The amplitude of alternating soil temperatures and its duration, photoperiod, red/far-red quanta in the light spectrum and nitrate availability have all been implicated in determining the flux between dormancy states. Mechanisms of dormancy clearly prohibit plant species from 'literally putting all their eggs in one basket' and constitute an important adaptive strategy for the persistence of 'weedy' species.

The presence of a seedbank in and on the soil surface is obviously also a consequence of dormancy in the seed population and 'the other side of the coin' from seedling emergence. Fig. 1.2 illustrates the range of types of seedbank that have been proposed for herbaceous vegetation and by comparison with Fig. 1.1 and with knowledge of the longevity of seed in the soil it is possible to infer the type of seedbank possessed by weed species. Those such as onion couch, barren brome and fireweed display transient

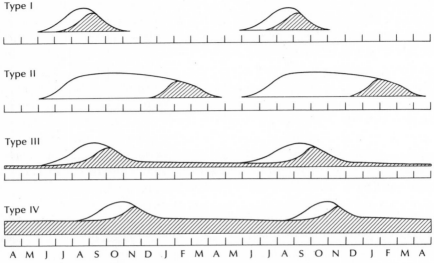

Fig. 1.2 Types of seedbank proposed for herbaceous vegetation. Shaded areas: seeds capable of germinating immediately after removal to suitable laboratory conditions. Unshaded areas: seeds viable but not capable of immediate germination. Type I: annual and perennial grasses of dry or disturbed habitats. Type II: annual and perennial herbs colonizing vegetation gaps in early spring. Type III: species mainly germinating in the autumn but maintaining a small persistent seedbank. Type IV: annual and perennial herbs and shrubs with large persistent seedbanks. After Thompson and Grime (1977).

seedbanks — Type I. These species possess little dormancy and the ability to germinate over a wide range of temperatures and the seedbank is typically exhausted after autumn germination. Dog's mercury, that is sometimes a persistent horticultural weed, also possesses a transient seedbank but one that is present only in winter (Type II). The distinction between Types III and IV is based on the relative proportion of seeds that remain viable within the seedbank over seasons. Representative species of Type III include thale cress and annual meadow-grass although this latter species along with heather and rushes may also exhibit long term persistent seedbanks (Type IV).

Dispersal

The importance of effective dispersal has been alluded to earlier in considering the incorporation of alien weeds into a flora. The pattern of seed dispersal or 'seed shadow' around individual terrestrial plants resulting from passive dispersal, or even of propagules with structures aiding wind dispersal, show that the majority of seeds are dispersed in relative close proximity to the parent, although this is not to dismiss the few that by chance will be spread to a considerable distance. Long distance dispersal of many weed species results from the activities of man. Rayless mayweed was first recorded in 1871 and prior to 1900 was only very locally abundant. Between 1900 and 1925 however it spread in a spectacular way along roadsides to all English counties. E.J. Salisbury records that the ribbed fruits which bear no pappus are effectively transported in mud adhering to wheeled vehicles. Farm machinery and associated implements in all probability play an essential role in the active transportation of weed seeds locally and among farms. Hooked or awned seeds (as of cleavers and wild oat) may easily be transported this way as may horticultural weeds when container grown plants are moved amongst nurseries. Studies on the movement of seed in the soil by cultivation practices in arable farming however have shown that seed in the soil is only moved up to 1–2 m from point of first landing. Such dispersal is probably achieved by bulk soil movement. In pastures, grazing stock play a pivotal role in distributing seed of both weed and forage species. A single cow has been estimated to distribute 900,000 viable seeds per annum in dung which itself may provide a suitable site for seed germination.

Growth and seed production

Traits 3–8 (Table 1.4) reflect the developmental characteristics of a plant species that ensures reproductive success. The time between germination and the production of mature seed varies amongst species and may be as little as 6 weeks in some species (shepherd's purse and nettle) whilst in others is prohibited until a critical size is reached and environmental cues for flowering are experienced (as in some biennial species). Some of the most abundant weeds have no specialist requirements for flowering and are found in flower at any time during the year — this is true of chickweed, groundsel, annual meadow-grass and gorse. One feature of habitat unpredictability may be wide fluctuations in the resources required for plant growth — for instance water and soil nutrients. Studies of weeds sown in monocultures at very high densities such that resources are extremely limiting per plant, have shown that even very small individuals may produce progeny. Common poppy and wild oats display this ability. Such plasticity is commonly held to be an attribute of 'weediness'.

Day length is by far the most important cue for flowering in both weeds and crops. Some species flower only when day lengths exceed some critical

Table 1.5 Production of propagules (seeds or fruits) by some common weeds. Data reflect the upper range in production of isolated plants in field conditions

Species	Botanical name	Propagule production (thousands)
Wild oat	*Avena fatua*	1–3
Groundsel	*Senecio vulgaris*	1–2
Common chickweed	*Stellaria media*	2–3
Shepherd's purse	*Capsella bursa-pastoris*	3–5
Greater plantain	*Plantago major*	13–15
Common poppy	*Papaver rhoeas*	14–20
Prickly sow-thistle	*Sonchus asper*	21–25
Canadian fleabane	*Conyza canadensis*	38–60
Hard rush	*Juncus inflexus*	200–250

duration (long day plants, e.g. greater plantain, common poppy) whilst others show the reverse, requiring short days. Relatively few British weeds flower in the short days of autumn — Canadian fleabane and corn caraway being two. Flowering may also be cued by a progressive change (e.g. long days following short days) and in some species (e.g. sheep's sorrel there may be a vernalization (exposure to low temperature) requirement. Near simultaneous flowering within a species population is of obvious importance if the species is outbreeding and predictable seasonal stimuli provide stable cues. Moreover in annual species cues for flowering are ecologically correlated with cues for seed germination; thus exposure to (winter) low temperatures to break seed dormancy may be associated with a long day flowering requirement. In perennial or biennial species such correlations may be uncoupled.

All other matters being equal (see below) a species that is very fecund and produces large numbers of seeds is likely to become abundant. Table 1.5 illustrates the ability of the more prolific weed species when grown as near isolated plants in ideal conditions.

Vegetative regeneration

Many of the most serious weeds in Britain exhibit forms of 'vegetative reproduction'. This broad term encompasses two sets of morphological and anatomical features that enable regrowth from plants other than by seedling establishment. On the one hand there are species such as bracken, common couch, field bindweed and ground elder which exhibit clonal growth and a plant architecture that in geometric arrays, spaces buds (or strictly meristems) from which new adult plants may establish. On the other hand there are those species which through possession of branched or simple tap roots can survive repeated damage and regenerate new growth.

Plants with creeping stems or roots typify clonal growth forms. Creeping stems have a series of nodes from which scale leaves arise and each scale encloses a lateral bud which may remain dormant or alternatively can grow on to form either branches of the stem or aerial shoots. Creeping stems may occur above or below ground as in creeping cinquefoil and common couch respectively. All creeping roots on the other hand are subterranean and display both horizontal and vertical growth. Creeping roots are distinguished by the absence of scale leaves. In addition the root apices never turn upwards and develop into aerial shoots, and can also produce adventitious buds at any point along their axis in contrast to creeping stems.

The annual growth cycle of these perennials varies and Fig. 1.3 illustrates the main periods of foliage growth and flowering in some important weeds. In some species, coltsfoot and bracken, above ground foliage dies back and the plant overwinters below ground whilst others exhibit continuous growth so long as temperatures permit (creeping soft-grass). Species also vary in the time when new aerial shoots emerge. In yarrow and common couch new shoots are produced before the old stems die and growth of creeping stems is renewed in the following spring and summer. In arable land the shoots of perennial sow thistle and creeping thistle appear late in spring.

In open undisturbed ground the spread of clonal growth forms can be extremely extensive and rapid. Up to 12 m in a single year has been recorded for creeping thistle and 2 m for common couch. In grassland however where grasses have a suppressive effect on growth, creeping rooted weeds are rarely of importance although creeping thistle is known to persist over long periods within swards.

Weeds that regenerate from tap-roots occur in several genera and include species of Umbelliferae, *Rumex* and *Taraxacum*. Perennial tap roots provide overwintering carbohydrate reserves for foliage and flowering in the coming growing period. Species with tap roots vary in their ability to regenerate when fragmented though most are fairly prolific. Docks will regenerate new shoots from the uppermost part of the tap-root (10 cm) alone whereas any part suffices in dandelions. Undoubtedly the reason why plants with clonal growth have assumed significance as weedy species is their ability also to regrow from cut parts. Physical fragmentation of

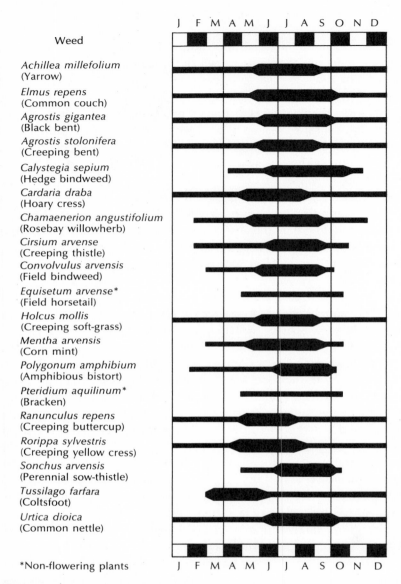

Fig. 1.3 The periods of growth, when green foliage is present (thin lines) and of flowering (thick lines) for some common perennial weeds.

creeping stems releases buds from dormancy and new shoots may rapidly establish as exemplified in common couch.

Land cultivation practices that fragment perennial plants and hence aid the propagation of new separate individuals act as an important force of interspecific selection. Their intensive repeated use however can serve to exhaust the bank of dormant buds so long as growing plants are destroyed before new creeping stems and roots are formed.

'Vegetative reproduction' in a species then is an important trait that may well confer 'weediness'. This however is not to belie the importance of contributions from seed reproduction to growth of these species. Table 1.6 describes the relative roles of the two.

Weed population ecology

The abundance of a weed species within a crop at a particular time is a momentary statement of demographic processes undergone by the species concerned. Explaining why a weed species may have increased in abundance (or declined) requires a knowledge of the factors that regulate the size of a weed population. By knowing how these factors, whether natural or man-managed, operate or interact in applying a brake to population growth: (a) weed control practices can be assessed and interpreted; and (b) the role of particular biological traits can be assessed.

A simple diagrammatic life table which can be applied to most weeds that annually renew themselves from a bank of buried seeds is given in Fig. 1.4. If weed control practices are sufficient to ensure that no seedlings survive to become seed producing adult plants then the weed population will decline according to the overall rate of loss of seeds from the soil bank. The major causes of losses to this bank will be seed germination, seed death *in situ* and predation. Detailed studies (see below) have shown that there is a constant death risk to seed in the soil, that is characteristic to a species under a particular management regime. Since seed buried at depth tend to remain in enforced or induced dormancy seed populations decay at a slower rate than corresponding ones near the soil surface.

The fraction of seeds that germinate from the bank and establish seedlings ($g.e$) in a growing season is commonly small ($0.001-10\%$). Apart from this important observation, of greater significance in the overall dynamics of the weed population is the proportion of those plants that survive (s) to return viable seed to the seedbank. Weed mortality may depend upon a host of factors but two obvious important biological features are species characteristics that enable escape from phytotoxic chemicals and competitiveness. Moreover as noted earlier the potential fecundity of a weed species (F) is important in determining the size of the multiplicative event in the cycle that offsets losses accrued earlier.

Fig. 1.4 also illustrates how a very simple equation of population

Table 1.6   Characteristics of some important perennial weeds

Species	Reproductive parts and overwintering state	Depth of vegetatively reproductive parts*	Reproduction by seed
Achillea millefolium (Yarrow)	Stolons; terminal rosettes of leaves overwinter	Very shallow	Very important
Aegopodium podagraria (Ground elder)	Rhizomes with dormant underground buds; aerial parts die	Shallow	Unimportant
Agrostis gigantea (Black bent)	Rhizomes with dormant underground buds; aerial shoots overwinter	Shallow	Very important
Agrostis stolonifera (Creeping bent)	Aerial creeping stems that overwinter	Above ground	Importance unknown
Allium vineale (Wild onion)	Offset bulbs and bulbils that overwinter	Aerial or very shallow	Rarely produced
Armoracia rusticana (Horseradish)	Fleshy tap root that overwinters; leaves die	Deep	None produced
Arrhenatherum elatius (Onion couch)	Bulbous shoot bases that overwinter (some forms only)	Very shallow	Very important
Calystegia sepium (Hedge bindweed)	Rhizomes that overwinter; aerial shoots die	Deep	Rarely produced
Cardaria draba (Hoary cress)	Creeping roots; small rosettes of leaves overwinter	Deep	Important
Cirsium arvense (Creeping thistle)	Creeping roots that overwinter; shoots die	Deep	Occasionally produced
Convolvulus arvensis (Field bindweed)	Creeping roots that overwinter; shoots die	Deep	Important
Elymus repens (Common couch)	Rhizomes with dormant underground buds; aerial shoots overwinter	Shallow	Moderately important
Equisetum arvense (Field horsetail)	Rhizomes with tubers that overwinter; aerial shoots die	Deep	Non-seeding plant

Table 1.6 Cont.

Species	Reproductive parts and overwintering state	Depth of vegetatively reproductive parts*	Reproduction by seed
Mentha arvensis (Corn mint)	Rhizomes; aerial shoots die	Shallow	Very important
Oxalis spp.	Bulbils, tap roots and rhizomes; leaves die	Shallow	Important in some species
Poa trivialis (Rough meadow-grass)	Short stolons; a few leaves overwinter	Above ground	Very important
Polygonum cuspidatum (Japanese knotweed)	Rhizomes, dormant underground buds; aerial shoots die	Shallow	None produced
Polygonum amphibium (Amphibious bistort)	Rhizomes, dormant underground buds; aerial shoots die	Shallow	None in arable
Pteridium aquilinum (Bracken)	Rhizomes; leaves die	Deep	Non-seeding plant
Ranunculus repens (Creeping buttercup)	Procumbent stems: a few leaves overwinter	Above ground	Very important
Rumex crispus *R. obtusifolius* (Docks)	Tap roots; rosette of leaves overwinter	Very shallow 7–10 cm	Very important
Sonchus arvensis (Perennial sow-thistle)	Creeping roots; aerial shoots die	Very deep	Important
Taraxacum officinale (Dandelion)	Fleshy tap roots; few leaves overwinter	Shallow	Important
Tussilago farfara (Coltsfoot)	Rhizomes; leaves die	Very deep	Important
Urtica dioica (Common nettle)	Rhizomes; short green shoots overwinter	Very shallow	Very important
Veronica filiformis (Slender speedwell)	Stems creeping on the surface	Above ground	None produced

* Depth may vary considerably; the categories are intended as an approximate guide only. 'Very shallow' indicates depths down to 15 or 25 cm, 'shallow' down to 30 or 45 cm, 'deep' down to 1 m and 'very deep' down to 3 m or more

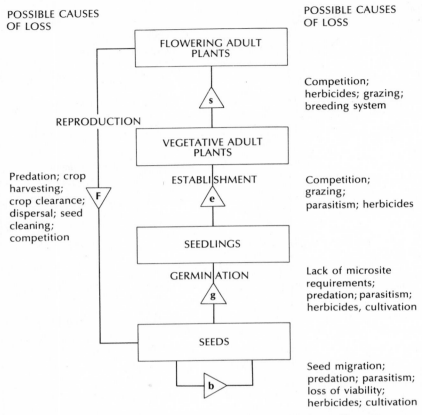

POSSIBLE CAUSES
OF LOSS

POSSIBLE CAUSES
OF LOSS

FLOWERING ADULT
PLANTS

s

Competition;
herbicides; grazing;
breeding system

REPRODUCTION

VEGETATIVE ADULT
PLANTS

ESTABLISHMENT

e

Predation; crop
harvesting;
crop clearance;
dispersal; seed
cleaning;
competition

F

Competition;
grazing;
parasitism; herbicides

SEEDLINGS

GERMINATION

g

Lack of microsite
requirements;
predation; parasitism;
herbicides, cultivation

SEEDS

b

Seed migration;
predation; parasitism;
loss of viability;
herbicides; cultivation

Fig. 1.4 A diagrammatic life table for a weed species reproducing by seed but with a persistent seedbank. The transition probabilities b, g, e and s are the proportions (range 0–1) of individuals that move from stage to stage, or remain in the seedbank. F is the seed production of an individual adult plant. Typically the model is worked over a timescale corresponding to a generation of growth on a unit area basis.

growth can be built up. If the time period t to $t + 1$ is the duration of a cycle of population growth then the size (S) of the weed population at census point $t + 1$ will be the sum of the gains resulting from the seed production of surviving adult plants and the fraction of seeds that persist in the buried seedbank. Algebraically describing the transitions in Fig. 1.4,

$$S_{t+1} = g.e.s.F.S_t + b.S_t,$$

where S_t and S_{t+1} are the respective sizes of the seed population at the beginning and end of the growth cycle, the rate of change in population size over that period being S_{t+1}/S_t. Theoretically in a seasonally constant environment, if the losses from the seedbank $(1 - b)S_t$ are not exactly matched by the gains from seed production $g.e.s.F.S_t$, only two long term outcomes can occur — either the population will increase geometrically or it will decline geometrically (in theory to extinction). It is then necessary to recognize that it is the abundance or density of the weed population itself that regulates the rate of increase of the population at particular stages during the life cycle. Studies over the last 10 years have indicated that many transitions during the life cycle of weed species may be subject to negative *density dependent* regulation, with the exception of mortality within the buried seedbank. Notably seed production per plant declines with increasing density of surviving adult plants. Similarly the likelihood of survivorship declines from seedling to adult plant but over an entirely different density range (Fig. 1.5). Recognizing the fact at least that one transition probability is negatively density dependent introduces a stabilizing component to the dynamics of the population. For instance a lowering of seed production per individual plant in response to increasing adult plant density will lead over subsequent generations to successive reduction in population rate of increase and the population will approach an equilibrium size (Q in Fig. 1.6). This equilibrium will be stable since any displacement away from it will be balanced by corresponding changes in seed production.

There are very few weed species for which the relationship between population growth rate and density (Fig. 1.6) is known. However the figure serves to illustrate the underlying nature of regulatory systems intrinsic to weed populations (response A) and the possible responses in relation to a companion crop (response B) and to chemical control (response C).

The utility of this simple approach lies in the ability to describe the effect of control practices on any weed species which displays discrete generations. Mathematical description of these responses points to ways of forecasting levels of weed infestations.

Whether weed populations are at or near equilibrium population levels in practice is a matter of conjecture. Such levels would only be achieved with common cropping and management practices over several seasons and then would be subject to seasonal variation. For annual weeds in arable crops, it is more likely that the accumulated buried seed population is a reflection of past management practices *in toto*. In grasslands, however, where consistency in grazing management can occur for long periods, long-term monitoring has indicated that populations of both grasses and perennial herbs (e.g. buttercups) fluctuate around a mean size (Fig. 1.7) and in the long term show considerable stability in size.

Chapter 1

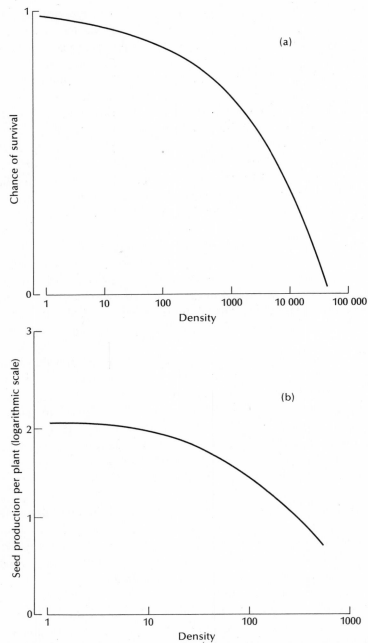

Fig. 1.5 The influence of plant density on: (a) plant survivorship; and (b) on seed production per plant. Relationships are idealized and densities are of notional arable weeds.

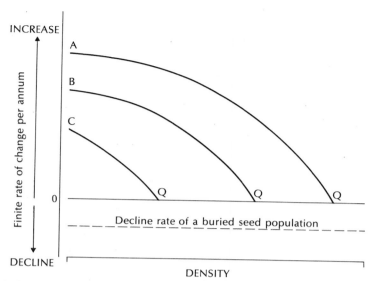

Fig. 1.6 Idealized population growth rate of a weed in relation to its own population density: A on its own; B in competition with a companion crop; and C experiencing crop competition and control from a chemical herbicide. Q = equilibrium size; scales are logarithmic.

Describing the population ecology of weeds that are not simple annuals by life tables (Fig. 1.4) requires much greater detail and usually the incorporation of additional stages based on the size or possible age of plants in the population. They have only recently begun to be constructed. Fig. 1.8

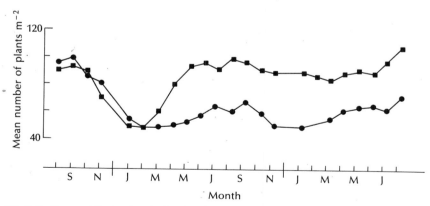

Fig. 1.7 Seasonal fluctuations in the abundance of two pasture grasses under sheep grazing. ■ = *Creeping bent*, ● = Yorkshire fog.

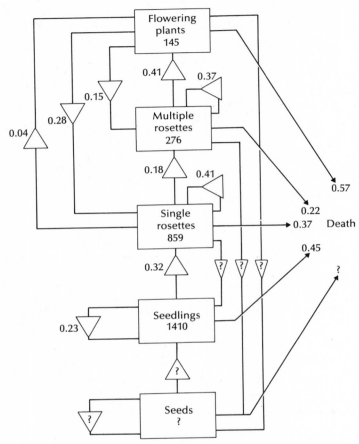

Fig. 1.8 A flux diagram for ragwort in grassland. Population sizes are mean densities
100 m^{-2} with transitions per annum. After Forbes (1977).

shows the complexity that may result when perennial structures, in this
case rosettes have to be monitored and overlapping generations of plants
occur. The task is even more daunting when plants with clonal underground growth forms are attempted — for instance yellow nutsedge or
common couch! Even with annual species that show episodic germination
like wild oat each group of emerging seedlings (cohort) can be followed
according to Fig. 1.4 resulting in a complex description of the life cycle.

Life tables, then, provide a constructional framework upon which to
examine the effects of management practices on the abundance of weed
species and Fig. 1.4 indicates some of the regulatory forces that may

influence transition probabilities. The magnitude of some effects may be such that weed species occur at such low densities particularly in arable and horticultural land that intrinsic (density dependent) regulation is largely absent. Under such conditions rapid changes in population size are to be expected when such regulatory controls are altered.

Weed biology and management

The integrated ecological approach

It would be gratifying to be able to compare population growth rates for individual weed species for a whole range of management practices ideally following Fig. 1.6. However, the speed with which weed control practices have changed both in relation to chemical and cultural control procedures has largely precluded long term population studies of all but a few persistent weed species. The detailed effects of specific management practices have been most intensively studied for arable weeds at specific stages in their life cycle. Generalizations about the effect of management practices upon species remain fraught with difficulty and often at best are inferential because of limited data, but autecological comparisons can provide some insight.

Table 1.7 illustrates some of the ecological traits that may be important in determining the persistence and abundance of three important grass weeds in winter sown cereals. All three occur on a wide range of soil types although blackgrass is favoured by heavier soils. The germination ecology and the resultant pattern of seedling emergence is a primary key to understanding the success of these weeds.

As commented earlier, sterile brome increased in abundance with the extension of reduced tillage practices particularly in the late 1970s and early 1980s. Isolated plants are prolific seed producers but the species has no persistent seedbank. Populations renew annually from the previous season's crop and the absence of innate dormancy means that seed germination occurs when there is sufficient soil moisture. This characteristic coupled with the inability of seedlings to emerge from the soil at plough depth constrains the establishment 'niche' to late summer and early autumn from seed populations close to and on the soil surface. Germination in light is episodic and dependent on rainfall. In wet summers where the weed has been abundant it is not uncommon to see dense infestations of seedlings of sterile brome in the stubbles of the previous crop. Husbandry systems that promote early sown winter cereal varieties and reduced cultivations there-

Table 1.7 Comparison of the ecological traits of three annual grass weeds of winter wheat in the UK. Data are approximate and best considered as relative ranges

Species	Germination pattern	Seed longevity (months)	Loss of viable seed from the seedbank — all causes (% per annum)	Maximum depth for seedling emergence in the field (cm)	Seedlings established per annum (% viable seedbank)	Weed density in the crop at which weed mortality noticeable (m^{-2})	Winter kill of seedlings (%)	Seed production of an isolated weed plant in the crop	Seed dissemination relative to the time of crop harvest*
Blackgrass	Episodic Oct–Mar	72+	60–80	3–5	10–20	500–1000	20–40	300–800	B; D
Wild oat	Extended episodic Oct–May	72–102	40–60	20–25	5–20	3000–5000	45–60	100–1000	B; D
Sterile broom	Restricted July–Dec	<12	95–100	3–7	40–90	5000	3–10	1000–2000	B

* B = before; D = during

fore encourage the germination of the species and recruitment of seedlings may often occur as a single flush in autumn at an early stage in the life of the crop.

In marked contrast, both blackgrass and wild oat exhibit a periodicity of seedling emergence usually over an extended time span which reflects the flux amongst seed dormancy states intrinsic to the species. In wild oat these fluxes are known to be regulated in a complex manner. In consequence, the establishment 'niche' is widened and both species may become weeds of spring sown cereals as well. Whilst the depth from which seedling recruitment can be achieved in blackgrass is limited to the surface soil layers wild oat seedlings may appear from reserves buried below normal cultivation depth.

The ability to recruit seedlings from the seedbank over an extended period enables the weed population to offset losses of autumn established plants that may have occurred during winter in cereal crops. Detailed studies have shown that those cohorts of plants recruited in spring can make important contributions to population growth in both species. This appears to be particularly more important in wild oat which may suffer significant mortality resulting from mesocotyl breakage with frost heavage of soil. Similarly, seedlings recruited late in the cropping cycle may escape autumn and early spring herbicidal control and ultimately set seed to make a small but significant contribution to population growth.

Inspection of Table 1.7 also indicates that all three species may establish dense plant populations (>500 seedlings m^{-2}) within a wheat crop, before any weed mortality which is due to density becomes noticeable. All three species have a high potential seed output but the fraction of seed returned to the seedbank effects build-up significantly. In sterile brome although grain contamination can occur, dissemination of seed is often largely complete by the time of crop harvest. Contrastingly significant proportions (20–30%) of seed of both blackgrass and wild oat may be removed by combining.

Autecological comparisons such as these are necessarily limited to inferences about the relative importance of features within the life cycle of a weed. Typically they ignore the competitiveness of the weed with the crop (see below) — an important characteristic of a weed species. At best they focus attention on the likely responses of a weed species to major shifts in management practices. It is clear from this analysis that alterations in soil cultivation methods (ploughing versus reduced tillage) and crop sowing (spring versus winter) are likely to reduce the abundance of blackgrass and particularly sterile brome but less so wild oat.

Seedbanks

The species composition of living vegetation is typically a poor reflection of the species diversity of propagules buried in the soil. Both the magnitude and composition of these banks reflects past and present weed management practices as well as the crops grown. Comparatively there are between 100 and 1000 seeds m^{-2} in forest soils, 100 and 1 000 000 m^{-2} in grassland and in arable soils 1000 and 100 000 m^{-2}. Table 1.8 illustrates the range in size of some seedbanks that have been recorded.

Seed size and shape, the changing pattern of dormancy within the species and the nature of cultivation practices are all important in determining the size of the seedbank. The small seeds of meadow-grasses and dicotyledonous species such as poppy and chickweed enter the soil profile readily on dispersal being aided by activities of invertebrates. Behavioural studies have shown that earthworms are selective in taking seeds according to seed shape, size and surface texture. The seed of some grass weed species possess trichomes and hygroscopic awns that aid burial but typical undisturbed soil profiles display a concentration of seeds in the upper layers which diminishes with depth.

Cultivation practices on the one hand act to bury seeds and on the other to promote germination of dormant seeds. Fig. 1.9 indicates that not only does the frequency of cultivations exhaust the buried seedbank at differing rates but also that this decline is specific to species; chickweed declining at a rate of 56% per annum in contrast to 28% for fat-hen in soil cultivated four times a year.

In arable cropping the fate of weed seeds may be markedly dependent upon straw and trash clearance as a prelude to seedbed preparation. Detailed studies of wild oats have shown that the influence of stubble cul-

Table 1.8 Numbers of viable weed seeds in horticultural and arable soils in UK, sampled to a depth of 15 cm

Crop or cropping sequence	Herbicide application (+/−)	Range in number of viable seeds m^{-2} (thousands)
Cabbage/leeks/Brussels sprouts/peas	−	2.5–47
Vegetables	+	0.3–24
Cereals	−	12.8–43.9
Cereals	+	8.4–15.5

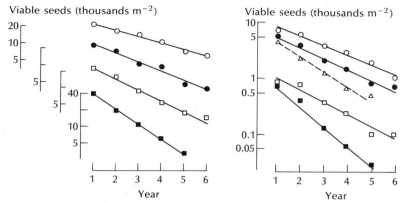

Fig. 1.9 The decrease in the number of viable weed seeds in soils under various cultivation treatments. *Left*: All species together: undisturbed soil, top 23 cm (○); cultivated twice a year, top 23 cm (●); cultivated four times a year, top 23 cm (□); cultivations involved in rotation of vegetable crops, top 15 cm (■). *Right*: Species-specific rates of decrease on plots dug four times a year. ○ = fat-hen, ● = shepherd's purse, △ = annual meadowgrass, □ = black nightshade, ■ = chickweed. Redrawn after Roberts and Dawkins (1967).

tivations and straw burning interact in a subtle way to determine the input of seed to the seedbank. Substantial seed losses (up to 85%) have been observed post-harvest when seeds remain on the soil surface throughout autumn. Such losses are most likely due to post-germination mortality and predation. Stubble cultivations to aid straw and trash decomposition reduce these losses substantially by seed burial. The influence of straw burning is twofold — the direct killing of seeds when temperatures remain high for sufficient duration (*c.* 200 °C min⁻¹) and breaking of innate dormancy which is possibly due to the physical effect of burning. Since burn temperatures vary considerably between and within straw swathes seed populations are treated differentially. Overall burning may decrease the total viable seeds by 32% but increase the numbers of non-dormant seeds by about 10-fold. One study showed that seed decline was 32% when straw was burnt, 67% when cultivations were delayed and 73% when straw was burnt and cultivations delayed. Even though absolute losses occurred in seed number, the effect of burning was reflected in the seedling populations in the following crop in spring. The least seedlings were present in the absence of burning and stubble cultivations and the most where straw was burnt and the stubble cultivated.

The depth of cultivation has important implications not only for seed survival but also on the vertical distributions of seed in the soil. Mould-board ploughing may initially result in an uneven placement of seed as a result of bulk downward movement of soil but over successive seasons tends to 'even-out' the distribution of dormant seeds within the profile. In a comparison, chisel ploughing resulted in no seeds being found below a depth of 10 cm and 60% being in the top 0.5 cm whereas mouldboard ploughing gave 43% in the top 5 cm, 37% at 5–10 cm depth and 20% in the 10–18 cm layer.

Few studies have addressed the rates of decline of buried seeds at differing depths but they are likely to be very different. In blackcurrant plantations the percentage loss of groundsel seed per annum at 7 cm was 38% in comparison to 60% in the top 1 cm layer. Rotary cultivation once a year which disturbed but did not invert the soil profile tended to equalize the seed distribution, the difference between the layers being 42% to 30% respectively.

Seedling and plant survivorship

At first sight the cultivation of seed beds in horticulture and agriculture prepares the soil surface for both weeds and incoming crops alike. Yet weed species, and perhaps particularly dicotyledonous ones, often have specific requirements for germination and establishment, which may be shown experimentally by manipulation of the habitat. Three species of plantain, *Plantago lanceolata*, *P. major* and *P. media* display strikingly different establishment rates if the soil surface is modified by compaction or small objects shade the surface where seeds have been sown. In physical terms it is not always easy to characterize the exact conditions of the microsite and moreover individual seed lots show differing requirements. Alternating conditions of light and temperature (amplitude and frequency) are in many cases critical cues for germination. On newly reclaimed polder soils creeping thistle seeds germinate in response to a maximum at high (30°C) and strongly alternating (10–28°C) temperatures.

The presence of associated vegetation may also dramatically reduce the chance of successful establishment of weed seedlings, particularly in grassland. 'Gap detection' in vegetation results from changes in the light quality and quantity receipts of seeds leading to the breaking of dormancy.

Even in optimal seedbed conditions for crops not all weed seedlings establish in the seedbed. Establishment success in weed populations is noticeably variable and points to a range of natural causes. The experimental addition of fungicides, insecticides and molluscicides have all been shown to reduce the risks of mortality and predation and it may well be that natural agents of regulation take a very significant toll of seedling populations.

Soil cultivations have the dual effect on stoloniferous and rhizomatous species of fragmenting perennating structures and releasing buds from internal dormancy. Whilst herbicides have largely superseded the use of repeated tillage to break up rhizomes and stolons for subsequent desiccation on the soil surface, this remains an effective method of control. The use of strike beds and bastard fallows in organic farming provide similar examples of the timeliness of cultivation effect upon seedling mortality.

Clearly the most noticeable effects of management practices upon weed seedling and plant survival are in the choice of crop and husbandry practice. Crop rotations exert very considerable regulation upon populations of individual weed species and the value of cleaning crops in rotations is well known. It is however the combined effect of crop and associated husbandry practice that causes the death of weeds and selects the species composition of survivors. Thus the weed flora of a root crop is strikingly different from that of cereals although some species will be common to both particularly if spring sown crops follow each other in a rotation. Chickweed, charlock and *Polygonum* species can be important in this category.

Weed–crop competition

An implicit assumption in all studies of competition in mixtures of plants is that species are competing for the same limiting resources for growth — water, nutrients and light — and that the outcome of competition is the unequal acquisition of those resources by the competing species. This is then reflected in the relative yields of the components of the mixture and may be seen as yield loss of the crop component when weeds are present. Frequently weed species are considered to be competitive because they show vigorous growth in crops and do indeed reduce crop yield. Two forms of competition have been recognized — *exploitation* competition and *interference* competition. In the former, one species rapidly gains all of a limiting resource for growth at the total expense of the other(s) in the mixture, whilst interference competition refers to the contest for resources typically over an extended period of growth. In natural populations of plants exploitation competition is uncommon (but large forest trees can suppress progeny in close proximity very extensively by shading). However in weed–crop associations exploitation competition may occur when a weed species smothers a slow growing crop. Such situations are, however, almost always a consequence of poor management.

Studies in this decade have begun to clarify the nature of interference competition and Fig. 1.10 shows the commonly observed relationship be-

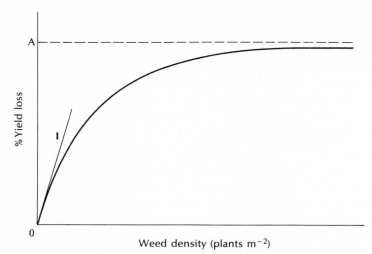

Fig. 1.10   The relationship between yield loss and weed density in arable crops. In the region of **I** yield loss is linearly proportional to weed density. A is the maximum yield loss. Scales of axes are linear.

tween weed density and crop yield loss. (Some workers prefer to invert the curve and view crop yield diminishing with increasing density from a weed-free maximum.) Such responses (rectangular hyperbolae) arise when weeds are sown experimentally at a range of densities into a constant density of a crop (an additive design) and loss in crop yield measured in relation to weed-free yields. Studies have clearly shown that over the initial, low, range of weed densities crop yield loss tends to be linearly proportional to sown weed density but that yield loss achieves a maximum at high densities when intraweed competition becomes increasingly important. These curves can be described by two parameters — the percentage yield reduction per unit weed density for very low weed infestations and the upper percentage yield loss at very high infestations. Both these parameters will depend upon other variables such as crop density, relative time of emergence of weed and crop and soil type and as such are useful for comparative assessments. The approach may be extended to considering a variable crop density or competition between two weeds in a constant crop density.

Whilst this type of analysis does not point to the resources for growth that may be limiting it has suggested new approaches to studying the competitive effects of weeds. The first is that at low weed densities and for defined management regimes species comparisons can be made in terms of 'crop equivalents': (a) the numbers of weeds required to cause an equivalent crop yield reduction; and (b) the development of economic threshold models for weed control practices. Both developments are in their infancy but hold promise for objective analysis of the competitive influence of weeds and the deployment of economically rational control measures.

The concept of thresholds in weed control is intuitively appealing and the theory is straightforward, but its application intrinsically more difficult. The economic threshold is defined as the weed density at which the cost of control in a single cropping season is returned in savings in production. These savings extend over the whole crop production system and include yield gains, reduced costs of harvesting operations and the lowering of the effects of crop damage (direct contamination or spoiling) by weed presence. The economic threshold is however based upon a consideration of costs, likely profits and losses in the current cropping season only. Thus the acceptance of an economically justifiable level of a weed infestation in one season may, through weed population growth, lead to high infestations in the following year. It is clearly naive to manage according to the economics of only the present cropping period, assuming it is indeed possible to achieve. There is therefore a need to calculate the economic optimum threshold which is the density which must be achieved long term to maximize profits. This in turn requires knowledge of the population dynamics of weed species under particular management regimes (Fig. 1.6).

Calculations have shown that for two scenarios involving post-emergence herbicides in weed control in winter cereals the economic optimum threshold in spring for wild oat was in the range 5–50 seedlings m^{-2} and 20–50 seedlings m^{-2} for blackgrass. These specific estimates are however of limited use in themselves since they rely on a series of assumptions — not least constancy, the efficacy of control practices, and the predictability of seasonal yield losses. The natural extension of these ideas is however to examine the variance in the economic optimum threshold and to calculate a *safe* threshold which should not be exceeded. Underpinning the whole approach requires a detailed understanding of the combined effect of management practices on the population dynamics of mixtures of weed species and the seasonal variation in their competitive interactions one with another and with the crop.

Application in the field requires a census of the size of the weed population at a relevant time, and knowledge of its future population growth rate. It has been suggested that measures of weed abundance should be taken at crop harvest when weed species are most conspicuous. This evaluation has to be applied at the field level, taking into account the patchiness of the weed. Despite all these complications the approach has been implemented in the Federal Republic of Germany.

Biological control

Biological weed control is based on the observation (as Fig. 1.4 reiterates) that natural enemies are of prime importance in limiting the distribution and abundance of plants. Spectacular range expansions have typically occurred when plant species have been introduced into areas outside their natural distribution and hence acquired the status of 'weediness'. For this reason biological weed control has been and still is primarily concerned with the control of naturalized weeds by importing exotic organisms. By 1982 approximately 86 naturalized weed species have been the targets of around 200 organisms in worldwide control programmes. In contrast only 25 native weed species have been subjected to biological control. There have been noticeable successes as evidenced in Australia by the control of prickly pear with *Cactoblastis cactorum*, a moth of South American origin and of skeleton weed by the rust fungus *Puccinia chondrillina*.

The identification of both the target weed species for biological control and the agent itself is complex and subject to many constraining factors. The need for host specificity, the underlying dynamics of the interaction of weed and control agent (insect or pathogen) in determining the level of control achieved long term, and the persistence of the control agent when the target is patchy are major questions to be addressed in every control programme. The objective of biological weed control is not the eradication of weeds but the reduction and long term stabilization of weed density to non-injurious levels. Vegetatively perennating weeds frequently suggest themselves as potential targets but as Fig. 1.11 indicates the introduction of a grazer, in this case the cinnabar moth *Tyria jacobeae* on ragwort, may result in fluctuations in the abundance of the weed as cycles of defoliation followed by regeneration occur. Such oscillations may also interact with other natural control agents. To date there have been no examples of the successful use of biological control agents on terrestrial weeds in the UK. A number of candidates do, however, present themselves at first sight, for instance bracken and rhododendron. In the former case research for possible agents is under way (see Chapter 2).

Concluding remarks

This introduction to the *Handbook* has taken two contrasting approaches to introducing the biology of weeds. On the one hand it is clear that the

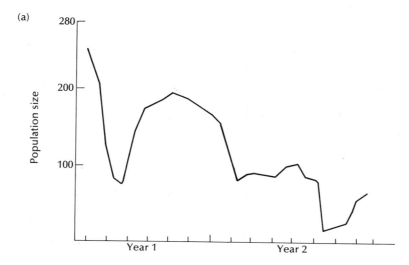

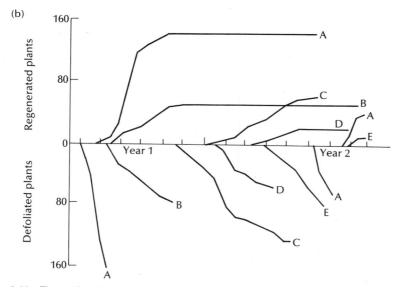

Fig. 1.11 Fluctuations in the abundance of ragwort: (a) changes in population size (rosettes 20 m^{-2}); and (b) cumulative effects of various defoliating agencies and subsequent regeneration of plants, resulting from: A cinnabar moth; B small mammals; C winter climate; D rabbits; and E drought and other causes. After van der Meijden (1971).

possession of certain biological characteristics has the potential to predispose a species to exhibiting weediness. On the other hand the habitat resulting from land use management which acts as the template selecting individual species must be evaluated. Consideration of one without the other is unfruitful and it is understanding the ecology of a weed species and the specific role of individual biological characteristics in determining the fitness of a species that is the ecological challenge of weed science.

Further reading

Subject matter that constitutes the biology of weeds occurs in many scientific disciplines. The following constitute a brief list of source material that may provide useful starting points for further reference.

Altieri, M. & Liebman, M. (eds) (1988) *Weed Management in Agroecosystems: Ecological Approaches*. CRC Press. 371pp.

Auld, B.A. Menz, K.M. & Tisdell, C.A. (1987) *Weed Control Economics*. Academic Press, London, 177pp.

Baker, H.G. (1984) The evolution of weeds. *Annual Reviews of Ecology and Systematics*, **5**, 1–24.

Baker, H.G. & Stebbins, G.L (eds) (1965) *The Genetics of Colonising Species*. Academic Press, New York.

Begon, M. & Mortimer, M. (1986) *Population Ecology*. Blackwell Scientific Publications, Oxford. 201pp.

Cousens, R. (1985) A simple model relating yield loss to weed density. *Annals of Applied Biology*, **107**, 239–252.

Fletcher, W.W. (ed.) (1983) *Recent Advances in Weed Research*. Commonwealth Agricultural Bureaux, Farnham Royal. 266pp.

Forbes, J.C. (1977) Population flux and mortality in a ragwort (*Senecio jacobaea* L.) infestation. *Weed Research*, **17**, 387–391.

Mortimer, A.M. (1984) Population ecology and weed science. In Dirzo, R. & Sarukhan (eds) *Perspectives on Plant Population Ecology*, pp. 363–388. Sinauer, Massachusetts.

Roberts, H.A. & Dawkins, P.A. (1967) Effect of cultivation on the numbers of viable weeds seeds in the soil. *Weed Research*, **7**, 290–301.

Salisbury, E. (1964) *Weeds and Aliens*, 2nd edn. Collins, London. 384pp.

Thompson, K. & Grime, J.P. (1979) Seasonal variation in seed banks of herbaceous species in ten contrasting habitats. *Journal of Ecology* **67**, 893–921.

van der Meijden, E. (1971) *Senecio* and *Tyria* (Callimorpha) in a Dutch Dune area: A study on an interaction between a monophagous consumer and its host plant. In den Boer, P.J. & Gradwell, G.R. (eds) *Proceedings of the Advanced Study Institute on 'Dynamics of Numbers in Populations*, pp. 390–404. Centre for Agricultural Publishing and Documentation, Wageningen.

2 / The evolution of weed control in British agriculture

J.A.R. LOCKHART, A. SAMUEL* AND M.P. GREAVES[†]

*Corinium House, The Avenue, Cirencester, Glos. GL7 1EJ; *The Royal Agricultural College, Stroud Road, Cirencester, Glos. GL7 6JS; and [†]AFRC IACR Long Ashton Research Station, Long Ashton, Bristol BS18 9AF*

Introduction

Britain in the 16th century was emerging from a period of famine, pestilence and war which had reduced the population to a bare three million. Much of the agriculture was based on a three-field system under which the holdings of individual farmers were unfenced, intermingled and small. From the 17th century onwards the movement of the rural population to industry in the towns created both an urban market for agricultural produce and reduction of labour on the land. The needs of the time accelerated the change from production for subsistence to production for sale. These trends coupled with the progressive enclosure of land within finite boundaries, legalized by Act of Parliament provided the conditions for structural and technical change. The 18th century was a period of innovation in agriculture. New methods of husbandry, including a systematic approach to weed control, moulded the pattern of crop production in the two centuries that followed. The Industrial Revolution of the 18th and 19th centuries affected the techniques of crop production little, but greatly assisted the ease and efficiency with which they could be accomplished.

43

The increased speed and reduced cost of transport facilitated the movement of goods on and off the farm and brought imports of cheap food to compete with the home produce. In the 20th century fluctuations in world trade caused by wars have brought periods of food surplus and of scarcity during which British agriculture has experienced changing prosperity.

It is the purpose of this chapter to describe the evolution of methods of weed control. From the 18th century onwards there was an increasing awareness that efficient weed control was essential to successful crop production, and the husbandry of those days increasingly incorporated control measures into all systems of land and crop management. Success depended on the operation of three measures, the rotation of crops, the disturbance of the soil by cultivation and the removal of weed seed from crop seed. Until the coming of chemical herbicides these three were central features of the art of husbandry.

Crop rotation

Prior to the 17th century the most common use of arable land in the southern half of Britain was cereals interspersed with fallow. In the north the system involved a permanently cropped infield and an extensive outfield cropped in patches and then allowed to fall back to grass. These simple alterations in cropping were concerned with the maintenance of soil fertility and the control of cereal diseases. The weeds on the fallows were grazed and apart from reducing the amount of shed weed seed the fallow did little to keep weeds under control. When 'open field' gave way to 'enclosure' and a greater freedom for the individual farmer to practise his own system, agriculture moved increasingly towards a balanced husbandry involving the production of a variety of crops. The rotation, traditionally a conserver of fertility, became the key to efficient husbandry with weed control as one of its major objects.

The principle of rotation

The growing of a succession of different crops on a field varied the competitive environment of the weed population. The working of the land in spring, the date of seedbed preparation, the seasonal presence or absence of the crop canopy, the time of harvest and subsequent cultivation all changed from year to year, so that no one weed species could benefit from a consistently favourable environment and thereby gain dominance.

In addition 'cleaning' crops such as turnips or potatoes which allowed mechanical weed control during their active growth were balanced against 'fouling' crops (usually the cereals) which did not. An effective rotation struck a balance that involved living with weeds which were never absent but seldom serious and usually under control.

The growth and decline of rotational husbandry

The introduction of turnips, potatoes and clover, the development of interrow cultivation and the establishment of the Norfolk four-course rotation (roots, barley, seeds, wheat) had led by the early 18th century to a new concept in which crop and animal husbandry were interdependent. The rotations which were developed in the 19th century were modifications of the Norfolk four-course or were combinations of it with leys of several years duration. While adhering to the basic principles of rotation, farmers chose crops and cropping sequences suitable to their soil, climate and production requirements.

Prior to the development of chemical fertilizers in the late 19th—early 20th century, farming achieved technical efficiency by using rotations containing cleaning crops of turnips and/or potatoes separating the cereals and containing leys of one or more years. The essentials for success were fertility from clover and livestock and weed control from the cleaning crops. The introduction of fertilizers removed the need for the ley on soils not subject to structural problems and paved the way for rotations limited to cereals and row crops.

A further change occurred with the expansion of the area under sugar-beet from the beginning of the 20th century. It provided a substitute for the fodder roots and enabled the arable farmers in the east to use a rotation based wholly on cash crops, an addition to their income which proved particularly useful from 1920 onwards. This rotation is in use today and may also include other cash crops such as peas, oilseed rape and carrots. A high level of efficiency in weed control was reached on mixed and arable farms in the late 1940s when the rotations of the 19th and 20th centuries were still in use and farmers were in addition obtaining the benefits of fertilizers and the newly developed herbicides. Thereafter, increased economic pressure and a substantial reduction in the labour force compelled specialization. The intensive arable farms have continued with varied rotations but with much reduced labour. Elsewhere selective herbicides and the combine harvesting made possible an increased area of cereal crops and a reduction in the labour-consuming row-crops. On many farms cropping now follows no set rotation. Changes are made only when disease or uncontrollable grass weeds force the issue. Over much of the Midlands and the North, cereals alternating with ley are the commonest form of land use. On land growing mainly cereals, despite the prevalence of grass weeds, farmers have, by increased use of herbicides, avoided the need to return to rotational husbandry for weed control. The major swing from spring-sown cereals to more productive winter cereals which began in the 1970s and is continuing in the 1980s has increased the problems posed by grass weeds. Even so, new herbicides, and techniques of using them, have enabled many farmers to sustain a programme of continuous autumn-sown crops. From time to time, however, a need may arise for specific tillage or introduction of a spring crop to control particular weeds. On farms with a high proportion of autumn-sown crops, the adoption of minimum tillage or direct drilling to speed up sowing has entailed a greater commitment to herbicides for weed control. The introduction of large areas of oilseed rape, and its effective herbicides, has helped to control some weeds, e.g. barren brome.

The rotations which were developed in the 19th century and which were modifications of the Norfolk four-course, attempted to avoid the use of the unprofitable fallow. However, potatoes and turnips, the two crops which allowed repeated hoeing for weed control, were not suited to heavy land. In consequence the fallow and the partial fallow were and still are used on the heavier clay soils; the technique employed is to put the soil into clods so that perennial weeds were desiccated in dry summer weather. Originally associated with soil fertility, and the subject of controversy as to its value in the 19th century, the fallow was used mainly for weed control by the 20th century. In the 1960s some 80 000 ha in Great Britain were fallowed, declining to 40 000 ha by 1987 (Table 2.1).

The movement of the soil by cultivation

The mouldboard plough was first used in Britain before the Roman occupation and with it came the beginning of cultivation as it is on our farms today. Over the intervening centuries the equipment and the techniques of cultivation have been reshaped and improved but the basic objects have remained essentially three: (a) the inversion of the top soil so as to bury trash and provide a clean surface; (b) the loosening of the soil surface to create a seedbed for the new crop; and (c) the control of weeds.

Cultivation was and is a major means at the farmer's disposal of

Table 2.1 Growth and decline of rotational husbandry

1000 BC–8th century	Cereals the main crop, some beans and peas; fallows of various duration for fertility conservation
8th–16th century	Open field system widespread in lowlands (autumn crop, spring crop and fallow or cereal, fallow) each farmer having many scattered strips. Also Scottish run-rig system — permanent infield and extensive outfield alternately cropped and rested
13th–19th century	The enclosure movement in progress; main periods of activity 16th and 18th centuries
17th century	Introduction of turnips, red clover and potatoes
18th century	Inter-row cultivation with greater flexibility of cropping led to the development of sound rotations with 'cleaning' crops and cereals eliminating the fallow on all but the heavy land. Four-course rotation of roots, barley, seeds, wheat
19th century	Development of rotations based on the Norfolk four-course principles to suit the environment and also the market provided by an expanding urban population *Examples* six-course: oats, roots, oats, 3-years' grass on dairying and stock farms mainly in the wetter and upland areas six-course: oats, potatoes, wheat, roots, barley, 1-year grass and clover six-course: rape, wheat, seeds, wheat, oats, wheat, in Lincolnshire fenland five-course: wheat, oats or barley, roots, barley, 1-year grass and clover, in the fertile areas of the drier east four-course: wheat, fallow, wheat, beans, for heavy land
20th century	Sugar-beet increasingly established as a combined cleaning and cash crop
Mid-20th century	Increased use of fertilizers and herbicides caused a decline in traditional cleaning crops. Area devoted to cereals increased
1970s	Major switch from spring- to autumn-sown barley and increase in winter oilseed rape, with associated need for minimal tillage or direct drilling to increase the speed of sowing and output per man. Increased specialization in cereals in certain areas
1980s	Surplus cereal production and lower profit margins; increasing areas of other combinable crops, e.g. oilseed rape, peas, beans, linseed; grants for cereal areas to be set-aside; fallowing may increase; increasing interest in organic farming methods

preventing weeds from competing with his crops. The ways in which cultivation achieves weed control are:

Burial: the covering of weeds by soil so that they die

Cutting: the severance of the aerial parts from the roots or rhizomes, usually below ground level

Stimulation: the provision of a soil environment that encourages dormant seeds or buds to change into an active and therefore vulnerable state

Desiccation: rhizomes or roots are brought to the soil surface where they may be exposed to a dry atmosphere, or to dry soil

Exhaustion: continuous stimulation of dormant buds on perennial weeds coupled with a denial of photosynthetic activity leading to the exhaustion of carbohydrate reserves and the death of meristems

The plough

As the basic tool of soil cultivation, the mouldboard plough has traditionally been used to start the sequence of soil movements leading to a seedbed. In the control of weeds its functions are to bury seeds and weeds growing on the surface, to loosen the soil so that other implements can attack weeds, and to bring to the surface dormant weed seeds which then germinate and can be killed by subsequent cultivations.

The objects, the first direct and the others indirect, have not altered since the introduction of the mouldboard plough 2000 years ago but considerable ingenuity has been used to improve the efficiency with which they are accomplished, particularly in the last three centuries.

The first ploughs were made of wood. Wooden mouldboards were progressively replaced by wood plated with iron and in the 18th century by cast iron, and in more recent times with high quality steel. All led to a progressive improvement in the inversion of the furrow. Comparable developments in shares and coulters also made a significant contribution.

The heavy ploughs of the Middle Ages required as many as eight oxen to draw them slowly through clay soil. Changes in design and the new manufacturers of the 18th and 19th century led to a plough suitable for two- or four-horse teams. The industrial revolution then brought the steam plough. These were progressive but not striking advances. A change of major significance came with the tractor in the 20th century; multi-furrow ploughs, which could travel at speeds greater than had previously been possible came into widespread use. The importance of these tools to weed control lies in the speed and timeliness with which the operation can be carried out. On the other hand, the rapidity with which seedbeds can be prepared can prevent the killing of seedling weeds that may be achieved when more time lapses between operations. The quality of ploughing has also suffered and poor inversion of the soil may not adequately suppress perennial weeds.

The single-furrow rigid plough which was capable of throwing the soil in one direction only called for considerable expertise in opening and finishing the 'lands'. When these 'one-way' ploughs were equipped with several bodies in the 20th century, opening and finishing required even more skill. It became common to see weeds which had been improperly buried, growing in lines across the field. In the 19th century, steam ploughing was carried out by reversible ploughs drawn by cables; in this system, the soil was inverted progressively across the field and 'opening and finishing' was unnecessary. Such ploughs were used mainly on the peat and clay soils. The introduction of the hydraulic linkage from 1946 onwards again brought the reversible plough into prominence and made possible a more complete burial of weeds.

The systematic approach to husbandry that came in the 18th century brought with it a recognition that for satisfactory cultivations after ploughing, whether for weed control or for sowing, there were particular requirements in the quality of ploughing. These requirements necessitated different plough designs. From the early 19th century onwards a profusion of novel ploughs were publicized. Their significance in weed control can be illustrated by the two extremes. The 'digger' plough developed about 1885 had a bluff elliptical body that turned the furrow slices quickly thereby breaking them better for subsequent spring cultivations; it did not, however, always provide good burial. In contrast, the 'lea' or long breasted body was used for turning unbroken slices; in the ploughing of grassland, its use in association with skim coulters achieved a complete inversion of the old sward. Similarly it provided excellent burial on arable land in the autumn and left the ridged finish necessary for effective weathering and the sowing of seed by hand. Against a background of a rapidly decreasing labour force since 1945, the tractor with its speed and power allowed a continuation of the laborious practice of ploughing. But in recent years many farmers have turned to reduced cultivation systems or direct drilling which avoid the need for ploughing (see pp. 52 and 69, and Tables 2.1 and 2.2).

Harrows and cultivators

In the Middle Ages the plough, used repeatedly, was in many situations the only implement employed in seedbed preparation but since the 18th

Table 2.2 Development and decline of ploughing

17th century	Heavy wooden ploughs drawn by up to eight oxen gradually replaced by lighter wheeled horse ploughs
18th century	Parts made increasingly of wrought and cast iron; introduction of cast iron shares and coulters
19th century	General improvements in design, introduction of depth wheels, reversible ploughs for use on sloping land. Development of steam traction on large arable farms
20th century	Progressive substitution of the tractor for the horse associated with increased number of furrows turned, and increased speed and depth of operation. Decline in the use of steam. Use of tractor hydraulic depth control and reversible ploughs. In recent times, a decline in ploughing on arable farms

century a further sequence of operations such as cultivating, grubbing, harrowing or dragging has been involved. Intended primarily for the creation of seedbeds, the functions of these operations in weed control have traditionally been to pull roots and rhizomes to the surface, to stimulate seeds to grow so that they may be destroyed and to bring about exhaustion and damage by the burial and breakage of young shoots.

The working component of harrows and cultivators has since mediaeval times been a spike or tine held more or less vertically and dragged through the top soil. The earliest harrows were wooden spikes driven into frames made from the branches of trees. The use of cast iron made it possible to use forward-curved tines which facilitated penetration into the soil and also helped to drag the roots and rhizomes of weeds to the surface. In general cultivators are heavier and have their tines wider apart than harrows. They penetrate deeper but do not move the surface of the soil so thoroughly.

Harrows are used to kill weeds during seedbed preparation and robust crops, such as beans, cereals and potatoes have traditionally been harrowed after emergence and before the small weeds have developed a firm anchorage in the soil. The cultivator is used at two seasons for weed control: in spring and autumn to drag rhizomes to the surface when they may be pulled into heaps by harrows; and in the autumn on cereal stubble to break up the soil surface and to encourage weeds to germinate. With these implements, as with so many others, the 19th and 20th centuries brought improved efficiency but no change in the basic concept.

The motor tractor, an early 20th century introduction but one which did not substantially replace the horse until the 1940s, greatly increased the efficiency of cultivating and harrowing in three ways. Higher speeds caused more breakage and burial of weeds. It became possible to carry out the operation when soil conditions and climate were optimal, and tractor 'power-drive' allowed methods of soil disturbance other than that caused by dragging tines.

Rotary cultivators have their origin in the middle of the 19th century when several patents were taken out for novel machines which used the rotary principle for the breaking of clods. But the machines which were powered by steam were too heavy and expensive to be economic and it was not until the arrival of the modern tractor that widespread use of the rotary principle became feasible. The particular value of the rotary cultivator in weed control is its ability to cut rhizomes of perennial weeds into short lengths, thereby stimulating the growth of dormant buds. At the same time the fast movement of the blades breaks or damages any leaves and stems that are present, stopping photosynthesis and allowing desiccation. The machine is an effective agent of weed destruction by combining stimulation with exhaustion but is slow in travelling over the ground. Reciprocating and rotary power harrows have further extended the weedkilling abilities of conventional harrows.

The hoe

Ploughs, cultivators and harrows have in common that they disturb the whole of the soil surface over which they travel and they can be used only when the ground is free of a crop or when the crop is of such vigour as to withstand rough treatment. The control of weeds in the majority of growing crops was, until the development of herbicides, achieved by hand pulling or by the hoe. The important features of the hoe are: (a) its suitability for use in the growing crop because it can be guided accurately by hand; (b) it disturbs the soil only shallowly, hence uproots small annual weeds and cuts large annual and perennial weeds below the surface; and (c) its efficient use is impossible or difficult on wet soil.

Until the mid-20th century there were two main types of hoe: hand hoes and those drawn by horse or tractor and guided by hand (steerage or side hoes). The latter were an essential part of the 'row-crop' system of crop production developed in the 18th century by Tull and others and called 'the horse-hoeing husbandry'. Planting the crop in parallel rows allowed a horse to pull the hoe (guided by a man) between the rows and so kill the weeds growing on the major part of the soil surface. The remaining weeds growing within the rows were killed by hand-hoeing. The success of row-cropping led to a change from the broadcasting to the drilling of seed and made possible the use of rotations of cleaning and fouling crops.

Little needs to be said about the hand hoe except that its weight, shape and cutting performance were greatly improved by the introduction of steel in the mid-19th century. The horse hoe started as an assembly of blades drawn by a horse to weed one inter-row at a time. But with the improvements in design and materials brought by the Industrial Revolution it became possible in the early 19th century for one man and one horse to hoe several crop rows. Taking the country as a whole there was little change until the widespread acceptance of the tractor from 1945 onwards. During the post-war period, tractor-mounted steerage hoes, initially developed in the USA came into use in Britain. At first they were mounted on the rear of the tractor and required two men to cover four rows, but gave way to mid-mounted hoes with which one man could cover up to eight rows. Other refinements such as independent hydraulic control for undulating ground were also introduced.

The combination of hand- and horse-hoeing was in keeping with the values of the 19th century but in the 20th century manual labour became increasingly scarce and costly. The agricultural engineering industry sought to reduce hand work by developing systems of weed and crop plant removal within the row by means of weeders, down-the-row-thinners and gappers.

Electromechanical and electrochemical thinners have been developed for sugar-beet, and the former were commercially available. All such machines have so far suffered from the inability to distinguish between the crop and weed plant growing side-by-side. It is this weakness which has emphasized the value of selective herbicides.

Changes in cultivation

Since 1960 there has been increased interest in and use of systems involving reduced cultivation in which mouldboard ploughing does not occur and the seedbed is obtained by several shallow passes of a tine or disc cultivator. The invention of machines for direct-drilling crops without prior cultivations (except where required to bury ash after straw burning) has enabled this practice to develop on large areas of suitable soils. The main crops involved are cereals, oilseed rape and some fodder crops. The absence of soil inversion, or, in the case of direct-drilling, any major disturbance of the soil changes the weed populations requiring control (see Chapter 1). These simplified methods of seedbed preparation are perhaps the most significant cultural pattern changes in arable agriculture in Great Britain for over 200 years. They rely almost entirely on chemicals to control weeds and volunteer plants. By contrast, a small but increasing number of farmers are reverting to organic farming. This necessitates a return to many of the old laborious methods of controlling weeds by cultivations and hand-weeding.

Prevention of weed dispersal

Crop seed

When man first harvested and sowed seed he introduced a new process for weed dispersal, since the weed seeds which contaminated the crop seed were preserved and then sown in an environment favourable to germination and growth. At first this merely helped to ensure survival of the contaminating species within their habitat but, with the expansion of shifting cultivation and then of seed barter and sale, the area of dispersal gradually extended. In the 19th century a large scale trade in seeds grew up between the continents and weed seeds could move great distances along the trade routes of the world.

Farmers' efforts to reduce the growth of weeds in their crops by means of rotations and cultivations did much to reduce the numbers of weeds able to set seed and so the quantity of seed harvested. But these activities were not enough to ensure a seed crop free from weeds.

The cleaning of threshed seed by movement of the air dates from ancient times. Winnowing in its simplest form (such as is still practised in primitive agriculture in many tropical countries) consists of throwing the seed in the air so that the wind currents carry away the chaff and light weed seeds, leaving the heavier material to fall back to the ground. By the late 18th century manually-operated fanners and hand winnowers were widespread. Later winnowers were incorporated into threshing machines. Subsequent developments in separation have exploited other physical differences between crop and weed seeds. Sieves or screens separated material of different external dimensions. Indented cylinders are now widely used to exploit differences in length.

Machinery was introduced in the early 1920s for the magnetic separation of rough and smooth textured seeds after coating with powdered iron filings; this played an important part in controlling the distribution of dodders in clover seed. The gravity separator, initially developed to purify and grade ore, uses differences in the specific gravity of seeds. Electrostatic separation, first developed for minerals, has recently been used to exploit differences in the conducting properties of seeds. Another recent development is the use of a photo-electric cell for separation on the basis of colour. Machines with conveyor belts made from Velcro type material are being used for the successful separation of wild oats from other cereal grains.

As increasingly sophisticated methods of seed cleaning have been developed it has become possible to eliminate all but those species with seeds closely resembling the seed of the crop that is acting as host. It is necessary to ensure that such weed seeds are not harvested with the crop, either by growing the crop on land free of the offending weed, or by eliminating them in the field before harvest. Administrative measures have become necessary to reduce the risk of contaminated seed being sown on clean land.

Reliable information on the type and extent of seed contamination can be obtained only from accurate seed analysis. The first seed testing service started in Saxony in 1869. In 1871 the Royal Agricultural Society of England appointed a consulting botanist to examine seed for its members. Subsequently the Society laid down standards which it recommended should be demanded by its members when purchasing seed.

Legislation to control the distribution of weed seeds in crop seed is of recent origin in Britain. Official seed testing stations were set up in 1914 in Scotland and in 1917 for England and Wales. Prior to 1914 there had been a campaign for the control and improvement of seed quality and after the outbreak of war it gained the support of the Ministry of Agriculture's Food Production Department, resulting in the introduction of the Seed Testing Order. This legislation later became permanent, as the Seeds Act 1920.

The requirement for declaration of information to a buyer concerning percentage purity and germination resulted in a marked improvement of seed quality. It also became an offence to sell or sow seed containing more than 5% by weight of certain injurious weed seeds; this particular provision had little effect because it was difficult to enforce and the permitted level was high. However, another clause, requiring the declaration of the presence of seeds of dodders in a sample, together with the introduction of improved cleaning machinery, resulted in the virtual elimination of these species in English clover seed. When new seeds regulations were introduced in 1961 the principle of declaration by the seller of the number of weed seeds in a sample was extended and a revised list of injurious species introduced.

Voluntary seed crop inspection and certification schemes, although primarily designed to produce seed to high varietal purity, have also played an important part in restricting weed seed distribution.

Table 2.3   Development of seed quality control

1869	First seed testing service established at Tharandt, Saxony
1871	Royal Agricultural Society appointed a consulting botanist to examine seeds. Subsequently the Society laid down standards to be demanded by members buying seeds
1914	Establishment of Scottish Seed Testing Station
1917	Official Seed Testing Station was set up in England
1920	The first Seeds Act
1947	Introduction of Cereal Field Approval Scheme by the National Institute of Agricultural Botany
1957	The various herbage seed certification schemes consolidated under the National Certifying Authority
1961	New Seeds Regulations under Seeds Act 1920 to replace those made in 1922
1964	Plant Varieties and Seeds Act
1974	Seeds Regulations under 1964 Act introducing EEC standardized system of seed quality control

The European Economic Community has a standardized system of seed quality control which must be operated by all member governments. Amongst the other provisions, detailed standards are set for the maximum contents of seeds of contaminating species in officially drawn and tested samples of prescribed size. The requirements were brought into effect in the UK from 1 July 1974 by a series of Seeds Regulations made under the Plant Varieties and Seeds Act 1964, and it is an offence to sell seed which does not comply with the conditions (Table 2.3).

Crop residues

Recent research on wild oats has indicated the significance of forms of weed dispersal which must have operated for a long time within farms but which became more widespread with the development of modern long-distance transport. Seeds of weeds become trapped in straw bales, or emerge from the harvesting process mixed with feed grain for consumption by cattle or game; and although the mortality of such seed is considerable a proportion of it may find its way back to the land in farmyard manure and other cleanings. However, not all such seed dispersals are bad: it has long been a practice of grassland farmers in hill areas to feed hay on land to be improved in order to introduce seeds produced on more fertile land.

Table 2.4 Major events in the development of herbicides before 1941

1896	France	{ Copper sulphate first used for selective weed control in
1898	Britain	{ cereals
1901–19	Europe and USA	Ferrous sulphate, sulphuric acid, sodium chlorate used as herbicides
From 1930	Britain	Substantial acreage of cereals sprayed annually with sulphuric acid
1932–33	France	Dinitro-phenols and cresols patented and used for weed control in cereals

Farm machinery

While nearly everything that moves onto a farm or field may carry small quantities of weed seeds with it, the combine harvester has been identified as a major agent of weed dispersal, seeds being trapped in the mechanism and subsequently released. The change in cereal harvesting methods from the reaper and binder to the combine harvester which occurred from 1940 onwards greatly enhanced this form of dispersal. There is a code of practice which is concerned with the cleaning of combines (National Wild-oat Advisory Committee publication, May 1975. Obtainable from the Ministry of Agriculture, Fisheries and Food — MAFF).

Chemical weed control

The first selective herbicide in British agriculture was copper sulphate used to control broad-leaved weeds in cereals. Other mineral salts were also used and later sulphuric acid. Then the value of dinitro-phenols was discovered (Table 2.4). Chemical weed control, however, developed rapidly only after the discovery in 1941 that the salts of the chlorinated phenoxy-acetic acids were selectively herbicidal. The chemicals hitherto used in cereal crops did not have the efficiency and safety of these new discoveries.

The development of modern herbicides

During the late 1930s research workers in Britain and elsewhere were searching for chemicals which would regulate the growth of plants. With the outbreak of war in 1939 these efforts became secret. Unknown to each other, workers at Jealott's Hill Research Station and at Rothamsted Experimental Station were studying the salts of the chlorinated phenoxyacetic acids and both reported the remarkable herbicidal activity of some of these chemicals to the Agricultural Research Council in November 1942. Thereafter development trials with MCPA and 2,4-D — which were the most active compounds discovered — were undertaken by the Council, the Norfolk Agricultural Executive Committee and Imperial Chemical Industries Ltd. The control of charlock in wheat, oats and barley was the main objective but many other dicotyledonous weeds were shown to be sensitive. With the return of peace the original findings together with the subsequent experiences were published, as were the results of independent work in the USA which had led to the development of 2,4-D. Experimental work also started in Britain in 1941 on the development of DNOC which had been used for weed control in France since 1933. The first carbamate herbicide, propham, was discovered at Jealott's Hill in 1941. MCPA, 2,4-D, DNOC and propham were the forerunners of the present array of selective herbicides.

From 1950 onwards industrial research began to produce many new herbicides. Britain, the USA and Switzerland were first in the field, but since then all the industrial nations have become involved. In consequence British agriculture has been the recipient of a continuously expanding range of herbicides.

The immediate impact of the growth-regulator herbicides was on the cereal crop and, to a much lesser extent, on peas and grassland. Throughout the 1950s both the range of herbicides and their use on cereals expanded steadily. By 1960 a large proportion of the cereal crop was sprayed. The discovery of mecoprop, dichlorprop, 2,3,6-TBA, dicamba and ioxynil increased the number of weed species that could be controlled. As a result of the regular use in cereals of herbicides active only against dicotyledonous weeds and consequential continuous cereal growing, annual and perennial grass weeds increased.

The decade of the 1970s brought increased awareness of the growing danger posed by grass weeds in cereals and other crops, common couch, wild oats, blackgrass, barren brome and onion couch being the principal problems; increased research has led to new herbicides and new techniques for their control. Thus herbicides such as amitrole, paraquat, and tri-allate became important.

In the 1960s and 1970s several wild oat herbicides were introduced, and the substituted ureas (metoxuron, chlorotoluron and isoproturon) became widely used in cereals — especially where blackgrass was a problem. These herbicides controlled both grass and dicotyledonous weeds and into the 1980s were increasingly used in mixtures where improved broad-leaved weed control was necessary; other dual-action chemicals such as pendimethalin have been increasingly used in the 1980s in cereals. Some of the newer herbicides, notably the sulphonylureas and imidazolinones are very active at rates as low as a few tenths of a gram ha^{-1}. The large increase in the area growing oilseed rape in the 1970s and 1980s has resulted in several herbicides being introduced for this crop, e.g. carbetamide, propyzamide and metazachlor.

New herbicides, many soil-acting, have brought a wider diversity of provision for crops such as beans, peas, potatoes, sugar-beet, carrots, onions, celery, brassica and fruit, and as a result there has been a considerable reduction in soil cultivation for weed control. Glyphosate has proved very useful for controlling most weeds between crops and for pre-harvest use in cereals, peas and oilseed rape. Herbicides such as alloxydim-sodium, sethoxydim, fluazifop-butyl and quizalofop-ethyl can be used to control grass weeds selectively in many broad-leaved crops.

This decade has seen the fruition of the direct-drilling and reduced cultivation techniques of growing crops — made possible by herbicides. Apart from the simplification of production of market-garden crops, the introduction of effective herbicides has encouraged the farm-scale production of these crops, e.g. Brussels sprouts, carrots, onions. The production of soft fruit has also benefited from the use of herbicides.

With increasing problems from numbers of weed species developing on farms it has become necessary to use two or more herbicides to control the full weed spectrum and this has often involved mixing of the herbicides in the sprayer tank and requiring special care and advice. In many cases, mixing herbicides gives an enhanced or synergistic effect, but in a few cases it can be antagonistic; many herbicide mixtures are now available as proprietary products.

Some herbicides such as the nitrophenols have been withdrawn from use because of their mammalian toxicity, and others are having tighter restrictions imposed on their use. A few instances of weed resistance to herbicides are now being found, including a strain of blackgrass resistant to chlorotoluron and diclofop-methyl in Essex, and some groundsel resistant to simazine (see Table 2.5 & Chapter 8).

Table 2.5 Approximate dates of the introduction of organic herbicides and formulated herbicide mixtures into commercial use in the UK

Year	Single herbicides	Two-component formulations	Multi-component formulations
1943	DNOC	—	—
1945	MCPA 2,4-D	—	—
1949	Dinoseb-ammonium Cresylic acids Mineral oils	—	—

Table 2.5 Cont.

Year	Single herbicides	Two-component formulations	Multi-component formulations
1950	TCA	—	—
1953	—	2,4-D/2,4,5-T	—
1954	MCPB	—	—
1955	2,4,5-T Pentachlorophenol	—	—
1956	Simazine Dinoseb-amine	MCPB/MCPA	—
1957	Dalapon Mecoprop Monuron 2,4-DB	2,3,6-TBA/MCPA	—
1958	Atrazine Barban	2,4-DB/MCPA Chlorpropham/diuron Chlorpropham/fenuron	—
1960	Aminotriazole (amitrole) Propham Sodium monochloroacetate 2,4-DES 2,4,5-TB Diuron	Endothal/propham Borate/monuron Borate/simazine Mecoprop/2,4-D	Propham/chlorpropham/ fenuron Propham/diuron/MIPC
1961	Dichlorprop Diquat Di-allate Chlorpropham	2,4-DB/2,4-D Dicamba/MCPA	—
1962	Pyrazon (chloridazon) Paraquat	Dichlorprop/MCPA Sulfallate/chlorpropham	—
1963	Chlorthiamid Desmetryne Dimexan Linuron Pentanochlor	MCPA/2,4-D Propham/diuron Fenoprop/MCPA Paraquat/diquat	2,4-D/monuron/sodium chlorate Dalapon/2,4-D/amitrole
1964	Monolinuron Chloroxuron Tri-allate	Ioxynil/dichlorprop Dichlorprop/2,4-D Prometryn/simazine Fenoprop/2,4-D Dalapon/MCPA Maleic hydrazide/2,4-D Mecoprop/fenoprop	MCPA/mecoprop/ dicamba/2,3,6-TBA
1965	Prometryn Nitrofen	Pentanochlor/ chlorpropham	Ioxynil/MCPA/ Dichlorprop

Table 2.5 Cont.

Year	Single herbicides	Two-component formulations	Multi-component formulations

	Lenacil
	Fenoprop
	Dichlobenil
	Metobromuron
	Bromacil
		Ioxynil/MCPA
		Ioxynil/mecoprop
		Borate/bromacil
			Benazolin/2,4-DB/
			  MCPA
			MCPA/2,4-DB/2,4-D
			Dimexan/cycluron/
			  chlorbufam

1966	Ametryne
	Morfamquat
	Dinoseb-acetate
		Linuron/monolinuron
		Mecoprop/dicamba
		Dichlorprop/mecoprop
		Sodium chlorate/
		  bromacil
		Sodium chlorate/diuron
		Mecoprop/2,4-DB
			Mecoprop/MCPA/
			  dichlorprop/2,4-D

1967	Propachlor
	Picloram
	Trifluralin
	EPTC
		Linuron/chlorpropham
		Bromoxynil/MCPA
		Medinoterb acetate/
		  propham
		Chloridazon/chlorbufam
		Dinoseb-acetate/
		  monolinuron
			MCPA/dicamba/
			  mecoprop
			Endothal/propham/
			  medinoterb-acetate

1968	Phenmedipham
	Terbutryn
	Methabenzthiazuron
	Asulam
		Monolinuron/paraquat
		Picloram/dichlorprop
		Methoprotryne/simazine
			Ixoynil/bromoxynil/
			  dichlorprop
			Atrazine/amitrole/
			  2,4-D
			Bromacil/2,4-D/sodium
			  chlorate
			Benazolin/MCPA/
			  dicamba
			Benazolin/MCPA/
			  MCPB

1969	Metoxuron Chlorbromuron Aziprotryne	Flurecol/MCPA Ioxynil/linuron Amitrole/diuron Bromoxynil/mecoprop	Chlorpropham/fenuron/ monolinuron Amitrole/dalapon/diuron

1970	Ioxynil
	Alachlor
	Terbacil
		Linuron/trietazine
		Dinoseb/MCPA
		Buturon/isonoruron
		2,4-DES/chlorpropham
		Metoxuron/simazine
		TCA/fenuron
		Dichlobenil/fluometuron
			Amitrole/diuron/2,4-D
			Maleic hydrazide/2,4-D/
			  chlorpropham
			Amitrole/dichlorprop/
			  diuron/MCPA
			Chlorpropham/propham/
			  diuron

1971	Bromoxynil
	Chlorotoluron
	Propyzamide
		Bromoxynil/ioxynil
		Sodium chlorate/
		  monuron
			Dalapon/diuron/MCPA
			Dicamba/2,4,5-T/
			  mecoprop

Table 2.5 Cont.

Year	Single herbicides	Two-component formulations	Multi-component formulations
	Cyanazine Chlorfenprop-methyl	Terbutryn/terbuthylazine Bromofenoxim/ mecoprop Aziprotryne/simazine Picloram/bromacil Amitrole/simazine Paraquat/diuron	
1972	Benzoylprop-ethyl Diphenamid Methazole	Nitrofen/neburon Sodium chlorate/atrazine Dichlobenil/dalapon Trietazine/simazine Dinoterb/mecoprop Cyanazine/MCPA	Amitrole/simazine/ MCPA MCPA/dichlorprop/ benazolin Picloram/paraquat/ diuron Dichlorprop/MCPA/ dicamba Dichlorprop/benazolin/ dicamba Ioxynil/bromoxynil/ dichlorprop/MCPA
1973	Benzadox Cycloate Metribuzin Bentazone	Bentazone/dichlorprop Lenacil/propham Bentazone/MCPB Cyanazine/linuron Linuron/lenacil	Dichlorprop/MCPA/ mecoprop Dicamba/MCPA/ mecoprop Barban/MCPB/ dichlorprop/mecoprop Dicamba/2,4,5-T/2,4-D
1974	Difenzoquat Carbetamide Chlorthal-dimethyl Glyphosate Ethofumesate		2,4-D/dichlorprop/ MCPA
1975	Isoproturon Flamprop-isopropyl Hexazinone	Bromofenoxim/ terbuthylazine Dichlobenil/bromacil Cyanazine/atrazine Amitrole/atrazine	—
1976	Diclofop-methyl Dinitramine	3,6-dichloropicolinic acid (clopyralid)/mecoprop Clopyralid/benazoline carbetamide/dimefuron metoxuron/mecoprop	Clopyralid/dichlorprop/ MCPA Metoxuron/simazine/ barban
1977	Flamprop-methyl Metamitron Oxadiazon	Chlorthal-dimethyl/ methazole Chloridazon/di-allate Cyanazine/mecoprop	Ethofumesate/propham/ fenuron/chlorpropham Ioxynil/bromoxynil/ mecoprop Mecoprop/asulam/MCPA

Table 2.5 Cont.

Year	Single herbicides	Two-component formulations	Multi-component formulations
1978		Trifluralin/napropamide Trifluralin/linuron	2,3,6-TBA/dichlorprop/ mecoprop
1979	Pendimethalin Clopyralid Fosamine-ammonium	Chloridazon/ ethofumesate Dichlorprop/bromoxynil Nitrofen/linuron Atrazine/sodium monochloroacetate Bromacil/diuron	Isoproturon/ioxynil/ bromoxynil Ioxynil/bromoxynil/ mecoprop/linuron Sodium monochloroacetate/ dalapon/MCPA
1980	Alloxydim-sodium Butam	Atrazine/dalapon Clopyralid/propyzamide Bifenox/linuron	Chloridazon/ chlorpropham/ propham/fenuron Bentazone/MCPB/ MCPA
1981	Pyridate Endothal Fluazifop-butyl	Simazine/methazole Bifenox/mecoprop Isoproturon/tri-allate Chloridazon/lenacil	Clopyralid/mecoprop/ bromoxynil Bromoxynil/MCPA/ mecoprop Propham/fenuron/ chloridazon
1982	Metazachlor Triclopyr	Clopyralid/bromoxynil Cyanazine/MCPA Chlorsulfuron/ methabenzthiazuron Isoproturon/trifluralin	Clopyralid/ioxynil/ mecoprop Ioxynil/mecoprop/ bromoxynil/MCPA Linuron/trietazine/ trifluralin Isoproturon/ioxynil/ mecoprop
1983	Napropamide	Metoxuron/simazine Cyanazine/clopyralid Paraquat/simazine Napropamide/simazine	Bromoxynil/ioxynil/ benazolin Chlorsulfuron/ioxynil/ bromoxynil Triclopyr/dicamba/ mecoprop Dichlorprop/linuron/ MCPA Benazolin/ioxynil/ mecoprop Amitrole/diquat/ paraquat/simazine Bromoxynil/ioxynil/ trifluralin

Table 2.5 Cont.

Year	Single herbicides	Two-component formulations	Multi-component formulations

1984	Fluroxypyr
	Isoxaben
	Quizalofop-ethyl
		Fluroxypyr/clopyralid
		Chlorotoluron/bifenox
		Chlorsulfuron/
		  metsulfuron-methyl
		Fluroxypyr/ioxynil
		Trifluralin/terbutryn
		Terbutryn/linuron
		Phenmedipham/
		  clopyralid
		Chlorthal-dimethyl/
		  diphenamid
		Chloridazon/propachlor
			Fluroxypyr/ioxynil/
			  bromoxynil
			Bentazone/dichlorprop/
			  MCPA
			Ethofumesate/ioxynil/
			  bromoxynil
			Isoproturon/bromoxynil/
			  ioxynil/mecoprop

1985	Metsulfuron-methyl
	Imazapyr
		Terbutryn/prometryn
		Triclopyr/clopyralid
		Propachlor/
		  chlorthal-dimethyl
			Dicamba/triclopyr/2,4-D
			Fluroxypyr/clopyralid/
			  ioxynil

1986	Imazamethabenz-methyl	Diflufenican/isoproturon
		Thiameturon-methyl/
		  metsulfuron-methyl
			Fluroxypyr/ioxynil/
			  bromoxynil/clopyralid

1987		Isoproturon/bifenox
		Fluroxypyr/bromoxynil
		Pendimethalin/
		  isoproturon
			Isoproturon/bifenox/
			  mecoprop
			Ioxynil/bromoxynil/
			  benazolin/mecoprop
			Bentazone/2,4D-B/
			  cyanazine
			Ioxynil/dicamba/
			  dichlorprop

1988		Isoproturon/isoxaben
		Methabenzthiazuron/
		  isoxaben
		Phenmedipham/
		  ethofumesate
		Phenmedipham/lenacil
		Terbuthylazine/atrazine	

1989		Isoproturon/cyanazine Isoproturon/metsulfuron- methyl Trietazine/terbutryn	

Herbicides introduced first as single chemicals are also given when they occur later in formulated mixtures. When first introduced in a formulated mixture, they are listed as individual chemicals only if and when they are subsequently marketed as such. No account has been taken of activators, adjuvants or safeners and inorganic herbicides are noted only when formulated with an organic herbicide. Some of the herbicides and mixtures listed were developed for industrial or forestry use only. A few of the herbicides listed have been withdrawn from the market

The application of herbicides

Some wheeled sprayers — usually horse-drawn — were in use in the 1930s
to apply herbicides, notably sulphuric acid and therefore with wooden
tanks. However, the only piece of equipment commonly available for the
application of herbicides on farms in the 1940s was the fertilizer distributor
and in consequence MCPA and 2,4-D were first used mainly as dusts.
Liquid formulations were of minor importance and spraying was first seen
as mainly a contractor's operation. The introduction to Britain of the low-
volume spraying machine costing about £50 in 1948 made spray application
by farmers both easy and cheap. The number of ground crop sprayers
purchased increased rapidly from 4200 in 1950 to 57 500 in 1966 after which
the increase in numbers proceeded more slowly. Although the numbers of
sprayers has not increased in recent years there has been an increase in
individual sprayer capacity and an increase in boom width up to more than
18 m on some machines, and so there has been a steady increase in the area
that can be sprayed each day. Many improvements have been introduced
to facilitate safer and more efficient spraying.

    With the increased diversity of crops being sprayed and the increased
sophistication of herbicide formulations involved there has come a variety
of equipment for special purposes. For row crops there are band sprayers
to apply herbicides over the crop row and not between the rows. Low pressure nozzles and 'vibrojets' were promoted to reduce the likelihood of
drift. Granule applicators have come into use on cereal farms. There is a
multiplicity of small applicators for garden, small-holding and orchard use.
Roguing gloves for applying herbicide to weeds such as wild oats in cereals
can reduce hand-roguing time. Selective applicators utilizing rollers and
rope-wicks may be used to control weeds which are taller than the crop,
e.g. weed beet in sugar-beet and thistles and other weeds in grassland.
Controlled droplet application using rotary atomizers and electrostatic
spraying offer possible ways of reducing the amount of chemical required
and the likelihood of drift. Glyphosate is now commonly used to control
green perennial weeds in cereals, peas and oilseed rape applied 1–3 weeks
before harvest; high-clearance wheels and a sheet under the tractor minimize damage to the crop. Low ground-pressure vehicles are now available
for spray application when soils are too wet and soft for conventional
tractor sprayers to operate; self-propelled sprayers are now also available
(see Chapter 5). Repeated low dose applications of herbicides are now
commonly used — especially for post-emergence weed control in sugar-
beet.

Biological weed control

The use of biological agents to control weeds is almost as old as agriculture itself. Without realizing it, many farmers have, from the earliest times, reduced weed problems by establishing strong stands of their crops and thus have exploited crop competition as a weed control measure. In this, they have pre-empted modern agricultural practice. Similarly, grazing animals have been used for centuries as a means of suppressing unwanted vegetation. They are still used in the Somerset levels where, in autumn and winter, cattle are allowed into basket-willow beds to graze off grass and other weeds and keep them in check until the willows regrow in the following year and repress them by competition. Sheep are used to control bankside growth on rivers and drainage channels in many parts of the UK.

Biological control, as it is presently recognized, was only developed comparatively recently and has evolved to produce two principal tactics. The longest established classical tactic is aimed, generally, at plants that have been introduced into a new area and achieved weed status as a result of a lack of natural enemies and favourable growth conditions. Control agents, especially insects and fungal pathogens, are sought at the geographical origins of the weed and introduced, one or more times, in the weed's new site. Successful introductions establish, reproduce and spread. In the process the agent reaches levels which are sufficiently high to reduce the weed population density to acceptable levels, while maintaining sufficient host material to allow survival of a low population of the control agent. Thus, future control potential, should it be required, is ensured.

As a result of the colonial activities of Europeans, most of the targets of classical biocontrol are weeds of European origins now producing problems elsewhere. However, more recently, plants introduced into Britain, often as ornamentals, are presenting potential targets for biocontrol. Amongst these is Japanese knotweed which is becoming a serious problem along waterways in several areas.

The classical biocontrol tactic has produced some notable successes. The control of species of *Opuntia* cactus in Australia by larvae of a moth (*Cactoblastis cactorum*) from Argentina and the control of St John's worts in the USA by leaf-eating beetles are good examples. Skeleton weed has been successfully controlled in parts of Australia with a rust fungus (*Puccinia chondrillina*), as has the wild bramble in Chile. Most recently, the control of aquatic weeds by Chinese grass carp (*Ctenopharyngodon idella*) has had some success (see Chapter 18). Currently, efforts are being made to control bracken with insects imported into the UK.

The optimum situations for the use of classical biocontrol tactics are stable ecosystems, such as are found in waterways, rangelands or permanent grasslands, where relatively uniform, dense stands of weed occur. This is generally not true of modern agriculture and a second tactic, the bioherbicide tactic, has been developed to exploit the advantages offered by biological control. This is directed at indigenous weeds in the same way as chemical herbicides. While, in theory at least, it is possible to use a wide range of biocontrol agents in this way, micro-organisms, especially plant pathogenic fungi, are generally the favoured agents. For this reason, the terms 'microbial herbicide' or 'mycoherbicide' are used commonly.

The essence of this tactic is that biocontrol agents are sought among the target weeds, especially in undisturbed systems where their development is predictably most likely. The selected agents are produced in bulk and then applied to the weed infestation, usually annually, in the same way as a herbicide. By this means, each weed plant is inundated with the agent so that any natural resistance or environmental constraint is overcome. Once the weed has been controlled, the control agent population rapidly declines to its natural very low level.

A particular feature of mycoherbicides is their generally high level of selectivity and narrower spectrum of activity as compared with many chemical herbicides. Often, only one species is controlled. Although this may be attractive from an environmental point of view, in practical agricultural terms it presents a drawback which may be overcome by mixing two or more microbial species or by mixing with herbicides to increase the spectrum of activity. Both approaches have been used successfully in the USA. A major advantage of mycoherbicides at present is their low cost. Current estimates suggest costs of research, development and registration of a mycoherbicide may be as low as one-tenth of those for chemicals. Hence it may be economic to develop selective control agents specifically for small acreage crops.

Since the first attempts to use the technique in the late 1960s in the USA, two products have been placed on the market and several more are undergoing final commercial trials prior to registration. More particularly, there is a considerable amount of research under way which promises to produce many more 'mycoherbicides'. For example, in the UK, a major programme has been developed at the Agriculture and Food Research Council (AFRC) Institute of Arable Crops Research, Long Ashton Research Station, near Bristol. Since this programme started late in 1986, several indigenous pathogens, effective against a range of temperate climate weeds of arable crops, have been isolated. At present these pathogens have successfully passed through primary selectivity screens and have been included in field trials. There are strong hopes that the great potential shown by these organisms, especially several effective against cleavers and field bindweed, will be realized via patent application and commercial development in the near future.

Recently, emphasis has been given to developing more effective microbial herbicides by the use of genetic engineering techniques. This has enormous potential for improving their virulence and hence effectiveness and for producing herbicide and fungicide resistant strains which can be used more easily in integrated pest control strategies. However, there is a considerable suspicion regarding the consequences of releasing such organisms into the environment and progress in this area may be constrained as a result. Similar suspicions are developing about the release of non-engineered organisms. There is no evidence, after nearly 20 years of use of such organisms in the USA, for one organism involving tens of thousands of acres, that any mutational or other change, resulting in harm to non-target species, has occurred.

The present position

To understand the events of the past 40 years it is necessary to appreciate the basic weakness of the traditional forms of weed control. They were essentially aids to control; they could reduce but not prevent weeds growing with the crop. When weed removal was necessary machines could ease the work but in the end only a person with a hoe could select and remove all the weeds growing beside the crop plant. As long as manual labour was abundant and cheap in relation to the value of the produce the system could work efficiently. But from 1914 onwards the expense of labour-consuming tasks made them undesirable and stimulated interest in alternatives. In the farmer's and grower's hands selective herbicides have provided a substitute for the observational powers of man in distinguishing between the crop and weed plant growing side by side.

After the Second World War, most cereal crops were very badly infested with aggressive weeds such as yellow charlock, poppies and fathen. When the herbicide MCPA was introduced the effects were dramatic — up to 100% control and yield improvements of 30% or more in many cases. However, the herbicide had to be used every year for seven or more years to control the weeds growing from the large numbers of seeds which had accumulated in the soil during the war years. During the 1950s charlock started disappearing and weeds such as cleavers and chickweed, which were resistant to MCPA, became serious problems and CMPP (mecoprop) was introduced to control them. Later, other weeds such as the polygonums, speedwells, mayweeds, pansies and grass weeds increased to occupy the bare spaces between the cereal plants, and so more herbicides were developed to control them. In the 1950s dinoseb proved to be a very useful herbicide for controlling broad-leaved weeds in peas and made it possible to sow the crop in narrow rows which gave much higher yields.

The fruits of this revolutionary development were not gained during the 1940s and in the early 1950s because the initial advances in chemical weed control had occurred mainly in the cereal and pea crops in which hand labour for weed control was already no longer used. These crops were, in any case, considered to be weedy crops which had to be offset by a balance of cleaning crops in rotation. The coincidental spread of mechanical harvesting, which eliminated a major labour demand, opened the way for an expansion in their production. It was not until selective herbicides were developed for crops grown in rows from 1960 onwards that they, in turn, moved towards freedom from the restrictions of labour scarcity and cost.

Yield increase resulting from weed control is not the only criterion to be considered. For example, some climbing weeds such as bindweeds can cause lodging and harvesting problems; some can reduce the saleability of crops such as wild oats or wild onion bulbils in cereals, and black nightshade in peas; some grass weeds can carry cereal diseases; and a few weeds, if not controlled, can soon build up to a serious problem (e.g. wild oats).

The decade of the 1960s was an important one in demonstrating both the limitations of chemical weed control and the revolutionary possibilities opened up by herbicides. The availability of effective herbicides enabled extended sequences of cereals to be grown, but repeated use of chemicals active against dicotyledonous weeds only led to steady increase in the prevalence of grass weeds. These have been contained, though at increased cost, with the development of new chemicals. It is apparent that herbicide technology can ensure efficient crop production in the face of changing weed problems only through continuing research and production of new herbicides. Experience of the impact of herbicides on production methods in intensive crops such as vegetables has shown that growers become wholly dependent on chemicals for weed control and are therefore vulnerable to failures in supply or inadequacy of performance. There is also an increasing problem of providing chemical weed control in crops of minor area which do not justify the costs of herbicide development by the chemical industry.

Herbicides not only provide a modern alternative to the hoe but they allow crops to be grown without any soil disturbance, with consequent economy of labour, tractor power, fuel and capital investment as well as soil and moisture conservation. Very substantial changes have occurred in farmers' approaches to the cultivation of their soils. On land being prepared for cereals, mouldboard ploughing has been replaced on many farms by the use of tined implements for quicker and more shallow work, but the increasing problem of straw disposal has forced a return to ploughing on some farms. The practice of direct-drilling crops without prior cultivation increased to 278 000 ha during 1980 but has since been decreasing on the less suitable soils.

Straw-burning is widely regarded as essential for direct-drilling and this can create problems where soil-acting herbicides are required to control black-grass and other weeds since the herbicide may be adsorbed by the carbon in the ash and so become less effective. To comply with the Code of Practice for straw-burning it is usually necessary to cultivate the field to bury the ash — this helps to provide a better seedbed and overcome the herbicide adsorption problem. Ploughing every third or fourth year with complete inversion of the furrow slices buries and eliminates the small seeds of annual grasses. 'Tramlines' are now commonly used as sprayer wheel tracks throughout the season, and are especially useful for late application of herbicides in cereal crops.

Chemical weed control has also had an important impact on production methods of fruit and vegetables. Vegetable production methods evolved from hand-weeded beds to row crops suited to horse-drawn cultivation in the 18th and 19th centuries and are now reverting to bed systems and/or narrow rows but with the application of modern knowledge on density, spatial arrangements etc. which have made a major contribution to yield and quality of vegetables. The opportunity of avoiding soil cultivation in fruit also has allowed considerable improvements in production methods.

Throughout the centuries the methods of British agriculture have been influenced by the external pressures of national and world events. In the 1970s, energy costs increased rapidly leading to high inflation and world food surpluses disappeared. In the late 1970s, new developments in methods of growing crops (especially winter cereals), and the introduction of better varieties, herbicides, fungicides, growth regulators and so on, resulted in greatly increased yields; this trend continued through the 1980s resulting in very considerable surpluses of cereals and an urgent need to reduce production by grant-aided set-aside areas, or by reducing fertilizer inputs, or in other ways. Organic farming is a very effective way to reducing production and the area involved is increasing to cater for those who believe that food crops grown with the aid of chemicals are of inferior quality to those grown with organic manures and no pesticides; weed control in organic farming can be very difficult and expensive in labour. The methods of crop production in British agriculture will continue to evolve and so will methods of weed control. The development of biological control agents is simply one evolutionary step in this process. As factors such as the development of herbicide resistance, increasing costs of herbicides, and increasing environmental constraints on chemicals come to bear more strongly, so the impetus for development of biological agents will be stronger. In particular, the development of microbial herbicides (mycoherbicides) will undoubtedly accelerate as a result of a considerable increase in interest by commercial concerns.

Co-ordination of the new technology

The growth of weed control in Britain has been stimulated and guided by the British Weed Control Council and its successor the British Crop Protection Council (BCPC), both bodies being composed of representatives of the State's research and advisory organizations, chemical industry, contractors, merchants and farmers. By conferences, this *Handbook*, the *Pesticide Manual*, other publications and by public pronouncements, the Council has exerted its influence on events and has assisted similar developments in Europe.

State participation

Because of the nature of the substances used as herbicides, their importance and the need to ensure public safety, the State has been involved intimately in their development. This involvement has been by regulation, research, information and advice.

The voluntary schemes which have served well in the past — the Agricultural Chemicals Approval Scheme (ACAS) which was a guide to technical performance, and the Pesticide Safety Precautions Scheme (PSPS) which dealt with safety aspects, have now been replaced by The Control of Pesticides Regulations from 6 October 1986 (under the Food and Environment Protection Act of 1985); only provisionally or fully approved herbicides may now be advertised, sold, supplied, stored or used (see Fig. 2.1 & Chapter 21).

The main State effort in research is now based in the Institute of Arable Crops Research at the Weed Research Division, Long Ashton Research

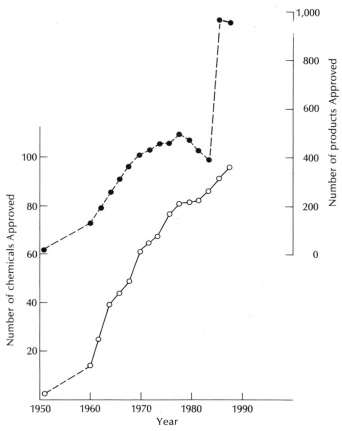

Fig. 2.1 Annual total of herbicide chemicals and products 'Approved' by the Agricultural Chemicals Approval Scheme (up to 1985) or under the Control of Pesticides Regulations 1986. ● = number of 'Approved' products, ○ = number of 'Approved' chemicals. Between 1976 and 1985, many products were cleared for use by the Pesticide Safety Precautionary Scheme (PSPS) but not 'Approved'. This accounts for the drop in products 'Approved' and the dramatic increase in numbers in 1986 when only 'Approved' products could be sold.

Station, at Broom's Barn near Bury St. Edmunds, and for aquatic weeds at Reading. Specialized research on weeds and weed control is also undertaken at state-aided AFRC institutes such as the Institute of Horticultural Research and the Scottish Crop Research Institute, by the Natural Environment Research Council's Institute of Terrestrial Ecology, by independent bodies such as the Processors and Growers Research Organization and the Norfolk Agricultural Station, and by universities. Development work on herbicides, and especially on the cost-effectiveness of control programmes, is undertaken by the Agricultural Development and Advisory Service (ADAS), on farms and on Experimental Husbandry Farms and Horticultural Stations.

Commercial participation

The agro-chemical industry has played a major role in the revolution brought about by chemical methods of weed control; in addition to discovering, developing and marketing the herbicides, it has also been largely responsible for initiating the technologies for using them safely and effectively; the concept of direct-drilling and minimal cultivation techniques are notable examples. The British Agrochemicals Standards Inspection Scheme (BASIS) is an independent scheme set up in 1978 at the behest of Government for distributors and manufacturers of pesticides, and contractors and advisors involved in crop protection (see Chapter 21).

Calendar

A summary of events governing the growth of weed control in the UK since the 1920s is given in Table 2.6.

Table 2.6   Events governing the growth of weed control

1920	Industry-published specifications of composition for lead arsenate, lime sulphur, nicotine and one or two other materials in use at the time, providing standards to which firms could work
1942	Voluntary scheme for approval of proprietary pesticides agreed between government and industry (Crop Protection Products Approval Scheme)
1941–42	Formation of a research team at Imperial College financed by the Agricultural Research Council. The team moved to Oxford University in 1945, and was established as the Agricultural Research Council Unit of Experimental Agronomy in 1950
1951	*Toxic Chemicals in Agriculture*. Report to the Minister of Agriculture and Fisheries of the Working Party on Precautionary Measures against Toxic Chemicals used in Agriculture. HMSO, London, 1951
1952–53	Agriculture Poisonous Substances Act and resulting regulations. Formation of Weed Control Joint Committee
1953	*Toxic Chemicals in Agriculture. Residues in Food*. Report to the Minister of Agriculture and Fisheries, Health and Food and to the Secretary of State for Scotland of the Working Party on Precautionary Measures against Toxic Chemicals used in Agriculture. HMSO, London, 1953
	First National Weed Control Conference at Margate.
	First Report of the Recommendations Sub-Committee (forerunner of this *Handbook*)

Table 2.6 Cont.

	Formation of the Council of the National Weed Control Conference, later re-named the British Weed Control Council (BWCC)
1955–57	Publication of Reports of BWCC Recommendations Committee in booklet form
1955	*Toxic Chemicals in Agriculture. Risks to Wild Life.* Report to the Minister of Agriculture, Fisheries and Food and to the Secretary of State for Scotland, of the Working Party on Precautionary Measures against Toxic Chemicals used in Agriculture. HMSO, London, 1955
	Informal notification of pesticides by industry
1956	Establishment of British Weed Control Conference on a biennial basis
	Publication of first *British Weed Control Handbook* (paperback).
1957	Notification of pesticides by industry based on formal agreement with the Association of British Manufacturers of Agricultural Chemicals (Notification of Pesticides Scheme)
1958	Publication of the first hardback edition of the *British Weed Control Handbook*. Formation of International Research Group on Weed Control at Ghent, Belgium, forerunner of the European Weed Research Council
1960	Formation of European Weed Research Council and publication of journal *Weed Research*. Creation of the Agricultural Council's Weed Research Organization, Crop Protection Products Approval Scheme revised and re-named Agricultural Chemicals Approval Scheme (ACAS)
1964	Advisory Committee on Pesticides established. Notification of Pesticides Scheme renamed Pesticides Safety Precautions Scheme (PSPS)
	Initiation of the *Annual Reviews of Herbicide Usage* (1–4 for BWCC, 5th onwards for BCPC)
1965	Supplementary Report of the Research Committee on Toxic Chemicals, Agricultural Research Council, London. (Specifically on herbicides.)
1967	*Review of the present safety arrangements for the use of toxic chemicals in agriculture and food storage.* Report by the Advisory Committee on Pesticides and other Toxic Chemicals. HMSO, London, 1967
1968	Major expansion of this *Handbook* into two volumes with a basis for regular updating
1969	*The Collection of Residue Data.* A report by the Advisory Committee on Pesticides and other Toxic Chemicals covering a report by its Working Party on the collection of Residue Data. HMSO, London, 1969
1970	Third Report of the Research Committee on Toxic Chemical Agricultural Research Council, London, 1970
1972	Poisons Act 1972 and Poisons Rules. HMSO, London
	Air Navigation Order, 1972 and Aerial Application Certificate (Permission), Civil Aviation Authority
	Deposit of Poisonous Waste Act, 1972
	From 1972, with Britain's application for membership of European Economic Community, increasing moves to harmonize procedures relating to use of herbicides
1973	The Institute of Terrestrial Ecology (ITE) established
1974	Control of Pollution Act, 1974
1975	Health and Safety (Agriculture) (Poisonous Substances) Regulations (1975). HMSO, London
	Formation of the European Weed Research Society (EWRS) to replace European Weed Research Council

Table 2.6 Cont.

1978	British Agrochemical Supply Industry Scheme Ltd (BASIS) incorporated, later changed to British Agrochemical Standards Inspection Scheme (BASIS)
1979	Royal Commission on Environmental Pollution, 7th Report — *Agriculture and Pollution.* HMSO, London, 1979
	Formation of International Weed Science Society with support from EWRS and other organizations
	Advisory Committee on Pesticides, *Review of the Safety for Use in the UK of the Herbicide 2,4,5-T.* MAFF, London, 1979
1980	Advisory Committee on Pesticides, *Further Review of the Safety for Use in the UK of the Herbicide, 2,4,5-T.* MAFF, London, 1980
1980s	Publication of specialized *Crop Protection Handbooks* by BCPC
1985	Food and Environment Protection Act
1986	Control of Pesticides Regulations made under the Food and Environment Act 1985
1986	First edition of *Pesticides* (1986) (Approved products under the Control of Pesticide Regulations 1986) — the 'Blue Book.' HMSO Reference Book No. 500

3 / The properties of herbicides

R.J. HANCE AND K. HOLLY*

*Consultant, Brook Hill, Woodstock, Oxford OX7 1XH;
and *Consultant, 36 Sunderland Ave, Oxford OX2 8DX*

Introduction

Over 200 compounds, that is active ingredients, with effects on plant growth are commercially available worldwide although not all are sold in every country. The distinction between those that control weeds directly (herbicides) and those whose action is to regulate crop growth and hence may affect weeds indirectly by changing crop–weed competition (plant growth regulators) is occasionally blurred but this chapter is confined to the former as far as possible. An exception is that a short section on safeners is included. The emphasis is on compounds used in the UK but others with important uses elsewhere are also mentioned. In this chapter herbicides are classified so as to try to show chemical relationships between compounds, with tables and text dealing with chemical structures, chemical and physical properties, and relevant environmental behaviour.

Herbicides may, however be classified in other ways, for example on the bases of selectivity, method of use or mode of action. Therefore, information on these topics is included and it is appropriate to discuss first the main features of these methods of classification.

Selectivity

Total or non-selective treatments aim to kill all vegetation so they are used before sowing a crop, immediately pre-harvest or in areas such as railway tracks, paths and industrial sites. Selective treatments are intended to control undesired plants (weeds) without seriously affecting those which are being cultivated (crop). Selectivity may be due to differences in retention, uptake, movement, metabolism and biochemical action between crop and weed, or in timing and method of application.

Whilst this classification can be useful for *treatments* it may not be valid for *compounds* since, with suitable rates, methods and timing of application, a 'non-selective' herbicide may sometimes be used selectively and vice versa.

Method of use

Herbicides may be applied directly to the plant foliage or to the soil in which it is growing.

Those applied directly are said to have *contact* action when only the treated part of the plant is affected, and *translocated* activity when the herbicide enters the plant and moves within it to the site(s) of action elsewhere.

Soil-applied herbicides generally affect germinating weeds and so must persist in the soil for a period if they are to be effective; hence such treatments are often referred to as *residual*.

Both foliar and soil treatments may be classified by the relation between time of application and stage of crop development:

Pre-sowing treatment: applied before the crop is sown
Pre-emergence treatment: applied between crop sowing and emergence
Post-emergence treatment: applied after crop emergence

Mode of action

The biochemical and physiological effects of herbicides are considered in detail in Chapter 7 and there herbicides are classified on this basis. Sometimes it may be convenient for limited purposes to group herbicides using a mixture of criteria. For example it is not uncommon to find terms like 'photosynthesis inhibitors', 'germination inhibitors', or even 'post-emergence grass-killers' although each group contains several classes of chemical and biochemical modes of action.

Types of herbicide treatment

Non-selective treatments

In cropping situations non-selective treatments are widely used to clear ground for cultivation or, in the case of minimum tillage/direct drilling systems, for sowing the next crop. For this purpose foliage applied compounds with no or very short residual activity are required, glyphosate and paraquat being good examples. This sort of procedure could equally be classified as a selective pre-sowing treatment. Herbicides are also sometimes used to kill vegetation, both crop and weed, immediately before harvest to make the operation easier.

Where an area is to be kept free of vegetation for some time, usually for a non-agricultural purpose, chemicals with a residual action are needed, such as diuron or simazine. If the land must first be cleared of plants an initial treatment with a foliage applied material such as aminotriazole or paraquat is applied or a dual purpose compound such as sodium chlorate is used.

Selective treatments

Pre-sowing treatments

Contact foliage treatments are used pre-sowing (or pre-planting) to kill annual weeds before planting or sowing the crop. *Translocated foliage treatments* are usually needed to control perennial weeds, such as common couch or creeping thistle in this way. For spring sown crops, this operation may be carried out during the previous autumn. Usually the selectivity of such treatments relies solely on timing.

Soil-applied pre-sowing or planting treatments involve compounds that are absorbed by weed seeds, roots, or shoots before emergence and may be used for both annual and perennial species. These treatments are usually most effective when the compounds are mixed into the soil by cultivations and this is essential for the more volatile materials. For annual weed control in an annual crop this mixing can be combined with seedbed preparation. Timing of the application depends on the dose required to achieve weed control, the persistence of the chemical and the tolerance of the crop toward it.

Perennial weeds often require dose rates that may cause crop damage so that long intervals that last several months between application and sowing may be necessary in such cases, whereas in circumstances of good selectivity, usually involving annual weeds, sowing can succeed immediately after treatment. Thus the correct balance must be struck between the various factors.

Crop safety is generally the overriding consideration so that weed control is likely to suffer if there is any uncertainty. However, there are many situations where this use of soil-applied chemicals allows weeds to be attacked that are difficult to control with post-emergence treatments (which are usually considered to be more reliable).

Pre-emergence treatments

Contact foliage pre-emergence treatment is the application of contact herbicides of negligible residual activity to weed seedlings that develop between sowing the crop and its emergence. This strategy requires a weed population that largely germinates in a single flush and the availability of an effective contact herbicide. Subsequent cultural or chemical weed control operations will be needed if the crop is not an effective competitor with weeds that emerge later. The method carries the risk that, because the treatment must be made during a short time interval, bad weather can prevent the operation so that the crop emerges in a weedy seedbed.

The system was developed for slow germinating crops such as *Allium* spp. and root crops where the interval between sowing and crop emergence allows large numbers of weeds to germinate. These are sprayed ideally 2–3 days before the crop starts to appear. It can, however, be used in crops which germinate more rapidly by using 'the stale seedbed' technique in which drilling the crop is delayed until the weeds have emerged and the herbicide is applied within a day or so. It is generally successful but there may be problems on poorly structured soils where the seedbed is prone to deterioration.

Translocated foliage treatments are not commonly applied pre-emergence because most translocated herbicides have some residual activity. Therefore they can only be used if the crop is tolerant so the compound can be used more effectively post-emergence.

Residual treatments use compounds that persist in the soil and so can control weeds for some time. To be effective the herbicide must remain near the soil surface in the region where most weeds germinate. Therefore the crop must be tolerant or it must be sown below this layer which is usually of the order of 5 cm or less. There is, however, a large number of crops that meet one or other of these criteria. Also in row crops it may be possible to apply the treatment only between the rows or to protect the crop with a band of an adsorbent, usually, charcoal placed just above the seed. The availability of soil-applied herbicides is affected by the weather, especially rainfall, and the physical and microbiological properties of the soil, topics which are treated in detail in Chapters 6 and 9. Consequently rates of application may depend on soil type, and herbicidal performance can be erratic.

Post-emergence treatments

Post-emergence treatments are the most widely used in Britain and there are examples in most agricultural and horticultural situations.

Contact foliage treatments are generally used to kill annual weeds. In most cases both crop and weed are sprayed so selectivity is based on either bio-chemical differences or differential retention between crop and weed. Formulation, volume rate and drop size of the spray, weather conditions and growth stages of crop and weed can be critical (see Chapters 4, 5 & 6). Selectivity is sometimes affected by using directed sprays or by shielding the crop mechanically in some way.

Translocated foliage treatments are used for both annual and perennial weeds. Selectivity in this case usually results from physiological or bio-chemical differences between weed and crop or sometimes from selective application. The action may take several days or even weeks. In some cases the weed is suppressed rather than killed but this is still useful if it eliminates or reduces effective weed competition with the crop, particularly if it is accompanied by inhibition of seeding. Translocated herbicides are generally less sensitive to application factors than contact materials but they may be affected by the weather.

Residual post-emergence treatments are uncommon in annual crops except with compounds that have foliar activity as well, for example isoproturon, but they are used extensively in perennial crops. Frequently directed sprays or granule formulations are used to maximize the amount of material that reaches the soil and reduce the risk of crop damage. Most soil-applied compounds are not effective on weeds that have passed the very early seedling stage so the ground must be weed-free before spraying and a preliminary weed control treatment with a contact herbicide or a cultivation is often necessary. With perennial crops it is often convenient to apply a residual herbicide while the crop is dormant.

The ways in which herbicides may be used are summarized diagram-

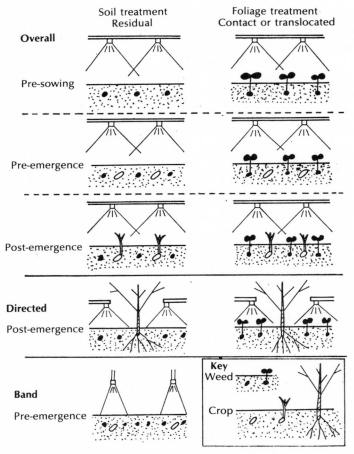

Fig. 3.1 Situations in which herbicides may be used for weed control in crops.

matically in Fig 3.1. These categories of use are by no means rigidly distinct. In particular many foliage-applied compounds have significant soil activity and some residuals have foliar activity. The balance between activities can often be shifted considerably by formulation. Also there are many commercial products that contain mixtures of herbicides that work in different ways.

Types of herbicide

The rest of this chapter gives brief descriptions of the properties and main uses of the most important herbicides available commercially in the UK and elsewhere. Some of historical or scientific interest are also included. More detailed information is to be found in *The Pesticide Manual* (ed. C.R. Worthing) and *The UK Pesticide Guide* (ed. G.W. Ivens), both published by the British Crop Protection Council, the latter jointly with CAB International. Inclusion in this chapter does not imply that a chemical has approval for use in the UK under the Food and Environment Protection Act 1985 (see Chapter 21). No statement should be regarded as a recommendation in any sense.

Compounds are grouped on the basis of common chemical functional groups so that related compounds appear in the same table. They are listed under the common names approved by the British Standards Institution (BSI), or chemical name if an item is considered by BSI not to need a common name, or by manufacturer's code number. Common names adopted by the International Standards Organization (ISO), the American National Standards Institute (ANSI) and the Weed Science Society of America (WSSA) are listed where they differ from the BSI name.

The tables summarize chemical structures, molecular weights (and weights of related salts and ions), solubility in water, vapour pressures and where relevant pK_a values (the negative logarithm of the acid dissociation constant K_a; it is also the value of the pH at which 50% of the acid is ionized). An indication of the type of activity is also included: S = selective; NS = non-selective; F = foliar activity; So = soil activity; C = contact action; and T = translocated. It is important to recognize that these indications give only an approximate guide to the main activity; the categories are not clearly distinct and the absence of any particular indicator letter does not mean that a compound is totally devoid of the relevant activity.

The toxicity ratings which were included in previous editions of the

Handbook have been omitted on the grounds that they gave only a partial indication of the potential hazards and hence could be misleading.

It is not possible to list the classes of compounds in an order based on the hierarchy of main functional groups used in IUPAC nomenclature because in some cases the common functional group is not the most senior in IUPAC terms. For example sulphonylureas with carboxyl groups are named as benzoic acids while those without are usually named as benzenesulphonylureas but it is clearly useful to group them together. Therefore the sequence does not follow an objectively logical system. To a large extent the order is similar to that of previous editions so as to maintain continuity but there are some changes where it seemed sensible to put classes of compounds in a sequence that reflects chemical or biological similarities. The sequence is set out in the contents list of this chapter.

Inorganic salts

The use of inorganic materials (Table 3.1) has a long history but they have largely been replaced by modern organic compounds.

Sodium chlorate still retains a substantial use for total weed control in non-crop areas. It is most active when applied to the foliage of actively growing plants causing scorch and chlorosis. Its mode of action is not well established but it has been suggested that the chlorate ion competes with nitrate for metabolic sites, although this would not explain the symptoms produced. Mammalian toxicity is low but there are hazards with its use because when dry it renders organic material, such as plant residues and

Table 3.1 Inorganic salts

Name and structure	Mol wt	Water sol	VP	Activity
Ammonium sulphamate [AMS (WSSA)] $NH_2SO_2.ONH_4$	114.1	$2.16\,kg\,l^{-1}$ (25°C)	—	NS, F, C
Borax* $Na_2B_4O_7\,10H_2O$	381.4	$51.4\,g\,l^{-1}$ (20°C)	—	NS, So
Ferrous sulphate $Fe\,SO_4\,7H_2O$	278.0	$266\,g\,l^{-1}$ (10°C)	—	S, F, C
Sodium chlorate $NaClO_3$	106.4	$790\,g\,l^{-1}$ (0°C)	—	NS, F, So, C

* Other hydrates and borates have also been used

timber, highly flammable. For this reason it is often formulated with a fire retardant. It does not usually persist in the soil for more than a year. Leaching is a major route of loss but this does not normally cause problems because the activity of the chlorate ion is intrinsically low and so is greatly reduced by dilution.

Ferrous sulphate still finds a use as a constituent of lawn sand, mainly to kill moss, although some higher plants are also damaged. It causes severe scorch but has little or no soil activity.

Ammonium sulphamate (AMS) has activity against some herbaceous and woody weeds.

Halogenated alkanoic acid derivatives

The members of this group (Table 3.2) are derivatives of simple alkanoic acids.

Sodium chloroacetate (or monochloroacetate), also known as SMA, is a contact herbicide used post-emergence to control a wide range of seedling annual weeds in some brassica crops, leeks, onions, and fruit. Dalapon is a translocated and soil-acting compound which can be used pre-planting and pre- or post-emergence to control mainly grass weeds, both annuals and perennials, in many broad-leaved crops. It is also used to control reeds and sedges in and near water as well as for pasture improvement and total weed control in uncropped situations. TCA is soil-acting and is used pre-planting or pre-emergence mainly to control grasses such as common couch, wild oat and volunteer cereals in brassicas, notably kale and rape, peas and sugar-beet.

Table 3.2 Salts of halogenated alkanoic acids

Name and structure	Mol wt	Water sol	VP	Activity
Sodium chloroacetate (BSI & ISO) $ClCH_2CO.ONa$	116.5*	$850\,g\,l^{-1}$ 20°C	—	S, F, So, C
TCA (BSI & ISO) [TCA-sodium (WSSA)] $Cl_3C.CO.ONa$	185.4*	$1.2\,kg\,l^{-1}$ Room temp.	—	S, So, T
Dalapon-sodium (BSI, ANSI & WSSA) $CH_3.CCl_2.CO.ONa$	165.0*	$502\,g\,l^{-1}$ 25°C	—	S, F, So, T

* Subtract 22 to give the mol wt of the free acid

These compounds are readily absorbed by both roots and shoots and cause formative effects (such as increased tillering), growth inhibition, chlorosis and leaf necrosis. Dalapon and TCA also affect leaf wax formation in some species. Dalapon moves in both the symplast and apoplast whereas TCA seems to be largely confined to the apoplast. They are detoxified only slowly, if at all, by the plant. These compounds have been shown to affect a range of biochemical processes including carbohydrate, lipid and nitrogen metabolism but it is thought likely that these are secondary responses brought about by primary effects on the structures of proteins. Changes in these structures could affect enzyme activity and membrane permeability and so bring about the observed effects.

Sodium chloroacetate and dalapon persist in the soil for only a few weeks and TCA for a little longer.

(Aryloxy)alkanoic acids

This group of herbicides (Table 3.3) was developed in the early 1940s simultaneously in Britain and the USA following the discovery of MCPA and 2,4-D. It includes both 2-(aryloxy)alkanoic acids (phenoxy- and pyridyloxy-) and also precursors of phenoxyacetic acids that are converted to the corresponding acids *in vivo*. The propionic acid derivatives contain a chiral centre and only the $(R)(+)$ isomers are herbicidally active. The usual commercial formulations are the racemates but it is now possible to prepare the active isomer alone. The names of single isomers are given the suffix *P* if they rotate polarized light clockwise (as with mecoprop-*P*) and *M* if the rotation is anti-clockwise.

All exert a characteristic growth regulator/hormone type of activity on most broad-leaved species, both annual and perennial, while graminaceous species are generally tolerant. The propionic acid derivatives are more active than the acetic acids on weeds such as common chickweed, cleavers and the *Polygonum spp.* Triclopyr and 2,4,5-T are effective on many woody species. All these compounds can enter both roots and shoots and are readily translocated in herbaceous plants.

The initial symptom on broad-leaved plants is severe epinasty of leaves, stems and petioles. Stems can swell followed by splitting and root primordia may develop above ground level. Leaves can be deformed with veins developing closely together as a result of inhibition of their expansion. Necrosis usually develops slowly. Grasses and cereals can show symptoms, their leaves resembling those of onions, if the herbicide is applied at a very early growth stage.

Table 3.3 2-(aryloxy)alkanoic acids and precursors

Name and structure	Mol wt	Water sol	VP	Activity
MCPA (BSI, ISO, WSSA)	200.6 pK$_a$ 3.07	825 mg l^{-1} Room temp.	200 nPa 21°C	S, F, T
2,4-D (BSI, ISO, WSSA)	221.0 pK$_a$ 2.64	620 mg l^{-1} 25°C	53 Pa 160°C	S, F, T
2,4,5-T (BSI, ISO, WSSA)	255.5	150 mg l^{-1} 25°C	700 nPa 25°C	S, F, So, T
Dichlorprop* (BSI, ISO, WSSA)	235.1	350 mg l^{-1} 20°C	—	S, F, T
Mecoprop* (BSI, ISO, WSSA)	214.6	620 mg l^{-1} 20°C	—	S, F, T
Fluroxypyr (BSI, ISO, WSSA)	255.0	900 μg l^{-1} 27.7°C	14 μPa 25°C	S, F, So, T
Triclopyr (BSI, ISO, ANSI, WSSA)	256.5 pK$_a$ 2.68	440 mg l^{-1} 25°C	168 μPa 25°C	S, F, T

Table 3.3 Cont.

Name and structure	Mol wt	Water sol	VP	Activity
MCPB (BSI, ISO, WSSA)	228.7 pK_a 4.84	$44\,mg\,l^{-1}$ Room temp.	—	S, F, T
2,4-DB (BSI, ISO, WSSA)	249.1 pK_a 4.8	$46\,mg\,l^{-1}$ 25°C	—	S, F, T

* The properties refer to the racemic mixture; only the (+) enantiomer is significantly herbicidal

The primary modes of action are thought to involve nucleic acid and protein metabolism as well as enzymes affecting cell wall plasticity but no clear details have been established. Interference with respiration and intermediary metabolism has also been observed but they are probably secondary effects. Detoxification involves ring hydroxylation and conjugation with glucose and amino acids. Decarboxylation does not seem to be of primary importance.

The precursors of phenoxyacetic acids, 2,4-DB and MCPB, can be used in some legume crops. This is because susceptible plants convert them to 2,4-D and MCPA by β-oxidation but legumes lack the necessary enzymes.

These compounds are usually formulated as salts or esters. The esterifying alcohol must be fairly large to reduce the volatility of the product and so avoid vapour drift damage to susceptible adjacent crops. Except for 2,4,5-T and triclopyr, they are not very persistent in the soil.

Arylcarboxylic acids and their derivatives

The first herbicidal derivative of an aromatic acid to be introduced was 2,3,6-TBA in 1954. There are members of this group based on benzoic acid (2,3,6-TBA, dicamba, chloramben), picolinic acid (chlopyralid, picloram) and terephthalic acid (chlorthal-dimethyl) (Table 3.4).

All except chloramben and chlorthal-dimethyl are primarily post-emergence although they do have some soil activity. They are usually used as a component of mixtures to control many annual and some perennial broad-leaved weeds in cereals and grass seed crops and, in the case of dicamba, in grassland and turf as well. Clopyralid is also used, both alone

Table 3.4 Arylcarboxylic acids and their derivatives

Name and structure	Mol wt	Water sol	VP	Activity
Chloramben (BSI, ISO, ANSI, WSSA)	206.0	$700\,mg\,l^{-1}$ 25°C	930 mPa 100°C	S, So, T
Dicamba (BSI, ISO, ANSI, WSSA)	221.0	$6.5\,g\,l^{-1}$ 25°C	45 μPa 25°C	S, F, So, T
2,3,6-TBA (BSI, ISO, WSSA)	225.5	$7.7\,g\,l^{-1}$ 22°C	3.2 Pa 100°C	S, F, So, T
Clopyralid (BSI, ISO, WSSA)	192.0 pK_a 2.33	$9\,g\,l^{-1}$ 20°C	1.6 mPa 25°C	S, F, So, T
Picloram (BSI, ISO, ANSI, WSSA)	241.5 pK_a 3.6	$430\,mg\,l^{-1}$ 35°C	82 μPa 35°C	S, F, So, T
Chlorthal-dimethyl (BSI, ISO, WSSA)	332.0	$<500\,mg\,l^{-1}$ 25°C	<67 Pa 40°C	S, So

and in mixtures, for beet and brassica crops. Chloramben is basically a soil-acting compound and is used to control annual grasses and broad-leaved weeds in soya beans, groundnuts and ornamentals. Chlorthal-dimethyl is a residual herbicide used mostly in mixtures in many vegetables, fruit, ornamentals and turf.

Although their methods of use differ, all these compounds can enter leaves and roots but chloramben and chlorthal-dimethyl are translocated little, if at all, whereas the others move both symplastically and apoplastically, and accumulate in areas of high metabolic activity. The benzoic and picolinic acids cause similar symptoms of growth-regulator or hormone activity as the aryloxyalkanoic acids and are thought to have similar primary effects on nucleic acid and protein metabolism. A number of other systems seem to be affected in consequence. Detoxification of these compounds by plants is usually by conjugation with sugars or by ring hydroxylation; decarboxylation does not seem to be very important.

Chlorthal-dimethyl is neither translocated nor detoxified to any great extent in the plant. It inhibits cell division, apparently by stopping mitosis at the metaphase. Cells may become multi-nucleate and nuclear material can occupy most of the cell. They usually become abnormally large and irregular so that affected shoots and roots are swollen although they do not grow in length. At high rates of application there is virtually no germination.

Dicamba and chloramben have a short soil persistence and ester and amide formulations may be used to reduce mobility. The other compounds have moderate to long (notably picloram) persistence. Residues of 2,3,6-TBA and picloram persist in the straw of treated crops and have caused problems if such straw was composted or used in farmyard manure.

Esters of 2-[4-(aryloxy)phenoxy]alkanoic acids

These compounds (Table 3.5) control grass weeds by both foliar and soil uptake. They are used in many broad-leaved crops, onions and, in the case of diclofop, in wheat and barley. They are effective against such important grass weeds as common couch, ryegrasses, wild oat, barnyardgrass, goosegrass, *Leptochloa, Panicum,* and *Setaria* but annual meadow-grass is quite resistant.

These compounds contain a chiral centre and the $(R)(+)$ enantiomer is the most active herbicidally. They are sold as the racemates and as the active isomer in the cases of fluazifop-*P* and propaquizafop (the *P* isomer of quizalofop).

Diclofop-methyl is an auxin inhibitor and affects fatty acid synthesis; the action of the others is presumably similar. In contrast to the arylalanine esters, both the free acid and methyl ester of diclofop are active although their relative potencies seem to vary with the species. Thus selectivity is not based on hydrolysis but probably on rate of detoxification which involves hydroxylation and conjugation with sugars. In *Avena fatua* conjugation mainly occurs without prior hydroxylation and it is possible that the con-

Table 3.5 Esters of aryloxyphenoxyalkanoic acids

Name and structure	Mol wt	Water sol	VP	Activity
Diclofop-methyl (BSI, ISO, ANSI, WSSA)	341.2	3 mg l^{-1} 20°C	34 µPa 20°C	S, F, So, T
Fenoxaprop-ethyl (BSI, ISO, ANSI, WSSA)	361.8	0.9 mg l^{-1} 25°C	19 µPa 20°C	S, F, So, T
Fluazifop-butyl (BSI, ISO, ANSI, WSSA)	383.4	1 mg l^{-1} 20°C	55 µPa 20°C	S, F, So, T
Haloxyfop-2-ethoxyethyl (BSI, ANSI, ISO, WSSA)	433.8	2.7 mg l^{-1} 20°C	86 µPa 25°C	S, F, So, T
Quizalofop-ethyl (BSI, ISO, ANSI, WSSA)	372.8	0.3 mg l^{-1} 20°C	40 µPa 20°C	S, F, So, T

Properties refer to racemic mixtures; (R) (+) enantiomer is usually the active entity and in some cases is marketed separately (see text)

jugate subsequently releases diclofop and so acts as a pool of active herbicide. Thus it is hydroxylation not conjugation that seems to be the critical process of detoxification.

The activity of these compounds is aided by moderate persistence in the soil.

Nitriles and their precursors

There are two groups of nitriles with quite different properties (Table 3.6).

Dichlobenil, a chlorobenzonitrile, and chlorthiamid, which is converted

Table 3.6 Nitriles (and precursor)

Name and structure	Mol wt	Water sol	VP	Activity
Dichlobenil (BSI, ISO, ANSI, WSSA) CN / Cl—[ring]—Cl	172.0	$18\,mg\,l^{-1}$ 20°C	67 mPa 25°C	S, So, T
Bromoxynil (BSI, ISO, ANSI, WSSA) CN / Br—[ring]—Br / OH	276.9 pK$_a$ 4.06	$130\,mg\,l^{-1}$ 25°C	—	S, F, C
Chlorthiamid (BSI, ISO) CS.NH$_2$ / Cl—[ring]—Cl	206.1	$950\,mg\,l^{-1}$ 21°C	130 μPa 20°C	S, So, T
Ioxynil (BSI, ISO, WSSA) CN / I—[ring]—I / OH	370.9 pK$_a$ 3.96	$50\,mg\,l^{-1}$ 25°C	—	S, F, C

to dichlobenil in the soil, are soil-applied although some foliar uptake can occur. They control not only germinating annual weeds but also buds and shoots arising from underground parts of perennials. They are used in fruit crops, forestry, non-crop areas, and some aquatic situations. Dichlobenil is also used in ornamentals.

Dichlobenil is translocated in the xylem but the rate varies with the species. Its primary mode of action is now thought to be by inhibition of cellulose biosynthesis. It also affects protein and RNA synthesis. The parent compound has little effect on phosphorylation or electron transport but its 3- and 4-hydroxy derivatives, which are produced in the plant, are quite potent uncouplers of oxidative and photosynthetic phosphorylation. Symptoms include swelling of tissues such as root hairs, brittleness of stems and sometimes dark coloured foliage. Detoxification seems to be by conjugation with sugars of the hydroxy-derivatives.

Dichlobenil can persist for some 6 months in the soil. It is metabolized to 2,6-dichlorobenzamide and then to the acid. It is appreciably volatile so must be incorporated into the soil if a granule formulation is not used.

The other group are the hydroxybenzonitriles (HBNs), bromoxynil and ioxynil, which are contact herbicides. They control annual broad-leaved weeds and are used, usually in mixtures with phenoxyalkanoic acids, in cereals. Ioxynil alone is used also in *Allium* crops and alone or in a mixture with ethofumesate and bromoxynil in new grass leys.

The primary site of action of the HBNs is in the chloroplast. Affected plants are rapidly scorched and show chlorosis. Like dichlobenil, they are hydroxylated in the plant and the 3- and 4-hydroxy derivatives uncouple oxidative phosphorylation and photophosphorylation. They have little residual soil activity but they are metabolized to the corresponding acids in the soil which, if they enter a plant, show auxin activity similar to the other halogenated benzoic acids.

Amides

As a group (Table 3.7), the amides are primarily soil-acting and control annual grasses and some broad-leaved weeds. However, propyzamide is also toxic to some perennials, notably common couch and sheep's sorrel. These compounds are mainly used in broad-leaved crops, tree and bush fruits but isoxaben is used in cereals. Diphenamid and propyzamide are also used in strawberries and some forestry situations. Napropamide and propyzamide can be used in oilseed rape and the uses of propyzamide also include some legumes, brassicas, sugar-beet and lettuce.

The mode of action seems to involve cell division or enlargement in some way. Symptoms include stunting of roots and shoots with grasses often failing to emerge from the coleoptile. Leaves may be temporarily darker green, especially with propyzamide. The primary detoxification mechanisms are thought to involve conjugation with sugars and dealkylation. The amides are relatively persistent in soil.

Anilides

Herbicides derived from aniline can conveniently be considered in three groups (Table 3.8(a)–(c)).

Table 3.7 Amides

Name and structure	Mol wt	Water sol	VP	Activity
Diphenamid (BSI, ISO, ANSI, WSSA)	239.3	$260\,mg\,l^{-1}$ 27°C	—	S, So
Isoxaben (BSI, ISO, ANSI, WSSA)	332.4	$1-2\,mg\,l^{-1}$ 25°C	—	S, So
Napropamide* (BSI, ISO, WSSA)	271.4	$73\,mg\,l^{-1}$ 20°C	$530\,\mu Pa$ 25°C	S, So
Propyzamide (BSI, ISO, WSSA)	256.1	$15\,mg\,l^{-1}$ 25°C	$11.3\,mPa$ 25°C	S, So
Tebutam (BSI, ISO, ANSI, WSSA)	23.4	Low	—	S, So

Structures:

Diphenamid: CHCO.N(CH₃)₂ — written as CHCO.N(CH$_3$)$_2$ attached to two phenyl rings.

Isoxaben: OCH$_3$, CO.NH, O, N, CH$_3$, C(CH$_2$CH$_3$)$_2$, OCH$_3$

Napropamide: CH$_3$CHCON(CH$_2$CH$_3$)$_2$, O, naphthalene ring

Propyzamide: Cl, CH$_3$, CO.NH–C–C≡CH, CH$_3$, Cl

Tebutam: CH$_2$–N.CO.C(CH$_3$)$_3$, CH(CH$_3$)$_2$

* Properties refer to racemate; the (R)(−) enantiomer is more active than the (S)(−)

2-chloroacetanilides

These compounds are used pre-emergence to control annual grasses and some broad-leaved weeds in mainly broad-leaved crops such as rape, brassicas, groundnuts and soya bean. Alachlor, propachlor and metolachlor are also used in maize, pretilachlor in rice and propachlor in strawberries. Some are also used in cotton.

The 2-chloroacetanilides are germination inhibitors and any seedlings

Table 3.8 (a) Anilides (2-chloroacetanilides)

Name and structure	Mol wt	Water sol	VP	Activity
Acetochlor (BSI, ISO, ANSI, WSSA) CH_3 / CO.CH$_2$Cl / N / CH$_2$OCH$_2$CH$_3$ / CH$_2$CH$_3$	269.8	223 mg l^{-1} 25°C	—	S, So
Alachlor (BSI, ISO, ANSI, WSSA) CH_2CH_3 / CO.CH$_2$Cl / N / CH$_2$OCH$_3$ / CH$_2$CH$_3$	269.8	242 mg l^{-1} 25°C	2.9 mPa 25°C	S, So, T
Butachlor (BSI, ISO, ANSI, WSSA) CH_2CH_3 / CO.CH$_2$Cl / N / CH$_2$O(CH$_2$)$_3$CH$_3$ / CH$_2$CH$_3$	311.9	23 mg l^{-1} 24°C	600 µPa 15°C	S, So, T
Metazachlor (BSI, ISO) CH_3 / CO.CH$_2$Cl / N / CH$_2$—N—N / CH$_3$	277.8	17 mg l^{-1} 20°C	49 µPa 20°C	S, So
Metolachlor (BSI, ISO, ANSI, WSSA) CH_2CH_3 / CO.CH$_2$Cl / N / CHCH$_2$OCH$_3$ / CH$_3$ CH$_3$	283.8	530 mg l^{-1} 20°C	1.7 mPa 20°C	S, So
Pretilachlor (BSI, ISO) CH_2CH_3 / CO.CH$_2$Cl / N / CH$_2$CH$_2$O(CH$_2$)$_2$CH$_3$ / CH$_2$CH$_3$	311.9	50 mg l^{-1} 20°C	133 µPa 20°C	S, So (paddy)
Propachlor (BSI, ISO, WSSA) CO.CH$_2$Cl / N / CH(CH$_3$)$_2$	211.7	613 mg l^{-1} 25°C	30.6 mPa 25°C	S, So, T

Table 3.8(b) Anilides (esters of *N*-arylalanines)

Name and structure	Mol wt	Water sol	VP	Activity
Benzoylprop-ethyl (BSI, ISO, WSSA)	366.2	20 mg l^{-1} 25°C	4.7 µPa 20°C	S, F, T
Flamprop-methyl (racemate) Flamprop-M-methyl (*R*)isomer (BSI, ISO)	335.8	35 mg l^{-1} 22°C	47 µPa 20°C	S, F, T
Flamprop-isopropyl (racemate) Flamprop-M-isopropyl (*R*)isomer (BSI, ISO)	363.8	10 mg l^{-1} 20°C	32 µPa 20°C	S, F, T

that do emerge are stunted or malformed. Cell division is inhibited but whether by interference with a regulatory or a synthetic process is not known although effects on protein synthesis, lipid synthesis, ion transport and membrane function have been demonstrated. Detoxification is apparently by dealkylation and conjugation. Their soil persistence varies from a few weeks to several months.

Table 3.8 (c) Anilides (miscellaneous)

Name and structure	Mol wt	Water sol	VP	Activity
Diflufenican (BSI, ISO) 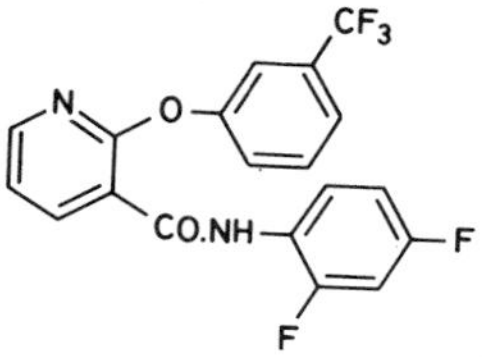	394.3	$50 \mu g\,l^{-1}$	—	S, F, T
Mefluidide (BSI, ISO, ANSI, WSSA) 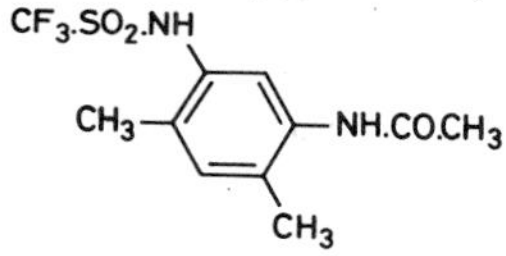	310.3	$180\,mg\,l^{-1}$ 23°C	<13 Pa 25°C	F, T
Naptalam (BSI, ISO, WSSA) 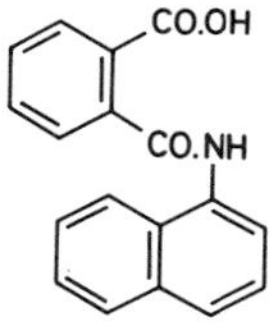	293.1	$200\,mg\,l^{-1}$ Room temp.	<133 Pa 20°C	S, So .
Pentanochlor (BSI, ISO, WSSA) 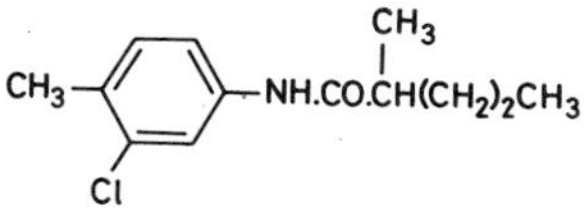	293.7	$8-9\,mg\,l^{-1}$ Room temp.	—	S, F, So, T
Perfluidone (BSI, ISO, ANSI, WSSA) 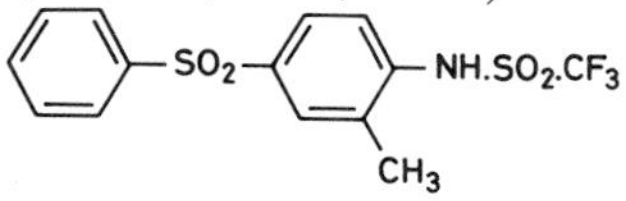	379.7	$60\,mg\,l^{-1}$ 22°C	<1.3 mPa 25°C	S, So
Propanil (BSI, ISO, WSSA) 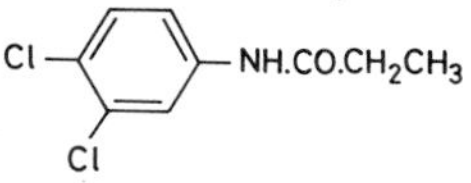	218.1	$225\,mg\,l^{-1}$ 25°C	12 mPa 60°C	S, F, So, C

Diflufenican structure: CF$_3$-substituted phenyl, O, pyridine N, CO.NH, 2,4-F phenyl.

Mefluidide structure: $CF_3.SO_2.NH$ — benzene ring with CH_3, CH_3 and $NH.CO.CH_3$.

Naptalam structure: CO.OH, CO.NH, naphthalene.

Pentanochlor structure: CH_3-, CH_3-benzene with Cl, $NH.CO.CH(CH_2)_2CH_3$ with CH_3.

Perfluidone structure: phenyl-SO_2-benzene-$NH.SO_2.CF_3$ with CH_3.

Propanil structure: Cl-, Cl-benzene-$NH.CO.CH_2CH_3$.

Esters of N-arylalanines

These compounds can also be regarded as 2-substituted propanoic acids
of which the carbon atom in the 2 position is a chiral centre. Benzoylprop-
ethyl is the racemic mixture but flamprop-isopropyl is now sold only as the
$(R)(-)$enantiomer (flamprop-M). They are used for the post-emergence
control of wild oats in cereals and, in the case of benzoylprop-ethyl, also
in field beans, oilseed rape and mustard.

The esters are hydrolyzed in the plant to the corresponding acids which
are the active principles. They inhibit auxin action in some way and treated
plants are not often killed but stunted, so that they do not compete successfully with the crop. Because of the anti-auxin activity they can an-
tagonize the action of auxin-type herbicides such as 2,4-D and dicamba.
Selectivity depends on relative rates of hydrolysis and subsequent detoxification which is primarily by conjugation with sugars but some debenzoy-
lation occurs. Benzoylprop-ethyl causes yield loss in barley and so is not
used in this crop. Although used post-emergence, these compounds do
have some soil activity and can persist for several weeks.

Miscellaneous anilides

The herbicides in this group cover a wide range of activities and uses.

Diflufenican is a pre- or post-emergence herbicide for use in winter
cereals to control broad-leaved weeds. It causes chlorosis and bleaching so
its action is presumably on photosynthetic electron transport. In the UK it
is used in a mixture with isoproturon.

Naptalam is a soil-acting pre-emergence herbicide for the control of
annual broad-leaved and grass weeds in crops such as groundnuts, soya
beans and potatoes. It is thought to be an auxin inhibitor and seedlings, if
they germinate at all, are stunted and malformed. The compound is probably unique in that it interferes with geotropic and phototropic responses.
Its soil persistence is of the order of a few weeks.

Propanil and pentanochlor are primarily foliar applied compounds
active against annual grass and broad-leaved weeds. Propanil is used in
potatoes and rice and pentanochlor in a variety of mainly broad-leaved
horticultural crops. Both act primarily by interfering with photosynthetic
electron flow but propanil has a range of activities and has been reported
to interfere with root and coleoptile growth.

Mefluidide and perfluidone, though structurally similar are quite different in their biological properties. Mefluidide is mostly used as a growth

regulator to suppress the growth of grass, trees and ornamentals and to increase the sucrose content of sugar-cane. Its mode of action has not yet been identified. Perfluidone is a soil-acting pre-emergence herbicide which controls many grasses, broad-leaved weeds and yellow nutsedge and is used in cotton, tobacco, sugar-cane, rice and groundnuts among others. Its mode of action is obscure although effects on photophosphorylation and IAA-oxidase activity have been reported.

Phenols

These compounds (Table 3.9) are of little more than historic interest as they have now been mostly replaced by compounds with better selectivity and lower mammalian toxicity. DNOC, introduced as a herbicide around 1933 was one of the first organic herbicides.

The dinoseb salts are basically contact materials which can control broad-leaved weeds in cereals, grass, legumes and potatoes. Pentachlorophenol is a general herbicide and preharvest defoliant.

Table 3.9 Phenols

Name and structure	Mol wt	Water sol	VP	Activity
Dinoseb (BSI, ISO, ANSI, WSSA)	240.2 pK_a 4.62	$100\,mg\,l^{-1}$ Room temp.	—	S, F, C
DNOC (BSI, ISO, WSSA)	198.1	$130\,mg\,l^{-1}$ $15°C$	$14\,mPa$ $25°C$	S, F, C
Pentachlorophenol (BSI, ISO) (WSSA name PCP)	266.3 pK_a 4.71	$20\,mg\,l^{-1}$ $30°C$	$16\,Pa$ $100°C$	F, C

The nitrophenols uncouple oxidative phosphorylation and so prevent the formation of ATP during respiration with consequent secondary effects on lipid, protein and RNA synthesis. This occurs in animals, insects and micro-organisms as well as plants so these compounds are general poisons. Dinoseb also inhibits photosynthesis. Detoxification is probably via reduction of the nitro groups and conjugation with glucose.

Diphenyl ethers

Compounds in this group (Table 3.10) discovered so far are all nitrodiphenyl ethers. They are usually applied pre-emergence to control germinating annual weeds, both grass and broad-leaved. They are mostly used in broad-leaved crops including brassicas, soya bean and cotton but some are used in rice and others are, or have been, used in wheat.

The nitrodiphenyl ethers can be absorbed by both leaves and roots but they are not translocated extensively. Compounds with 2-,4- or 6-substituents, which include all those so far sold as herbicides, require activation by light. This light-dependent action involves inhibition of chloroplast energy transfer and photosynthetic electron transport leading to free radical formation, lipid peroxidation and membrane disruption. The site of action is not, however, apparently the same as for compounds like the ureas and triazines (see Chapter 7). Detoxification mechanisms vary with the compound. For example, nitrofen appears to form conjugates with lipids and lignin; fluorodifen is cleaved at the ether linkage following reduction of the nitro group; bifenox undergoes ring hydroxylation and is then conjugated.

These compounds persist in the soil for several months at normal application rates. The use of some members of the group has been restricted or stopped altogether because although acute mammalian toxicity is low their chronic effects give rise to some concern.

2,6-dinitroanilines

Most of these compounds (Table 3.11) have been developed for specific, mainly broad-leaved, crops including brassicas, cotton, groundnuts, peppers, soya bean, tobacco and tomatoes. Mixtures containing trifluralin are used in winter wheat and barley in Europe although elsewhere, notably in Australia, carryover of trifluralin residues can cause damage to wheat.

Because they are germination inhibitors, all these compounds are applied to the soil. Some of them are sufficiently volatile and photolabile

Table 3.10 Diphenyl ethers

Name and structure	Mol wt	Water sol	VP	Activity
Acifluorfen (BSI, ISO, ANSI, WSSA) CO.OH CF_3—⟨⟩—O—⟨⟩—NO_2 Cl	361.7	$120\,mg\,l^{-1}$ 23–25°C	$133\,\mu Pa$ 20°C	S, F, So, C
Bifenox (BSI, ISO, ANSI, WSSA) $CO.OCH_3$ Cl—⟨⟩—O—⟨⟩—NO_2 Cl	342.1	$350\,\mu g\,l^{-1}$	$320\,\mu Pa$ 30°C	S, So
Fluorodifen (BSI, ISO, ANSI, WSSA) CF_3—⟨⟩—O—⟨⟩—NO_2 NO_2	328.8	$2\,mg\,l^{-1}$ 20°C	$9.3\,\mu Pa$ 20°C	S, F, So, C
Nitrofen (BSI, ISO, WSSA) Cl—⟨⟩—O—⟨⟩—NO_2 Cl	284.1	$0.7–1.2\,mg\,l^{-1}$ 22°C	$1.06\,mPa$ 40°C	S, So
Oxyfluorfen (BSI, ISO, ANSI, WSSA) OCH_2CH_3 CF_3—⟨⟩—O—⟨⟩—NO_2 Cl	361.7	$100\,\mu g\,l^{-1}$ 25°C	$26.7\,\mu Pa$ 25°C	S, So

that they must be incorporated into the soil immediately after application.

They can enter roots and shoots but translocation is limited. They inhibit cell division probably by interaction with the microtubular system. Several other processes including the various syntheses and oxidative phosphorylation can also be affected. If germination does occur, the roots and shoots are severely stunted and the meristematic tissue may swell. Inhibition of lateral roots is a characteristic symptom as is the failure of grasses to merge from the coleoptile. Detoxification reactions can include oxidation, *N*-dealkylation and reduction of nitro-groups.

Table 3.11 Nitroanilines

Name and structure	Mol wt	Water sol	VP	Activity
Butralin (BSI, ISO, ANSI, WSSA) $(CH_3)_3C$—ring with NO_2, $NHCHCH_2CH_3$ (CH_3), NO_2	295.3	$1\,mg\,l^{-1}$ 24–26°C	$17\,mPa$ 25°C	S, So
Ethalfluralin (BSI, ISO, ANSI, WSSA) CF_3—ring with NO_2, N (CH_2—C=CH_2 with CH_3)(CH_2CH_3), NO_2	333.3	$200\,\mu g\,l^{-1}$ 25°C	$11\,mPa$ 25°C	S, So
Nitralin (BSI, ISO, WSSA) CH_3SO_2—ring with NO_2, $N[(CH_2)_2CH_3]_2$, NO_2	345.4	$600\,\mu g\,l^{-1}$ 22°C	$1.2\,\mu Pa$ 20°C	S, So
Oryzalin (BSI, ISO, ANSI, WSSA) H_2NSO_2—ring with NO_2, $N[(CH_2)_2CH_3]_2$, NO_2	346.4	$2.4\,mg\,l^{-1}$ 25°C	$<13\,\mu Pa$ 30°C	S, So
Pendimethalin (BSI, ISO, ANSI, WSSA) CH_3—ring with NO_2, $NHCH(CH_2CH_3)_2$, CH_3, NO_2	281.3	$300\,\mu g\,l^{-1}$ 20°C	$40\,\mu Pa$ 25°C	S, So
Trifluralin (BSI, ISO, ANSI, WSSA) CF_3—ring with NO_2, $N(CH_2CH_2CH_3)_2$, NO_2	335.3	$<1\,mg\,l^{-1}$ 27°C	$14\,mPa$ 25°C	S, So

These compounds do not apparently interact with animal tubulin and mammalian toxicity is low but toxic nitrosamine impurities have been found in some formulations. Persistence in the soil varies from a few weeks upwards depending on the compound, the formulation and whether or not the material is incorporated. Degradation pathways, and hence rates, are different in aerobic and anaerobic conditions.

Carbamates

The first carbamate herbicide to be introduced was propham in 1945 and other simple carbamates appeared over the next 15 years. The sulphonyl carbamate, asulam became available in 1968 and the biscarbamates, phenmedipham and desmedipham during the following decade.

The activities and actions of this group are quite diverse (Table 3.12). The biscarbamates are primarily foliar-actings; asulam is usually regarded as being in this category but also has appreciable soil activity, while the others are mainly soil acting.

Table 3.12 Carbamates

Name and structure	Mol wt	Water sol	VP	Activity
Chlorbufam (BSI, ISO)	223.7	540 mg l^{-1} 20°C	2.1 mPa 20°C	S, So
Chlorpropham (BSI, ISO, WSSA)	213.7	89 mg l^{-1} 25°C	—	S, So
Propham (BSI, ISO, WSSA)	179.2	>32 mg l^{-1}	—	S, So
Carbetamide (BSI, ISO, ANSI, WSSA)	236.3	3.5 g l^{-1} 20°C	—	S, F, So

Chlorbufam structure: ring—Cl; —NH.CO.OCHC≡CH with CH$_3$

Chlorpropham structure: ring—Cl; —NH.CO.OCH(CH$_3$)$_2$

Propham structure: ring—NH.CO.OCH(CH$_3$)$_2$

Carbetamide structure: ring—NHCO.O—C(H)(CH$_3$)—CO.NHCH$_2$CH$_3$

Table 3.12 Cont.

Name and structure	Mol wt	Water sol	VP	Activity
Desmedipham (BSI, ISO, ANSI, WSSA)	300.3	$9\,mg\,l^{-1}$ Room temp.	400 nPa 25°C	S, F, C
Phenisopham (BSI, ISO)	342.4	$3\,mg\,l^{-1}$ 25°C	665 nPa 25°C	S, F, So, C
Phenmedipham (BSI, ISO, ANSI, WSSA)	300.3	$4.7\,mg\,l^{-1}$ Room temp.	1.3 nPa 25°C	S, F, C
Asulam (BSI, ISO, ANSI, WSSA) pK$_a$ 4.82	230.2	$4\,g\,l^{-1}$ 20–25°C	—	S, F, So, T

Desmedipham structure: NHCO.OCH$_2$CH$_3$ ... NHCO.O

Phenisopham structure: NHCO.OCH(CH$_3$)$_2$ ... NCO.O ... CH$_2$CH$_3$

Phenmedipham structure: NH.CO.OCH$_3$ ... NH.CO.O ... CH$_3$

Asulam structure: H$_2$N—⟨ ⟩—SO$_2$.NH.CO.OCH$_3$

The soil acting materials are largely used to control grass weeds in certain broad-leaved crops but some important broad-leaved weeds are also susceptible to some of them. For instance common chickweed can be controlled by chlorpropham and to some extent by propham. The biscarbamates control broad-leaved weeds post-emergence in beet crops and in the case of phenisopham, in cotton. Asulam controls some perennials, notably docks in grassland and bracken on hill and grazing land and in forestry.

The primary site of action of asulam, chlorpropham and propham seems to be on the microtubule organizing centre so that cell division is affected with consequent deformation of roots, shoots and buds. Also foliage is usually darker green. Various synthetic processes are also inhibited including, in the case of asulam, that of folic acid. The biscarbamates interfere with photosynthetic electron flow which produces the usual symptoms of chlorosis and even scorch.

Selectivity seems to depend on different rates of detoxification which

Table 3.13 Thiocarbamates

Name and structure	Mol wt	Water sol	VP	Activity
Butylate (BSI, ISO, WSSA) $[(CH_3)_2CHCH_2]_2NCO.SCH_2CH_3$	217.4	46 mg l^{-1} 20°C	170 mPa 25°C	S, So
EPTC (BSI, ISO, WSSA) $[CH_3(CH_2)_2]_2NCO.SCH_2CH_3$	189.3	375 mg l^{-1} 24°C	4.5 Pa 25°C	S, So
Molinate (BSI, ISO, WSSA) NCO.SCH$_2$CH$_3$	187.3	880 mg l^{-1} 20°C	764 mPa 25°C	S, So (paddy)
Thiobencarb (BSI, ISO, ANSI, WSSA) $(CH_3CH_2)_2N.CO.SCH_2$—⟨ ⟩—Cl	257.8	30 mg l^{-1} 20°C	—	S, So (paddy)
Di-allate* (BSI, ISO, WSSA) $[(CH_3)_2CH]_2NCO.SCH_2$ $C=C$ Cl/H (E); $[(CH_3)_2CH]_2NCO.SCH_2$ $C=C$ H/Cl (Z)	270.2	14 mg l^{-1} 25°C	20 mPa 25°C	S, F, So, T
Tri-allate (BSI, ISO, WSSA) $[(CH_3)_2CH]_2N.CO.SCH_2C=CCl_2$ (with Cl on central C)	304.7	4 mg l^{-1} 25°C	16 mPa 25°C	S, So, T

* Figures refer to mixture of (E) and (Z) isomers

usually involves hydroxylation and conjugation to glucose or in some cases lignin and possibly glutathione. The biscarbamates may also undergo hydrolysis.

Thiocarbamates

The first member of this group (Table 3.13) to be introduced was EPTC in 1954. They control weeds by inhibiting development soon after germination or by inhibiting bud formation in underground storage organs, so they are applied to the soil. Some, notably EPTC, di-allate and tri-allate, are sufficiently volatile to require incorporation or granular formulations. Those used in rice (EPTC, molinate, thiobencarb) are applied to the irrigation water.

Di-allate and tri-allate are specific against wild oat and blackgrass but the others basically control annual grasses and broad-leaved weeds and, in the case of EPTC, common couch as well. They are mostly used in broad-leaved crops but tri-allate is used in small grain cereals, maize and onions, EPTC and butylate are used in maize, in addition to those used in rice mentioned above.

Uptake seems to be greater through the emerging shoot than the root. Several physiological processes can be affected including respiration, protein synthesis and photosynthesis but their main action seems to be interference with fatty acid and wax formation. The primary effect is thought to be on cell elongation. At high doses grasses often fail to emerge from the coleoptile or the leaf may emerge through the base of the cole-optile with the leaf tip enclosed in a loop. If the plant emerges from the soil, leaves are usually darker than normal and wax structure and/or cuticle thickness is altered. Consequently any other compound applied to the leaf penetrates more easily and because transpiration is increased xylem mobile pesticides may be taken up more extensively by the roots. Thus carried-over residues may affect the selectivity of herbicides applied to a following crop, although only EPTC, di-allate and tri-allate are likely to persist in the soil for more than about 3 months.

Detoxification in the plant is largely by conjugation with glutathione and cysteine. In some cases this is preceded by oxidation to the corresponding sulphinylformamides ('sulphoxides') which may be more phytotoxic than the parent molecule.

Ureas

A large number of urea derivatives (Table 3.14) with a variety of uses has been developed since the introduction of monuron in 1952. They are basically active through the soil and kill seedlings of most annual weeds soon after they begin to photosynthesize but some, for example linuron, isoproturon and metoxuron, can be taken up significantly through foliage and effective post-emergence formulations can even be produced. The range of selectivities is extensive and is reflected in the crops in which they are used which include cereals (chlorotoluron, isoproturon, metha-benzthiazuron), cotton (fluometuron), potatoes (linuron, monolinuron) and strawberries (chloroxuron) to name but a few, while some compounds

Table 3.14 Ureas

Name and structure	Mol wt	Water sol	VP	Activity
Chlorotoluron (BSI, ISO) CH_3—⬡—$NH.CO.N(CH_3)_2$ (with Cl)	212.7	$70\,mg\,l^{-1}$ 20°C	$4.8\,\mu Pa$ 20°C	S, F, So, T
Chloroxuron (BSI, ISO, ANSI, WSSA) Cl—⬡—O—⬡—$NH.CO.N(CH_3)_2$	290.7	$4\,mg\,l^{-1}$ 20°C	$239\,nPa$ 20°C	S & NS, F, So, T
Diuron (BSI, ISO, ANSI, WSSA) Cl—⬡—$NH.CO.N(CH_3)_2$ (with Cl)	233.1	$42\,mg\,l^{-1}$ 27°C	$410\,\mu Pa$ 50°C	S & NS, So, T
Fenuron (BSI, ISO, ANSI, WSSA) ⬡—$NH.CO.N(CH_3)_2$	164.2	$3.85\,g\,l^{-1}$ 25°C	$21\,mPa$ 60°C	S & NS, So, T
Fluometuron (BSI, ISO, ANSI, WSSA) ⬡—$NH.CO.N(CH_3)_2$ (with CF_3)	232.2	$105\,mg\,l^{-1}$ 20°C	$66\,\mu Pa$ 20°C	S, F, So, T
Isoproturon (BSI, ISO) $(CH_3)_2CH$—⬡—$NH.CO.N(CH_3)_2$	206.3	$55\,mg\,l^{-1}$ 20°C	$3.3\,\mu Pa$ 20°C	S, F, So, T
Linuron (BSI, ISO, ANSI, WSSA) Cl—⬡—$NH.CO.NOCH_3$ (with Cl, CH_3)	249.1	$81\,mg\,l^{-1}$ 24°C	$2.0\,mPa$ 24°C	S, F, So, T
Methabenzthiazuron (BSI, ISO) (WSSA, methibenzuron) benzothiazole—$N.CO.NHCH_3$ (with CH_3)	221.3	$59\,mg\,l^{-1}$ 20°C	$133\,\mu Pa$ 20°C	S, F, So, T

Table 3.14 Cont.

Name and structure	Mol wt	Water sol	VP	Activity
Metoxuron (BSI, ISO)	228.7	$678 \, mg \, l^{-1}$ 24°C	4.3 mPa 20°C	S, F, So, T
Monolinuron (BSI, ISO, WSSA)	214.6	$735 \, mg \, l^{-1}$ 25°C	6.4 Pa 65°C	S, So, T
Monuron (BSI, ISO, ANSI, WSSA)	198.7	$230 \, mg \, l^{-1}$ 25°C	67 µPa 25°C	S, So, T
Tebuthiuron (BSI, ISO, ANSI, WSSA)	228.3	$2.5 \, g \, l^{-1}$ 25°C	267 µPa 25°C	S & NS, So, T

Metoxuron: CH_3O—⟨ring⟩(Cl)—$NH.CO.N(CH_3)_2$

Monolinuron: Cl—⟨ring⟩—$NH.CO.NOCH_3$ (CH_3)

Monuron: Cl—⟨ring⟩—$NH.CO.N(CH_3)_2$

Tebuthiuron: $(CH_3)_3C$—⟨thiadiazole ring: S, N—N⟩—$N.CO.NHCH_3$ (CH_3)

(such as diuron) can be used for total weed control in uncropped areas.

Most of the ureas move in the xylem but there are sometimes large species differences in rates of absorption and transport which, together with different rates of detoxification, form the basis for selectivity. The primary mode of action is to inhibit photosynthetic electron transport in Photosystem II (see Fig. 7.1 in Chapter 7). This leads to the production of a range of powerful oxidants which damage membranes, pigments and so on, causing rapid destruction of the cell. Symptoms are thus chlorosis, and yellowing at low doses and/or low light intensities and a bleaching effect and desiccation when conditions (of light, humidity, dose, soil conditions) are most favourable to the action. Detoxification reactions in the plant include *N*-demethylation (usually the initial step), *N*-demethoxylation, ring hydroxylation (which can lead to the formation of conjugates with carbohydrates or peptides) and hydrolysis to the corresponding aniline.

In keeping with the wide variety of activities the persistence in the soil of members of this group ranges from a few weeks to more than a year.

Sulphonylureas

The sulphonylureas (Table 3.15), introduced around 1980, control a wide range of annual broad-leaved and some grass weeds in small grain cereals,

Table 3.15 Sulphonylureas

Name and structure	Mol wt	Water sol	VP	Activity
Bensulfuron-methyl (BSI, ISO, ANSI, WSSA)	410.4 pK_a 5.2	$2.9–120 \, mg \, l^{-1}$ pH5–7 25°C	2.8 pPa 25°C	S, F, So, T
Chlorimuron-ethyl (BSI, ANSI, WSSA)	414.8 pK_a 4.2	$11–1200 \, mg \, l^{-1}$ pH5–7 25°C	492 pPa 25°C	S, F, So, T
Chlorsulfuron (BSI, ISO, ANSI, WSSA)	357.8 pK_a 3.6	$60–7000 \, mg \, l^{-1}$ pH5–7 25°C	3.1 nPa 25°C	S, F, So, T
DPX L5300	396	$28–280 \, mg \, l^{-1}$ pH4–6 25°C	36.0 µPa 25°C	S, F, So, T
DPX-M6316 (WSSA, thiameturon)	387.4 pK_a 4.0	$260–2400 \, mg \, l^{-1}$ pH5–6 25°C	17.3 nPa 25°C	S, F, So, T

Bensulfuron-methyl structure: $CO.OCH_3$ — $CH_2.SO_2.NH.CO.NH$ — pyrimidine with OCH_3, OCH_3

Chlorimuron-ethyl structure: $CO.OCH_2CH_3$ — $SO_2.NH.CO.NH$ — pyrimidine with Cl, OCH_3

Chlorsulfuron structure: $SO_2.NH.CO.NH$ — triazine with OCH_3, CH_3; benzene ring with Cl

DPX L5300 structure: $CO.OCH_3$ — $SO_2.NH.CO.N$ — CH_3 — triazine with OCH_3, CH_3

DPX-M6316 structure: thiophene with S, $CO.OCH_3$ — $SO_2.NH.CO.NH$ — triazine with OCH_3, CH_3

Table 3.15 Cont.

Name and structure	Mol wt	Water sol	VP	Activity
Metsulfuron-methyl (BSI, WSSA, draft, ISO, ANSI)	381.4 pK_a 3.3	$1.1-9.5\,g\,l^{-1}$ pH 5–7 25°C	332 pPa Extrapolated to 25°C	S, F, So, T

| Primisulfuron-methyl (BSI, WSSA, ISO, ANSI) | 468.3 | $0.7\,g\,l^{-1}$ pH7 20°C | 975 pPa 20°C | S, F, So, T |

| Sulfometuron-methyl (BSI, ISO, ANSI, WSSA) | 364.4 pK_a 5.2 | $8-70\,mg\,l^{-1}$ pH5–7 25°C | 73.2 fPa 25°C | S, F, So, T |

| Triasulfuron (BSI) | 401.8 | $1.5\,g\,l^{-1}$ pH7 20°C | 99.7 pPa 20°C | S, F, So, T |

rice and soya bean. They are remarkable for their high potency; application rates are of the order of $5-20\,g\,ha^{-1}$, and also for wide margins of selectivity, differences in tolerance of up to 20 000-fold have been reported.

They can enter through both roots and aerial parts so can be used pre- or post-emergence. They are readily translocated by both susceptible and tolerant plants. The mode of action seems to be by inhibiting acetolactate synthase (acetohydroxyacid synthase) the first enzyme in the synthetic pathways for valine, leucine and isoleucine. This has a number of consequences including a reduction in nucleic acid synthesis which causes effects in the meristem.

Selectivity appears to be based on relative rates of detoxification. The major pathway is thought to be via hydroxylation of the phenyl ring (this metabolite is still phytotoxic), followed by conjugation with glucose and then hydrolysis to the glycoside of the benzenesulphonamide.

These compounds are mobile in the soil as they are not extensively adsorbed. Times for 50% disappearance are of the order of a few weeks for metsulfuron-methyl and months for chlorsulfuron, with the rate of loss being slower in alkaline soils. Although rates of breakdown are not particularly slow, because of the wide range of tolerance referred to above, susceptible crops (notably sugar-beet) can be damaged by residues that are very small by conventional standards so appropriate care must be taken with rotational crops.

Imidazolinones

This group (Table 3.16) was introduced in the early 1980s for the control of annual and perennial grass and broad-leaved weeds. Imazamethabenz-methyl is used post-emergence in cereals but is active through the soil, imazaquin and imazethapyr are used pre- and post-emergence in soya beans while imazapyr is used in plantation crops, for brush control and for total weed control in non-crop areas.

All can enter plants through roots and foliage and have the same site of action as the sulphonylureas, that is, they inhibit acetolactate synthase. Selectivity depends on differences in rates of metabolism. With imaza-methabenz-methyl, which is an ester, susceptible species hydrolyze it to the acid (the phytotoxic principal) more rapidly than tolerant ones. The others are acids or salts and selectivity depends on speed of detoxification. When used pre-emergence, seedlings emerge but growth stops at the cotyledon stage. Post-emergence application causes growth to stop and meristems may become necrotic.

Breakdown rates in the soil vary, imazapyr being extremely persistent. Since sensitive crops may be damaged by very low residue levels rotational crops must be selected with care.

Pyrimidines (uracils)

These compounds (Table 3.17), the first of which was introduced in 1963, are primarily soil-acting although some foliar entry can occur. They control

Table 3.16 Imidazolinones

Name and structure	Mol wt	Water sol	VP	Activity
Imazamethabenz-methyl (BSI, ISO)	288.4	$1.3-2.2\,g\,l^{-1}$ 25°C	—	S, F, So, T

m-isomer   *p*-isomer

Name and structure	Mol wt	Water sol	VP	Activity
Imazapyr-isopropylammonium (BSI, ISO, ANSI, WSSA)	320.4	$620-650\,g\,l^{-1}$ 25°C	—	S, F, So, T

Name and structure	Mol wt	Water sol	VP	Activity
Imazaquin (BSI, ISO, ANSI, WSSA)	311.3	$60\,mg\,l^{-1}$ 25°C	—	S, F, So, T

Name and structure	Mol wt	Water sol	VP	Activity
Imazethapyr (BSI, ISO, ANSI, WSSA)	289.3	$1.3\,mg\,l^{-1}$ 15°C	—	S, F, So, T

germinating annual weeds and certain perennials. Bromacil is used for total weed control and selectively in such crops as cane fruits, citrus and pineapple; lenacil is used in beet crops (mainly pre-emergence), ornamentals, some cane fruits and strawberries; terbacil is used in apples, asparagus, citrus, peaches, mint, strawberries and sugar-cane.

They are all readily taken up by the roots and translocated in the xylem to the leaves where they interfere with photosynthetic electron transport,

Table 3.17 Pyrimidines

Name and structure	Mol wt	Water sol	VP	Activity
Bromacil (BSI, ISO, ANSI, WSSA)	261.2	815 mg l^{-1} 25°C	33 µPa 25°C	NS, So, T
Lenacil (BSI, ISO, ANSI, WSSA)	234.3	6 mg l^{-1} 25°C	—	S, So, T
Terbacil (BSI, ISO, ANSI, WSSA)	216.7	710 mg l^{-1} 25°C	62.5 µPa 29.5°C	S, So, T

apparently in the same way as the ureas. They also inhibit the induction of nitrate reductase by molybdenum and nitrate in the leaf in contrast to the ureas and triazines which may increase the activity of this enzyme. Detoxification in the plant involves hydroxylation and conjugation with glucose. They are of moderate to long persistence in the soil.

Pyridazines

These compounds (Table 3.18) have little in common except the pyridazine ring in their structures.

Chloridazon is applied pre- or early post-emergence, alone or in mixtures, to control annual weeds in beet crops. When taken up by the roots it is translocated in the xylem to the leaves but little translocation occurs following leaf entry. It interferes with photosynthetic electron transport in the same way as the ureas. The tolerance of beet species is due to their

Table 3.18 Pyridazines

Name and structure	Mol wt	Water sol	VP	Activity
Chloridazon (BSI, ISO) (ANSI, WSSA, Pyrazon)	221.6	400 mg l^{-1} 20°C	<10 μPa 20°C	S, F, So, T
Norflurazon (BSI, ISO, ANSI, WSSA)	303.7	28 mg l^{-1} 25°C	2.8 μPa 20°C	S, F, So, T
Pyridate (BSI, ISO, WSSA)	378.9	1.5 mg l^{-1} 20°C	133 nPa	S, F, C

ability to detoxify it by conjugation with glucose. It can persist in the soil for several weeks.

Norflurazon is applied pre-emergence to control annual grasses and sedges in cotton, stone and pome fruits, nuts and cranberries. It inhibits carotenoid synthesis with a consequent bleaching of the tissue caused by the singlet oxygen produced during photosynthesis which is normally quenched by β-carotene. Norflurazon also inhibits the dehydrogenation of linoleic acid with consequent effects on membrane structure.

Pyridate is a post-emergence herbicide for the control of annual broad-leaved weeds, notably cleavers and pigweed in small grain cereals, maize and oilseed rape. Its mode of action does not seem to have been established.

1,3,5-triazines

A large number of amino- 1,3,5-triazines has been produced since the introduction of simazine in 1956 (Table 3.19). Two of the carbon atoms carry substituted amino groups. The third has a chloro-, or methylthio-group which is denoted in the common name which ends *-azine* for chloro,

Table 3.19 1,3,5-triazines

Name and structure	Mol wt	Water sol	VP	Activity
Chloro-compounds:				
Atrazine (BSI, ISO, ANSI, WSSA) Cl, N, NHCH$_2$CH$_3$ / N, N / NHCH(CH$_3$)$_2$	215.7	30 mg l^{-1} 20°C	40 µPa 20°C	S & NS, So, T
Cyanazine (BSI, ISO, WSSA) CN / Cl, N, NHC(CH$_3$)$_2$ / N, N / NHCH$_2$CH$_3$	240.7	171 mg l^{-1} 25°C	213 nPa 20°C	S, F, So, T
Simazine (BSI, ISO, ANSI WSSA) Cl, N, NHCH$_2$CH$_3$ / N, N / NHCH$_2$CH$_3$	201.7	5 mg l^{-1} 20°C	810 nPa 20°C	S & NS, So, T
Terbuthylazine (BSI, ISO, ANSI, WSSA) Cl, N, NHC(CH$_3$)$_3$ / N, N / NHCH$_2$CH$_3$	229.7	8.5 mg l^{-1} 20°C	150 µPa 20°C	S, So, T
Methylthio-compounds:				
Ametryn (BSI, ISO, ANSI, WSSA) CH$_3$S, N, NHCH$_2$CH$_3$ / N, N / NHCH(CH$_3$)$_2$	227.3	185 g l^{-1} 20°C	112 µPa 20°C	S, F, So, T
Aziprotryne (BSI, ISO) (USA, Aziprotryn) CH$_3$S, N, NHCH(CH$_3$)$_2$ / N, N / N$_3$	225.3	55 mg l^{-1} 20°C	267 µPa 20°C	S, So, T

Table 3.19 Cont.

Name and structure	Mol wt	Water sol	VP	Activity
Desmetryn (BSI, ISO, WSSA) CH_3S ... $NHCH(CH_3)_2$ ... $NHCH_3$	213.3	580 mg l^{-1} 20°C	133 µPa 20°C	S, F, So, T
Prometryn (BSI, ISO, ANSI, WSSA) CH_3S ... $NHCH(CH_3)_2$ ... $NHCH(CH_3)_2$	241.1	33 mg l^{-1} 20°C	133 µPa 20°C	S, F, So, T
Terbutryn (BSI, ISO, ANSI, WSSA) CH_3S ... $NHC(CH_3)_3$ ... $NHCH_2CH_3$	241.4	25 mg l^{-1} 20°C	128 µPa 20°C	S, So, T

or *-tryn* for methylthio. The corresponding methoxy compounds (end-ing *-meton*) are also herbicidal but are no longer of commercial importance. The 1,3,5-triazines control many annual weeds in a range of crops and are usually applied pre-emergence. The major use is in maize (simazine, atrazine and terbuthylazine) but others are used in small-grain cereals (terbutryn, cyanazine), and many other crops including brassicas, peas, potatoes and fruit crops. In addition some are used for total weed control at higher rates.

They are usually taken up by the roots and then translocated to the aerial parts but some can enter the leaf. Their main action is to interfere with photosynthetic electron transport in the same way as the ureas but the molecular site of attachment is not entirely the same (see Chapter 7 & 8). Differences at this site seem to be responsible for some types of resistance to these compounds (Chapter 8) but generally selectivity is based on de-toxification processes. Dechlorination with simultaneous hydroxylation is probably the major process by which maize detoxifies the chlorotriazines but in most other species dechlorination, demethoxylation or demethyl-thiolation involves conjugation with glutathione. Dealkylation of the amino substituents can also occur.

Table 3.20 Triazinones

Name and structure	Mol wt	Water sol	VP	Activity
Hexazinone (BSI, ISO, ANSI, WSSA)	252.3	$33\,g\,l^{-1}$ 25°C	$27\,\mu Pa$ 25°C (extrapolated)	S, F, C
Metamitron (BSI, ISO)	202.2	$1.8\,g\,l^{-1}$ 20°C	$13\,mPa$ $c.$ 70°C	S, So
Metribuzin (BSI, ISO, WSSA)	214.3	$1.2\,g\,l^{-1}$ 20°C	$<1.3\,mPa$ 20°C	S, F, So, T

Soil persistence ranges from <3 months (aziprotryne, cyanazine, ametryne for example) to >6–9 months (atrazine, simazine).

1,2,4- and 1,3,5-triazinones

These are all contact and residual materials for the control of annual weeds and in the case of hexazinone, some perennials as well (Table 3.20). Metribuzin is used for potatoes, soya bean, asparagus and some other crops; metamitron in beet crops and hexazinone in young forest trees.

They interfere with photosynthetic electron transport in the same way as the ureas and triazines. Selectivity seems to be based on rates of detoxification, at least for metribuzin and metamitron where deamination is a major step but metribuzin can also undergo *N*-glycosylation and sulphoxidation.

Metamitron persists in the soil for a few months, metribuzin and hexazinone rather longer (6 months or more).

Bipyridinium compounds

The first of these quaternary ammonium compounds (Table 3.21) was introduced in 1957. They are post-emergence, foliar acting compounds

Chapter 3

Table 3.21 Bipyridinium compounds

Name and structure	Mol wt	Water sol	VP	Activity
Diquat dibromide (BSI, ISO, ANSI, WSSA)	344.0 (ion 184.2)	$700\,\mathrm{g\,l^{-1}}$ 20°C	—	NS, F, C
Paraquat dichloride (BSI, ISO, ANSI, WSSA)	257.2 (ion 186.3)	Very	—	NS, F, C

with no soil activity under normal circumstances because of their strong and extensive adsorption by soil, particularly the clay mineral fraction. Paraquat and diquat are generally non-selective but tolerance has been bred into some crops (see Chapter 8). Paraquat is used to control a wide range of annual broad-leaved weeds in such situations as stubble cleaning and grassland destruction prior to cultivation or direct drilling, in plantations, orchards and forestry. It is also used as a component of mixtures for total weed control. Diquat is used for the pre-harvest desiccation of a number of seed crops and potatoes and for the control of some aquatic weeds.

Paraquat and diquat cause rapid scorch and desiccation of treated foliage the speed of which usually limits translocation. They act by accepting electrons from Photosystem 1 (see Fig. 7.1, Chapter 7) to produce a radical ion which is reoxidized back to the original ion by molecular oxygen with the production of the superoxide radical (O_2-). The superoxide radical is a powerful oxidant which attacks plant tissues itself but also generates other active oxygen species, singlet ($^1O_2$) and triplet ($^3O_2$) oxygen, hydrogen peroxide (H_2O_2) and the hydroxide radical ($\cdot OH$) which do the same.

The acute mammalian toxicity of paraquat and diquat is very high and no specific antidote is available.

Miscellaneous heterocyclic compounds (Table 3.22)

Amitrole is a compound which can enter both roots and foliage and is translocated in both phloem and xylem so can be used in foliar sprays and in soil acting formulations. It has little selectivity and is used against annual and perennial weeds especially common couch and *Rumex* spp.

Table 3.22 Miscellaneous organic compounds

Name and structure	Mol wt	Water sol	VP	Activity
Amitrole (BSI, ISO, ANSI, WSSA)	84.08	$280\,g\,l^{-1}$ 25°C	—	NS, F, So, T
Benazolin (BSI, ISO, WSSA)	243.7	$600\,mg\,l^{-1}$ 20°C	396 nPa 20°C	S, F, T
Bentazone (BSI, ISO) (WSSA, Bentazon)	240.3	$500\,mg\,l^{-1}$ 20°C	$<10\,\mu Pa$ 20°C	S, F, C
Difenzoquat methyl sulphate (BSI, ISO, ANSI, WSSA)	360.4 (ion 249.3)	$765\,g\,l^{-1}$ 25°C	—	S, F, T
Ethofumesate (BSI, ISO, ANSI, WSSA)	286.3	$110\,mg\,l^{-1}$ 25°C	$86\,\mu Pa$ 25°C	S, F, So
Fluridone (BSI, ISO, ANSI, WSSA)	329.3	$12\,mg\,l^{-1}$ pH7	$13\,\mu Pa$ 25°C	S, So, T

Table 3.22 Cont.

Name and structure	Mol wt	Water sol	VP	Activity
Methazole (BSI, ANSI, WSSA)	261.1	1.5 mg l^{-1} 25°C	133 µPa 25°C	S, F, So
Oxadiazon (BSI, ISO, ANSI, WSSA)	345.2	700 µg l^{-1} 20°C	133 µPa 20°C	S, F, So

(docks), mainly prior to planting or in orchards and noncrop areas. Its main action seems to be to inhibit carotenoid synthesis and hence reduce the ability of the plant to quench singlet oxygen produced during photosynthesis, so the characteristic symptom is bleaching. Ammonium thiocyanate is a synergist for amitrole activity and is often included in formulations. Detoxification seems to be primarily by conjugation with amino acids and sugars, especially serine and glucose.

Benazolin is used in mixtures to control annual dicotyledon weeds in cereal crops and grass. It has auxin activity and behaves in the same way as the phenoxyalkanoic acids.

Bentazone is a post-emergence contact herbicide used alone to control annual dicotyledons in legumes, linseed, potatoes and narcissi and in mixtures for undersown cereals and grass. It also has uses in rice and soya bean. It acts primarily by interfering with photosynthetic electron transport on the reducing side of Photosystem 2. Selectivity is based both on differential retention, absorption and translocation and on detoxification which involves hydroxylation and conjugation with sugars.

Difenzoquat is a post-emergence herbicide which is specific for the control of wild oats in cereals (only in some cultivars of wheat and triticale) and ryegrass seed crops. It also provides control of powdery mildew. Difenzoquat apparently affects cell division and elongation in the apical meristem as the symptom it produces is prolific production of lateral tillers which are themselves often inhibited.

Ethofumesate is a pre-emergence herbicide used, often in mixtures, to control annual weeds in beet crops, grass seed crops, leys and strawberries. Uptake from the soil is mainly through the shoot in the case of grass species and the root in broad-leaved species. The mode of action is unclear but based on the symptoms (grasses often fail to emerge from the coleoptile and broad-leaved species do not develop beyond the cotyledon stage) it seems it involves the meristem.

Fluridone is a pre-emergence herbicide for the control of annual and some perennial weeds in cotton. It is absorbed by roots and shoots and is translocated readily except in cotton, the only crop which so far has shown tolerance. It inhibits carotenoid synthesis so produces the characteristic bleaching symptoms.

Methazole is used pre- and post-emergence to control annual weeds in cotton and some horticultural crops. It is metabolized in the plant to produce 3,4-dichlorophenyl-3-methylurea (DCPMU) and 3,4-dichlorophenylurea (DCPU) so its symptoms and mode of action are similar to those of the ureas, but there is also some carbamate-like activity on cell division. Selectivity appears to rest on the rate at which the various metabolic reactions occur. It can persist in the soil for 6 months or more.

Oxadiazon is a pre- and post-emergence herbicide used against annual and some perennial grass and broad-leaved weeds in rice, fruit and ornamental crops, being particularly effective against bindweeds. Post-emergence applications produce symptoms similar to the diphenyl ethers so presumably involves radical formation after interaction with carotenoid pigments. Used pre-emergence it reduces germination or causes leaf trapping or stunting, indicative of meristematic activity. It can persist in the soil for several months.

Oximes

Alloxydim-sodium and sethoxydim-sodium are translocated post-emergence herbicides for the control of a range of grass weeds, including volunteer cereals but excluding annual meadow-grass, in broad-leaved and *Allium* crops. They do have some soil action as well (Table 3.23).

Mechanism of action is uncertain but the free acid is thought to be the active principal. Mitosis is disrupted so the development of roots and shoots is inhibited, probably as a result of interference with lipid synthesis. Chlorosis and sometimes anthocyanin formation may also occur in the leaves. Detoxification seems to be primarily by dealkoxylation of the substituted hydroxylamine group to leave the free amine. Rate of detoxification may be the basis of selectivity as there is no clear relationship between activity and spray retention, uptake and translocation.

Table 3.23 Oximes

Name and structure	Mol wt	Water sol	VP	Activity
Alloxydim-sodium (BSI, ISO)	345.4	>2 kg l^{-1} 30°C	133 µPa 25°C	S, F, So, T
Sethoxydim-sodium (BSI, ISO)	327.5	25–4700 mg l^{-1} pH4–7 20°C	—	S, F, So, T

Organophosphorus compounds

There are several herbicides containing phosphorus and their chemistry and biological properties are various (Table 3.24). Several are phosphorodithioates; bensulide and piperophos are used pre-emergence in cotton and in the case of the latter, maize, rice and soya beans; butamifos and anilofos are used in rice. Glufosinate, a phosphinate, is non-selective, controlling both mono- and dicotyledons. Fosamine-ammonium, a phosphonate, is used for the control of woody plants and brush in forestry and non-crop areas. Field bindweed and bracken are also controlled. Symptoms often do not appear until the new growing season when buds fail to develop and fronds of bracken and shoots of bindweed may be deformed and striated. It is not translocated in some woody species where it can be used as a chemical trimmer.

The most important organophosphorus herbicide is undoubtedly glyphosate. It is a foliar-applied translocated compound with essentially no soil activity. It is active against almost all annual and perennial species so is non-selective, although its activity on some tropical grasses and sedges may be limited by restricted penetration. It is used pre-planting or drilling of field crops, on uncropped land, in some aquatic situations and as a directed spray in plantation crops and forestry. It is also used prior to

Table 3.24 Organophosphorus compounds

Name and structure	Mol wt	Water sol	VP	Activity
Bensulide (BSI, ISO, WSSA)	397.5	$25\,mg\,l^{-1}$ 20°C	$<133\,\mu Pa$ 20°C	S, So
Butamifos (BSI, ISO)	332.4	$5.1\,mg\,l^{-1}$ 20°C	$84\,mPa$ 27°C	S, F, C
Fosamine (ammonium) (BSI, ISO, ANSI, WSSA)	170.1	$1.79\,kg\,l^{-1}$ 25°C	$530\,\mu Pa$ 25°C	F, C
Glufosinate (ammonium) (BSI, ISO)	198.2	Very soluble	Low	NS, F, C
Glyphosate (BSI, ISO, ANSI, WSSA)	169.1	$12\,g\,l^{-1}*$ 25°C	—	NS, F, T

Bensulide structure:
$$\text{—SO}_2\text{NHCH}_2\text{CH}_2\overset{\overset{\displaystyle S}{\|}}{\text{S}}\text{P(OCH(CH}_3)_2)_2$$

Butamifos structure:
$$\text{—OP NH CHCH}_2\text{CH}_3$$ with NO$_2$, CH$_3$, S, O CH$_2$CH$_3$ substituents

Fosamine structure:
$$\text{NH}_2\text{CO–}\overset{\overset{\displaystyle O}{\|}}{\underset{\underset{\displaystyle ONH_4}{|}}{P}}\text{–OCH}_2\text{CH}_3$$

Glufosinate structure:
$$\text{CH}_3\overset{\overset{\displaystyle O}{\|}}{\underset{\underset{\displaystyle ONH_4}{|}}{P}}\text{CH}_2\text{CH}_2\underset{\underset{\displaystyle NH_2}{|}}{\text{CH}}\text{CO.OH}$$

Glyphosate structure:
$$\text{HO.COCH}_2\text{NHCH}_2\overset{\overset{\displaystyle O}{\|}}{P}\text{(OH)}_2$$

* Isopropylammonium and sesquisodium salts very soluble

harvest in small grain cereals to control some perennial weeds. It is readily absorbed by most species and is highly mobile in the phloem and probably also in the xylem. It produces symptoms of chlorosis and necrosis and leaves of regrowing perennials are often malformed or striated and multiple shooting is common. Glyphosate has a number of physiological and biochemical effects in the plant but it seems likely that the initial action is to inhibit the enzyme enol pyruvyl shikimate-3-phosphate synthase (EPSP synthase, see Fig. 7.4, Chapter 7) which disrupts the production of aromatic amino acids and affects levels of other enzymes, notably phenylalanine lyase (PAL). Various phenolic compounds and ammonia accumulate in the tissue. Glyphosate is detoxified only slowly in plants by degradation to aminomethylphosphonic acid (AMP). This is also the pathway of breakdown in the soil, where it occurs quite rapidly.

Table 3.25 Organoarsenic compounds

Name and structure	Mol wt	Water sol	VP	Activity
Dimethylarsinic acid (Draft BSI & ISO) (WSSA, cacodylic acid) $(CH_3)_2\overset{\underset{\textstyle \parallel}{O}}{As}.OH$	138.0	$2\,kg\,l^{-1}$ 25°C	—	NS, F, C
Methylarsonic acid (Draft, ISO, WSSA, MAA) $CH_3\overset{\underset{\textstyle \parallel}{O}}{As}(OH)_2$	140.0	Very	—	S, F, C
Sodium hydrogen methylarsonate (WSSA, MSMA) $CH_3-\overset{\underset{\textstyle \mid}{\underset{\textstyle OH}{}}}{\overset{\overset{\textstyle O}{\parallel}}{As}}-ONa$	162.0	$1.4\,kg\,l^{-1}$ 20°C	—	S, F, C
Disodium methylarsonate (WSSA, DSMA) $CH_3-\overset{\underset{\textstyle \mid}{\underset{\textstyle ONa}{}}}{\overset{\overset{\textstyle O}{\parallel}}{As}}-ONa$	183.9	$279\,g\,l^{-1}$	—	S, F, C

Organoarsenic compounds

These compounds (Table 3.25) have limited uses none of which is current in the UK. Methylarsonic acid (MAA) and several of its salts (sodium hydrogen methylarsonate, MSMA, disodium methylarsonate, DSMA and calcium methylarsonate, CMA) are used for the control of fingergrasses in turf and other annual weeds in lawns, cotton, citrus and non-crop areas. They also control some perennials, notably johnsongrass and *Cyperus* spp. Dimethylarsinic acid (cacodylic acid) is a contact material used as a desiccant in cotton and in pasture renovation. Because of its desiccant action it is not translocated extensively but the methylarsonates are translocated in both phloem and xylem. The sites of action are unclear; although effects on photosynthesis and respiration have been reported they are not altogether consistent with the symptoms of growth inhibition and abnormal cell division. Detoxification probably proceeds through conjugation with sugars and amino acids. Mammalian toxity is much lower than that of inorganic arsenic compounds. They have no soil action but soil

Table 3.26 Fumigants

Name and structure	Mol wt	Water sol	VP
Chloropicrin (BSI, ISO) Cl_3CNO_2	164.4	$2.27\,g\,l^{-1}$ 0°C	3.2 kPa 25°C
Dazomet (BSI, ISO, WSSA)	162.3	$3\,g\,l^{-1}$ 20°C	37 μPa 20°C
Methyl bromide CH_3Br	94.9	$13.4\,g\,l^{-1}$ 25°C	101 kPa 4.5°C
Methyl isothiocyanate CH_3NCS	73.1	$7.6\,g\,l^{-1}$ 25°C	2.7 kPa 20°C
Metham-sodium (BSI) (ISO, metam-sodium; WSSA, metham) $CH_3NH.CS.SNa$	129.2	$722\,g\,l^{-1}$ 20°C	

residues are finally converted to arsenates which have similar properties to phosphates so some would be likely to appear in crops.

Soil fumigants

Chloropicrin, methyl bromide and methyl isocyanate, together with two compounds which generate it in the soil, metham-sodium and dazomet, are used as soil sterilants or fumigants to control soil pests including weed seeds (Table 3.26). They are general toxicants so must be handled with care and planting of treated soil must be delayed until they have been completely dissipated.

Safeners

Alternative names for these compounds (Table 3.27) are crop protectants and antidotes but the term 'antidote' should not be encouraged to avoid possible confusion with medical antidote. The availability of a safener increases the range of crops that may be treated and allows higher dose rates to be used so that the weed spectrum may be extended. It may also

Table 3.27   Safeners

Name and structure	Mol wt	Water sol	VP
Cyometrinil (Z isomer) (BSI)	185.2	$95\,mg\,l^{-1}$ 20°C	4.7 mPa 20°C
Fenclorim (BSI, ISO)	225.1	$2.5\,mg\,l^{-1}$ 20°C	11.9 mPa
Naphthalic anhydride (Draft ISO)	198.2	IOW	—
Oxabetrinil (BSI, draft ISO)	232.2	$20\,mg\,l^{-1}$ 20°C	520 µPa 20°C
R-25788* (WSSA, Dichlormid) $Cl_2CHCO.N(CH_2CH=CH_2)_2$	208.1	$5\,g\,l^{-1}$ 20°C	800 mPa 25°C

* *N,N*-diallyl-2,2-dichloroacetamide

increase the likelihood of the development of herbicide resistant weeds if used injudiciously (see Chapter 8).

Naphthalic anhydride is used as a seed dressing on maize to increase tolerance to EPTC and on rice to safen against molinate and alachlor. Experimentally it has shown promise with other crop/herbicide combinations.

R-25788 (*N,N*-diallyl-2,2-dichloroacetamide, WSSA name, dichlormid) is used to protect maize against EPTC and can be used either as a seed dressing, or mixed with the EPTC formulation.

Cyometrinil is used as a seed dressing to protect sorghum from the effects of amides and oxabetrinil protects sorghum from metolachlor/triazine mixtures. Fenclorim protects rice from pretilachlor and is used in a mixture with the herbicide.

The mechanisms of action of these compounds are not altogether clear but in most cases are presumed to involve enhancement of detoxification. Certainly R-25788 can increase glutathione content and glutathione-S-transferase activity in treated plants.

Further reading

Ivens, G.W. (ed.) (1989) *The UK Pesticide Guide.* BCPC Publications, Bracknell & CAB International, Wallingford.
Worthing, C.R. (ed.) (1987) *The Pesticide Manual.* 8th edn. BCPC Publications, Bracknell.

4 / The principles of formulation and application

E.S.E. SOUTHCOMBE AND D. SEAMAN*

Schering Agrochemicals Ltd, Chesterford Park Research Station, Saffron Walden, Essex CB10 1XL; and *ICI Agrochemicals, Jealott's Hill Research Station, Bracknell, Berks. RG12 6EY

Formulation principles

Purpose of formulation

There are many classes of herbicides and they cover a wide range of chemical and physical properties. They can be solids or liquids with various melting and boiling points. They may be stable or unstable and be soluble or insoluble in water.

Herbicides are applied to weeds, soil or water at rates from a few tens of grams to a few kilogrammes of active ingredient per hectare. To enable this small amount of material to be distributed over a large area, the herbicide is supplied to the farmer as a formulation, which provides him with the chemical in either a very fine form or as a concentrate giving a fine form when diluted for spraying, in order to get an even cover of a large number of herbicide-containing sources over the target.

A satisfactory formulation must be safe and easy for the farmer to use and must not deteriorate in any way on storage. The formulation composition can affect the biological activity and selectivity of a herbicide and the preferred properties need to be built into the formulation during its development by the chemical company.

Types of formulation

Herbicide formulations can be divided into two main categories:
1 Concentrates for dilution with water.

2 Products to be applied undiluted.

The most important formulation types for dilution with water are:

Soluble concentrates	SL	Suspension concentrates	SC
Wettable powders	WP	Water-dispersible granules	WG
Emulsifiable concentrates	EC		

The principal formulation supplied ready-for-use is:

Granules GR

The International Coding System used above was devised by GIFAP (International Group of National Associations of Agrochemical Manufacturers).

Choice of formulation

The principal factors affecting the choice of formulation are the properties of the herbicide, the impact of formulation type on biological expression and the handling of the product by the farmer. Chemicals which are soluble in water in their own right or as salts offer a simple way for formulation as soluble concentrates. Liquid pesticides which are immiscible in water can normally be formulated as emulsifiable concentrates and solid pesticides as wettable powders. Solids, which have low solubilities and are chemically stable in water, can be formulated as suspension concentrates, which are dispersions of fine particles of the pesticide.

Water-dispersible granules are an alternative to wettable powders and are much safer and easier to handle as they are free flowing and free of dust. Granules are preferred for application to soil where it is desirable to hold back the chemical for a period. Examples are for volatile herbicides which otherwise would need to be incorporated into the soil to minimize volatilization.

Composition and properties

Wettable powders are prepared by dry grinding and blending the herbicide, a filler such as clay, a dispersing agent to suspend the herbicide and a wetting agent to wet out the powder into the spray water. The particle size of the herbicides is about $10–20\,\mu m$ unless the formulation has been air-milled in which case it can be as fine as a few micrometres. The fairly coarse particle size can lead to lower foliar activity than other formulations and to wash-off on to the soil.

Emulsifiable concentrates consist of the liquid herbicide, solvents and emulsifiers. Solids which are sufficiently soluble in solvents can be formulated in this way also. When added to spray water, fine emulsions are spontaneously formed, giving a droplet size of only a few micrometres. As the chemical is in solution in a lipophilic solvent, it is readily available to penetrate leaves, so emulsifiable concentrates can be a highly active presentation. An alternative to emulsifiable concentrates for liquid herbicides, which are chemically stable in water, is as a preformed emulsion (EW). This form reduces or removes the solvent component and can, for some chemicals, reduce the dermal toxicity of the formulation.

Suspension concentrates are suitable for herbicides which have a solubility in water below a few hundred parts per million and are chemically stable in this medium. They are prepared by grinding a suspension of the coarse pesticide in the presence of a dispersing agent until the herbicide is <5 μm. To prevent settling of these fine particles to form a compact sediment, anti-settling agents are included. The finer particle size compared with wettable powders can lead to greater biological activity.

Water-dispersible granules can be prepared by extrusion, agglomeration or spray drying. They have similar components as wettable powders. The granules are a few millimetres in diameter or less, and on addition to spray water disperse to release the herbicide particles.

Granules are prepared by impregnation or coating of preformed granules with the herbicide and by extrusion or agglomeration of a herbicide mixture with a carrier. They can be broadcast or applied in bands on to soil, into paddy rice or rivers and watercourses by hand or using suitable application equipment.

Mixed formulations

In order to obtain the desired spectrum of weed control, many formulations are sold which contain more than one herbicide. These may be formulations described above but on some occasions it becomes necessary to formulate these mixtures in a novel way. An example is where a solid pesticide is required in admixture with a water-immiscible liquid pesticide. One possible formulation is as a wettable powder where the liquid is absorbed on a porous carrier. A more recent solution is as a suspo-emulsion (SE) which is a mixture of a suspension and an emulsion. Another possibility is as a suspension of the solid in the emulsifiable concentrate of the liquid pesticide.

Tests on formulations

To be satisfactory, formulations must have long shelf lives to ensure that, when the farmer uses the product, it is easy to use and performs satis-factorily. There must be minimal physical and chemical change in products over a period of a couple of years. Powders should not cake, nor should the wetting and dispersing properties deteriorate. Suspensions can separate to give a clear layer provided they mix back readily but they must not form a hard sediment which will be difficult to re-suspend. Water-dispersible granules and ready-for-use granules should remain free flowing and dust free.

Products are tested in soft and hard waters to cover the range of spray waters found in practice. Wettable powders and suspensions should not sediment quickly nor should emulsions cream and coalesce rapidly, other-wise it is possible that the concentration of herbicide in the spray varies during the spraying period. Each product is tested in combination with other formulations recommended as tank-mixes to ensure that no serious physical interactions occur which might lead to inhomogeneous sprays.

Adjuvants

Adjuvants are materials added to the formulation to improve the biological effect. They can be built into the formulation as part of its total com-position or added to the spray tank. The most common examples are surface-active agents, commonly called 'wetters' or 'wetting agents', and oils.

Wetting agents: These are detergent-like chemicals. They operate by aiding spray retention, spreading the deposit and increasing the uptake of the herbicide through the leaf cuticle. They are used at rates of a few hundred to a few thousand milligrams per litre in the spray at similar concentrations to that of the herbicide. Frequently they are built into the formulation but a number of products are available for tank-mixing.

Oils: Oils are applied at higher rates and have similar effects although less is known about their mode of action. They are applied at rates from 0.25% to a few percent and are used as tank-mixes. The products used at lower rates contain substantial amounts of surface-active agents so may combine the effects of oils and surface-active agents. The ones used at higher rates contain up to 3% emulsifier to enable the oil to be mixed into water.

Sticking agents: Some pesticide deposits on foliage are readily washed off by rain. Where a persistent effect is required, a sticking agent may help. These are film-forming materials applied as tank-mixes to stick or coat the herbicide onto the leaf. Their use is limited as it has been difficult to demonstrate that they work reliably and it is possible that the sticking agent can reduce the activity of a herbicide by locking it into the film.

Controlled-release formulations

The two principal examples are granules and micro-capsules. Controlled-release formulations are useful where a herbicide lacks persistence. Standard granules can be to some extent controlled release as the chemical is not immediately available. It needs to be washed out or to volatilize from the granule. The release can be further slowed down by coating the granule with a film of material through which the chemical must diffuse.

Micro-capsules (CS) are tiny packages of herbicide held within a plastic film (Fig. 4.1). They are spherical and are made by a process called 'interfacial polymerization' which creates a plastic surface on the herbicide droplet. They can be made as very fine capsules of a few micrometres, or larger up to about 30 μm. The herbicide must pass through the surface plastic coating to exhibit its effect. There are examples of quite volatile herbicides, which by micro-encapsulation can be applied as a spray instead of as a granule. The micro-capsule formulation process produces a suspension of capsules which can be formulated in the same way as a suspension concentrate.

Wrapping the chemical up in this way can reduce its toxicity and make the product safer to use. Micro-encapsulation reduces the release of a chemical and can extend its persistence. The initial availability of the herbicide is reduced so whether an overall improved effect is achieved depends on whether a more persistent toxic dose is both required and achieved.

Delivery of herbicides

The most common way to apply herbicides is by spraying the formulated product diluted in water as a cloud of droplets directed towards the target to produce an evenly distributed deposit. The concentration of the product in the spray mixture is usually less than 10% and the volume of application between 100 and 400 litre ha^{-1} depending on the product and method of application. In some cases, such as CDA spraying, volumes can be as low as 20 litre ha^{-1} with concentrations up to 25%.

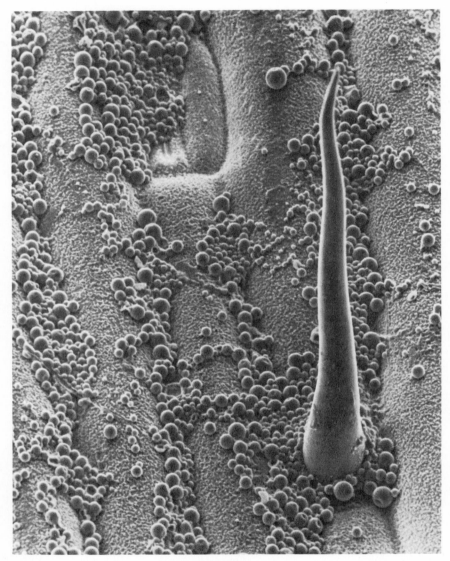

Fig. 4.1   A scanning electron micrograph (SEM) of a dried-down deposit of an experimental wild oat selective herbicide formulated as a microcapsule. A leaf prickle hair, stomata and epicuticular wax can also be seen in the picture. SEM courtesy of ICI Agrochemicals.

Target surface

It is clearly important to understand the relationship between the target surface, the spray mixture and the spray cloud. The effectiveness of the deposit depends on all three factors.

The surface of leaves can have a significant effect on the way in which the spray droplets are deposited. Although most leaf surfaces are covered in a wax layer, it is the quantity and nature of this layer which determines whether the droplets will be retained or reflected from the surface. In general terms it has been shown that micro-rough surfaces, whether waxy or dry, tend to reflect droplets, whereas smooth surfaces allow good retention.

Retention

The physico-chemical characteristics of the product can be influenced by the formulants used in the herbicide. The effect of these is often to reduce the surface tension which is a good way to improve retention on difficult surfaces. This aspect is normally taken into consideration during the design of the formulation. However in recent years many surfactant and polymeric adjuvant products have been marketed which attempt to improve effectiveness of herbicides. This may be to increase the spreading of each drop across the surface or to aid the penetration of the active ingredient into it. Care must be taken to ensure that the adjuvants are compatible chemically and physically with the herbicide and do not reduce crop safety.

Another way to improve the retention of droplets onto difficult target surfaces, such as young vertical grass weed leaves, is to use finer sprays with smaller droplets. This technique must be used with care as it may increase the risk of drift. In such cases the correct choice of nozzles producing the optimum spray quality is increasingly being indicated on labels by the herbicide manufacturers (see p. 143).

The spray quality required depends very much on the nature of the herbicide. Highly-systemic materials may work well with a less complete cover of the target than that required by contact-acting products. In dense foliage the product may need to achieve good penetration to reach underlying weeds and this may be better achieved with faster medium-quality sprays at medium volume rates than with swirling fine sprays at low volumes. Penetration can be assisted by air currents produced either by external ducts or within the nozzle itself, or by deflecting the crop with ropes under the boom, though this latter technique is more applicable to late applied fungicides.

Retention can be improved by the use of electrostatic techniques to charge the spray droplets so giving them a positive attraction to earthed plants. However the charged droplets tend not to penetrate into canopies well and may need air or other assistance.

Residual herbicides sprayed directly onto the soil, paths or other such surfaces need to be applied to a surface free from obstructions which could result in uneven deposits, such as cloddy seedbeds. Herbicides with a volatile component applied either as sprays or granules must often be incorporated into the upper layers of the soil soon after application. Many herbicides used in such situations can be applied as relatively coarse-quality sprays which reduce the risk of drift.

Application methods

Herbicides may be applied in a variety of ways depending on the target, scale of treatment and the product. Spraying is the most common method followed by the use of granules. Herbicides can be applied to the soil or to foliage, either pre- or post-planting or emergence depending on the nature of the product. They may be applied as broadcast treatments or in spots or bands. The preferred choice of method is often indicated by the product label.

Conventional sprayers

Portable hand-operated sprayers

A very wide range of small, hand-operated sprayers of around 5 litre capacity is available, but for serious agricultural and horticultural work, knapsack sprayers with tank capacities of around 15 litre are used (Fig. 4.2). Plastic tanks with a small hand-operated pump mounted in their base are widely available. The pump is incorporated with a small pressure vessel which evens out pulsations and makes continuous pumping unnecessary unless a large nozzle, or multiple nozzles, are being used. Pressures of several bars can readily be generated.

The need for pumping during spraying can be eliminated by the use of sprayers in which a tank can be pressurized when it is filled with liquid. The spray liquid is put into the tank and air then pumped in above, either by a pump or from a compressed air supply. While spraying the pressure de-

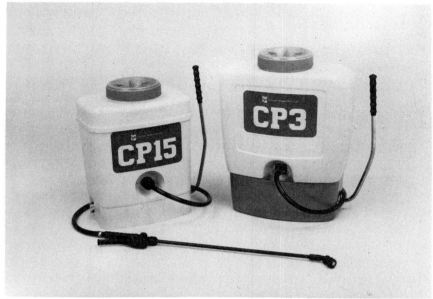

Fig. 4.2 Two typical knapsack sprayers. Photograph courtesy of Cooper-Pegler Ltd.

creases so the tank may need to be recharged with air before all the liquid has been sprayed. In one type of compression sprayer, the air may be retained in the sprayer which is recharged by pumping in the spray liquid. Although these pressurized sprayers are manufactured with an adequate safety factor, it is advisable to subject them to an annual hydraulic pressure test. Testing, with suitable precautions, should be at twice the working pressure. In use, the recommended working pressure should never be exceeded.

These sprayers can be fitted with a trigger valve on a lance at the end of which is usually a single nozzle. The use of deflector or solid cone nozzles is recommended to give, respectively, a well-defined band or spot treatments of individual weeds. Flat-fan nozzles are best for applying an even deposit to wider areas. It is wise to use a small pressure-control valve or a gauge on the lance to keep the pressures down to <2 bar (200 kPa). Short booms are available for treating small plots.

Knapsack and compression sprayers can be used to treat the area around trees or shrubs in nurseries and small plantations or in non-agricultural areas such as paths. They may have to be used for larger areas where the unevenness or slope of the land precludes the use of a wheeled vehicle.

Knapsacks emptying by gravity are used in forestry for the treatment of frill girdles.

Portable-powered sprayers

Knapsack sprayers incorporating a small petrol engine and a fan to provide an air blast are widely used for the application of insecticides and fungicides to trees and bushes, but since they are designed to produce very fine drops in a stream of air they are unsuited to most herbicide applications. They have been used in forestry for applying certain herbicides to brushwood and for similar treatments.

Where the hand-pumped knapsack has inadequate capacity, herbicides can be applied by small motor-driven pump units which may be mounted in a portable frame or mounted integrally with a tank on a wheeled chassis. A diaphragm pump is most often used on these small units and the output is only a few litres per minute. Tank size can be from 20 to 200 litre but in the larger sizes the power unit must be used to assist in propulsion or the whole unit must be placed on a vehicle. The spray is usually manually directed from a spray lance at the end of a length of hose but fixed spray booms may also be used. Such equipment is used on horticultural holdings, and is suitable for use in the treatment of weeds in industrial sites or similar places where irregular and relatively small areas have to be sprayed.

Tractor-powered sprayers

The majority of sprayers in use in agriculture are mounted on the tractor three-point linkage, but this imposes a limit on the tank capacity. The pump is usually mounted on the power take-off of the tractor and connected to the spraying unit by suction and delivery hoses. The smallest sprayers may be fitted with roller-vane pumps. Diaphragm pumps are more popular, for although the higher pressures available from such pumps are certainly not required for herbicide application, separation of the spray fluid from the moving parts gives a much greater pump life. The pump output must be sufficient to maintain pressure at the nozzles, especially at high spray volumes and with wide booms, and also to provide adequate return to tank for efficient agitation.

These sprayers range from 200 to 1000 litre in tank capacity, about $50-150$ litre min^{-1} pump output, and $6-18$ m in boom width. Most of the work required of farm sprayers (in particular weed control in cereals) can satisfactorily be done at spray volumes of $100-300$ litre ha^{-1} and the choice of

Fig. 4.3   Tractor mounted boom sprayer. Photograph courtesy of E. Allman & Co. Ltd.

tank capacity is usually decided by the area to be sprayed and therefore rate of work required (Fig. 4.3).

Trailed sprayers are used on larger farms. Their operation is, in principle, no different from that of the popular tractor-mounted sprayer, but because the work load may be greater, their output and capacity are larger with tank capacities of 1500–2000 litre, pump outputs of 200 litre min^{-1} and booms up to 24 m wide. For the larger farmer and contractor, performance is of paramount importance. Minor savings in capital expenditure by the purchase of a 'cheaper' unit can rapidly be more than lost by breakdowns. For these situations very large self-propelled sprayers fitted with advanced electronic and hydraulic control systems are available.

Booms: The accuracy of deposition depends on the nozzles being at a correct height above the crop and travelling over it at the correct speed. The more uneven the ground, the greater the swing of boom carrying the spray bar is likely to be. The wider the boom, the more the tips can swing in relation to the ground. Wide booms also make swath matching more difficult as the meeting point is further from the operator. On the other hand, such booms mean less wheelings and quicker work.

While very little can be done about the limitations of uneven ground, improved design of the boom mounting can reduce the transfer of movement from the vehicle to the boom and so decrease both fore-and-aft and up-and-down movement of the boom. Excessive movements can given rise to considerable variations of dose. A stable boom is essential particularly when reduced volumes of spray are applied.

A few sprayers have wide booms mounted on the front of the tractor. These are used either with the largest possible linkage-mounted tank, or with twin tanks saddle mounted on each side of the tractor engine, to give up to 1200 litre capacity. Many sprayers are fitted with hydraulically-operated booms operated from the tractor cab.

Controls: The control of the output of spray and its pressure is often provided by remote means from within the tractor cab. It is also possible to fit instruments to monitor speed, pressure and output. Furthermore, these can be integrated to provide a means for the automatic control of output, usually by varying spray pressure, over small speed changes.

Agitation of spray mixture: Agitation of the liquid in the tank is necessary to maintain satisfactory suspension of wettable powders or emulsions. Mechanical agitators are costly and are used only on the largest sprayers. 'Hydraulic' agitation, that is the re-circulation of some of the pumped liquid, is satisfactory provided the return flow is adequate and is suitably arranged in relation to the tank shape. It is very important to ensure that the material is well mixed before spraying starts, whether it be a liquid or solid formulation, and particularly in the case of a suspended powder, to maintain adequate agitation until the tank is emptied. If the sprayer has to be stopped, owing to weather or mechanical breakdown, long enough for powder to settle to the bottom of the tank, the agitation on some sprayers may be incapable of properly re-mixing the spray.

Frothing may occur with some materials if the return flow does not enter at the bottom of the tank or if a mechanical stirrer is breaking the liquid surface in the tank. This problem is not usually serious on modern sprayers unless there is an air leak into the pump suction. In this case a fine foam will be produced which will make it impossible to obtain the correct spray pressure until the fault is rectified.

Aircraft equipment

The limitations imposed in order to avoid drift or over-shooting the sprayed area have rather precluded the use of aircraft for herbicide application in the UK apart from forestry and some specialized applications to cereals. Work is being undertaken to improve the placement of spray from the air in overseas crops where there is greater economic pressure to use this method and some of the developments may widen the potential for herbicide application from the air in this country. Much more sophisticated spraying equipment can be used on aircraft than could be justified for farm equipment.

There are certain statutory requirements laid down for aerial spraying in a document entitled *The Aerial Application Certificate — Requirements and Information* obtainable from The Civil Aviation Authority (CAA). The National Association of Agricultural Contractors can provide guidance and documentation relating to aerial spraying.

Special sprayers

Various adaptations can be made of the basic sprayer mechanisms by replacing the conventional boom by other nozzle arrangements. A common arrangement is for 'band spraying' where individual nozzles are arranged to cover a strip only a few centimetres wide, commonly 18 cm. This is usually done to conserve chemical by obtaining weed control only in a crop row, using mechanical means to deal with inter-row weeds. For this purpose, pre-emergence treatments can be applied at the time the crop is sown by attaching nozzles to the seeder units. The nozzles, giving an even deposit across the band are designed for band spraying but their height above the seedbed must be carefully controlled to avoid over-dosing by contraction of the band width. Sometimes a triangular-type deposit pattern is preferred to reduce the dose along the edges of the band where it overlaps the hoed ground.

Special arrangements of nozzles, sometimes combined with guards to protect crop plants, are used for weed control along rows of trees, bushes and canes in plantation and nursery crops. For widely-spaced orchard trees, nozzles have been arranged on swinging arms to treat circles around the tree base which is protected by a U-shaped shield. Other arrangements of nozzles on special booms are needed for treatment of verges, railway lines etc., but in most cases the operation is essentially the same as normal crop spraying except that the dose must be related to the target area; this may often not be flat and there may be a much denser herbage cover than is usual in farm crops.

Spraying equipment can be fitted to a wide variety of vehicles ranging from small all-terrain motor-cycles, through low-ground-pressure vehicles, pick-up trucks, to multi-purpose self-propelled units. The use of special vehicles often allows spraying operations to be carried out in seasons and on ground conditions which would prohibit normal tractors, so enabling optimum application timing.

There are a few specialized sprayers for use on railways and roadsides where more elaborate means of dose control are introduced. These have high-output nozzles arranged so that the swath width can be continuously varied as the work proceeds by turning them on and off individually. Compound nozzles (a cluster of nozzles projecting spray to various distances) are sometimes used in place of a boom. To deal with speed variations imposed by traffic and other conditions, some sprayers of this type have used a fixed output of water with herbicide metered in from a pump whose output is proportional to the speed. Long distances may be travelled at high speed and large quantities of chemical and water carried.

Nozzles

Hydraulic nozzles produce droplets by the formation of a sheet of liquid which quickly becomes unstable, forming ligaments which break down into droplets. Typically, a wide spectrum of droplet sizes is produced and research work continues to find ways to produce narrower spectra, eliminating the extreme small and larger size fractions.

Work also continues on methods to measure and describe droplet spectra. A number of sophisticated instruments, most using laser principles, exist but there is little agreement between the data they produce. Traditionally, droplet spectra have been described by the mathematically-derived figure, volume median diameter (VMD) which gives a single diameter weighted towards the larger, higher-volume droplets. Number median diameter (NMD) gives a similar figure weighted towards the more numerous smaller droplets. One way to refer to spectrum width is to use the VMD:NMD ratio, typically >10 for hydraulic nozzles and <2 for rotary atomizers depending on the accuracy of the NMD measurement, which is variable.

Types

There are three types of hydraulic nozzles used to apply herbicides which can be fitted to any of the above types of spraying equipment (Fig. 4.4).

Flat fan nozzles: These are the most common. They produce a fast, well-defined spray useful for penetrating canopies. Those giving a triangular

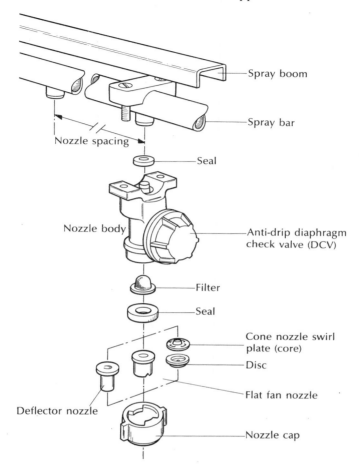

Fig. 4.4 Typical components of hydraulic nozzle assembly.

deposit pattern, must be used on boom sprayers to give an even overall deposit, whilst those giving a rectangular or 'evenspray' pattern are for use with single-nozzle sprayers such as knapsack or band sprayers. Low-pressure nozzles operated at 1 bar (100 kPa) give a coarser spray more suited to use with pre-emergence products.

Hollow cone nozzles: These tend to give a more turbulent spray cloud often with a finer spray which gives a better coverage of foliage. Their uneven deposit pattern makes it difficult to achieve a uniform overall deposit when fitted to boom sprayers.

Deflector nozzles: These are often known as 'anvil' or 'impact nozzles' and produce a sheet from the impact of a stream of liquid onto a flat surface. They operate at lower pressure and produce a 'coarse' spray with a rather uneven deposit pattern. They are suitable for use with simple nozzle lances and also often used to apply liquid fertilizer.

Nozzles are generally fitted to lances or booms by means of a body containing an anti-drip device usually known as a 'diaphragm check valve' (DCV). They must be protected with good filters fitted either directly behind each nozzle or in the hoses feeding each boom section. Nozzle caps often connect to the body by means of a quarter-turn bayonet fitting so enabling nozzles to be changed quickly and easily.

There are a number of parameters which affect the performance of nozzles:

1 *Output*: A function of hole size and pressure. Smaller nozzles produce finer sprays.

2 *Pressure*: Normally between (1.5 and 4 bar (150 and 400 kPa). Pressures above this are not useful. Higher pressure produces finer sprays.

3 *Spray angle*: With flat fan nozzles these are normally 80° or 110°. Wider angles produce finer sprays but allow the boom to be carried closer to the ground.

4 *Filters*: Filters are usually found at the output from the tank to protect the pump and either in the hose feeding the booms or fitted directly into the nozzle bodies. A mesh size of 50 mesh inch^{-1} (300 μm aperture) is typically used.

5 *Material*: Nozzles may be made in brass, plastic, stainless steel or ceramic — in rough order of increasing cost and wear resistance. All nozzles should be renewed at least once yearly.

6 *Height*: The best height to set the boom is so that fans or cones from

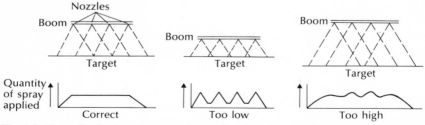

Fig. 4.5 Spray distribution patterns under a series of nozzles with triangular distribution, and the effect of distance between nozzle and target. The deposit is represented by the area below the thick line.

Table 4.1 BCPC nozzle code

BCPC code	Description	Output l min^{-1}	Rated pressure bar (kPa)
F80/1.27/3	80° Fan	1.27	3 (300)
F110/1.59/3	110° Fan	1.59	3 (300)
FE80/1.16/3	80° 'Even spray'	1.16	3 (300)
FLP80/1.30/1	80° Low pressure fan	1.30	1 (100)
HC/1.43/3	Hollow cone	1.43	3 (300)
D/1.12/1	Deflector	1.12	1 (100)

alternate nozzles meet just above the target, or the crop if spraying underlying weeds (Fig. 4.5).

BCPC Nozzle Code

Nozzles are identified by number, letter or colour codes. Unfortunately there is no uniformity between the codes used by different manufacturers and none indicates the output of the nozzle in metric units. The *BCPC Nozzle Code* provides a universal means of describing all such nozzles in terms of type, spray angle and output. Spray quality has not been included as this can vary from make to make. The examples in the Table 4.1 show how the *Nozzle Code* is used.

BCPC nozzle-selection system

BCPC have introduced a system for the selection of nozzles to help the agro-chemical supplier to indicate not only what dose and spray volume to use but also the type of spray quality or range of droplet sizes best suited to the product. This enables the optimum spraying conditions for the product to be specified and discourages the use of unnecessarily fine or coarse sprays which may be environmentally or biologically undesirable.

Nozzles are tested to standard protocols and classified according to their spray quality into categories of 'Fine', 'Medium' or 'Coarse' (Fig. 4.6). In addition there are two additional categories of 'Very Fine' and 'Very Coarse' for exceptional uses such as fogging machines or liquid fertilizers.

Information on the categories for common nozzles is available from the suppliers of nozzles and spraying equipment. Agro-chemical suppliers are using phrases such as 'apply as "Medium" spray (as defined by BCPC)' on their labels, unless they wish to specify actual nozzles by name.

(a)

(b)

(c)

Fig. 4.6 Examples of BCPC 'fine', 'medium' and 'coarse' sprays. Photograph courtesy of Lurmark Ltd.

The system can be very easily adopted by users in practice. At its simplest, all that is needed are two sets of nozzles — one giving a 'Fine' spray and the other a 'Medium' spray. With careful choice, it may be only necessary to change the nozzle and pressure and using the bayonet-fitting caps common to many sprayers these days this is a quick and simple job.

Spinning disc spraying equipment

Principles

Droplets more evenly-sized than those produced by nozzle orifices can be achieved by using a spinning disc nozzle (Fig. 4.7). The mean droplet size is adjusted by the speed of rotation of the disc and the flow rate of the liquid. Smaller droplets are produced as the speed of rotation is increased.

Droplets are produced individually at very low flow rates. As flow rate increases, some ligaments are formed which produce main drops and also much smaller satellite droplets. When sufficient liquid is fed, only ligaments are formed and further increases in the flow enlarge each ligament and, with it, droplet size. Care must be taken not to overfeed the disc as the ligaments will then join to form a sheet and droplet-size control is lost.

Chapter 4

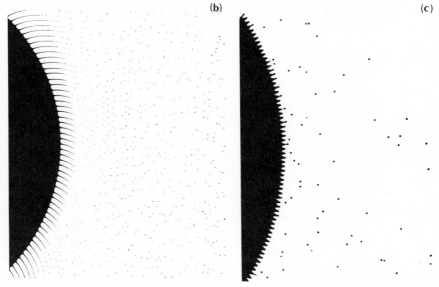

Fig. 4.7 Drop formation: (a) sheet and drop formation from an hydraulic pressure nozzle; (b) ligament break-up and drop formation at the periphery of a spinning disc; and (c) drop formation at the periphery of a spinning disc with a serrated edge. Illustration courtesy of AFRC Institute of Engineering. (Not to scale.)

Equipment with spinning discs allows the use of spray volumes in the $10–60$ litres ha^{-1} range so less time is spent refilling the sprayer and more time used actually spraying. This can be very important in the UK where weather conditions, particularly rain and strong winds, can reduce the number of days suitable for spraying. Alternatively, the lower-volume rate permits the use of a smaller tank and reduces soil compaction. The use of spinning disc equipment has been associated with low-ground-pressure transport vehicles for spraying, particularly in the autumn and winter months. The reduction in proportion of the small droplets in the spray may allow operation in slightly higher wind speeds.

The use of such equipment which restricts the range of droplets in a spray is often known as 'controlled droplet application' (CDA). It has been found to be particularly useful with open targets such as with the application of pre-emergence herbicides to soil; there are sometimes problems in achieving penetration into dense crops.

The use of a very narrow range of droplet sizes has made demands for the optimum droplet size for different crop/product situations to be identified and this has proved to be most difficult, hence the need for sprays with a reasonable mix of droplet sizes.

The use of such low spray volumes is often below the minimum volumes recommended for most products, so the *Code of Practice* covering reduced volume applications must be followed (see p. 149).

Hand-carried equipment

Herbicide is gravity fed through an interchangeable orifice on to a small plastic saucer-shaped or flat disc usually with teeth around the edge (Fig. 4.8). The disc is rotated by a small d.c. electric motor powered by dry-cell batteries carried in the sprayer handle or in a separate holder. Disc speed is governed to maintain the correct droplet size and for herbicide application most units operate at 2000 r.p.m. producing droplets of about 250 µm in diameter. Some sprayers can restrict the emitted spray to a limited area to the front.

The hand-held unit is useful for treating edges of fields, fence lines and irrigation ditches, in orchards and forestry. Normally $10-20$ litre ha^{-1} is applied.

Some suppliers can offer a range of herbicide products specially formulated and packaged ready for use with their equipment. This gives considerable cost and safety advantages.

Tractor-mounted equipment

Rotary atomizers can be fitted to the booms of tractor-mounted or trailed sprayers, usually at about 1 m spacing. The drive to the atomizers may be

Fig. 4.8. Hand-carried spinning disc sprayer for herbicide application. Photograph courtesy of Micron Sprayers Ltd.

electric, cable or hydraulic. The atomizers may be a series of discs on a common shaft or a larger single cup-shaped disc. Spray volume rates of 20–60 litre ha^{-1} are general with sprays corresponding to the BCPC 'Medium' category.

Great care must be taken to avoid damage of the discs by ground contact and to calibrate and control disc speed, flow rate and forward speed.

Reduced volumes

There is increasing interest in the use of spray volume rates lower than those previously given in product labels, especially to increase work rates and timeliness of application. Some spraying techniques can only function at low volumes, such as CDA, and although there are several labelled-product recommendations, most use of this equipment would be outside the label. Recent proposals made by the registration authorities in the UK allow the use of volumes down to 1/10th the label minimum volume rate providing certain conditions are observed. These are given in detail in the official *Code of Practice on the Agricultural and Horticultural Use of Pesticides*. The main conditions are that the user accepts responsibility for efficacy, that the products are not rated under certain toxicity categories or otherwise not restricted for such use, that certain additional protective clothing is worn, that spray quality restrictions are observed and that all other conditions on the label are adhered to. This is an important development as it allows the use of a much wider range of application equipment than has been possible in the past.

Granule applicators

Granular formulations of herbicides are used in industrial areas, in the control of water weeds and in forestry, agriculture and horticulture. They may be applied as spot or band treatments or overall (Fig. 4.9). Hand, knapsack, band and overall applicators are available.

Whatever the size and type of the applicator, it will consist of a hopper, metering mechanism and distribution device. The smallest equipment in spot treatment of individual weeds may consist of a hand-held container fitted with a device allowing a small volume of granules to escape at each operation down a rigid tube reaching from the bottom of the container to the weed.

Fig. 4.9 A pneumatic granule pesticide applicator. Photograph courtesy of Horstine-Farmery Ltd.

Equipment available for applying granules in bands can be used either on its own or in association with a seed drill to lay down bands of granules containing a suitable herbicide simultaneously with the sowing operation. A satisfactory metering device consists of a fluted rotor driven from a land wheel so that its revolutions are directly geared to the distance covered. In this way, the application remains even, regardless of variations in forward speed. Granules fall by gravity first into the flutes and then after discharge down the distribution tubes.

The dose can be regulated either by changing the width of the rotor or its speed of revolution in relation to the forward speed. Where only narrow bands are treated it is adequate to let the granules fall under gravity down a tube with a fitting to spread them over the desired width. Such equipment may be used manually, as a knapsack applicator, or mounted on a tractor or implement.

To treat large areas where a broadcast distribution is required it is usual to distribute the granules by using an airstream. The granules are metered, very much in the same way as in gravity machines, into an airstream produced by fan units driven by the tractor p.t.o. The granules are conveyed to an impact nozzle (acting in a manner similar to the liquid deflector nozzle) which spreads the granules uniformly over a band of up to 1.5 m. The width of the swath is dictated by the number of outlets and their spacing.

Knapsack applicators working on the pneumatic principle, through the agency of a small motor, may be used where tractor-mounted applicators are impractical, as in some forestry and industrial situations. The granules are metered by gravity through an orifice into the airstream. Accuracy depends on the operator and the results may be less satisfactory than with tractor-mounted equipment.

The combined application of fertilizer and herbicides through granule applicators is not satisfactory because of segregation of the different granule sizes during application. This can occur in the hopper, during metering and in the air flow. For this reason, herbicide granules themselves are manufactured to within close limits of density and size.

Other considerations

Selective application

One method of application which does not easily fit into any of the above categories is that of 'selective application' (sometimes known as 'Selap') using various designs of equipment to transfer herbicide to the weed by direct contact. Reflecting their designs these types of equipment are often called 'weed-wipers', 'rollers', 'wicks' or 'ropes'.

As a general rule, herbicide is metered onto the absorbant surface which is brought into contact with the weed surface. The technique is mainly used to control weeds which grow taller than the main crop and relies on products which are sufficiently translocated to enable weeds to be killed from only partial contact with the foliage. The key advantages are economy of use of herbicides and placement of herbicide only onto the weed and not onto the crop, ground or into the environment. However there are only a limited number of suitable crop/weed situations and few herbicides with the appropriate properties.

Drift

It is clearly important to reduce the possibility of drift from herbicide applications to a minimum to prevent damage to neighbouring crops and to the environment. There are a number of ways in which this can be done:
1 Avoiding fine quality sprays.
2 Avoiding low spray volume unless medium or coarse sprays can be produced.
3 Using CDA/rotary atomizer equipment, providing biological activity is satisfactory and the appropriate droplet size selected.
4 Avoiding unsuitable weather conditions, such as calm with temperature inversion, turbulence on warm sunny days and strong winds.
5 Using anti-drift additives, though these must be compatible with the herbicide and not create too coarse a spray.

Handling of chemicals

As with all uses of pesticides it is essential to handle the concentrated product with care, wearing all the protective clothing prescribed for the product. As most contamination takes place when transferring the product from container to the spray tank, the use of various filling devices now found on many sprayers can significantly reduce this problem. These devices range from simple suction probes, to induction hoppers at low level on the sprayer, to complete closed-fill systems.

Chemical handling problems can also be reduced by the use of improved formulations such as water-dispersible granules (WG) or water-soluble packs, where these are appropriate to the product.

Acknowledgements

Thanks are due to E. Bals of Micron Sprayers Ltd and to A. Peck of Horstine Farmery Ltd for their assistance in checking relevant parts of the text, and to R. Robinson of Rhône-Poulenc Group for his valuable help and comments.

Further reading

BCPC (1988) *Nozzle Selection Handbook*. BCPC Publications, Bracknell.

CAA *The Aerial Application Certificate — Requirements and Information*. CAA Cheltenham.

MAFF (1988) *Code of Practice for the Agricultural and Commercial Horticultural use of Pesticides (revised draft)*. MAFF Publications, Lion House, Willowburn Estate, Alnwick, Northumberland NE66 2PF.

Matthews, G.A. (1979) *Pesticide Application Methods*. Longman, Harlow.

5 / The methodology of application

T.H. ROBINSON

Ciba-Geigy Agrochemicals, Whittlesford, Cambridge CB2 4 QT

Introduction — objectives of the operator

The saying 'A spray is only as good as the person who applies it', will remain as true in the future as it has done in the past. The way in which a herbicide is manufactured, tested and supplied is meticulously monitored by the manufacturer and the registration authorities right up to the time when it is delivered to the end user, at which point the success or failure of that herbicide depends upon the quality of its application.

The spray operator is responsible for the fate of that herbicide. For a safe and effective application he must apply the correct dose of product in the recommended volume of liquid at the recommended spray quality, at the optimum time. This must be carried out with the minimum risk of contamination to the operator, the environment and the general public. Following the application of a pesticide, the spraying machinery and protective clothing must be decontaminated in a safe and effective manner.

The importance of the sprayer operator to the perception that the general public has of agrochemicals and spraying cannot be over-emphasized. Of all the activities from the initial development of a herbicide, it is the actions of the operator before, during, and after applying the product which the general public are most likely to observe at first hand.

153

The principles of herbicide application are the same for all types of herbicide formulation and application equipment. The operator is trying to apply the correct quantity of product evenly over a unit area of ground in a manner that is most acceptable in producing effective weed control with the minimum off-target wastage.

Owing to the similarity of the principles, aspects such as calibration and maintenance apply equally to all types of sprayer, from the largest self-propelled machine to the smallest knapsack, and even to band sprayers and granule spreaders.

Interpreting the label

The following instructions on herbicide labels are now mandatory:
1   Field of use (e.g. whether the product is for use within broad categories such as agriculture, horticulture, home garden, etc.).
2   The crops, plants or surfaces on which the product may be used.
3   The maximum dose rate.
4   The maximum number of treatments.
5   The latest time of application.
6   Any limitation on area or quantity allowed to be treated.
7   Any statements about operator protection or requirements for operator training.
8   Any statements about environmental protection.

There may also be other specific prohibitions relating to individual products which will need to be obeyed. Any breach of these or the other statutory conditions of approval relating to use constitutes a criminal offence. A herbicide product label should always be read thoroughly before the product is bought or used. Labels also carry other advisory information which users are encouraged to follow to obtain the best results.

Preparing the ground-crop sprayer

Preparation of the sprayer commences at the end of the previous season when the machine is put into storage. It should have been triple-rinsed internally using washing soda on the second rinse, and finally flushed through with a small quantity of 30% anti-freeze.

The outside of the sprayer must be washed down thoroughly and notes made of any faults or alterations that require attention before the new season. Nozzles and diaphragm check valves (DCVs), should be removed and stored in the filter basket.

A sufficient time before starting the new season's spraying, the following checks should be carried out:

1 Re-assemble all sprayer components.

2 Check chassis, tyres and wheels, and boom for wear and damage. Repair as necessary.

3 Check that power take-off shaft (p.t.o.) is of the correct length for the tractor–sprayer combination and that the guard is intact and secure.

4 Check condition of all wearing parts on the sprayer and lubricate according to the manufacturer's recommendations.

5 Check tank for security of mounting.

6 Dismantle filters to check for cleanliness and condition. Replace all 'O'-rings to avoid leakage problems in season.

7 Check oil level in the pump and make sure it does not contain water. Water turns the oil milky, and indicates a diaphragm failure. On grease-lubricated diaphragm pumps, diaphragm failure is signalled by water pouring out of the bottom of the pump.

8 Check that the pressure gauge returns to zero and is undamaged.

9 Fill the sprayer with water.

10 Run the sprayer at normal p.t.o. speed and with all nozzles spraying increase the pressure up to 5 bar (500 kPa) pressure. If the pressure will not reach 5 bar there is either an air leak in the suction system, or the pump is defective (see Table 5.1 on page 173). This test does not apply to centrifugal pumps which are unable to attain a pressure greater than 3 bar (300 kPa).

11 Re-set the pressure to the expected operating pressure, which is normally in the range of 2–3 bar (200–300 kPa) and check the spray for pulsation. If the spray pulsates adjust the air pressure in the pump pulsation damper. Follow the manufacturer's instructions, but generally a pressure of 2 bar (200 kPa) is recommended. The capacity of the damper chamber is very small so it must be filled by a hand pump or a foot pump. An industrial compressor is too harsh and may damage it.

12 Check that all switchgear operates as intended, and that by-passes for individual boom sections are correctly adjusted such that the pressure remains constant when individual boom sections are switched off. If nozzles drip when the boom is turned off, the diaphragm check-valves require servicing.

13 Remove tank lid and check that the vent is not blocked.

14 Check that no air bubbles are circulating in the tank. Air bubbles indicate an air leak on the suction side of the system.

15 When everything is working as the manufacturer intended the sprayer is ready for calibration. However, before moving on, it is worth emphasizing that many problems associated with sprayers during the spraying season, stem from over-tightening plastic components. Virtually all plastic components such as DCVs and filters should only be moderately hand-tightened. Over-tightening invariably leads to leakages and frequently causes damage to the components. An over-tightened DCV can also restrict the flow of liquid to that nozzle.

Calibrating the sprayer

First read the product label for information specifying spray volumes, forward speed, nozzle type, spray quality and operating pressure. Where all this information is given, the operator's task is simple. He must fit the recommended size and type of nozzle, and calibrate for the recommended forward speed and nozzle output. Where less information such as spray volume range only is given, the operator must work from basic principles.

Choose spray volume rate

The volume rate on the label is recommended in litres of water per hectare with lower and upper limits. The volume rate must be chosen within that range, taking into account:
1   Any other information on the label.
2   Previous experience.
3   That the lowest recommended volume rate will give the highest work rate.
4   That the maximum capacity of the pump might limit work rate.
5   That certain crop situations such as a dense canopy may require the higher end of the volume range.

Measure speed

1   Carry out a trial run to establish a forward speed which gives an acceptable level of boom bounce and yaw, and a gear which gives a p.t.o. speed of about 540 r.p.m.
2   Carry out a speed check over 100 m, using gear and p.t.o. speed as above. Measure the time taken, in seconds, to cover this distance.
3   Establish the forward speed from the formula:

$$360 \div \text{Time (seconds)} = \text{Speed (km h}^{-1}).$$

Calculate nozzle output

1 Measure and record nozzle spacing, in metres.

2 Establish and record the output per nozzle required to achieve the intended volume of application, using the formula:

$$\text{Volume of application} \times \text{Speed} \times \text{Nozzle spacing} \div 600 = \text{Nozzle output}$$
$$(\text{litre ha}^{-1}) \qquad (\text{km h}^{-1}) \qquad (\text{m}) \qquad (\text{litre min}^{-1})$$

Select nozzle

Refer to nozzle manufacturer's data charts or cards, or the BCPC Nozzle Selection Handbook, and select type and size of nozzle that will provide the calculated nozzle output and the spray quality required. Set pressure to recommended level.

Check nozzles

1 Fit the nozzles and check spray patterns and alignment visually. Replace any rogue nozzles.

2 Compare the output of individual nozzles by use of a nozzle flowmeter (e.g. Jetchek) or a recording jar. Replace nozzles with $>\pm5\%$ variation from the average.

Calibrate sprayer

1 Using a calibrated vessel, measure the output from four nozzles, at least one from each boom section, and compare with the calculated nozzle output.

2 If the output of these four nozzles differs by a small amount from the calculated output, alter the pressure and repeat the calibration; if the output differs by a large amount, re-check calibration and calculations and change nozzle size if necessary.

Calculating the doses

Pesticides are formulated either as solids or liquids and their doses are specified as kg ha^{-1} or litre ha^{-1} of the formulated product. Where possible, chemical manufacturers package their products in unit area packs usually of 1 ha or 2 ha to simplify the job of measuring the correct quantity of pesticide into the spray tank, and reduce the risk of errors. However, this is not always possible, particularly when a product has several recommended doses.

The quantity of product to be required per load is calculated by dividing the tank capacity (litres) by the application rate (litre ha^{-1}) and multiplying the resulting figure by the dose rate (litre ha^{-1} or kg ha^{-1}), e.g.:

$$
\begin{aligned}
\text{Capacity of tank} &= 1000 \text{ litre ha}^{-1} \\
\text{Application rate} &= 200 \text{ litre ha}^{-1} \\
\text{Therefore area treated per load} &= \frac{1000}{200} \\
&= 5 \text{ ha}
\end{aligned}
$$

If the dose rate = 7 litre ha^{-1}, then the quantity of herbicide required for each tank load will be:

$$5 \times 7 = 35 \text{ litre}$$

If only a part load is required, the quantity of pesticide added must be proportionately less.

Mixing the product

Whether filling the sprayer direct or from a pre-mix tank the following procedure should be observed:

1   Read the label before opening the chemical container, and follow the recommended mixing procedure.

2   Ensure that weather, field, and crop conditions comply with the instructions for use and safe spraying practice.

3   Only use Approved products and tank mixes.

4   When applying tank-mixes, pay particular attention to the order of mixing.

5   Wear the protective clothing stated on the label.

6   Part-fill the sprayer with water; some products such as water-soluble bags require a minimal amount of water in the sprayer for maximum agitation, whereas others require almost the maximum amount of water for the greatest possible dilution.

7   Start the pump and check that the agitation is working correctly.

8   Measure into the sprayer the calculated amount of products, using a low-level filling device where fitted, to avoid risk from glugging and splashing.

9   Add each product separately to the sprayer tank and allow it to completely disperse before adding the next product.

10   Wash out containers with clean water.

11 Continue to agitate while adding the remainder of the water to the required amount. Avoid creating excess frothing by over-agitation and adding water from too great a height.

12 Wash any spilt pesticides off the sprayer and containers.

13 Wash spilt pesticides off impermeable clothing such as gloves, boots, aprons and face shields.

14 Close or cover all used and unused pesticide containers, and store in a secure place to prevent theft, misuse or contamination.

15 Before getting into the tractor cab remove any protective clothing not required by law, except boots and coveralls, and store them in the tractor locker, not inside the cab.

Field procedure

Accurate marking out is a prerequisite to accurate spraying. Commonly-used marking systems include tramlining with the drill, foam marking and flags. Tramlining with the drill is the most accurate and convenient system providing the drilling is done accurately in the first instance. Foam markers are particularly useful for pre-emergence sprays, but are less reliable than tramlines, and are an extra factor on which the sprayer operator must concentrate while driving. This is a problem particularly with wide-boom sprayers. Portable flags are the least preferred method as they are time-consuming to set up accurately and impossible to see when the field has a rise in the middle.

It is good practice to allow two headland bout widths with a 12 m sprayer. The headland should be drilled to the same width as the sprayer headland, as the change in direction of the crop rows signals the sprayer operator when to turn on and off.

Where large obstruction such as ponds or pits exist in the middle of a field they should be drilled around and sprayed around as separate headland (Fig. 5.1).

On entering the field to be sprayed, the first task is to set the nozzles to the correct height above the crop. Having purged the spray line of air and water, the operator sprays out the headlands. At the corners the operator must stop spraying a bout width away from the edge of the field, reverse back into the corner then spray down the next field edge. This manoeuvre squares off the corner and avoids the localized under- and over-dosing associated with spraying round corners, during which the faster-moving outside boom under-doses while the slower-moving inside boom over-doses. Outside corners should be squared off in a similar manner.

Having sprayed the headland, the sprayer should work up and down

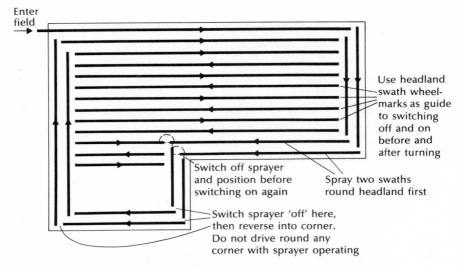

Enter field →

Use headland swath wheel-marks as guide to switching off and on before and after turning

Switch off sprayer and position before switching on again

Spray two swaths round headland first

Switch sprayer 'off' here, then reverse into corner. Do not drive round any corner with sprayer operating

Fig. 5.1 Ground crop sprayer — field procedure. From MAFF Booklet 2272.

parallel to the longest side of the field, or follow the drill rows if these are different.

Large obstructions have already been dealt with. Small obstructions such as trees and telegraph poles require a different approach. A sprayer cannot be sprayed round a telegraph pole without encountering the same under- and over-dosing problems associated with spraying round corners. A hydraulic folding sprayer should be driven right up to the telegraph pole, stop, reverse back, fold in the boom, drive past the pole, unfold the boom, reverse back to the pole and then continue spraying. The same procedure applies for manual folding booms except that the tractor has to make a wider loop to negotiate the telegraph pole and consequently damages more crop.

It is important to maintain a constant speed when spraying whether using a basic field crop sprayer or even one fitted with an automatic spray-regulating system. The basic sprayer relies on accurate maintenance of the forward speed and pressure for an accurate application rate, while automatic spray-regulating devices compensate for an increase in forward speed by increasing the operating pressure to maintain a constant application rate. Of course, increasing the pressure reduces the droplet size of the spray resulting in an increase in drift (Fig. 5.2(a) & (b)). This effect is more serious than frequently perceived. A doubling of the forward speed requires a four-fold increase in pressure at the nozzle to maintain a con-

Fig. 5.2 (a) The effect of increasing forward speed and pressure on spray quality. A sprayer applying 150 litre ha^{-1} at 9 km h^{-1} through Albuz red nozzles, at 2.5 bar pressure (250 kPa).

Chapter 5

Fig. 5.2 (b) Effect of increasing forward speed and pressure on spray quality. A sprayer applying 150 litre ha^{-1} at 18 km h^{-1} through Albuz red nozzles, at 10 bar pressure (1000 kPa).

stant application rate, e.g. a sprayer working at 3 bar (300 kPa) pressure and 6 km h^{-1} would require an operating pressure of 12 bar (1200 kPa) to maintain the same application rate at 12 km h^{-1}. The resulting increase in drift would be totally unacceptable and may lead to crop damage and poor weed control as well.

Further points to watch when spraying are:

1   Do not run out of spray half way across the field.
2   Keep a regular check on the nozzle while spraying. Always carry spare nozzles as replacements.
3   Do not blow through nozzles to clean them.
4   Whenever changing nozzles, wear the required protective clothing for handling the concentrate.

Have on hand as you spray:
An adequate supply of clean water
Soap
Towel
First-aid kit
Spare pair of protective gloves

Preparing the rotary atomizer

The preparation of the rotary atomizer sprayer is essentially the same as for the hydraulic ground-crop sprayer except for the nozzles which are replaced by the rotary atomizer heads. The drop size is dependent on flow rate of liquid to the head and rotational speed of the head. The uniformity of flow to the individual heads is easily checked using water. The rotational speed of the discs must be checked with a rev. counter.

Rotary atomizer sprayers are generally associated with very low liquid flows and concentrated spray mixtures. They are much more prone to blockages than a hydraulic ground-crop sprayer. The operator frequently cannot see the heads working from his position in the cab, so a blockage may go unnoticed for a considerable period of time. Good filtration and meticulous sprayer hygiene cannot be over-stressed for this type of equipment.

Calibration

Read label

Check the label on the chemical pack for recommended volume of application and drop size.

Check vehicle performance

1 Carry out a trial run to establish a forward speed which gives an acceptable level of boom bounce and yaw, and a gear which gives p.t.o. speed of about 540 r.p.m.

2 Carry out a speed check over 100 m, using gear and p.t.o. speed as above. Measure time taken in seconds to cover this distance.

3 Establish the forward speed from the formula:

$$360 \div \text{Time (seconds)} = \text{Speed (km h}^{-1}).$$

Calculate atomizer output

1 Measure and record atomizer spacing, in metres.

2 Establish and record the output per atomizer required to achieve the intended volume of application, using the formula:

$$
\begin{array}{ccccccc}
 & & & & \text{Atomizer} & & \text{Atomizer} \\
\text{Volume of application} & \times & \text{Speed} & \times & \text{spacing} & \div\, 600 = & \text{output} \\
\text{(litre ha}^{-1}) & & \text{(km h}^{-1}) & & \text{(m)} & & \text{(litre min}^{-1})
\end{array}
$$

Select rotational speed and metering orifice

Refer to the machine's handbook to select the correct rotational speed and size of metering orifice that will provide the calculated nozzle output and the drop size required. Set pressure to the recommended level.

Check atomizers

1 Fit the orifice plates and check spray patterns visually. Strip down and service any rogue atomizer heads.

2 Compare the output of all the individual heads by use of a measuring cylinder. Outputs should not vary between heads by $> \pm 5\%$.

Calibrate sprayer

Using a calibrated vessel, measure the output from four atomizers, at least one from each boom section, and compare with the calculated atomizer output. If the output differs from the calculated output, alter the pressure and repeat the calibration.

Field use calibration

Because the products applied through rotary atomizers are very concentrated, the output from one head must be checked with spray mixture when the first tank load is mixed up. If the output varies from the calculated output, alter the pressure and repeat the calibration.

Record for future use:

Orifice size fitted
Application volume
Spray pressure
Drop size
Rotational speed of heads (r.p.m.)
Tractor gear
Tractor speed
Tractor r.p.m.
Tractor wheel size

Granule applicators

There are two main types of granule applicators: those which are specifically designed for the application of pesticide granules, and the combined fertilizer/pesticide granule spreader. Before commencing the application of a herbicide with either type of granule applicator, it is important that the following factors are correct:

1 The machine is correctly assembled.
2 The spreader is fitted with the correct metering rollers for the type of product to be applied.
3 All parts must be in good working order: the booms horizontal, the tubing free from obstruction, the spreader plates entire and in their original shape, and the metering rollers functioning correctly without excessive wear.
4 All parts must be dry.
5 All parts requiring lubrication must be lubricated carefully, with no oil or grease getting into contact with the granules.

Calibration

Unlike aqueous sprays, the different size and density of herbicide granules means that granule spreaders must all be calibrated with the herbicide to be used. For many types of pesticide and spreader, the manufacturers have collaborated and suggest settings for calibration. These settings are, however, only suggestions and due to variations between individual machines and other factors such as wear of the metering device, a calibration check is still essential. The method for checking a granule spreader is essentially the same as for a field-crop sprayer and, for that matter, any other application device:

$$\text{Flow rate (kg min}^{-1}) = \frac{\text{Dose}}{\text{(kg ha}^{-1})} \times \frac{\text{Speed}}{\text{(km h}^{-1})} \times \frac{\text{Swath width}}{\text{(m)}} \div 600$$

With the granule spreader set up as per the manufacturer's recommendation and half-filled with product, the true forward speed is ascertained by measuring the time taken in seconds to cover a distance of 100 m at the intended spreading speed:

$$\text{Speed (km h}^{-1}) = 360 \div \text{Time to travel 100 m (seconds)}$$

The tractor speed and r.p.m. are recorded, and the metering mechanism set to the manufacturer's recommendations appropriate to the product, speed, and power take off r.p.m.

Most granule spreaders have two sets of metering rollers, each of which feeds product to either half of the machine. One roller is isolated and the other has a collecting tray placed beneath it, to collect the output of half the swath width of the machine. The fan must be either disconnected, or blanked off for calibration.

The machine is set in operation for a minute or more until the operator is sure that the metering mechanism is fully primed. The granules collected in the tray during the priming period are returned to the hopper.

The calibration is now checked by setting the tractor r.p.m. to that at which the speed test was carried out, then collecting and weighing the output of the granule spreader over 1 min.

For example, if the product had an application rate of 25 kg ha^{-1} and was to be applied through a 12 m spreader, a typical calibration might be as follows:

Time taken to travel 100 m in fifth gear at p.t.o. speed = 50 s.

$$\text{Forward speed} = 360 \div 50 = 7.2 \, \text{km h}^{-1}$$

Output (kg min^{-1}) required from half the width of the 12 m spreader:

$$25 \times 6 \times 7.2 \div 600 = 1.8 \, \text{kg min}^{-1}$$

The quantity of chemical delivered over the period of 1 min is collected, weighed, and if incorrect the machine is adjusted accordingly.

Maintenance and cleaning

After use, the granule spreader must be cleaned down as carefully as a sprayer. All unused granules and dust must be removed, as any accumulation of herbicide is a potential hazard when the machine is next used. Heavy deposits of dust and debris will also affect the flow of subsequent products through the machine. Granules must be stored in a cool dry environment. Damp granules flow erratically and cannot be applied accurately.

Band sprayers

The purpose of band spraying a row crop is primarily to economize in the amount of expensive herbicide used. In addition, on certain soil types liable to 'blow' it is possibly an advantage to retain some weed cover between the rows until such time as the crop is big enough to give some protection to the soil. At this stage the weed cover can easily, and safely, be removed by a tractor hoe. Band spraying is also considered a good practice in that it reduces to a minimum the amount of chemical put on the soil.

Pre-emergence band spraying is normally a simple operation in that each nozzle is attached to the rear of a seeder, and if properly adjusted, places a band of spray immediately over the line of seeds. The main problem is that the low tractor speeds required for satisfactory sowing necessitate the use of nozzles having correspondingly low outputs. Where wettable powders are used this can lead to an annoying frequency of nozzle filter blockage. If coarser filters are substituted this may lead to undue nozzle-tip blockages.

On a very long field it may be necessary to top-up the seeders at the end of every row. Similarly it may be necessary to top-up the band sprayer tank at the same time, and before it has emptied. If at this stage the tank is one-quarter full then the full amount of chemical for a tankful must not be added to the tank before re-filling with water, otherwise over-dosing will occur — only three-quarters of the normal amount should be added. An alternative is to keep a quantity of the chemical mixed at the correct strength in a tank on the headland, topping-up the sprayer from this. With this system the tank must be kept agitated to ensure that the correct proportion of chemical to water is maintained for each topping-up.

Post-emergence band spraying is a more skilled operation. The band sprayer must be matched with the original drilling. If the seed was sown in, say, five rows at a time, then the band sprayer must have five nozzles or a multiple of this figure. Moreover, the sprayer tractor must follow in the wheelings of the seeding tractor in order to ensure continual matching of the nozzles with the rows. Although the original wheelings may be visible from the headland, it is often impracticable to drive accurately along them when there is a heavy cover of weeds. It is essential to keep an eye on the lines of crop and drive accurately so that a nozzle is directly over each row all the time. Obviously a rear-mounted band sprayer makes this almost impossible to achieve. A front-mounted sprayer permits the driver to look straight along one row and keep the appropriate nozzle over that row. A pressure gauge should be mounted as close as possible to that nozzle so that the driver never needs to take his eye off the row he is following.

Post-emergence band sprayers should not be mounted under the tractor, behind the front wheels. The air from the tractor cooling fan is liable to disturb the spray pattern on the rows under the tractor.

Band-sprayer nozzles

Types of nozzle available for band spraying

1 *Nozzles giving an even dose across the band*: These even-spray nozzles are flat-fan nozzle tips very similar in appearance to those used for overall spraying. However, they must not be used for overall spraying as their spray pattern will give rise to severe striping. Generally, these tips, although giving an even volume of spray across the band, produce a relatively coarse spray at the edges of the band. This is quite satisfactory for pre-emergence work but is often unsuitable for post-emergence spraying — the coarser spray is liable to be reflected off a number of weed species. Where this happens poor control of difficult weeds can be seen at the edges of the band, making accurate hoeing later on much more difficult.
2 *Nozzles which give a reduced dose at the edges of the band, conventional flat fan nozzles*: Where a herbicide is in use which has a high margin of safety for the crop, this type of tip is favoured because it ensures adequate weed control in the crop row, where no hoeing can be carried out except by hand or using complex hoeing devices.
3 *Nozzles which give a higher dose at the edges of the band, e.g. hollow-cone spray nozzles*: These nozzles are to be recommended where there is some risk to the crop of damage or delayed growth when over-dosing

occurs. The slight over-dosing at the edges of the band makes for easier tractor hoeing later on, at the same time ensuring slight under-dosing on the crop-line. In addition, when the light falls on the crop from certain angles the spray from hollow-cone nozzles is often easier to see from the driver's cab, than that from a flat fan.

Nozzle output

Each nozzle must pass precisely the correct volume of liquid to ensure that some rows are not over-dosed at the expense of others. Furthermore, each nozzle must produce a good spray distribution across the band. A heavy streak or a light streak in the pattern close to, or over, the crop-line could give rise to crop damage or make hoeing difficult, respectively.

If any tips are worn or damaged they must be replaced by a complete matched set of nozzles. It is not advisable to replace one nozzle by a new one which is not matched to the relatively worn remainder. Nozzle outputs can be tested with the sprayer stationary, and this should be done before taking the sprayer out to the field. Band sprayer manufacturers normally can provide ready-matched sets of tips. A sprayer built on the farm should be designed to operate with readily-available matched sets of nozzles.

Band width

It is essential to maintain a constant correct band width. If the nozzles are mounted on a rigid bar, the slightest undulation of the ground will cause variation in band width and hence variation in dose. When spraying post-emergence, each nozzle should be mounted on its own wheel, to maintain it at a constant height above the soil (performing the similar function of a seeder unit, but the wheel running alongside the row of plants). As the wheel (or the seeder unit) will tend to sink into the soil a little, thereby narrowing the band width, it is always essential to check the band width in the field under the normal soil conditions. Cone nozzles will have to be raised to increase the width of band and vice-versa. The same technique can be used for flat-fan nozzles, or alternatively these can be angled to the direction of the row to reduce the band width.

Calibration

This is essentially the same as for the hydraulic ground crop sprayer.

Read label

Check the label on the chemical pack for recommended volume of application, spray quality, band width and nozzle type.

Measure speed

1 Carry out a trial run in the field to establish the optimum operating speed, and a gear which gives a p.t.o. speed of about 540 r.p.m.
2 Carry out a speed check over 100 m using gear and p.t.o. speed as above. Measure the time taken, in seconds, to cover this distance.
3 Establish the forward speed from the formula:

$$360 \div \text{Time (seconds)} = \text{Speed (km h}^{-1}).$$

Calculate nozzle output

1 Record the band width required for the product, in metres.
2 Establish and record the output per nozzle required to achieve the intended volume of application using the formula:

$$\begin{array}{c} \text{Nozzle} \\ \text{Volume of application} \times \text{Speed} \times \text{Band width} \div 600 = \text{output} \\ \text{(litre ha}^{-1}) \qquad \text{(km h}^{-1}) \qquad \text{(m)} \qquad \text{(litre min}^{-1}) \end{array}$$

Select nozzle

Refer to nozzle manufacturer's data charts or cards, or to ADAS advisory lists, and select type and size of nozzle that will provide the calculated nozzle output and the spray quality required. Set pressure to recommended level.

Check nozzles

1 Fit the nozzles and check spray patterns and alignment visually. Replace any rogue nozzles.
2 Compare the output of individual nozzles by use of a nozzle flow meter (e.g. Jetchek) or a recording jar. Replace nozzles with $>\pm5\%$ variation from the average.

Calibrate sprayer

1 Using a calibrated vessel, measure the output from four nozzles, at least one from each boom section, and compare with the calculated nozzle output.

2 If the output of these four nozzles differs by a small amount from the calculated output, alter the pressure and repeat the calibration; if the output differs by a large amount, re-check calibration and calculations and change nozzle size if necessary.

3 Obtain the correct band width by spraying water and at least 0.1% wetter over dry concrete or dry soil, and adjusting the height of the nozzles until the correct band width is obtained.

Record for future use:

Nozzle-tips fitted	Spray quality
Application volume	Tractor gear
Band width	Tractor speed
Nozzle height	Tractor r.p.m.
Spray pressure	Tractor wheel size

Band area

For the purposes of purchasing sufficient chemical for a band spraying operation it is necessary to calculate the band area to be sprayed:

$$\text{Band area} = \text{Field area} \times \text{Band width} \div \text{Row width}$$
$$\text{(ha)} \qquad \text{(ha)} \qquad \text{(m)} \qquad \text{(m)}$$

For example, a 15 ha field cropped with 0.5 m row spacing using a band width of 0.18 m requires sufficient chemical to spray:

$$15 \times 0.18 \div 0.5 = 5.4 \text{ ha of band}$$

Calibrating a knapsack sprayer

The method is fundamentally the same as for a ground-crop sprayer except that knapsack sprayers are generally less sophisticated and have few options for adjusting spray pressure. Some knapsacks are fitted with neither a pressure regulator or a gauge, and rely on the skill and consistency of the operator for a uniform application.

Calibration of a single lance nozzle

1   Read the herbicide label and sprayer handbook. These will provide the correct range of spray volumes for applying the herbicide and may advise a particular type and size of nozzle, or spray quality.

2   Check forward speed. Half-fill the knapsack sprayer with water, spray over a measured distance of 100 m at your normal speed and record the time taken to cover this distance. This exercise should be repeated at least three times and the average time taken to calculate the spraying speed:

$$360 \div \text{Time (seconds)} = \text{Speed (km h}^{-1})$$

3   Referring to nozzle charts, select a nozzle that should produce a suitable application rate consistent with constraints of spray quality, operating pressure and swath width.

4   Fit the nozzle to the sprayer and half-fill the sprayer with water plus 0.1% wetter, and spray over dry concrete. The swath width (m) of wet mark on the concrete is recorded, and can be adjusted by raising or lowering the nozzle height as necessary.

5   Check the flow rate (litre min^{-1}) through the nozzle by collecting the output from spraying for 1 min; record the pressure setting where appropriate.

6   Calculate the volume of application (litre ha^{-1}) by the formula:

Volume (litre ha^{-1}) = Flow rate (litre min^{-1}) $\times$ 600 $\div$ Swath width (m)
$\div$ Speed (km h^{-1})

7   Should the volume of application be outside of the range recommended on the herbicide label, it should be altered by fitting a different sized nozzle for a large deviation or altering the pressure for a small deviation.

8   *Example*: The herbicide label requires an application rate of 200–400 litre ha^{-1}. The nozzle fitted to the sprayer produces an output of 2.2 litre min^{-1} at 1.5 bar pressure (150 kPa). The swath width is measured as 1.2 m. Forward speed is 3.6 km h^{-1}.

Spray volume (litre ha^{-1}) = 2.2 $\times$ 600 $\div$ 1.2 $\div$ 3.6 = 306 litre ha^{-1}

9   Record:

Spray volume	Forward speed
Nozzle type	Nozzle height
Spray quality	Swath width

Fault finding

Table 5.1 Fault-finding and correction chart for sprayers

Fault	Probable causes	Remedy
Fails to spray when turned on	Nozzles assembled incorrectly	Re-assemble correctly — see manufacturer's handbook
	Outlet at bottom of tank blocked	Disconnect outlet pipe and clear
	Filter on suction side of pump completely blocked	Dismantle, clean and re-assemble
Sprays for a short time only after switching on	Air inlet to tank blocked	Clean vent hole, otherwise the tank may collapse
	Filter on suction side of pump blocked rapidly	Dismantle, clean and re-assemble — determine and remove cause of blockage
Spray is not even across the spray bar	Some nozzle filters or tips are becoming blocked	Remove, clean and refit correctly
	Nozzle tips are not all the same size	Check the number on each tip and change any wrong tips
	Nozzle tips may be worn, check output	Replace worn tips with new ones. Calibrate
	Nozzles at each end of the spray bar have a lower output	Check pressure at end of bar by replacing end nozzle with pressure gauge. If pressure is lower at end of bar, nozzle output is too large for the pump's capacity. Fit smaller tips or change pump, if worn
Pressure gauge reading going up; spray volume from nozzles decreasing	Nozzle filters blocking up gradually	Dismantle, clean and refit, check pressure has returned to normal
	If cleaning nozzle filters has no effect, gauge may be strained	Check that gauge returns to zero when spray turned off. If not, replace with new gauge.
Pressure gauge reading falling off	Main filter on suction side of pump blocking up	Dismantle, clean filter and replace
	If filter cleaning gives no improvement, nozzle tips may be worn	Replace tips with new ones of same size and make

Table 5.1 Cont.

Fault	Probable causes	Remedy
Pressure gauge reading falling off	If replacing nozzles gives no improvement pump may be worn	Replace with new or reconditioned pump
	Airlock in the pump	Take tension off relief valve spring and operate pump to allow air to escape through agitator tube
Spray fans or cones very narrow	Pressure too low	Check that pressure and output are within the range recommended for sprayer. Use smaller tips if necessary
	Pressure too low and air spluttering out of nozzles	Check that tank is not very nearly empty. If not there may be an air leak between the tank and the pump — or in the pump itself. Locate and repair leak
Coarse foam in the spray tank, on top of the liquid	Faulty agitation	If mechanical agitation is used this is too violent and is beating air into the liquid. If there is a return pipe above the level of the tank liquid, extend it to the bottom of the tank or deflect output against the tank more carefully
Very fine foam in the liquid in the tank	Air leak in system probably between the tank and the pump, or in the pump itself	Locate and repair leak
Spray fans or cones streaky when viewed against a dark background	Nozzle partly blocked by minute hairs or flakes	Remove tips and clean. Refit correctly and test
	Nozzle clean but still streaky — probably faulty or worn tip	Replace tip with new one of same size and make; test
Excessive pulsation of the spray pattern	Damaged pump inlet valve	Strip and repair pump
	Incorrect air pressure in the equalizing chamber	Set pressure to 2 bar (200 kPa) and adjust in small increments if necessary
Output of pump below expectation	Damaged exhaust valve	Strip pump and replace
Spray pressure drops as tank empties	Air entering tank outlet to the pump	Fit anti-vortex plates, re-direct jet agitator or reduce pump speed

Errors in applying herbicides

Efficient selective weed control depends, amongst other things, upon applying the correct dose of a herbicide uniformly to the target across the whole of the area to be sprayed, whilst minimizing off target losses.

Incorrect dose

Under-dosing presents only the problem of unsatisfactory weed control which can sometimes be rectified by further efforts on behalf of the grower.

Over-dosing is more serious; it is a criminal offence. It may also damage the crop and leave residues that are harmful to the following crop and the consumer. Over-dosing can be caused by spillage: all herbicides should be handled carefully. Spraying from a stationary sprayer, which often occurs when the spray-line is purged on the headland before starting off across the field with a fresh tank load, results in very heavy herbicide loads on the crop and soil. It can be minimized by the operator recording the time taken to purge the line, and using this information in practice.

Dribbles from nozzles cause localized over-dosing as do leaking joints and components on the sprayer and pipework that fouls the spray pattern, all of which are items of maintenance.

Over-dosing will occur if the tractor slows down when travelling uphill or through wheel slip, unless fitted with an electrical or mechanical system to compensate. Over-lapping of the spray swaths is a common cause of over-dosing; accurate bout matching and switching off the sprayer at the headland are essential.

Failure to observe herbicide mixing instructions is a frequent cause of heavy over-dosing at the start of spraying a new load and reduced weed control for the remainder. Mistakes in calculation or calibration will inevitably lead to the wrong dose being applied.

Table 5.2 Common errors which occur in spraying and their possible remedies

Observed error	Probable cause	Remedy
Longitudinal stripes	Nozzles too low	Adjust nozzle height
Longitudinal stripes	Pressure too low	Adjust pressure or remove blockage
Longitudinal stripes	Worn or damaged nozzles	Replace nozzles
Longitudinal stripes	Foam in spray liquid	Find cause of foaming
Short intermittent stripes	Spray boom roll or bounce	Reduce speed, check linkages for free movement
Intermittent stripes at boom end	Spray boom yaw	Reduce speed, check linkages for free movement
Uneven patchy results	Too fine a spray quality	Fit coarser quality nozzles
Uneven patchy results	Excessive wind at application	Don't spray

Herbicide drift

Drift must be minimized at all times. It is potentially damaging to neighbouring crops and hedgerow flora and fauna, and creates an impression of irresponsibility to the casual observer even in low risk situations. Drift may occur in three ways:

1 *Spray drift*: This is the result of the smaller drops in the spray being carried off-target by wind or convection currents.

2 *Vapour drift*: This occurs when the vapour from a volatile herbicide is carried away from the target area during or after spraying. It is most likely to occur in warm, still weather. A gentle breeze will tend to disperse the vapour to such an extent that there will be no hazard.

3 *'Blow'*: This is the movement by high wind of dried spray particles or of soil impregnated with the herbicide away from the area originally treated.

Spray drift

This is the most common form of herbicide drift. Growth regulator herbicides are the most dangerous from this point of view because quite small amounts, which can travel considerable distances, may be highly damaging. Contact herbicides are generally less damaging but will cause necrotic spotting on susceptible crops. Soil-applied herbicides which have no foliar activity have little risk of causing damage over a distance of more than a few metres. The amount of herbicide carried away is likely to be too small to have any effect when deposited on the soil.

Table 5.3 Wind speed and when to spray

Approx. airspeed at boom height	Beaufort scale (at height of 10 m)	Description	Visible signs	Spraying
Less than 2 km h^{-1} (less than 1.2 mph)	Force 0	Calm	Smoke rises vertically	Avoid spraying
2–3.2 km h^{-1} (1.2–2 mph)	Force 1	Light air	Direction shown by smoke drift	Avoid spraying
3.2–6.5 km h^{-1} (2–4 mph)	Force 2	Light breeze	Leaves rustle, wind felt on face	Ideal spraying
6.5–9.6 km h^{-1} (4–6 mph)	Force 3	Gentle breeze	Leaves and twigs in constant motion	Avoid spraying herbicides
9.6–14.5 km h^{-1} (6–9 mph)	Force 4	Moderate	Small branches moved, raises dust or loose paper	Spraying inadvisable

Spray drift can be reduced or prevented by one or more of the following methods.

1   Spray only when there is a gentle breeze blowing away from a susceptible crop. Avoid spraying in strong winds. Also avoid spraying in warm, still conditions (particularly evenings) when convection currents rise from fields and fine spray drops take a long time to settle.

2   The spraying of herbicides next to susceptible crops should, if possible, be carried out before these crops appear above the soil.

3   Avoid the use of fine sprays. Use the coarsest-quality spray recommended on the label. Operators of machines fitted with electrical or mechanical constant-volume devices must pay particular attention to their forward speed. A doubling of the forward speed results in a corresponding four-fold increase in spraying pressure which greatly increases the potential for drift. Conversely, such devices can be beneficially used to reduce the potential for drift by reducing the forward speed of the machine.

4   Keep the nozzles as close as possible to the soil, weeds or crop, whichever is the taller, consistent with the minimum recommended nozzle height.

5   Where a susceptible crop is above the ground and is on the down-wind side of a crop requiring treatment, the operator should leave an untreated strip of sufficient width along the edge. The untreated strip may be sprayed later when the wind is blowing away from the susceptible crop.

Vapour drift

Vapour drift is generally associated with the ester formulations of certain growth-regulator herbicides. It is best avoided by choosing alternative products for high-risk situations.

'Blow'

Damage by 'blow' was mainly associated with the now obsolete active ingredients DNOC and dinoseb. Where there is a risk of 'blow' onto a susceptible crop the operator must either spray before that crop has emerged, or leave a suitable width of untreated ground beside it.

Decontamination of sprayers and disposal of waste material

The question of disposal should be considered before herbicides are purchased. Quantities and pack sizes should be appropriate to the task in hand and the shelf life of the product. It is preferable to choose products and quantities that will not cause disposal problems wherever possible. Disposal problems of waste herbicides can be classified as follows.

Dilute end of tank surplus spray

The area to be sprayed should be accurately calculated so that mixing surplus spray is avoided. However, there will always be some excess mixture left in the tank.

The operator's first task is to read the product label to ascertain whether there are specific decontamination procedures, and follow these if provided. Where no specific procedure is present on the label, the operator should dilute the spray mix to the maximum volume of the sprayer, and spray out onto a local water authority approved soakaway. Where there is no approved soakaway, the user should carefully identify an area of non-cropped land of minimum wildlife value. This area must be sign-posted and fenced to exclude people and animals and it must be able to carry the quantity of diluted material that is to be sprayed, without run-off, without leaving puddles, and without risk to wildlife, watercourses, groundwater, public sewers, septic tanks or field drains. The local water authority will be able to advise on the siting and frequency of use of such a site.

The sprayer should be rinsed through a further two times with clean water, and it is advisable to incorporate washing soda in the first rinse when changing to a different type of pesticide or when heavy contamination of the sprayer's internal components is suspected.

When decontaminating the sprayer the correct protective clothing must be worn for the last pesticide applied. The sprayer must be thoroughly washed inside and out, with special attention paid to nozzles, filters and spray-lines. Sprayers should never be put away with chemical in the spray-lines as this will lead to damage of the components.

If the operator has the misfortune to be prevented from finishing a tank load owing to bad weather, the contents of the sprayer should be vigorously agitated at least three times daily. Failure to do so can result in serious damage to the sprayer. Sometimes it is possible to dispose of dilute pesticides on cropped areas where this activity is within the conditions of approval of the product.

Disposal of concentrated product

Where possible this should be returned to the supplier. However if the product is very old, or the label has come off the container, then the product must be disposed of by a reputable chemical waste-disposal contractor. It is an offence to store a pesticide which does not carry current approval. Users who find they have an unapproved product in store should dispose of it by a reputable chemical waste-disposal contractor. The waste-disposal authorities (the county councils in England and district councils in Scotland and Wales) will give advice on disposal matters and where appropriate the names of reputable chemical waste-disposal contractors.

Disposal of empty containers

After the contents of a container have been added to the spray tank, the container should be rinsed out and the rinsate added to the spray tank. The empty container must be stored in a secure used container compound, until it is disposed of.

Metal containers: These should be rinsed, punctured, flattened and disposed either by the local council or by burying to a depth of at least 0.8 m in a site on land occupied by the person disposing of the container without risk of pollution to surface or groundwater sources. The area should be marked and a record kept of the site and the materials buried.

Plastic and paper containers: These should be rinsed where appropriate and burnt. However, containers which have held one of the substances with the active ingredients listed below should on no account be burnt owing to the potential hazards of their vapour:

Benazolin
Clopyralid
2,4-D
2,4-DB
Dicamba
Dichlorprop
Fenoprop
MCPA
MCPB
Mecoprop
Oxadiazon
Picloram
Sodium chlorate
2,4,5-T
2,3,6-TBA
Triclopyr

Other such containers with pesticides or formulations marked as highly flammable, pyrotechnical devices and atomizable fluids should be disposed of as if they were metal. If waste is to be burnt the user should ensure that:

1 The fire is open and not within 15 m of a public highway, or where smoke will drift over a person's livestock, houses or business premises.

2 The containers have been opened and are placed on a very hot fire, a few at a time.

3 Care is taken to avoid breathing the smoke, the fire is under constant supervision and is extinguished before leaving.

Storage of herbicides

The farm herbicide store should:

1 Be secure against thieves and vandals with locked doors and bolted windows.

2 Be dry, well ventilated and protected from frost.

3 Be sited well away from public roads, private houses, livestock buildings, as well as stores for fodder, fertilizer, fuel or any combustible materials.

4 Be sited to prevent any pollution hazard to watercourses, ponds, ditches, surface catchment areas and bore-holes.

5 Have floors and walls which contain spillage or flooding at a level below the stored containers and also prevent spilt liquids from seeping into the ground.

6 Ideally have a separate washroom with hot and cold running water. Soiled protective clothing should be stored separately from clean protective clothing and laundered as soon as possible.

7 Provide easy access for the fire service and any other emergency vehicles.

8 Have a stocklist of all stored pesticides with separate copies in the farm office and on display in the washroom. One copy of the list should be readily available to the fire service.

9 Display a warning sign in a prominent position outside the store at 2 m above the ground.

10 Have adequate first-aid facilities.

Where <50 litre of liquids or <50 kg of powders and/or granules is to be stored, a ventilated and lockable steel cabinet would be adequate. As a general rule, one can expect a shelf life of at least two or three years if chemicals are correctly stored. Shelf life will largely depend on the stability of the active ingredient and its formulation, and can in some temperatures be much greater than three years. Chemicals should be date marked as they are put into store, and new bottles placed at the back of the shelves, to ensure that the old containers get used up first.

6 / Entry and transport of herbicides in plants

J.C. CASELEY AND A. WALKER*

*AFRC IACR Long Ashton Research Station, Long Ashton, Bristol BS18 9AF; and *AFRC Institute of Horticultural Research, Wellesbourne, Warwick CV35 9EF*

Introduction
Interception and uptake of foliage-
    applied herbicides
    Interception
    Retention
    Uptake into foliage
Availability and uptake of soil-applied
    herbicides
    Factors influencing the availability of
    herbicides in soil
Uptake by roots and other below-
    ground organs
    Differential uptake and selectivity
Transport of herbicides in the plant
    Short-distance transport
    Long-distance transport
    Transport and selectivity
    Weather and herbicide transport
Further reading

Introduction

Herbicides kill weeds by interfering with biochemical processes, such as photosynthesis, which take place in the symplast or living system of the plant. Thus it is a prerequisite for herbicide action that sufficient active ingredient enters the plant and is transported to the appropriate site of action.

This chapter reviews some of the principles concerning the progress of active ingredient from the surface of the plant to a distant site of action, discusses some of the factors which affect this process, and considers some aspects of selectivity which are not covered in Chapter 7. These include herbicide availability in the soil, weather conditions, plant factors such as growth stage, application and formulation. The approach which is taken here follows the pathway of movement of herbicides after they have been applied to foliage or soil and some of the major steps involved are shown in Fig. 6.1. The pathways are different initially for foliage-applied and soil-applied herbicides (Fig. 6.2) and the concentration of active ingredient in spray deposits on the foliage is usually far greater (1000–10 000-fold) than is found in herbicide solutions in contact with below-ground parts of the plant. For foliage-applied herbicides, these high concentration gradients across the cuticle enable rapid uptake of herbicide, thus conferring a certain degree of 'weatherproofing' to the treatments. With soil-applied residual herbicides, the period of time available for uptake is usually much longer.

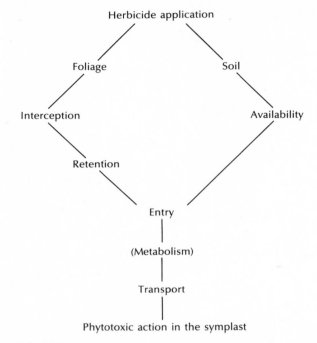

Fig. 6.1 Steps in herbicide action.

In this chapter, herbicide uptake by foliage and by underground parts of the plant are considered separately. However, it should be emphasized that while some compounds such as glyphosate enter the plant exclusively via the foliage and others such as simazine enter only via the soil, many post-emergence herbicides, e.g. chlorsulfuron, may to some extent enter the plant by both routes.

It has been estimated that for many herbicides, <1% of the dose on the surface of the plant reaches the site of action. While the distribution of herbicide on, and to some extent, within plants can be measured, our knowledge of the mechanisms involved in herbicide entry and transport and the losses sustained during these processes, tends to be scanty. For this reason and because of the diversity of chemicals, plants and environments, there is often insufficient information available to allow firm statements to be made and this must be remembered if some of the ideas expressed below are extended to field practice. Variables frequently interact with each other and, in consequence, deliberate alteration of one may not have a simple predictable effect.

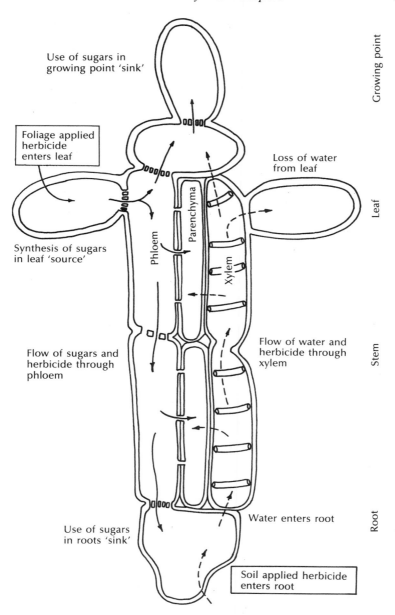

Fig. 6.2 Herbicide transport in the xylem and phloem. From Bonner and Galston (1952).

Interception and uptake of foliage-applied herbicides

For a foliage-applied herbicide, toxicity is likely to be related in part to the amount of spray retained. In some cases, the weed retains a much greater volume of herbicide than the crop and selectivity may be achieved without any other factor being involved. More often, however, differences in retention aid selectivity without being wholly responsible for it.

Interception

To be wetted, a leaf or stem must both intercept and retain spray droplets. In the extreme cases of directed sprays in perennial crops, and contact pre-emergence applications, the crop does not intercept any spray and provided there is no action through the soil, selectivity is complete. With overall sprays too, there are ways in which differential interception occurs. If a weed infestation covers a crop, the latter will intercept very little spray and herbicide treatments can be used which would not usually be selective. In other cases more subtle factors may be involved. The angle and arrangement of leaves can be very important. Many grass weeds and cereals, with erect leaves, present relatively small targets to a vertical spray and may, in consequence, receive relatively little spray per unit area of leaf.

Retention

Leaves vary in the proportion of intercepted spray which they retain and many factors influence this retention.

Plant features

Leaf angle may influence retention as well as interception. The momentum of drops and the angle at which they strike a leaf will clearly affect the chances of a spray drop adhering to a leaf rather than bouncing off.

The nature of the leaf surface is one of the most important factors controlling retention of spray droplets. For example, the shape and arrangement of wax particles (epicuticular wax) on a normal pea leaf produces a rough surface giving it a water-repellent character, with the result that larger water droplets have little adhesion and readily bounce or roll off. This was the main basis for selectivity of the no longer used dinitrophenol herbicides as post-emergence sprays in peas. The reduction of surface wax as a result of pre-planting treatment with TCA or dalapon leads to much greater retention and severe damage. The physico-chemical properties of the surface waxes are not the same from one species to another. The mealy surface on leaves of fat-hen, which is due to wax globules, is not nearly so water repellent as the wax on a pea leaf and this weed can be controlled in peas by using bentazon with MCPB.

Hairiness is another character which can influence retention. A downy leaf such as that of red clover is difficult to wet, but hairiness does not necessarily indicate poor retention; tomato leaves, for instance, are hairy yet readily wetted. When the hairs are both rigid and water-repellent the leaves are most difficult to wet.

The nature of leaf surfaces can vary as the plant ages. Young leaves may have an incomplete wax deposit and may retain more spray than mature foliage. Older leaves, however, may again become wettable if the wax surface becomes abraded. Younger, but not older leaves, are able to regenerate the wax deposit following abrasion. The trifoliate leaves of clover and lucerne are less retentive of spray droplets than are the cotyledons, hence the fact that herbicides such as MCPB must not be applied too early in these crops.

Where crop and weed plants have equal inherent tolerance to a herbicide, selectivity may be obtained when there is seasonal variation in growth. For example, in spring, white clover tolerates 2,4-D better than does dandelion, for the latter is growing vigorously but the clover has not begun active growth. Later in the summer the reverse may occur.

Application, formulation and weather factors

Small drops may be produced by small-aperture nozzles often used to achieve low-volume rates or by use of high pressures. These drops have less momentum than larger ones and also a higher surface area to volume ratio, so that the tendency to bounce or roll off the leaf is reduced and, of those which impinge, a much greater proportion is retained, especially on waxy surfaces. These characteristics are particularly important for achieving good retention on grasses such as wild oat. Rotating disc applicators produce a more uniform drop size and can be used to apply very low volume rates ($10-50$ litre ha^{-1}). Control of speed of rotation and liquid flowrate permits some selection of drop size to suit particular situations. A sufficient volume of liquid to dissolve or suspend the herbicide must be used and it should be noted that if the volume of spray solution is reduced and the same drop size maintained, there are fewer drops per unit area and this reduces the chance of small plants intercepting the spray. On the other hand if the number of drops per unit area is maintained and their size reduced, retention on difficult-to-wet surfaces will increase and problems with drift may be encountered, especially with drops of <200 μm in diameter. Some further implications of drop size, etc. are considered in Chapters 4 and 5.

The leaves of some plants, such as corn marigold or cereals, are hardly wetted at all by simple aqueous sprays, even at quite low volume with small drop sizes. Reducing the surface tension by the addition of wetting agent completely alters the surface dynamics of the spray drops and results in greater adhesion with the leaf surface and less bounce or roll-off. On easily wetted surfaces, however, the addition of a wetting agent may lead to the opposite effect where only a thin film of liquid is retained, the remainder running off. This type of run-off can occur, even in the absence of a wetting agent, at volumes above 500 litres ha^{-1}; with the presence of wetter it may occur from 250 litres ha^{-1} upwards.

Oil emulsions, suspensions of wettable powders and suspension concentrates usually behave in a manner comparable with aqueous solutions which contain wetting agent. Wetting agents, in addition to influencing the retention of the herbicide, may increase its entry and sometimes movement within the plant.

The characteristics of the plant can be influenced by the environment such that both interception and retention are affected. Crop density, or competition from weeds, can result in weeds becoming more erect in habit, thus leading to reduced interception.

The nature of the leaf surfaces, especially hairiness and wax deposits, may be influenced by the environment and this may lead to changes in retention. Common couch grown experimentally to the same stage of development under warm (16/10°C) compared with cool (10/6°C) conditions, retained twice the amount of glyphosate because the leaves from the warm environment were larger, more prostrate and less waxy.

Wind also may be important; waxy surfaces of leaves are abraded by rubbing together or by airborne sand or soil particles. Damage to the leaf surface may also result from rain, hail and frost.

The effect of precipitation soon after spraying depends to some extent on its intensity; heavy rain may wash the herbicide off the leaf, while slight rain, drizzle and dew may redistribute the herbicide, depositing some of it in areas where entry may be facilitated. In grasses, for example, the adaxial surface of the leaf sheath is only lightly covered by wax, the humidity is high, and uptake therefore may be more rapid than from the surfaces of the leaf blade. Absorption may also be enhanced by migration of the herbicide to the inner surfaces of the sheath.

Uptake into foliage

All aerial parts of a plant are covered by a cuticle which varies in composition and thickness over the surface of the plant, throughout the life of the plant and from one species to another. It is this cuticle which a herbicide must penetrate before it can meet and possibly interact with the live tissues of the plant. Unlike the underlying cell walls, the cuticle is water repellent, consisting of cutin with larger or smaller amounts of wax embedded within and projecting from it. If continuous, it would seal the plant from its environment preventing loss of water and entry of water-soluble substances. However, because slow but continuous evaporation of water can occur through the cuticle and water-soluble compounds are absorbed, particularly under conditions of high relative humidity, it has been argued that the cuticle may be somewhat sponge-like in structure. The pores of the sponge are thought to be open and full of water under conditions of high humidity but to close as the relative humidity is lowered. The stomata penetrate the leaf surface and are possible points for entry of applied solutions. However, most common surfactants are unable to reduce the surface tension of aqueous solutions sufficiently to allow stomatal entry, but recently some organo-silicone surfactants have been shown to facilitate glyphosate uptake into gorse by this route. Even then a cuticle, albeit thin, lines the sub-stomatal cavity of many if not all species. In fact, there probably exist two main routes along which compounds can cross the cuticle, one for fat-soluble materials which is permanent (the lipoidal route), the other for water-soluble materials (the aqueous route) but which is functional only under conditions of high humidity. Experiments using aqueous solutions containing surfactants and fluorescent dyes indicate preferential sites of entry associated with guard cells, hairs and, in broadleaved species, veins.

Provided an applied compound does not crystallize out on the leaf surface it will diffuse across the cuticle at a rate governed by its solubility in the region through which it is passing and the concentration gradient across that region. Penetration rates are directly proportional to the external concentration of these compounds and the rate of their removal from the inner surface of the cuticle.

Light stimulates penetration of at least the phenoxyacetic acid and benzoic acid groups of herbicides. This stimulation appears to involve biochemical processes and is most marked in young immature leaves. A sufficiently high internal level of naturally-occurring auxin is required for the stimulation to occur.

Temperature increases penetration probably by altering the physical characteristics of the cuticle and the velocity of certain physiological processes.

The pH of the applied solution can markedly affect penetration of certain herbicides. For example weak acids, which at pH values >5 exist in aqueous solution mainly as charged ions, cannot follow the lipoidal route through the cuticle but are restricted to the aqueous route. At lower pH values, a higher proportion of the dissolved acid is in the form of uncharged molecules which can diffuse along the lipoidal route. Penetration of such herbicides is therefore assisted by acidifying the applied solution but care must be taken to avoid lowering the pH to a level where damage to the plant surface or to the underlying tissues responsible for distributing the herbicide occurs.

Availability and uptake of soil-applied herbicides

An essential requirement of any soil-applied herbicide is that it should accumulate in weed seedlings in toxic concentrations from soil treated at an economic rate without risk of damage to the crop. The amount taken up over a period of time will be determined by the ability of the soil to supply the chemical to the plant and the ability of the plant to absorb it. Although this chapter is concerned primarily with the plant aspects of herbicide behaviour, discussion of how soil and weather factors interact to control availability and hence uptake is also relevant.

Factors influencing availability of herbicides in soil

Soil-acting herbicides are normally applied to the soil surface as a suspension or emulsion in water and this water usually evaporates rapidly to leave a deposit of herbicide on a dry soil surface. If herbicides remained in this state they would have little activity and weed seedlings which emerged from moist soil beneath the dry surface layer would not be controlled. Some moisture in the form of rainfall or irrigation after application is essential to 'activate' soil-applied herbicides and there are several reasons for this. Firstly, the herbicide must be brought into solution from its formulation since most herbicides are only available for uptake by weed seedlings when present in the soil water. Secondly, rainfall or irrigation is required to move the herbicide from the immediate soil surface into the top few centimetres of soil where most weed seeds germinate and seedlings have their roots. Finally, adequate rainfall is important to maintain high moisture contents in the surface soil layers and hence maximize herbicide up-take by the weeds. Some of the ways in which soil water content affects herbicide availability are summarized in Fig. 6.3. It is difficult to specify the amount of rain required for optimum activity since this will depend on the amount of water required to dissolve the herbicide, the ease with which the herbicide will move into the soil, and the extent to which the herbicide must be moved in order to be absorbed by the seedlings and hence achieve good weed control.

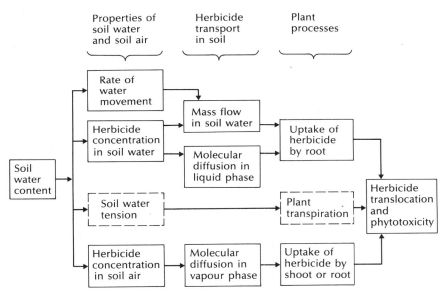

Fig. 6.3 Schematic diagram showing probable mechanisms by which soil water content influences the phytotoxicity of a soil-applied herbicide at water contents above the wilting point. From Moyer (1987).

There is little doubt that some herbicides (e.g. propachlor and related compounds) are absorbed mainly by the shoot or hypocotyl of germinating weeds. They will usually be most active when present in high concentration in moist soil above the germinating weeds. In fact, mechanical incorporation of these chemicals into soil can often lead to a reduction in weed-control performance presumably because this effectively reduces the concentration in the shoot zone of the weeds. It seems likely that excessive movement in the soil under the influence of heavy rainfall may also lead to occasional lack of weed control. With compounds of this type, activation before weed emergence is important because there is often little uptake via the stem once the seedlings have emerged. Other herbicides (e.g. simazine, lenacil) are absorbed primarily by the roots of weed seedlings and they must be moved further into the soil to achieve optimum performance. Adequate soil moisture is essential for uptake by the root, since roots actively search out water and will not explore dry soil. This latter type of compound is often activated after weed emergence has begun and provided that the weeds have not grown too large and become established with large root systems below the top few centimetres of soil they may still be effectively controlled. A final group of herbicides (e.g. napropamide, trifluralin, meta-mitron, atrazine, metribuzin) can be absorbed by root and shoot. The main prerequisite for activity of these compounds is adequate soil moisture, although trifluralin which is highly volatile, also shows activity via the gas phase of the soil. Because the requirements for movement of these compounds in soil are less critical, their activity is probably somewhat less variable than that of chemicals where specific distributions in the soil are required.

Although the amounts of rainfall required to give optimum performance are difficult to specify precisely, there is little doubt that some rain shortly after application is essential. This is because the normal cultivations for seedbed preparation stimulate weed seed germination and if the seedlings emerge and become established before the herbicide is activated by rainfall, they will often not be controlled. On the other hand, rainfall shortly after application will usually ensure that the herbicide is readily available to the seedlings as they develop and emerge from the soil. Young seedlings are generally more susceptible than older ones, if only because they are smaller and hence require less herbicide to kill them.

As mentioned earlier, most herbicides must be present in the soil solution before they can be absorbed appreciably by germinating weed seeds or the roots and shoots of seedlings. Therefore, those factors which control the concentration of herbicide in solution, or the availability of this solution to the plant, will influence activity of the compound. Assuming adequate rainfall, the two most important factors are soil texture and soil organic matter. The movement of water through the soil and the extent to which this results in redistribution of herbicide are controlled to a large extent by soil texture, i.e. the relative proportions of sand, silt and clay in the soil. Closely related to soil texture is soil structure, i.e. the extent to which the individual particles are bound together to form larger crumbs or aggregates. Light-textured sandy soils are generally structureless, and the particles are not aggregated together to form larger crumbs. These soils have low water-holding capacities, are freely drained, and water plus dissolved herbicide move in them relatively rapidly. Heavy textured soils on the other hand tend to hold more water and water infiltration and movement occur more slowly. They also tend to be more structured with well-defined aggregates of soil. This can lead to less efficient movement of herbicide, since a significant proportion of the water which enters the soil will move in the wider channels between soil aggregates with the herbicide remaining in finer pores within the crumbs or clods. Herbicide availability to weed seedlings in these soils may therefore be restricted.

The other factor which plays an important role in controlling the availability of herbicides to plants in soil is its organic matter content which, in mineral soils, is loosely related to soil texture. Organic matter is important because it effectively removes herbicide from the soil solution by adsorption and hence reduces the amount which is available for movement and subsequent uptake by the weeds. A fuller discussion of the processes of herbicide adsorption and mobility in soil, and of the factors governing the relationship between soil type and the practical recommendations for herbicide use are given in Chapter 9.

Uptake by roots and other below-ground organs

Once a herbicide has been moved or mixed into the soil and is available in the soil solution, those factors which influence its uptake into the plant come into play. In studies of the uptake of inorganic nutrients, three means have been distinguished by which ions arrive at the root surface: (a) interception, when the growing tip of the root makes contact with stationary ions; (b) mass flow, when the ion is carried passively to the root surface in the water which moves towards the root in response to transpiration; and (c) diffusion, when the ion is moved to the root surface through the soil solution along a concentration gradient. Of these, mass flow is probably the most important for non-volatile herbicides. Studies with a wide range of triazine, urea and uracil herbicides have shown that uptake from both nutrient solutions and soils can be related to the concentration of herbicide available in solution and the amount of water transpired by the plant. There is also evidence that, when a root system is absorbing water from the soil, the rate of transfer of herbicide to the root surface increases as the rate of water uptake increases. These changes in rate of water uptake can be caused by changes in the aerial environment, e.g. temperature or relative humidity, or by changes in soil water stress.

Once a herbicide has arrived at the root surface, its entry into the root may occasionally be independent of the entry of water, and uptake of some herbicides, e.g. 2,4-D may be associated with concurrent metabolic processes. With most herbicides, however, passive diffusion via the apoplast is the most likely uptake mechanism. Movement into the root with water is possible only as far as the Casparian strip which is a water-tight barrier separating the cortex and the stele. It has been suggested that passive movement across this barrier is only possible if the herbicide can dissolve in the suberised layer and diffuse through it. The lipophilic/hydrophilic balance of the specific herbicide molecule is then critical in determining the ultimate ease of transfer across the root. For this reason the apparent concentration of herbicide in the xylem is usually less than that in the external solution and the magnitude of the difference is closely related to the lipophilic/hydrophilic balance of the chemical, often indicated by its octanol/water partition coefficient. Highly lipophilic or hydrophilic compounds are not readily transferred across the root and hence concentrations in the xylem are often relatively low.

Although movement in the water phase of the soil is the main means by which herbicides arrive at the root surface, significant movement in the vapour phase can also occur in some circumstances. Diffusion in the vapour phase is rapid and is important in the movement of volatile herbicides in the soil. As well as providing a means of transport to the root surface, it is believed to be of special importance in the transfer to below-ground regions of the shoot and to newly-germinated seedlings whose uptake of water is insufficient to bring about significant transfer by mass flow.

The classical view that the roots of seedlings are largely responsible for the uptake of herbicides from soil has been modified since it is now known that some soil-applied herbicides can enter the parts of the shoot system that are underground. Entry into the underground shoot is thought to be essential for the full effectiveness of the thiocarbamate herbicides EPTC and tri-allate, and it may also contribute to the activity of other herbicides such as pendimethalin, trifluralin, dichlobenil and chlorpropham. All of these have appreciable volatility and hence contact between the herbicide and the underground shoot can be maintained via the gas phase of the soil. Even relatively non-volatile herbicides such as atrazine, linuron and metribuzin have been shown to cause damage as a result of direct entry into the shoot but adequate soil moisture in the surface layers of soil is essential for this to occur. In practice, it seems unlikely that uptake via the underground shoot alone can be responsible for the activity of non-volatile herbicides other than in very fine-textured soils in which good contact between the soil water and the shoot system can be maintained.

Differential uptake and selectivity

The selective phytotoxicity of soil-acting herbicides may result from differential uptake, differential response or a combination of the two. Only the former is discussed here as the latter is considered in detail in Chapter 7. Differences in uptake from soil are based primarily on differential interception. This is usually achieved by producing a high concentration of herbicide near the soil surface, where weed seedlings germinate and have their roots, with a much lower concentration in the root zone of the crop. Selectivity based on this difference is often called 'depth protection'. Because anything which causes downward movement of the herbicide increases the chances of crop damage and because all weeds are not shallow rooted, depth protection can rarely provide complete selectivity. It is probably of some importance for large-seeded, deep-sown crops such as peas and beans. Likewise it may play some part in the safety of herbicides used in plantation crops although these usually have some roots near the surface so there must be an element of physiological tolerance involved as well.

Depth protection can occur with herbicides which enter the shoot. The selectivity of tri-allate against wild oat in wheat can be improved by incorporating the herbicide only shallowly after sowing, leaving 2–3 cm of untreated soil over the wheat seed. This is because wild oat has a mesocotyl which raises the sensitive meristematic region of the seedling into the layer of treated soil very early during growth, whereas wheat has no mesocotyl and the sensitive tissues only enter the treated soil some time later when they are protected by the mature sheathing leaf bases. In this and other instances where depth protection is important in selectivity, special care is needed to ensure a uniform and accurate sowing depth of the crop. This can be particularly so in light, fluffy seedbeds where due allowance must be made for subsequent consolidation of the surface soil.

The selectivity of thiocarbamate herbicides between some broad-leaved and grass species is probably influenced by a situation which is the reverse of depth protection. While the sensitive meristematic region of the grass seedlings remains below ground exposed to the herbicide, the stem apex of the broad-leaved species emerges rapidly and is exposed relatively briefly to the treated soil.

Transport of herbicides in the plant

Herbicides may move in the plant along pathways which are essentially non-living (apoplast) or living (symplast) or both. The term 'apoplast'

refs to the cell-wall continuum with the xylem as its specialist transport
system; the symplast refers to the cytoplasmic continuum with the phloem
as its specialist transport system. The movement of herbicides in plants is
often considered in two separate categories: short-distance transport and
long-distance transport.

Short-distance transport

Penetration into leaves and uptake by roots have already been discussed
and both of these processes necessarily involve the movement of herbicides
over short distances in the plant. After penetration, further short-distance
transport is important for a number of reasons:
1   Since herbicides can only be applied to the surfaces of plants, transport
must occur if cells other than those at the immediate surface are to be
killed.
2   Since the long-distance transport systems are always internal, short-
distance transport from the point of entry is essential prior to further
redistribution in the plant.
3   When long-distance transport has been achieved, the chemical must
move away from the cells of the transport system to its ultimate site of
action.
    The exact mechanisms by which organic chemicals are loaded into and
unloaded out of the long-distance transport systems are not well understood and are the subject of considerable research interest at present.

Long-distance transport

Herbicides are often most useful when they are able to move within the
plant to cells, tissues and organs which are remote from the site of initial
uptake or penetration. This can be particularly important where control of
perennial weeds is concerned. Two systems are primarily responsible for
this long-distance transport — the xylem and the phloem.

The xylem system

This is the system in which water and dissolved mineral nutrients pass from
the root system to the transpiring surfaces of the plant, which in most
species are the leaves. Both naturally-occurring and foreign organic substances are moved in the xylem and there is little doubt that most soil-
applied herbicides taken up by the roots are initially transported to the
aerial parts of the plant in this system. The xylem is essentially non-living

and the process of transport within it is largely governed by environmental factors, although adsorption of herbicides onto the organic matrix of the walls and escape into the surrounding cells is possible. Water loss from a plant is determined by light, temperature, wind speed and humidity as well as the availability of water in the soil. It is reasonable to expect, therefore, that in non-drought conditions the rate of upward transport of a herbicide which has penetrated to the xylem in the root will be governed primarily by the rate of water loss from the leaves. However, as soil water becomes less readily available other factors may override the simpler ones which control transpiration. Under extreme conditions of soil water stress a reversal of the transpiration stream may occur and water present on the leaves may be absorbed and then pass down the plant to the root system. There have been reports that herbicides sprayed onto leaves under such extreme conditions have been transported rapidly and extensively to the subterranean organs of perennial weeds.

A special case occurs in respect of both paraquat and diquat. Owing to the rapid desiccating properties of these herbicides, a flow of water containing the herbicide may occur from treated leaves to other parts of the plant. This form of long-distance transport of herbicides which are essentially of contact action is known to occur in the field under severe conditions of water stress and it is possible that this form of transport may be more common than has been suspected previously.

A herbicide taken up by the roots and distributed normally in the xylem system will be transported primarily into the expanded leaves, an ideal pattern of distribution for any compound whose mode of action is inhibition of photosynthesis. On the other hand, unless some further redistribution takes place within the plant, this pattern is not satisfactory for a herbicide whose mode of action is associated with growth processes. In such cases redistribution must occur into the growing apices, a process probably involving phloem transport.

The phloem system

Although the precise mechanism of transport of sugars in plants is not understood, it is generally agreed that they are translocated in the living symplast of which the phloem forms the long-distance route. Sugars produced by photosynthesis in the green tissues of the plant (sources) are conveyed in the phloem to regions of the plant where growth and storage are taking place (sinks). Generally, it is assumed that loading sugars into the phloem is achieved against a concentration gradient by use of metabolic energy. In most circumstances, herbicides only move out of a treated leaf in the phloem and concentrations of herbicide and/or formulation components which interfere with phloem loading limit herbicide translocation. For example, localized damage following difenzoquat application restricts its systemic activity with more than 95% of the absorbed herbicide remaining in the treated leaf. Often slow development of phytotoxic symptoms, as found for example with glyphosate, is associated with more effective translocation of the herbicide.

Leaves located at different positions on the shoot usually translocate sugars to different sink areas. Thus lower leaves usually supply the roots and underground storage organs, whereas from upper leaves there is a tendency for sugars to be translocated to the shoot meristem and developing leaves. The strength of individual sources and sinks changes during the year in response to senescence of leaves, developmental changes in the plant such as flowering, seed formation and storage organ development.

There are many examples of systemic herbicides (MCPA, amitrole, chlorsulfuron, glyphosate, imazapyr and fluazifop-butyl) following the 'source–sink' distribution patterns outlined above which have implications with regard to herbicide performance. Very young leaves behave as sinks and thus are poor targets for systemic herbicides. Leaves approaching full development on young plants tend to export sugars (and herbicides) predominantly to the stem apex. As the plant grows, the export pattern changes more towards the roots and subterranean organs. It is at this stage that application of herbicide usually results in good control of perennial plants such as common couch and bracken. In contrast, early spring herbicide treatment when sugars are being translocated from storage organs to developing shoots, will damage the latter but not the buds on the underground rhizomes. During seminal reproductive development herbicide treatment of the foliage usually results in at least some herbicide accumulation in the seed. Consequently pre-harvest application of glyphosate should not be made until the grain has fully developed and the leaves of the cereal have become fully senescent.

In addition to plant growth stage, environmental factors also affect the flow of sugars in the phloem. Stress factors that slow down the rate of plant growth, such as low temperatures and drought, reduce sink strength and less herbicide tends to be translocated. Other factors, such as low light intensity, limit production of sugars in the leaves and reduce source activity, and can impair systemic herbicide activity. For these reasons, it is commonly recommended that systemic herbicides should be applied when weeds are in an active phase of growth.

While in transit, herbicides may leave the phloem system and move into other tissues where they may be metabolized or made unavailable for phytotoxic action by conjugation, or compartmentalized. In some cases herbicide may be lost from the plant in root exudates, e.g. imazapyr and glyphosate.

In conclusion, for herbicides which depend on the phloem system for long-distance transport, an understanding of the general physiology of the weed species is an essential preliminary to successful development of a herbicide treatment.

Phloem and xylem transport

The patterns of long-distance transport of herbicides cannot be explained on the basis of either phloem or xylem transport alone. A herbicide entering a plant through the roots and conveyed to the leaves in the xylem system has eventually to enter the symplast since the site of action for all herbicides is located in the living tissue of the plant. Thus all herbicides to some extent are able to diffuse across membranes and enter the phloem. However, many herbicides such as chlorotoluron are not exported from a treated leaf in the phloem. The reason for this is probably that the herbicide diffuses freely across membranes, but the flow of water in the xylem, the transpiration stream, is more rapid than the flow in the opposite direction in the phloem (Fig. 6.2). Thus the net direction of chlorotoluron movement in the vascular bundle will be in the direction of the transpiration stream, i.e. towards the margins of the leaf.

Most herbicides that remain in the phloem long enough to be translocated therein, have ionizable groups which can exist in a non-charged or charged form depending on the pH of the surrounding medium. In the apoplast, with a pH of around 5, some herbicides, such as imazapyr, are in a non-charged form and can readily diffuse across the membrane surrounding the symplast. Within the phloem the pH is higher (around 8) and the herbicides exist as anions which diffuse less readily across the membrane. This mechanism is known as the 'ion trap' and provides an explanation for the phloem mobility of many systemic herbicides. Because the xylem and phloem are in close juxtaposition in the vascular bundles (Fig. 6.2) transfer of some herbicides occurs between the two systems. In this case herbicides are capable of circulation within the plant; examples include amitrole, maleic hydrazide, glyphosate, picloram, imazaquin and dalapon.

Transport and selectivity

Although herbicides differ widely in their mobility in plants, there is only limited evidence that the internal transport mechanisms are sufficiently varied to allow any major degree of selectivity to be obtained. However, differences in efficiency of transport may contribute to a total complex 'package' of sources of selectivity. For example, the soil-applied photosynthetic inhibitor linuron is selective in carrots, but toxic to turnips. In carrots, most of the absorbed herbicide (or metabolites) remains in the root whereas in the sensitive turnip, most is translocated to the shoots. Root accumulation of linuron is probably due to its conjugation or compartmentalization therein with differences in translocation between the two species playing a secondary role in selectivity.

Following foliage application, dicamba translocation is primarily apoplastic in tolerant wheat and symplastic in susceptible buckwheat allowing accumulation in the meristem, its site of action. However, there is more inactivation of dicamba in wheat than in buckwheat so differential transport is a contributing factor to selectivity between these two species.

Several herbicides including diclofop-methyl and imazamethabenz-methyl are applied as esters. These readily enter the leaf, but are not loaded into the phloem. In susceptible species such as wild oat, the herbicide is rapidly deesterified to a free acid which is phloem mobile and accumulates in the target meristem, whereas in the tolerant wheat this conversion occurs very slowly and so confers selectivity. Metabolism that produces the phytotoxic molecule is sometimes called 'suicide metabolism'.

Weather and herbicide transport

There has been no extensive and systematic field study made of the effects of the environment on the transport of herbicides in plants. From laboratory work enough information has accumulated to show that there are direct and indirect effects of light, temperature, nutrient and water status on transport. Unfortunately, it is always difficult to separate environmental effects on transport from those on penetration of herbicides and many of the techniques which have been used have not discriminated between these two stages of the process.

Further reading

Bonner, J. & Galston, A.W. (1952) *Principles of Plant Physiology.* W.H. Freeman, San Francisco.

Hance, R.J. (1980) *Interactions between Herbicides and the Soil.* Academic Press, London.

Hess, F.D. (1985) Herbicide absorption and translocation and their relationship to plant tolerances and susceptibilities. In (ed.) Duke, S.O. *Weed Physiology, Vol.2, Herbicide Physiology,* pp. 191–214. CRC Press, Boca Raton.

Moyer, J.R. (1987) Effect of soil moisture on the efficacy and selectivity of soil-applied herbicides. *Reviews of Weed Science,* **3,** 19–34.

7 / The mode of action and metabolism of herbicides

A.D. DODGE

School of Biological Sciences, University of Bath, Bath BA2 7AY

Introduction

Herbicides are used to kill or to control weeds while, as far as possible, leaving crop plants unharmed. Most of those in use today kill plants by a subtle interaction with plant physiological or biochemical processes. They initiate a chain of events that eventually leads to the death of the plant. Such initial actions may involve the inhibition of photosynthesis, the generation or accumulation of toxic molecules, the failure to produce vital intermediary metabolites or the inhibition of plant growth. In some instances crop plants are not killed because of their ability to de-toxify potentially lethal molecules. Alternatively, in some instances susceptible plants die because they metabolize a non-toxic pro-herbicide to a toxic molecule. Various aspects of herbicide action and selectivity are now considered in more detail.

Herbicide action

Interaction with photosynthesis and respiration

The process of photosynthesis involves the conversion of light energy to chemical energy. Light is absorbed by the photosynthetic pigments, chlorophylls and carotenoids, that are integrated in lipoprotein thylakoid membranes within the chloroplast. Each leaf mesophyll cell probably contains hundreds of chloroplasts. The initial events of photosynthetic energy conversion are represented in Fig. 7.1. Here, in a simplified form, light energy is shown to be absorbed by two photosystems, 2 and 1, and electrons are

201

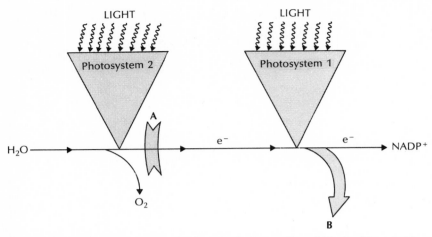

Fig. 7.1 Simplified scheme showing photosynthetic electron flow within the chloroplast from water to the natural acceptor $NADP^+$ (nicotinamide adenine dinucleotide). **A** = site of inhibition by acylanilides, ureas, hydroxybenzonitriles, phenylcarbamates, uracils, triazines and triazinones, etc. (see Table 7.1), and **B** = site of diversion of electron flow by bipyridiniums such as paraquat and diquat.

conveyed from water to reduce the acceptor $NADP^+$. Oxygen is released incidentally and NADPH and ATP, also generated within the chloroplast, are utilized for the incorporation of carbon dioxide into sugars and other molecules. It is evident that plants maintained in darkness will not grow, nor will plants in which photosynthesis is chemically inhibited. Table 7.1 shows a list of herbicides that work by inhibiting photosynthetic electron flow. This is achieved by interacting with a protein component of the chloroplast thylakoid membrane in the vicinity of photosystem 2, represented by **A** in Fig. 7.1. When chlorophyll absorbs light energy to activate electron flow from water (Fig. 7.1) it is excited to a so-called 'singlet state' (^{1}Chl). If the excitation energy is not utilized because electron flow is prevented, there is not only an increase in energy release as fluorescence, but also the transformation of ^{1}Chl to a longer-lived triplet state (^{3}Chl). The triplet form can interact in a damaging way with membrane lipids, but of more importance, it can excite oxygen to a singlet state ($^1O_2$) (Fig. 7.2). This highly damaging form of oxygen can interact with cellular lipids, proteins, nucleic acids and other molecules, and induce cellular disorganization and hence plant death. Although in the long term, an inhibition of photosynthesis will lead to plant starvation, the appearance of phytotoxic symptoms such as chlorophyll bleaching occurs quite rapidly and this is an indication of light-induced damage.

Table 7.1 Some herbicides inhibiting photosynthetic electron transport

Acylanilide:
Monalide
Pentanochlor
Propanil

Cyclic urea:
Methazole

Hydroxybenzonitrile:
Bromoxynil
Ioxynil

Phenylcarbamate:
Desmedipham
Phenisopham
Phenmedipham

Uracil:
Bromacil
Lenacil
Terbacil

Urea:
Benthiazuron
Buturon
Chlorbromuron
Chloroxuron
Chlorotoluron
Difenoxuron
Diuron
Ethidimuron
Fenuron`
Fluometuron
Isoproturon
Linuron
Methabenzthiazuron
Metobromuron
Metoxuron
Monolinuron
Monuron
Neburon
Siduron
Thiazafluron

Triazine:
Ametryne
Atrazine
Aziprotryne
Cyanazine
Desmetryn
Dimethametryn
Dipropetryn
Methoprotryne
Prometon
Prometryn
Propazine
Secbumeton
Simazine
Simatryn
Terbumeton
Terbuthylazine
Terbutryn
Trietazine

Triazinone:
Metamitron
Metribuzin

Miscellaneous:
Bentazon
Chloridazon

Other herbicides require to be activated by photosynthetic processes before their toxic action is realized. Of major importance in this category are the bipyridinium compounds, paraquat and diquat. These herbicides divert electron flow at the terminal end of photosystem 1, **B** in Fig. 7.1. The action of these herbicides is therefore dependent upon light to promote

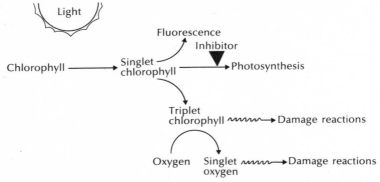

Fig. 7.2 Scheme showing the consequences of inhibiting photosynthetic electron transport and the diversion of excitation energy to generate damaging triplet chlorophyll and singlet oxygen.

electron flow and also oxygen to yield toxic radicals. Fig. 7.3 shows a sequence of events that occurs after the initial reduction of paraquat. The capture of one electron is followed by a transfer of the electron to oxygen. This gives rise to superoxide, which in turn produces hydrogen peroxide, and these two products interact in the presence of an iron catalyst to yield the highly damaging hydroxyl free radical. This, like singlet oxygen mentioned above, rapidly interacts with membrane lipids, amino acids of enzyme proteins and nucleic acids. The result is a rapid inactivation of cellular metabolism, and the breakdown of organized structure. This is manifest as rapid bleaching.

Other herbicides that require light for activation are some of the nitrodiphenylethers such as nitrofen, oxyfluorfen and acifluorfen. At present there is no consensus of opinion as to the exact mechanism of activation, but it could involve a direct electron donation to yield a toxic radical,

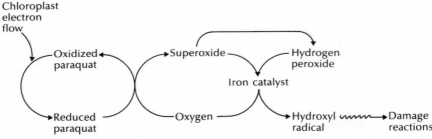

Fig. 7.3 Scheme showing the diversion of photosynthetic electron flow in the chloroplast by paraquat, with the consequent production of the hydroxyl free radical.

or possibly activation by a form of energy transfer from carotenoid or chlorophyll pigments.

In contrast to chloroplast electron transport, mitochondrial flow involves the oxidation of NADH through a series of carriers to reduce oxygen to water. This electron flow is indirectly linked to ATP generation. A number of nitrophenol herbicides such as dinoseb and DNOC, the benzonitrile dichlobenil, and also the hydroxybenzonitriles (in addition to their photosynthetic effect) sever the link between electron flow and phosphorylation, and are termed 'uncouplers'. A similar effect is achieved if the mitochondrial membrane is rendered more permeable, as occurs with perfluidone.

Inhibitors of biosynthesis

Amino acids

Amino acids are the building blocks of proteins, therefore the failure to produce certain acids will not only have an effect on the biosynthesis of enzymes, but also lead to the failure of plant metabolism in general.

One important biosynthetic route is the shikimate pathway that leads to the formation of the aromatic amino acids tryptophan, tyrosine and phenylalanine. A large number of secondary compounds, such as flavonoids, anthocyanins, auxins and alkaloids also arise from these amino acids. The failure of the shikimate pathway would therefore not only have a serious effect upon protein synthesis but also on the synthesis of compounds associated with growth regulation and defence. Glyphosate inhibits the pathway at the enzyme enolpyruvyl shikimate-3-phosphate synthase (Fig. 7.4). Glyphosate has been shown to be translocated rapidly to meristematic tissue and stem apices, especially underground rhizomes, where growth failure occurs.

Another group of protein amino acids is the branched-chain acids, leucine, isoleucine and valine. The synthesis of these acids is shown in outline in Fig. 7.5. The two related pathways in which acetolactate is produced from pyruvate, and acetohydroxybutyrate from threonine, are catalyzed by a common enzyme acetolactate synthase (acetohydroxy acid synthase). This is effectively inhibited by the sulfonylureas such as chlorsulfuron and metsulfuron, and also the imidazolinones such as imazapyr and imazaquin.

One further instance of the inhibition of amino acid synthesis is provided by glufosinate-ammonium. This molecule prevents ammonia combining with organic compounds by competing with the enzyme glutamine

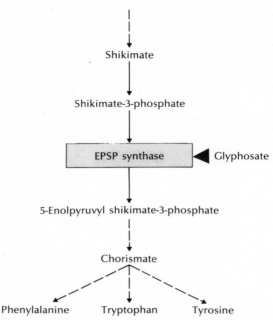

Fig. 7.4 Part of the biosynthetic pathway of aromatic amino acid synthesis, showing the site of action of glyphosate at EPSP synthase (5-enolpyruvyl shikimate-3-phosphate synthase).

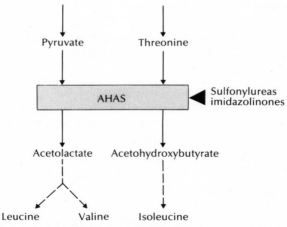

Fig. 7.5 Part of the biosynthetic pathway of branched chain amino acid synthesis, showing the site of action of some herbicides at the enzyme acetolactate synthase (acetohydroxy acid synthase).

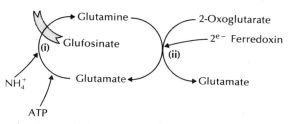

Fig. 7.6   Enzymes involved in the incorporation of ammonium into amino acids, showing the site of action of glufosinate. **(i)** glutamine synthetase, **(ii)** glutamate synthetase.

synthetase (Fig. 7.6). Although in the long run this would lead to a general failure to produce the essential amino acid glutamate, death is apparently more rapid due to the accumulation of toxic ammonia. This leads to cell damage, bleaching and death.

Carotenoids

The orange-coloured carotenoid pigments are not generally a conspicuous feature of leaves for their presence is masked by an approximately eight times greater concentration of chlorophylls. The role of carotenoids within the chloroplast is two-fold. Firstly, they act as extra light-absorbing pigments and secondly they act in a protective capacity. The damaging action of ^3Chl and $^1O_2$ was outlined above and both of these agents are quenched or inactivated by carotenoid pigments. It is quite possible that this protective role is essential in normal environmental conditions, let alone conditions of high light stress. Any diminution of carotenoid pigments is thus likely to have a potentially damaging effect upon chloroplasts. A number of herbicides including the pyridazinone norflurazon together with fluridone, fluometuron and diflufenican inhibit the so-called 'desaturase reactions' between phytoene and lycopene (Fig. 7.7). Amitrole among other effects, inhibits the cyclization stages between lycopene and α- and β-carotene. The overall effect of these herbicides is particularly evident in newly-developing tissue which is obviously bleached.

Lipids

Lipids are essential plant components, not only as major seed storage compounds, but also as constituents of membranes and cuticular waxes. A unifying feature of all lipids is the constituent fatty acids. These long-chain molecules are synthesized from acetyl coenzyme A. The action of an initial

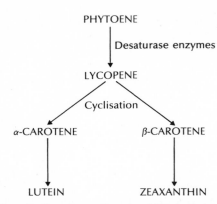

Fig. 7.7   Part of the biosynthetic pathway of carotenoid formation showing the sites of herbicide inhibition at the desaturase and cyclization enzymes.

carboxylase enzyme is followed by a complex sequence of 'fatty acid synthase' activity. Recent evidence suggests that the initial acetyl CoA carboxylase enzyme is inhibited by a number of herbicides including the aryloxyphenoxy propionics such as diclofop-methyl, haloxyfop, fluazifop-butyl, and fenoxaprop-ethyl. Furthermore, some oximes such as alloxydim and sethoxydim also function at a similar site. The failure to produce fatty acids and hence membrane lipids is particularly evident as necrosis in meristematic tissue, leading to a cessation of growth and death.

Fig. 7.8 summarizes the sites of action of lipid biosynthesis inhibitor herbicides and the consquences for membrane and cuticular lipids. The fatty acids of cuticular waxes are the so-called 'very long-chain acids', and the formation of these is specifically inhibited by thiocarbamate herbicides such as EPTC, di-allate and tri-allate. The failure to produce cuticular waxes will leave plants vulnerable to desiccation and pathogen attack.

Growth inhibition

Growth and development will obviously be affected by herbicides that prevent the biosynthesis of essential cellular molecules. However other compounds affect growth by inhibiting cell division, cell elongation, or by altering the balance of endogenous growth regulators.

Mitotic inhibitors

Cell division could potentially be prevented by a number of mechanisms including the failure of DNA synthesis or the disruption of micro-tubule

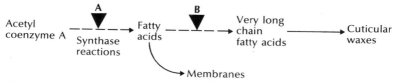

Fig. 7.8 Summary scheme of fatty acid biosynthesis showing **A** the site of action of aryloxyphenoxypropionics, alloxydim and sethoxydim at the enzyme acetyl CoA carboxylase, and **B** the site of action of thiocarbamates.

formation. The formation of micro-tubules of the mitotic spindle is a prerequisite for the separation of daughter chromosomes. The complex organization of these micro-tubules involves the aggregation of the globular proteins α- and β-tubulin at the micro-tubule organizing centre (MTOC) (Fig. 7.9).

A number of herbicides including the 2,6-dinitroanilines such as trifluralin, oryzalin, and carbamates such as asulam, barban, propham and chlorpropham affect either the formation of the microtubules or their organization on the MTOC. This will lead to a failure of cell division and hence growth.

Cell-elongation inhibitors

The *N*-arylalanine ester herbicides such as benzozylprop-ethyl, and flamprop-isopropyl are used to control certain grass weeds in cereals. Their exact mode of action has not been established. However, they appear to prevent cell elongation, and this would lead to stunted growth and thus a failure in the competition with crop plants. It is possible that the herbicides modify the site of action of the natural growth regulator, auxin.

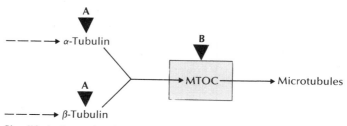

Fig. 7.9 Simplified scheme showing the formation of microtubules from tubulin-showing sites of herbicide action at **A** the formation of microtubules, and **B** at the MTOC (microtubule organizing centre).

Auxin-type herbicides

The well-known auxin or hormone type herbicides such as 2,4-D, 2,4,5-T, 2,3,6-TBA, MCPA and MCPB, have been used as effective herbicides for 40 years or more, yet their exact mode of action is unknown. Secondary effects such as stem enlargement, callus growth, secondary roots and leaf stunting are usually evident. These major structural changes are an indication of gross physiological and biochemical aberrations, brought about, it is assumed, by a massive and uncontrolled 'over-dose' of auxin-type compounds. We need to understand much more about the action of natural plant auxins before we can speculate with accuracy on the initial triggering events of these major biochemical and physiological changes.

Herbicide metabolism and selectivity

The spectacular selectivity shown by many herbicides may be due to a number of factors including differential uptake or movement. However, here consideration will be given of some important metabolic events that are the basis of selectivity. Fig. 7.10 outlines two major aspects. Selectivity may be due firstly to the metabolism of the *active* molecule to an *inactive* metabolite and secondly to the conversion of an *inactive* proherbicide to an *active* metabolite within the plant.

Herbicide inactivation

Herbicides are 'foreign' molecules and a number of important de-toxification processes may come into play within the plant to reduce their action. The usual sequence is an initial event such as hydroxylation that will render the molecule more water soluble followed by conjugation with sugars (glycosidation) or amino acids. These products could be removed from metabolic reactions by excretion within the plant cell vacuole.

Hydroxylation, almost certainly catalyzed by microsomal mixed-function oxidases, is a feature of the metabolism of bentazone (Fig. 7.11), the phenoxyacetic acids such as 2,4-D and diclofop. Chlorsulfuron is also

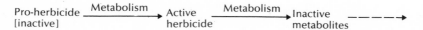

Fig. 7.10 Summary scheme showing some aspects of the metabolism of herbicides and proherbicides.

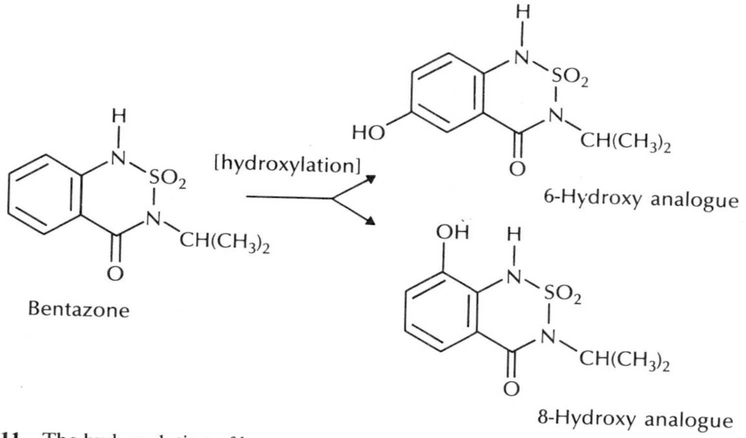

Fig. 7.11 The hydroxylation of bentazone.

hydroxylated, but this metabolite is herbicidally active, and inactivation is achieved by glycosidation.

Another type of inactivation occurs with metamitron (Fig. 7.12) and metribuzin which are deaminated almost certainly by a peroxisome-based deaminase enzyme. These two herbicides are photosynthetic inhibitors, and another, propanil, is de-activated in rice by an arylacylamidase enzyme that produces the inactive metabolites dichloroaniline and propionic acid (Fig. 7.13). With some other photosynthetic inhibitors, such as the phenyl-urea diuron, demethylation plays an important part in inactivation (Fig. 7.14). This reaction is possibly catalyzed by the microsomal mixed-function oxidase system, and may also precede hydroxylation and glycosylation.

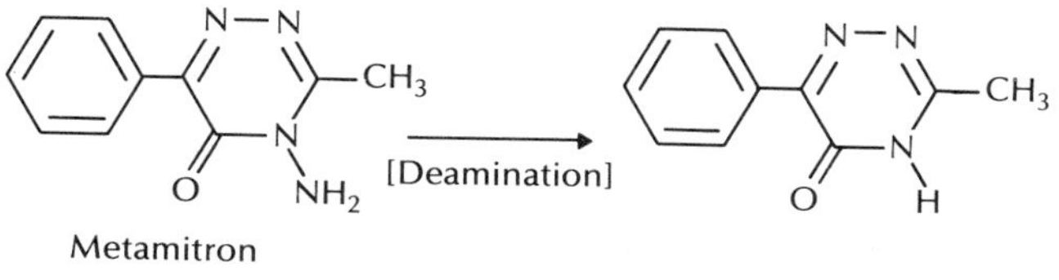

Fig. 7.12 The deamination of metamitron.

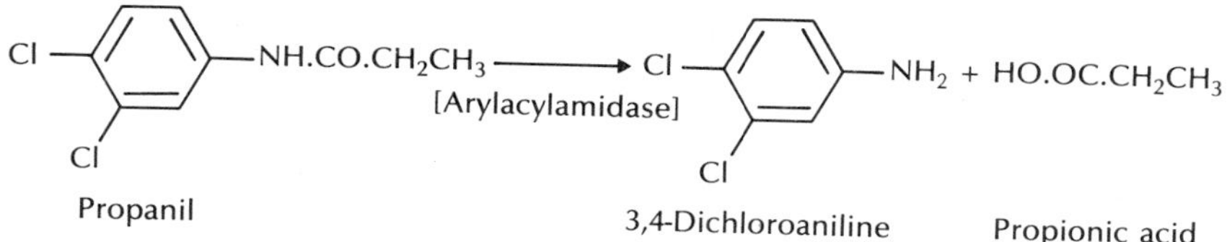

Fig. 7.13 The metabolism of propanil.

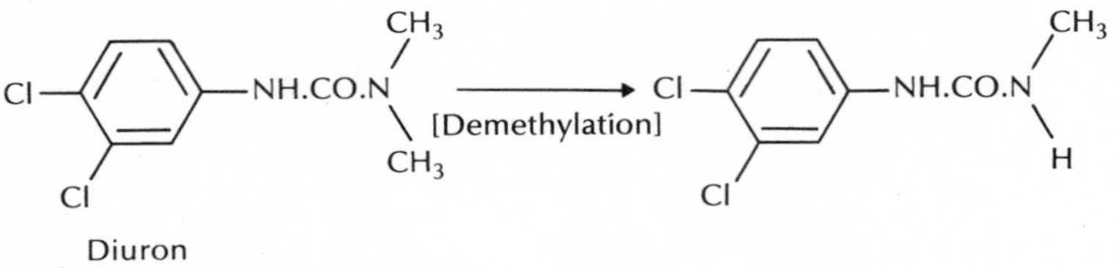

Diuron

Fig. 7.14   The demethylation of diuron.

The selectivity of the triazine herbicide atrazine in maize is due in part to the conjugation of the active herbicide with the tripeptide glutathione (Fig. 7.15). This is catalyzed by the enzyme glutathione-*S*-transferase. This important form of herbicidal inactivation may also occur to a limited extent in the inactivation of some diphenylethers, chloroacetanilides and possibly some sulphonylureas. In the case of the thiocarbamate EPTC, glutathione conjugation is preceded by sulphoxidation (Fig. 7.16). In this case, however, the S-oxide may be herbicidally more active than the original compound.

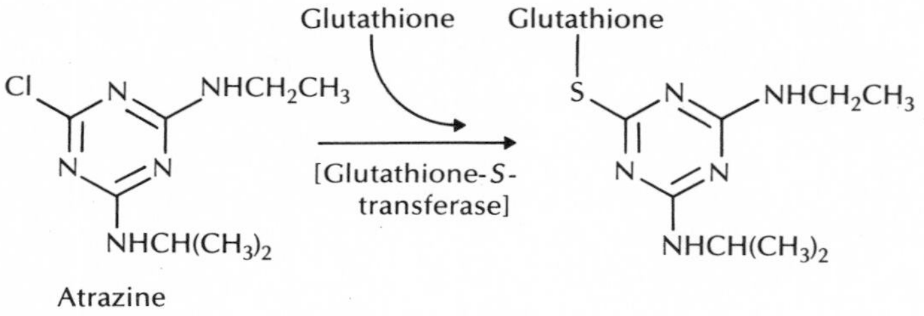

Atrazine

Fig. 7.15   The conjugation of atrazine with glutathione.

Herbicide activation

Herbicide activation is achieved by electron deviation or excitation energy transfer as mentioned above; however other metabolic forms of action of a pro-herbicide are well known. β-Oxidation is an important feature of the oxidation of fatty acids and its role is well illustrated in the conversion of the herbicidally-inactive MCPB to the active MCPA (Fig. 7.17). Other examples are provided by dichlobenil where hydroxylation results in the formation of a toxic hydroxylated metabolite and by the now obsolete metflurazon that is demethylated to the active carotenoid biosynthesis inhibitor norflurazon (Fig. 7.18).

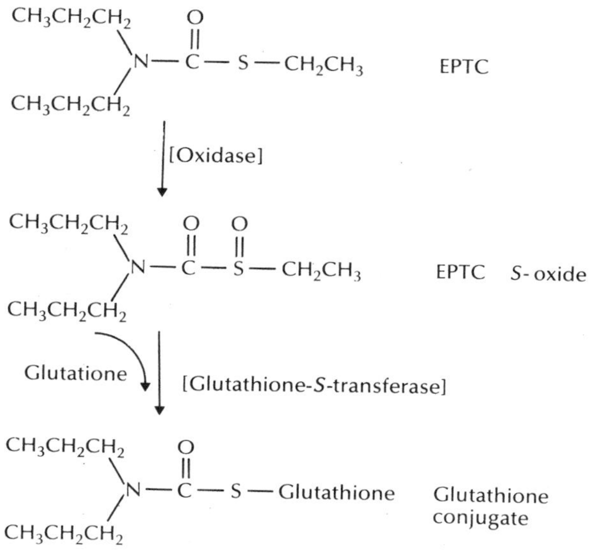

Fig. 7.16 The conjugation of EPTC with glutathione.

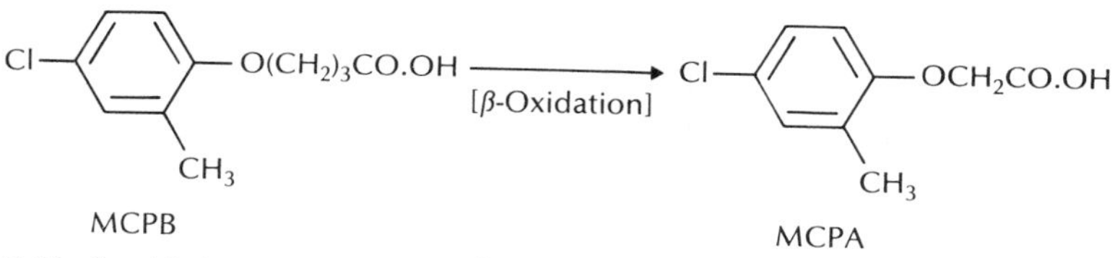

Fig. 7.17 β-oxidation of MCPB.

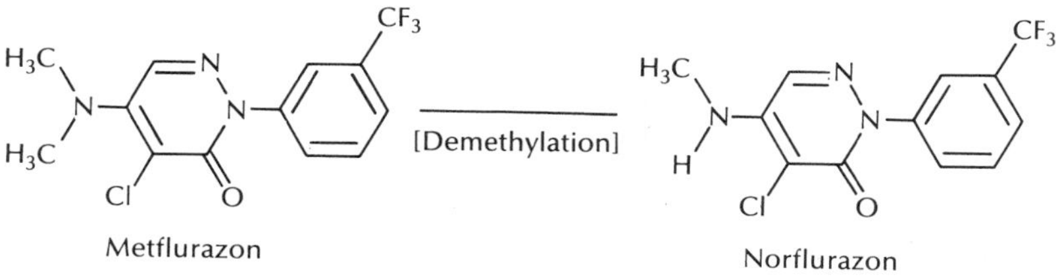

Fig. 7.18 The demethylation of metflurazon.

A number of herbicides that are esters are probably inactive unless hydrolyzed to the acid. This is the situation with benzoylprop-ethyl and related flamprop compounds (Fig. 7.19), the aryloxyphenoxypropionic herbicides such as fluazifop-butyl and also the isomeric imidazolinone herbicide imazamethabenz-methyl. In a number of instances the acid is

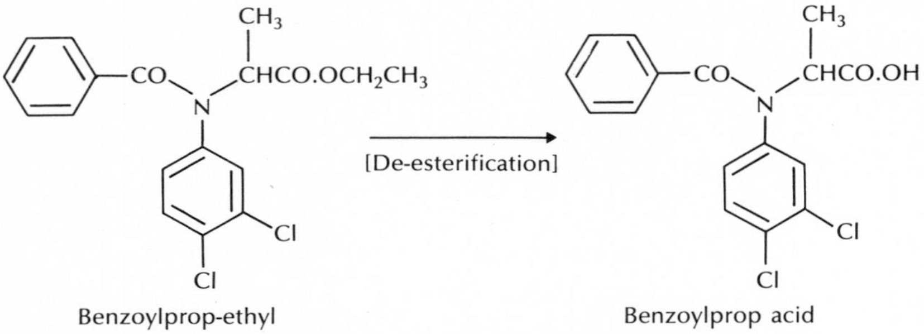

Benzoylprop-ethyl   Benzoylprop acid

Fig. 7.19 The de-esterification of the aryloxyphenoxypropionic herbicide benzoylprop-ethyl.

more mobile within the plant and therefore reaches the site of action more readily.

Herbicide modulation

Safeners are chemicals added to herbicides to modify or improve their action or selectivity. They are sometimes called 'antidotes' or 'protectants' but 'safeners' is preferable. They alter the metabolism of the herbicide within the crop plant. This is well illustrated by N,N-diallyl-2,2-dichloro-acetamide that will improve the selectivity of EPTC in maize by enhancing not only the level of glutathione (see above) but also the activity of the enzyme glutathione-S-transferase. Another herbicide, tridiphane (Fig. 7.20) could have the opposite effect because it is thought to inhibit the enzyme glutathione-S-transferase. This could be used as a herbicide synergist to prevent the metabolic removal of the active herbicide.

There is increasing evidence that some herbicide metabolism is achieved via microsomal mixed-function oxidases. Recent developments indicate that herbicide action could be modulated by activators of this oxygenase system that will promote de-toxification within crop plants or by inhibitors of oxygenase activity that could prevent herbicide breakdown within weeds.

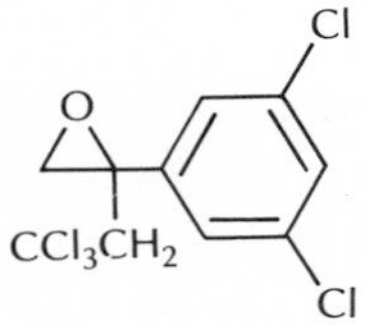

Fig. 7.20 Structure of tridiphane.

Further reading

Corbett, J.R., Wright, K. & Baillie, A.C. (1984) *The Biochemical Mode of Action of Pesticides*. Academic Press, London.

Dodge, A.D. (1989) *Herbicides and Plant Metabolism*. Cambridge University Press, Cambridge.

Hatzios, K.K. & Penner, D. (1982) *Metabolism of Herbicides in Higher Plants*. Burgess Publishing, Minneapolis.

Hutson, D.M. & Roberts, T.R. (1987) *Herbicides*. John Wiley, Chichester.

8 / The resistance of plants to herbicides

P.D. PUTWAIN

Department of Environmental and Evolutionary Biology, The University of Liverpool, PO Box 147, Liverpool L69 3BX

The significance of resistance to herbicides in weeds and crop plants

During the past 35 years, evolution of resistance to chemical pesticides has been recognized, first by entomologists and subsequently by plant pathologists, as a considerable threat to the chemical control of undesirable organisms. A substantial intellectual and financial investment has been made to devise methods to combat resistance or strategies to avoid the problem. Living organisms generally have enormous biological and biochemical adaptability so it was inevitable that eventually species of weeds would evolve resistance to herbicides. Farmers, the agrochemical industry and weed scientists are now confronting fundamentally the same kinds of problems arising from herbicide-resistant weeds that arose from the evolution of resistance to insecticides and fungicides. The main difference is that chronological time-scales are longer although in terms of number of generations, evolution of resistance in weed populations is probably similar to that required for the evolution of resistant insect populations.

Over the past two decades, in parallel with the increasing occurrence of herbicide-resistant weed strains there has been increasing interest by agricultural scientists, biotechnology companies and chemical industry in the possibility of creating herbicide-resistant crop varieties. Optimal cost-effective control of weeds could be achieved more readily if environmentally safe herbicides that kill a broad spectrum of weed species could be used in crops that are currently susceptible to them, that have been made resistant by molecular biotechnology. The registration costs connected with the introduction of new crop varieties would be greatly reduced in comparison with new herbicides and therefore some chemical companies have become involved in the creation of herbicide-resistant crops. However, there are potential drawbacks with this approach such as the eventual parallel evolution of resistant strains of weeds. Thus the nature of resistance to herbicides in both weed species and in crop plants is discussed in this chapter.

Definition of resistance

In the literature the term 'resistance' has a variety of meanings and therefore a precise definition is required for the purposes of this chapter. The World Health Organization defines resistance as a heritable, statistically demonstrable decrease in sensitivity to a chemical of a pest population relative to a 'normal' susceptible population. Resistance in weeds occurs as a consequence of the application of a herbicide and resistant populations evolve that are unharmed by a concentration of herbicide that completely kills all unselected populations. Natural tolerance or resistance that is due to the usual physiological characteristics of unselected plants should be clearly distinguished from 'evolved resistance'.

The difficulty with this concept of resistance is that there is a continuum of responses. At the susceptible end of the scale there is a natural 'background' variability in the sensitivity to herbicides within unselected populations. Some selected populations of a weed species or certain crop cultivars may exhibit substantially reduced susceptibility (or conversely enhanced resistance) which has often been given the term 'tolerance'. Tolerance is a response where there is partial survival or some reduction in growth at a herbicide concentration that would normally give 100% kill of a susceptible population. At one end of the continuum of sensitivity, 'tolerance' merges with 'resistance' and at the other end 'tolerance' merges with 'susceptibility'. Any division is somewhat arbitrary.

The global distribution of herbicide-resistant weed species

The first confirmed occurrence of a herbicide-resistant strain of a weed species was in the state of Washington, USA, in 1968 when a triazine-resistant biotype of groundsel was found in a tree nursery. Within a few years reports of 1,3,5-triazine-resistant weeds were common in the USA, Canada and western Europe. Subsequently in the late 1970s and during the 1980s, triazine-resistant weeds have become common in eastern Europe and Israel and now occur in New Zealand and Australia.

Worldwide, more than 50 plant species have evolved strains that are resistant to triazine herbicides. The majority of these are dicotyledons but 15 species of grasses have also evolved resistance including seven different grass species in Israel that occur in simazine-sprayed roadside habitats.

In almost all cases, persistent triazine herbicides were used alone on the same areas for 5–10 years or longer, in orchards, vineyards, plant nurseries, along rights-of-way and most importantly in maize crops. There have been a multitude of occurrences of triazine resistance in North America and Europe in widely separated locations within a few years. Many of these cases must have involved independent evolution of resistance at each site, although there has also been lateral geographic spread, sometimes extensive, for example along railroads in the USA. In Hungary, atrazine-resistant common amaranth apparently evolved in many different locations and by 1979, 75% of the maize fields contained the atrazine-resistant biotype. With the exception of velvetleaf, resistance to triazine herbicides is maternally inherited and is coded by a chloroplast gene. The rapid and multiple evolution of triazine-resistance in some species may have been enhanced by a nucleus-coded plastome mutator gene similar to that which has been detected in triazine-resistant black nightshade.

Multiple resistances involving triazines and other photosystem-2 inhibiting herbicides have recently occurred in a few places. In Hungary, atrazine-resistant biotypes of fat-hen that evolved in maize crops, also became resistant to chloridazon, when rotations were introduced that included sugar-beet. Resistance to pyridate then evolved after several treatments in maize crops. Fat-hen strains with triple herbicide resistance had evolved rapidly in just a few years. The rapidity of evolution of multiple resistance was attributed to the presence of a plastome mutator gene in the fathen populations. Cross-resistances have been found in Israel in triazine-resistant blackgrass and in bristle-spiked canarygrass that were also resistant to the urea herbicide methabenzthiazuron and to diclofop-methyl. The latter compound is unrelated to the triazines and the cross-resistance was unexpected.

Cases of resistance to herbicides other than triazines have been slower to occur. However in the past decade the evolution of resistance to a wider range of chemical structures has become more common in a variety of locations worldwide. Resistance to paraquat has evolved in ten different species. Three species of *Conyza* have evolved resistant biotypes: hairy fleabane (*C. bonaniensis* — Egypt), Canadian fleabane (*C. canadensis* — Hungary) and Philadelphia fleabane (*C. philadelphicus* — Japan). Paraquat resistance has also evolved in American willowherb in Belgium and the UK, annual meadow-grass in the UK, *Hordeum glaucum* and capeweed in Australia. Considerably enhanced tolerance to paraquat has also been reported in strains of annual meadowgrass and perennial ryegrass in the UK. There is evidence of moderate cross-tolerance to atrazine in Canadian fleabane and hairy fleabane. Although paraquat is a non-persistant herbicide, growers caused a high selection pressure by repeated application of the herbicide through each growing season, and thus paraquat resistance evolved despite the original expectation that it was unlikely to occur.

There are several cases of resistance evolving to other herbicides. In cotton fields in South Carolina and Alabama (USA), populations of goose-grass have evolved resistance to trifluralin in many different locations. In Australia, populations of annual ryegrass (*Lolium rigidum*) resistant to diclofop-methyl have been found in many widely-scattered locations representing concurrent evolution. Resistance to diclofop-methyl also occurs in wild oat. This is a serious development because there is cross-resistance to several unrelated groups of herbicides with different mechanisms of action such as chlorsulfuron, fluazifop-butyl, haloxyfopmethyl, alloxydimsodium, and dinitroaniline herbicides. There is a problem in finding alternative herbicides that will give satisfactory control of diclofop-methyl-resistant populations of annual ryegrass.

The occurrence and spread of herbicide-resistant weeds in the UK

Triazine-resistant strains of weed species previously controlled by simazine were the first to be reported in the UK in 1982 and have become a problem in ornamental plant nurseries, bush fruit plantations and orchards. The number of years of treatment with simazine prior to the appearance of resistant strains has varied from 4 years in a Sussex vineyard to 15 years in some fruit and ornamental crops. Use of simazine for 6–10 years has commonly been sufficient to cause evolution of resistance.

Groundsel, annual meadow-grass and American willowherb were the first species to evolve triazine-resistant biotypes in the UK (Table 8.1). Resistant strains of groundsel are known to be particularly widespread, although simazine resistance is probably also widely distributed within the

Table 8.1 Occurrence of herbicide-resistant weed biotypes in the UK confirmed by experimental tests

Herbicide	Cross-resistance	Weed	Year found	Geographic location	Crop
Simazine	Other 1,3,5-triazines	Groundsel	1981	Widespread. Midlands, East Anglia, South England	Plant nurseries Bush fruits Soft fruits
Simazine	Other triazines	Annual meadow-grass	1982	Midlands, Southern England	Orchards Plant nurseries
Simazine	Atrazine	Pineappleweed	1983	Luddington E.H.S., Warwickshire (1 location)	Experimental plots
Simazine	—	Canadian fleabane	1984	Probably in many locations in the South of England	Bush fruits Soft fruits
Simazine	—	American willowherb	1982	Probably in many locations in the South of England	Bush fruits
Chlorotoluron	Diclofop-methyl* Pendimethalin* Chlorsulfuron* Imazamethabenz and some others such as terbutryn	Blackgrass	1984	Essex, Cambridgeshire, Buckinghamshire, Oxfordshire, Suffolk, Warwickshire	Winter wheat
Mecoprop	—	Chickweed	1985	Vicinity of Bath, Avon	Winter wheat
Paraquat	—	Annual meadow-grass	1978	Berkshire (1 location)	Horticultural nursery
Paraquat	Oxyfluorfen* Pyridate*	American willowberb	1983	East Malling Research Station Kent (1 location)	Experimental plots

* Intermediate degree of resistance (tolerance)

other two species. Many populations of groundsel and annual meadow-grass are known to have evolved resistance quite independently in different locations and although there has been local spread of the resistant biotypes by passive dispersal of seed and attached to farm machinery, there has been no scientific investigation to evaluate the significance of local dispersal of resistant strains.

Other species that have evolved triazine-resistant strains in the UK are Canadian fleabane and pineappleweed. The resistant biotype of the latter species occurred on experimental plots at Luddington Experimental Horti-cultural Station and has been reported resistant to simazine and bromacil in East Anglia (not confirmed by experiment) but does not occur at any other location worldwide.

Evolution of resistance to herbicides other than triazines has also occurred recently in the UK. Populations of blackgrass that exhibit sub-stantial tolerance or virtual resistance to recommended application rates of chlorotoluron occur in the vicinity of Peldon (Essex). Other populations from Tiptree (Essex) and Faringdon (Oxfordshire) exhibit enhanced tole-rance. The chlorotoluron-resistant biotype is also cross-resistant to several other herbicides including diclofop-methyl, pendimethalin, chlorsulfuron and imazamethabenz. There is evidence that the evolution of greater mono-oxygenase activity could be responsible for imparting cross-resistance to these other herbicides with varied modes of action. The mixed-function oxidase inhibitor l-aminobenzotriazole is synergistic to isoproturon phyto-toxicity in the resistant Peldon population but had little effect in the susceptible population. This suggests that resistance to chlorotoluron may depend on rapid degradation and de-toxification of the herbicide. An important consequence of these findings is that there is the potential for herbicide resistance in blackgrass to become a widespread problem.

A second case of herbicide cross-resistance was reported in 1985 from an area north of Bath. Two populations of chickweed were resistant to recommended rates of the phenoxyalkanoic acid herbicide, mecoprop, and were also resistant to the closely-related herbicides MCPA and dichlor-prop. However, resistance did not extend to other growth regulator her-bicides, viz. benazolin, dicamba and fluroxypyr. Thus there are available to the farmer several effective alternative post-emergence herbicides.

The only other occurrences of resistance in the UK are to paraquat. A population of annual meadow-grass became resistant to paraquat at a Berkshire nursery where the herbicide had been used very frequently during each growing season for several years. In addition to this resistant population, increased tolerance to paraquat was also recorded in annual meadow-grass populations as a consequence of the frequent use of the herbicide in a hop yard near Worcester and in hops elsewhere in southern England. In 1983, resistance to paraquat was found in a single population of American willowherb that had received repeated applications on experimental plots at the East Malling Research Station, Kent. Paraquat resistance is likely to remain rare and spasmodic since resistance will probably only evolve if a grower maintains a very high selection pressure by repeated applications of the herbicide through the growing season for several years.

In the UK, it is generally recognized that the occurrence of triazine-resistant weeds in perennial crops is a nuisance rather than a serious problem, since there are several effective alternative residual herbicides available. However, there is normally a cost since the alternatives may be much (up to 10 times) more expensive than simazine. However, the appearance of resistance to chlorotoluron in blackgrass and to mecoprop in chickweed, combined with the evidence of multiple resistance in both cases, gives cause for greater concern. If the multiple resistances involve the evolution of greater mono-oxygenase activity, then resistance to a variety of major herbicides used in cereal crops with different chemistries and modes of action, could become widespread. The choice of alternative herbicides would be very limited.

Genetics of resistance in weeds and crop plants

The majority of the published evidence of herbicide resistance (in contrast to tolerance or reduced sensitivity) in weed and crop species suggests control by single nuclear genes (sometimes with modifier genes present). However, there is evidence that all types of inheritance are possible, not only nuclear genes with dominance, partial dominance or recessiveness but also polygenic inheritance and non-nuclear inheritance in plastid genomes.

Some examples of types of inheritance of herbicide resistance are given below; a full review is not appropriate.

Enhanced tolerance to atrazine in velvetleaf is inherited as a partially-dominant allele, a single nuclear gene imparting control of the formation of the glutathione conjugate of atrazine. Inheritance of triazine resistance in soybean is controlled by a single nuclear gene and metribuzin resistance in tomato is inherited as a dominant allele, a single gene but with modifier genes present. It is much more common for triazine resistance to be imparted by maternally inherited plastid genes.

Resistance to paraquat in fleabane is apparently owing to a single gene. A dominant allele appears to pleiotrophically control three oxygen-detoxifying enzymes. This is the only case of paraquat resistance where the genetics of resistance have been elucidated. Other examples of nuclear monogenic inheritance include: (a) resistance to barban chlorosis in barley controlled by a dominant allele; (b) sensitivity to metoxuron in winter wheat controlled by a recessive allele; (c) a dominant allele for resistance to 2,4-D in sorghum; (d) picloram-tolerant mutants in tobacco, isolated from cell-suspension cultures, involved dominant gene mutations in four out of seven lines; and (e) monogenic resistance to paraquat in haploid mutants of a fern, *Ceratopteris*.

Examples of polygenic inheritance include foxtail barley in which at least three complementary dominant genes are involved in conferring tolerance to siduron, and in maize several genes are involved in conferring tolerance to diclofop-methyl with high levels of heritability.

Quantitative inheritance due to polygenes is usually related to variation in herbicide tolerance or differential sensitivity rather than high levels of resistance. Quantitative inheritance with reasonably high levels of heritability occurs in paraquat-tolerant perennial ryegrass and in cucumber tolerant to chloramben methyl ester. Relatively low heritability was found in flax tolerant to MCPA and also in other cultivars tolerant to atrazine. Tolerance to simazine in wild cabbage (*Brassica oleracea*), oilseed rape and groundsel is also of low heritability.

The genetics and molecular biology of resistance to triazine herbicides in weeds has been intensively investigated in recent years. With the exception of velvetleaf, triazine resistance is maternally inherited as has been demonstrated in several species including Chinese cabbage (*Brassica campestris*), *Amaranthus* spp., fat-hen, black nightshade and annual meadow-grass. In the latter two species, there was evidence of paternal transmission of resistance via pollen although only in a very small proportion (0.1%) of hybrid progeny.

In a few weed species, e.g. fat-hen and annual meadow-grass, there are biotypes with intermediate characteristics of triazine resistance. The intermediate characteristic is maternally inherited similarly to full triazine resistance and it is probable that a chloroplast gene codes for the intermediate characteristics in addition to a gene coding for full-scale resistance.

Herbicide cross-resistance

Weeds become resistant to more than one class of herbicide either because there is independent mutation and selection by two or more herbicides to which the species has been exposed or because there is cross-resistance so that selection by one class of herbicide confers resistance to herbicides with different biochemical modes of action. The former has not occurred in the UK but in Hungary a remarkable series of multiple resistances has evolved. These have been mentioned previously, namely fat-hen co-resistant to atrazine/chloridazon, atrazine/pyridate and a biotype with triple resistance to the three herbicides. Another interesting case in Hungary involves a biotype of Canadian fleabane discovered in 1986 in a vineyard, that was resistant to both paraquat and atrazine. Since both of these herbicides had been applied repeatedly for a period of 10 years, it seems likely that independent mutations and parallel evolution was responsible.

Occurrence of cross-resistance in weeds is more common than double or multiple resistance caused by independent selection of mutants. Important examples include annual ryegrass resistant to diclofop-methyl and to sulphonylureas, and blackgrass moderately resistant to chlorotoluron and also showing some cross-resistance to pendimethalin, diclofop-methyl, imazamethabenz and chlorsulfuron. As already discussed, such cross-resistances may be due to the evolution of greater mono-oxygenase activity. If cross-resistance based on enhanced mono-oxygenase activity becomes more widespread, then this would have potentially serious implications for weed control in cereal crops since many commonly-used products would be susceptible to this type of biochemical degradation.

Some herbicide cross-resistances are readily explicable because the resistance extends only to chemically closely-related herbicides. In the UK, a good example is resistance to mecoprop in populations of chickweed. The weed was resistant to the related phenoxyalkanoic acid herbicides, MCPA and dichlorprop, but not to other growth regulatory herbicides, fluroxypyr, dicamba and benazolin.

Hairy fleabane that is resistant to paraquat also exhibits enhanced resistance to atrazine, acifluorfen and sulphur dioxide. These compounds cause the generation of toxic active oxygen species in plants as does paraquat (see Chapter 7). Resistant biotypes of hairy fleabane apparently have elevated levels of three enzymes that are involved in the detoxification of active oxygen. The partial cross-resistance can be explained by the enhanced activity of these enzymes.

An investigation of cross-resistance among herbicides that inhibit photosystem 2, at both the level of the intact plant and of photosynthetic electron transport, showed that a similar mutation conferring resistance occurs in triazine-resistant biotypes of the weeds, green amaranth, fat-hen, groundsel and the crop, canola (*Brassica napus*). There appeared to be a shared chlorophyll thylakoid membrane binding site for the triazine, atrazine and triazinone, uracil, pyridazinone and urea herbicides.

Multiple resistance and cross-resistance between chemically-unrelated herbicides have only appeared in weed species during the last few years. A more widespread occurrence in the future would pose a serious threat to cereal crop production in particular.

Biochemical and physiological mechanisms of resistance

An understanding of biochemical mechanisms of resistance to herbicides coupled with genetics and molecular biology of resistance is an essential baseline for progress in the genetic engineering of herbicide resistance in crop plants and also for the development of long-term strategies to avoid evolution of resistant weeds.

Three main types of resistance mechanism are discussed below. These are based on triazine resistance in many different weed species, glyphosate resistance in bacterial cultures and paraquat resistance in a few species of weed.

Resistance to 1,3,5-triazine herbicides

The site of inhibition of 1,3,5-triazine herbicides is a protein that forms part of the photosystem 2 complex located within the chloroplast thylakoid membrane. The protein regulates the electron transport between photosystem 2 and the plastoquinone pool. Triazine herbicides bind on to the membrane protein, block the flow of electrons and therefore inhibit photosynthesis.

Mechanisms that confer resistance to triazines will involve either a reduction in the concentration of the herbicide at the binding site or a change in the molecular structure of the binding site. In resistant biotypes the structure and composition of the membrane protein has been altered. The gene which encodes the binding site protein mutates so that the nucleotide sequence of the chloroplast DNA of a resistant biotype differs from a susceptible biotype by a single amino acid substitution. Black night-shade, fat-hen, groundsel and smooth pigweed all have the same amino acid transversion. Serine is substituted by glycine in the resistant biotype. This must be the cause of the change in the herbicide affinity of the binding site that results in no herbicide binding.

Resistant algal mutants have different amino acid transversions and since in some weeds there is variation between biotypes in sensitivity and degrees of cross-resistance to other herbicides, it is possible that other amino acid transversions occur in addition to that of serine for glycine.

Resistance to paraquat

Paraquat competes for electrons from the primary electron acceptor of photosystem 1. There is production of superoxide radical anions and other active oxygen species that cause rapid death of cell membranes (see Chapter 7).

Uptake of paraquat is similar in both sensitive and moderately resistant lines of perennial ryegrass and in hairy fleabane it penetrates rapidly into leaves and chloroplasts of the resistant biotype.

At least three enzymes are involved in the de-toxification of the active oxygen species generated by paraquat. These are superoxide dismutase, ascorbate reductase and glutathione reductase. In the resistant biotype, enhanced levels of these enzymes in the plastid may de-toxify the active oxygen species. However, this simple explanation of the resistance mechanism is probably inadequate since there is evidence that in hairy fleabane paraquat may be excluded from the site of action in the chloroplasts by an unknown mechanism of sequestration (perhaps binding of paraquat to cell wall components). In other species resistance may be conferred by inhibition of uptake and distribution of paraquat to the site of action. Thus it appears that the complexities of the mechanism of paraquat resistance are not yet fully understood. Investigation of the resistance mechanism will need to involve sensitive and resistant biotypes in the range of species that have evolved resistance.

Resistance to other herbicides

In addition to the triazines and paraquat, mechanisms of resistance in weeds (and in transgenic crop plants in some cases) have been investigated for several other herbicides. However, conclusions are often tentative and further work will be required before the resistance mechanisms are clearly elucidated. An understanding of glyphosate resistance is emerging quite rapidly. Work with bacteria, fungi and cell cultures of higher plants has shown that glyphosate is toxic because it inhibits the enzyme 5-enol-pyruvylshikimic-3-phosphate synthase (EPSPS) which catalyzes the conversion of shikimic-3-phosphate to 5-enolpyruvylshikimic-3-phosphate, a step in aromatic amino acid synthesis (see Chapter 7). Resistance to glyphosate in bacteria, in plant-cell cultures and in plants derived from cell culture appears to result from over-production of the enzyme EPSPS. A mutant gene encoding a glyphosate-resistant EPSPS has been isolated from a bacterium and introduced into tobacco cells. The plants that were regenerated from the transformed cells were two- to threefold more tolerant to glyphosate than control plants.

Resistance to chlorotoluron in blackgrass is less well understood but may be linked to de-toxification involving ring methyl oxidation (as occurs in wheat and barley) or N-demethylation. The P450 mixed-function oxidase inhibitor, 1-aminobenzotriazole, enhanced herbicide activity against the chlorotoluron-resistant biotype but not in the susceptible biotype.

The majority of recent research on biochemical mechanisms of resistance and metabolism is concerned with the pursuit of new herbicide-resistant crop cultivars rather than understanding resistance mechanisms in weed species. Knowledge of the mode of action of herbicides assists in the identification of genes that confer enhanced resistance. A typical example is the recent work on resistance to sulphonylurea herbicides which has shown the enhanced resistance in lines of bacteria, tobacco cell lines and seedlings of thale cress link to the enzyme acetolactate synthase. The genes that encode the wild type and mutant acetolactate synthase have been isolated. Rapid progress is being made in the conferring of herbicide resistance on crops. It is likely that an increasing knowledge of the biochemistry of resistance mechanisms and their genetic control will derive mainly from the work on the genetic engineering of resistance.

The population ecology of herbicide-resistant weeds and natural selection

A detailed understanding of the dynamics of resistant and susceptible phenotypes in weed populations is needed to measure the selective forces imposed by application of herbicides and to determine the potential for the evolution of resistance. Demographic studies also provide information on the relative ecological fitness of resistant and susceptible phenotypes and demonstrate the population decay rates of these phenotypes when a herbicide is temporarily withdrawn from use.

The whole plant life cycle should be examined since selection for, or against, evolution of resistance can act on dormant viable seed populations and also will operate through plant mortality, mainly on seedlings but also later in plant development and may operate on plant fecundity. The combination of these interacting processes will be a measurement of differential rates of increase or decrease in resistant and susceptible phenotypes in populations.

The first requirement for evolution of resistance to herbicides is that genetic variation is present in most natural populations of weeds and that heritability of resistance is relatively high. This must be so for triazines, bipyridiniums, aryloxyphenoxypropionates and probably for sulphonylureas. The duration and intensity of selection are also very important in determining rates of evolution of resistance. Selection exerted by herbicides is episodic and after one or two herbicide treatments there will often be scope for seedling recruitment later in a growing season when the phytotoxicity of a rapidly-degraded herbicide has disappeared. The same applies to more persistent herbicides although the window of selection for resistance will obviously be longer. Recruits that escape the selective effects of a herbicide will often contribute susceptible progeny to seed populations and reduce the proportion of resistant phenotypes. However, weed seedlings emerging later in the growing season often contribute few progeny and thus may have a very limited influence on selection pressure.

Another important influence will be the release of competition that occurs after herbicide treatment. If the density of the weed population is greatly reduced, both resistant and susceptible plants will experience less inter-specific competition and plasticity in plant growth may cause an increase in the fecundity of survivors. On the other hand, crop competition may reduce the fecundity of surviving weeds, particularly weakened susceptible plants, and therefore intensify the selective effect of the herbicide. In this situation the relative competitive fitness of resistant and susceptible phenotypes will be crucial in determining the proportion of resistant seed entering the soil seed population.

The majority of the evidence concerning the demography of herbicide-resistant weed populations and the duration and intensity of selection, comes from studies of triazine-resistant populations of groundsel. The population dynamics of groundsel have been examined in blackcurrant plantations with different experimental herbicide management of the weed. The control of weeds was either (a) by rotovation in spring (no herbicide, (b) by early spring simazine application, or (c) by spring application of simazine and directed application of paraquat in mid-summer.

In the unsprayed sites, seed production and dispersal of the susceptible biotype were restricted to plants arising from seedlings that germinated in April or May whereas in simazine-treated areas, the seasonal phenology of the susceptible biotype was altered so that it behaved as a winter annual. Susceptible seedlings that emerged during a hazard-free period in August

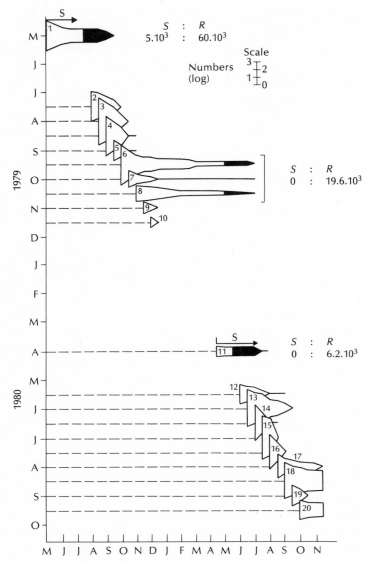

Fig. 8.1 Time tracks of cohorts in mixed populations of simazine-resistant and susceptible phenotypes of groundsel growing in experimental blackcurrent plantations. Two percent of the seed population sown in May 1979 were of the resistant phenotype. Solid areas in cohorts indicate flowering and seed dispersal. $S : R$ indicates proportions of susceptible and resistant phenotypes in progeny. Data are per m². The management regime was spring simazine application. 2.24 kg ha⁻¹. indicates time of application and duration of effect. From Mortimer (1983) by permission of CAB International.

and September established and survived over-winter to reproduce in the spring of the following year. This provides temporal escape from selection by simazine and is the only way in which the susceptible biotype can persist in simazine-treated areas apart from the occasional spatial escape.

The resistant biotype behaved as both a summer and a winter annual (Fig. 8.1) with partly overlapping generations in each 12-month period. However, over-wintering susceptible phenotypes had a higher survivorship and fecundity than resistant ones. They were apparently more fit (defined as the probability of leaving progeny) in an adverse environment.

An experimental study examined the survival and reproduction of resistant and susceptible phenotypes of groundsel when sown separately into patches within large areas with contrasting management of the weed flora. The area was rotovated and sprayed with glyphosate to kill the existing weed flora and plots were either (a) left undisturbed so that there was natural development of an undisturbed weed flora or (b) plots were kept free of all plants except groundsel by hand weeding, or (c) plots received a spring simazine treatment at a rate of $2.5\,kg\,ai\,ha^{-1}$.

In the undisturbed sites with intra-specific competition from a mixed species vegetation, all resistant cohorts of seedlings failed to survive and produce seed. Even the minor disturbance caused by hand removal of other weed species caused a higher mortality rate and lower fecundity in the resistant cohorts of groundsel. The mean fecundity per plant was 925 seeds in comparison with 1480 in the simazine-treated plants. The fitness of resistant biotypes in a competitive field environment was zero for the above-ground groundsel population; the seed population was not examined in this study.

A review of evidence from all the literature on herbicide resistance to compare the performance of resistant and susceptible biotypes demonstrated that relative fitness depended on the character measured and whether tested in monoculture or in mixture experiments. Realistic data measure comparative reproductive output in several different populations, involve mixture experiments in addition to monocultures and have resistant and susceptible biotypes originating in the same locality, preferably near isogenic.

Triazine-resistant biotypes were generally less fit than susceptibles, but in some experiments there was no statistical difference between the biotypes (e.g. triazine-resistant common amaranth, fat-hen) and in a few experiments the resistant biotype was more fit (e.g. certain triazine-resistant populations of fathen and bristle-spiked canarygrass).

A knowledge of the persistence and flux of resistant and susceptible seed is important in determining management strategies. A study of the dynamics of triazine-resistant and susceptible seed (achene) populations of groundsel showed that it is the flux of seed in the surface seedbank (0–2 cm) that is of crucial importance to the annual renewal of adult plant populations in simazine-treated soft fruit plantations. When seed was buried lower in the soil profile (7 cm), the rate of seed decline was substantially lower than at the surface and the rate of loss of seed of the resistant biotype was significantly less than the susceptible (Table 8.2). In undisturbed plots with vegetation cover, the mean half-life of a resistant seed population was 1633 days at 7-cm depth whereas that of the susceptible population was 1189 days. Under a soil rotovation treatment, overall mortality was greater, but at 7 cm, the resistant seed population had a longer half-life (884 days) than the susceptible (696 days).

The implication of these findings is that management practices that cause deep burial of seed, will result in more rapid depletion of the susceptible biotype than the resistant. If deep-soil cultivation subsequently returned resistant seed close to the surface, then use of triazine herbicides would lead to a rapid increase in the population of resistant biotypes.

Field studies that quantify selection by post-emergence herbicides with low persistence are rare. It has been suggested that selection may be weaker than is commonly assumed because susceptible phenotypes that escape herbicide treatment, either spatially or by seedling emergence after the herbicide phytotoxicity has disappeared, may through developmental plasticity produce more seeds per plant at low weed densities.

The dynamics of chlorotoluron-resistant and susceptible populations of blackgrass have been experimentally examined in a wheat crop. The intensity of selection for resistance in the field was substantial in full recommended rate treatments of chlorotoluron ($2.75 \, kg \, ai \, ha^{-1}$) and isoproturon ($2.1 \, kg \, ai \, ha^{-1}$).

Selection measured as plant fecundity (seeds m^{-2}) was generally more severe than measured as population mortality. Plants weakened by the herbicide treatment may have been further suppressed by crop competition. Depending on the sowing density of the weed, herbicide selection against the susceptible biotype based on mortality was -0.52 to -0.93 whereas based on plant fecundity selection against susceptibles was -0.82 to -1.00. Thus although some individuals escaped death, their fecundity was greatly reduced (Fig. 8.2).

In the absence of a herbicide, selection against the resistant biotype was inconsequential. There was little difference between resistant and sus-

Table 8.2 The decline of buried seed (achene) populations of groundsel in a blackcurrant plantation under three management regimes. Mean rates of decline are expressed as a half-life in days. Data in parentheses are the percentage loss per annum

Management regime	Seed burial at 1 cm depth		Seed burial at 7 cm depth	
	Simazine-resistant	Simazine-susceptible	Simazine-resistant	Simazine-susceptible
Soil rotovation	329(53.7)	454(42.7)	884(24.9)	696(30.5)
Spring simazine	245(64.4)	273(60.4)	986(22.6)	527(38.2)
Undisturbed control	501(39.6)	433(44.2)	1633(14.4)	1189(19.2)

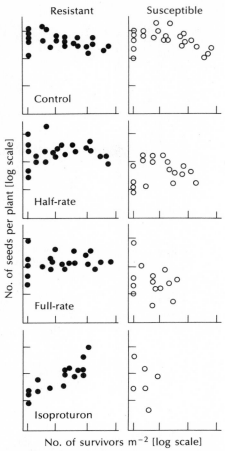

Fig. 8.2   The relationship between mean seeds plant[−1] and number of surviving plants at harvest in field populations of herbicide-resistant and susceptible biotypes of *blackgrass*, exposed to four herbicide treatments. Half- and full-rate refer to chlorotoluron herbicide, whilst isoproturon was applied at full-rate. Unpublished data courtesy of P.F. Ulf-Hansen.

ceptible plants although at the highest population density (1000 seeds sown m[−2]) mortality of the resistant biotype was higher.

   Given the paucity of experimental studies, it is not possible to make general conclusions or meaningful predictions concerning the factors that regulate the relative frequency of resistant and susceptible phenotypes in weed populations. Considerably more research is required to unravel the dynamics of herbicide-resistant and -susceptible populations and their responses to herbicides.

Management strategies to avoid or contain
herbicide resistance in weed populations

Long-term management strategies to avoid, delay or contain the eventual evolution of herbicide resistance are difficult to formulate when so little is known about the genetics of resistance in particular classes of herbicides and about the population dynamics of the weeds. Usually, only after herbicide resistance has appeared can sensible strategic decisions be made about how to prevent it.

If resistance is imparted by polygenes or gene amplification/duplication then high initial dose rates of the herbicide would delay the onset of resistance. The worst scenario would be for a grower to use initially relatively low doses of herbicide and then to increase dose rates when it became more difficult to kill the weed population. Selection would be intensified. However, even if a grower uses high initial doses, when resistance eventually evolved, the increase in frequency of resistant phenotypes will be very rapid and by the time a grower realizes, it will be too late to avoid a high population density of resistant individuals.

Single major gene resistances (particularly if a recessive allele) are more tractable to manage since the onset of resistance will be delayed by low herbicide doses. Application of herbicide at less than the full recommended rate, possibly involving the use of herbicide mixtures or synergists, is a practical management tactic to delay or avoid the evolution of resistance. The majority of herbicide resistances that have been investigated genetically appear to be major gene resistances. This is encouraging. Lower herbicide doses can be used if a synergist compound is available. An example is tridiphane which prevents grasses from metabolizing atrazine. This provides control of grasses at lower rates of atrazine and consequently lowers selection pressure on broad-leaved weed species, although selection intensity on grasses remains high. They would have to co-evolve resistance to atrazine and tridiphane and the occurrence of co-resistance would be delayed if genetic variation for resistance to tridiphane was more rare than for atrazine. Recent work has demonstrated that adding chelating agents can suppress paraquat resistance by removal of copper or zinc from superoxide dismutase or of copper from ascorbate peroxidase. It may be possible to achieve weed control with much lower rates of paraquat. However, the use of synergists in this way is a new concept with no practical applications at present as a management tactic to the onset of resistance.

Wherever possible, persistent herbicides should be avoided. This will reduce the window of selection to a minimum. It is probably not a co-incidence that the phenoxyalkanoic acids 2,4-D and MCPA have been in use for almost 40 years without any significant evolution of resistance and similarly the benzonitriles have a long history of use in cereal crops with no evidence of resistance appearing. Recently, predictions were made that widespread use of sulphonylurea herbicides that have considerable persistence will inevitably result in the evolution of resistance to this class of herbicide. Already there are reports of resistant weed populations appearing in the USA.

The rotation of herbicides in crop monocultures or alternatively more emphasis on the rotation of crops (and therefore herbicide rotation) are strategies that were originally proposed many years ago as the most practical approach to delaying (or avoiding completely) the evolution of resistance. In the UK, resistance to triazine herbicides (principally simazine) occurred long after it had evolved in North America and in many European countries although the chronological time-scale of usage was the same in the UK as elsewhere. It was argued that in the UK, growers had utilized a variety of herbicides in addition to simazine either in rotation or as an additional treatment in the same growing season. A consequence was greatly reduced selection pressure for triazine resistance. The majority of triazine-resistant weeds show reduced ecological fitness in the absence of triazine. Fitness (measured as seedling survival) of the resistant phenotype is further reduced by mechanical disturbance and by some other classes of herbicide. These factors would combine to reduce the probability of selection for resistance.

Triazine usage on maize in the Midwest area of the USA has only resulted in occasional instances of resistance. Although maize is a major crop in this area, it is generally grown in rotation with other crops and in a large percentage of the area, alachlor or metalachlor is applied for grass weed control in addition to atrazine. Although this has not been a deliberate management strategy for the avoidance of resistance, so far the approach has been successful whereas triazine-resistant weeds have become very common in maize monoculture elsewhere in North America and in Europe. Many growers may be reluctant to switch to crop rotations if investment in additional harvesting and crop processing machinery is required. There will be a long-term economic cost whether a grower develops a crop-rotation programme or tries to live with a resistant-weed problem.

In the case of perennial crops such as vine, soft fruits, top fruit orchards, tree nurseries and citrus fruits, crop rotation is clearly not possible and a strategy of using herbicide rotation and mixtures will be necessary. Mixtures must involve different classes of herbicide with different modes of action. The probability of evolving resistance independently to two or three classes of herbicide will be greatly reduced. There will probably be an economic penalty for the grower since alternative or additional herbicides will often cost more.

Chemical manufacturers, growers and their advisors will have to develop strategies that avoid or delay the onset of resistance by learning from previous examples of the evolution of resistance. As more information becomes available concerning the biochemistry of a compound's action, the genetics of resistance and the dynamics of resistant and susceptible phenotypes in field populations, it will become more feasible to predict the probability that resistance will evolve in a particular herbicide/crop combination.

Plant breeding and genetic engineering for herbicide resistance in crops

During the past 15 years there has been steadily increasing interest in the chemical industry and by some plant breeders in the development of crop cultivars that are resistant to herbicides. The chemical industry has become interested in the biotechnology of resistance in crops because the costs of introduction and registration of new conventional herbicides are continually increasing. In addition, herbicide-resistant crops may prolong the market share of well-established, low-cost herbicides. This approach would be reinforced if there is future withdrawal of hitherto valuable herbicides for environmental reasons. Herbicide-resistant crop cultivars could then provide greater flexibility in the choice of chemical weed control.

Four main approaches to the development of herbicide-resistant cultivars are discussed below.

Conventional plant breeding

In a few crops, herbicide resistance or tolerance has been detected either in the gene pool of the crop species or in a closely-related non-crop species. Conventional breeding programmes involving crossing, back-crossing, progeny testing and further selection have produced a few new resistant cultivars. A classic example is the metribuzin-resistant soyabean cultivar 'Tracy M'.

A population of perennial ryegrass that had evolved tolerance after repeated field application of paraquat has been used as the base material in a breeding programme to create new paraquat-tolerant cultivars of perennial ryegrass for amenity and agricultural uses. It was then possible to spray newly-established swards of perennial ryegrass with paraquat to prevent ingress of unwanted grasses and broad-based species. Unfortunately, the original amenity cultivars were insufficiently persistent (low ecological fitness) and further hybridization and breeding was necessary to improve the performance of the paraquat-resistant cultivars.

The development of triazine-tolerant cultivars of spring oilseed rape (canola), rutabaga (swede) and broccoli (*Brassica oleracea*) in Canada by conventional plant breeding illustrates most vividly the potential and the pitfalls of breeding herbicide-resistant crop cultivars. In Canada, canola is an important oilseed crop, but there are particular problems associated with its cultivation. Soil residues of triazines following maize may preclude canola production and there are several weed species including wild mustard and stinkweed where adequate control by existing registered herbicides is not possible. Adulteration of the oilseed crop by cruciferous weed seeds reduces oil quality. Using conventional plant-breeding methods, the triazine-resistant plastid genome was transferred from wild *B. campestris* (birds rape) initially into cultivated varieties of *B. campestris* (Polish rape) and subsequently into other cruciferous crops. The triazine-tolerant canola cultivars, 'OAC Triton', 'Tribute' and 'Triumph' were released to Canadian farmers in the mid-1980s. Weed-control benefits included not only control of wild mustard and stinkweed but also control of another serious weed, wild prosso millet using sethoxydim.

Unfortunately the triazine-resistant chloroplast genome mutant usually confers low fitness. The result is that triazine-tolerant canola cultivars have poorer agronomic performance, lower oil yield, delayed maturity and poorer seedling vigour than the equivalent triazine-sensitive varieties. Unless the weed problems of the crop are extremely severe, there is no point in breeding herbicide-resistant varieties. They must be as good agronomically as existing varieties. The justification for breeding herbicide-resistant cultivars may also be negated by the introduction of a new effective chemical herbicide that controls problem weeds. The impending introductions of a new selective sulphonylurea herbicide may reduce the advantages for weed control from the use of triazine-tolerant rapeseed cultivars.

Cell-culture selection and protoplast fusion

As a system for the selection of herbicide resistance, plant-cell suspension cultures have considerable advantages over intact plant selection. The growth of cell populations is very rapid (doubling every few days) and countless millions of cells may be uniformly treated with herbicides at high selection pressures.

Herbicide-resistant cell lines will only be useful for the eventual development of herbicide-resistant crop cultivars if there is a permanent heritable change that is genetically transmitted by plants regenerated from the cell culture. Sometimes resistance is not stable and is lost following regeneration of intact plants. The potential of herbicide-resistant cell lines was previously restricted since many crop species cannot be regenerated from cell lines, although recently rapid progress has been made in determining procedures for the regeneration of many species and cultivars of crop plants from cell cultures. It is now also possible to isolate genes for herbicide resistance from plant cell lines and use genetic engineering to create new transgenic crop plants. Thus the requirement for regeneration of plants from resistant cell lines is less important than in the past.

Tobacco is one of the easiest species to regenerate from cell culture and cell lines resistant to picloram and sulphonylureas have been selected. Sulphonylurea resistance was retained in field-grown regenerated plants and genetic crosses demonstrated that the resistant phenotype was owing to a single dominant or semi-dominant nuclear mutation. Recently, sulphonyl urea-resistant cell lines of soyabean have also been selected. Crop cultivars resistant to sulphonylurea herbicides are being developed commercially but reports of recently-evolved sulphonylurea-resistant weed populations cast doubt on the long-term usefulness of crop cultivars resistant to this chemical structure.

A commercial plant breeding programme is in progress in the USA to introduce an imidazolinone-resistant nuclear gene into maize. The gene originated in selected maize cell cultures that had >100-fold resistance to imidazolinones. The resistant alleles have been transferred from regenerated maize plants in a back-crossing breeding programme. There is also cross-resistance to sulphonylureas because the mode of action is similar to that of imadazolinones, involving inhibition of the enzyme acetolactate synthase (see Chapter 7).

When resistance has evolved in a plant species, it may be possible to transfer the resistance to another susceptible species which would not normally successfully cross-breed with the resistant species. This may be achieved using inter-specific protoplast fusions. It is only worth attempting when usual molecular procedures for genetic engineering are not feasible as for example with chloroplast inherited genes. Triazine resistance has been introduced into potatoes and tobacco using protoplast fusion. It is particularly important when transferring triazine resistance that genetic material is selected that does not endow reduced fitness in terms of the growth rates, seed yields or biomass of plants. Protoplast fusion will only have a limited potential for the production of new herbicide-resistant crop cultivars.

Genetic engineering of herbicide resistance

There has been rapid progress in recent years in the development of gene-transfer systems for higher plants. Consequently, genetic engineering of resistant crop cultivars has become a commercial goal for some biotechnology and large chemical companies. There has been a major research effort to engineer glyphosate resistance in transgenic plants. The research was stimulated initially because glyphosate inhibits the shikimate pathway enzyme EPSP synthase in both bacteria and in higher plants. Thus bacterial mutant genes encoding a resistant enzyme have been introduced into tobacco cells and into transgenic tomato plants. A plant EPSP synthase gene was isolated from a cell line of *Petunia*. Regenerated petunia plants (and transgenic tobacco plants) were shown to be moderately resistant to $0.9\,kg\,ai\,ha^{-1}$ glyphosate. However, there was not full expression of glyphosate resistance since the growth of petunia was reduced relative to unsprayed controls.

Recently, a mutant EPSP synthase gene has been located that encodes an enzyme that is 1000-fold less sensitive to glyphosate inhibition than the wild-type petunia enzyme. Tobacco plants that express the mutant gene are not injured by $0.9\,kg\,ai\,ha^{-1}$ of glyphosate. It seems that genetic engineering methods involving EPSP synthase show potential for the eventual development of glyphosate-resistant crop cultivars. However, it has been suggested that utilizing 'glyphosatases' might provide an alternative approach since glyphosate is biodegraded by soil micro-organisms.

Development of genetically engineered sulphonylurea resistance in crop plants is now a real possibility. Mutant acetolactate synthase (ALS) genes that confer resistance have been isolated from thale cress and tobacco cell lines. Transgenic tobacco plants have been produced that show substantial enhanced resistance to sulphonylurea herbicides. Commercial varieties of sulphonylurea-resistant tobacco, cotton and soyabean are currently being developed in the USA.

Other examples of genetic engineering of resistance that are being developed for incorporation in crop plants include the production of imidazolinone-resistant maize and transgenic plants expressing modest resistance to phosphinothricin and the detection of bacterial genes encoding enzymes capable of degrading bromoxynil. The gene confers enhanced resistance in transgenic tobacco and tomato plants.

Since the majority of chemical companies now own a plant-breeding firm, it appears likely that growers will be offered packages of compatible herbicide/crop cultivar combinations. However, it will remain relatively expensive to breed new cultivars that incorporate resistance to a herbicide, but are as good in terms of agronomic performance and quality of product, as other recommended varieties.

There are several potential problems arising from the extensive use of herbicide-resistant crop cultivars:

1 Resistance genes often reduce crop yield and sometimes also the quality of harvested crop products. This is not always the case and it will only be worthwhile introducing resistance genes into crops that do not cause unfavourable correlated responses.

2 Out-crossing between herbicide-resistant crop plants and closely-related wild relatives that are weed species is a risk that would have to be closely examined before a new herbicide-resistant cultivar was released to growers. Triazine resistance is maternally inherited and there is no risk of transfer from resistant cultivars to related weed species.

3 The phenomenon of cross-resistance has several implications for the development of herbicide-resistant crop cultivars. Introgression of crop plant genes which confer resistance to closely-related weeds will create additional economic problems for growers, if such genes also confer cross-resistance to other classes of herbicide. Herbicide-resistant crop volunteers that are weeds of other crops such as barley, potatoes or oilseed rape, will be more difficult to eradicate if they possess cross-resistance. There would be less economic incentive for plant breeders controlled by a chemical company to market resistant crop cultivars if they proved to be cross-resistant to other widely-used herbicides, since the marketing advantage of a specific compatible cultivar/herbicide package, would be undermined.

4 The other major potential drawback to the widespread use of herbicide-resistant crop cultivars is the obvious potential for the parallel evolution of resistant weeds. Recurrent use of resistant cultivars in combination with the appropriate herbicide will impose strong selection pressures on weed floras. When the biochemical site of action of a herbicide is controlled by genes encoding particular enzymes, they will be ideal selective agents. For example, the increased use of sulphonylurea and imidazolinone herbicides would surely result in the more widespread evolution of resistant weed populations. Resistance to sulphonylureas in weed populations has already been reported in the USA.

To conclude, it appears unlikely that herbicide-resistant crop cultivars will become widespread in the near future. Even when resistant genes have been successfully incorporated into plant-breeding programmes, herbicide-resistant cultivars will probably fulfill a specialist role in cropping programmes, rather than becoming mainstay varieties of major crops.

Acknowledgements

I would like to thank Dr P.F. Hansen and Dr D. Watson for permission to use their previously unpublished data.

Note added in proof

During the past 6 months new information has become available concerning occurrences of potentially serious resistances in weed. In 1987 and 1988 weed biotypes resistant to sulphonylurea herbicides and other acetolactate synthase inhibitors (e.g. imidazolinones), evolved in winter wheat growing areas in the USA and Canada. Prickly lettuce resistant to sulphonylureas was first discovered in Idaho after four years of commercial use of the herbicide, resistant Kochia occurs in North Dakota and several other states, resistant Russian Thistle occurs in Kansas and resistant chickweed occurs in Alberta. These are the first indications of what may become a major resistance problem in the USA, Canada and Europe within a few years.

In the UK there has been a considerable recent expansion in the occurrence of blackgrass showing partial resistance to chlorotoluron. Resistant populations have been detected in Buckinghamshire, Cambridgeshire, Essex, Lincolnshire, Oxfordshire, Suffolk and Warwickshire, involving 17 farms to date. It is probable that most of the occurrences were independent evolutionary events and further cases will almost certainly arise in the future.

Further reading

A.M. Mortimer (1983). On weed demography. In Fletcher, N.N. *Recent Advances in Weed Research*, ch. 2. CAB International, Wallingford.

9 / Herbicides in soil and water

D. RILEY AND D. EAGLE*

*ICI Agrochemicals, Jealott's Hill Research Station, Bracknell, Berks RG12 6EY; and *MAFF ADAS National Pesticide Residues Unit, Government Buildings, Brooklands Avenue, Cambridge CB2 2DR*

The effectiveness and optimum methods of using soil-applied herbicides depends on their persistence, distribution and movement. The environmental safety of all herbicides depends on their behaviour in both soil and water.

Soil

Processes of inactivation (Fig. 9.1)

Herbicides may be lost from soil either by physical removal of the unchanged molecule or by degradation. Their availability and thus activity, to plants and other organisms, is related to their concentration in the aqueous and/or vapour phases. Thus adsorption is included here as a mechanism of physical removal.

Adsorption

In soil, herbicides are distributed between the solid, liquid and gaseous phases. This is normally described by partition coefficients, although it is important to recognize that a true thermodynamic equilibrium rarely exists in soil. The most important is the adsorption coefficient, K_d, because the concentration of most herbicides in the soil water controls their biological availability and mobility. The K_d is defined as the concentration of pesticide adsorbed to the soil particles divided by the concentration in the equilibrium solution (Fig. 9.2); thus K_d values are the highest for strongly

243

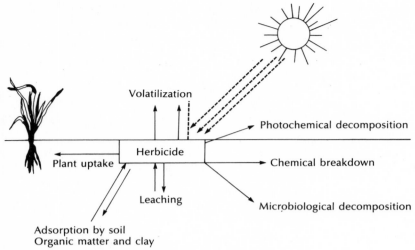

Fig. 9.1 Processes of herbicide inactivation.

adsorbed chemicals. If the adsorption isotherm is not linear, i.e. the value of K_d depends on the amount of pesticide added, then the Freundlich adsorption coefficient should be calculated using the equation:

$$\frac{x}{m} = kC^{1/n}$$

where x is the quantity adsorbed by mass m in equilibrium with concentration C; k and n are constants

Adsorption coefficients are determined by shaking the herbicides with dilute soil slurries until equilibrium is approached; this is normally achieved in a period ranging from less than one hour to a day. Desorption coefficients are measured by replacing the equilibrium solution with water and re-equilibrating. The adsorption process is not always completely reversible, particularly if the herbicide has been subjected to wetting/drying cycles under field conditions. A proportion of some herbicides, such as prometryn, can become so strongly adsorbed that they can only be released by destroying the binding sites, e.g. by refluxing the soil with strong alkali or high-temperature distillation. Such residues are referred to as 'bound' residues.

Many of the uncharged soil applied herbicides, such as triazines and ureas have adsorption coefficients in the range 1–20.

Herbicides which exist as anions at soil pH values are only weakly

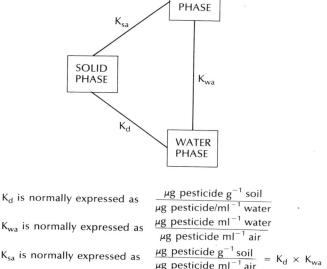

K_d is normally expressed as $\dfrac{\mu\text{g pesticide g}^{-1}\text{ soil}}{\mu\text{g pesticide/ml}^{-1}\text{ water}}$

K_{wa} is normally expressed as $\dfrac{\mu\text{g pesticide ml}^{-1}\text{ water}}{\mu\text{g pesticide ml}^{-1}\text{ air}}$

K_{sa} is normally expressed as $\dfrac{\mu\text{g pesticide g}^{-1}\text{ soil}}{\mu\text{g pesticide ml}^{-1}\text{ air}} = K_d \times K_{wa}$

Fig. 9.2 Partition coefficients of herbicides in soil.

adsorbed, unless they contain functional groups which bind to the soil, e.g. the P–O group in glyphosate is thought to bind to iron and aluminium sesquioxides. Examples of anionic herbicides which are weakly adsorbed by soil are 2,4-D, dicamba, dalapon and chlorsulfuron.

Uncharged molecules, such as triazines (at pH values >5) and the ureas are adsorbed by a combination of low-energy bonds, including non-specific forces of the van der Waals type, hydrogen bonds, and charge transfer. In addition, if the adsorption of a herbicide displaces water molecules from the surface, there is an increase in the entropy of the system which is therefore then more stable. This process of entropy generation is sometimes described as 'hydrophobic bonding'. The adsorption coefficient increases with the lipophilicity of the molecule, as measured by its octanol/water partition coefficient, K_{ow}. In some series of structurally-related molecules there is a good correlation between their lipophilicity and adsorption.

The adsorption coefficients of neutral and anionic molecules are usually directly proportional to the organic matter content of soils and the adsorption coefficient K_{oc} is less dependent on soil type than K_d, where K_{oc} is 100 times the K_d divided by the percentage organic matter in the soil.

Nevertheless, for a particular compound, K_{oc} can still differ several-fold between soils.

Soil organic matter and, in most temperate soils, the clay fraction have a nett negative charge. Positively-charged herbicides, such as the bipyridinium compounds and materials which can accept a proton at a suitable pH such as amitrole and the triazines, are strongly adsorbed by soil. They can be more strongly adsorbed than inorganic cations. Some cationic herbicides, such as divalent paraquat, are bound very strongly, owing to the formation of charge-transfer complexes, in addition to coulombic forces; the binding is so strong that it can only be released by refluxing the soil in strong acid to destroy the soil particles, particularly the clay minerals. These 'bound' residues have no residual activity in soil.

The herbicidal activity of fairly strongly adsorbed neutral molecules, such as trifluralin, depends on their relatively high vapour pressure. Thus the partition of the chemical between the gaseous and solid phases, K_{sa} (see Fig. 9.2) is more important than partition between the soil water and solid phase, K_d.

Considerable caution is needed when extrapolating data from laboratory studies done under near-equilibrium conditions to the field. In the field, chemicals are not uniformly distributed in the soil. Chemicals with a low solubility might also be present undissolved in the solid phase.

Degradation

Decomposition is the major route for herbicide disappearance from soil. To assess the effectiveness and safety of a herbicide, information is needed about both the pathways and rates of degradation. Photochemical degradation can be important for herbicides on plant surfaces but it is probably insignificant for most soil residues, particularly when they have been incorporated into the soil by soil cultivation or leaching. The main degradation mechanisms in soil are chemical and microbial. To estimate their relative importance, degradation in unsterile soil is compared with that in the same soil which has been sterilized by heat treatment or irradiation; however, the validity of such comparisons is debatable.

Many herbicides contain chemical groupings which are susceptible to hydrolysis but in aqueous solution at pH values similar to those in soils, such hydrolyses would be slow. In soil, however, many types of surface are present and there are many different species in solution so catalysis may well occur. If hydrolysis occurs in solution or is catalyzed by mineral surfaces in the soil, increasing soil organic matter will tend to slow the reaction through increasing adsorption and reducing the concentration of herbicide in solution. On the other hand, if the reaction is catalyzed by functional groups in soil organic matter, an increase in organic content of the soil may increase the rate of hydrolysis. It has been suggested that organic surfaces (particularly carboxyl groups) may catalyze the hydrolysis of some triazines. Degradation pathways can be different under aerobic and anaerobic conditions. For example, under reducing conditions nitro groups might be reduced to amines.

Microbial enzymes, both intra- and extra-cellular, are undoubtedly responsible for the degradation of many compounds in soil and organisms capable of metabolizing many of the common herbicides have been isolated and identified. Even when chemical degradation is responsible for some steps in the degradation pathway, micro-organisms are involved in further degradation steps, including mineralization to carbon dioxide. In some cases the herbicide probably acts as a substrate for micro-organisms which obtain energy from it. In others, microbial enzymes degrading natural substrates in the soil coincidentally carry out similar transformations on structurally-related 'foreign' molecules, a process known as 'co-metabolism'. Possibly simazine, monuron, diuron and other ureas and triazines fall into this group. Such compounds are usually of relatively long persistence.

Repeated treatment of soil with some herbicides can induce them to degrade subsequent applications more rapidly. The process by which the micro-organisms adapt to the herbicide is not fully understood. The phenomenon was first demonstrated and extensively studied for the phenoxyalkanoic acid herbicides in the laboratory. This was mainly of academic interest since they are primarily used post-emergence. However, in the USA, repeated treatment of some soils with thiocarbamate herbicides, such as EPTC, has resulted in their accelerated degradation and the development of 'extenders' to prevent accelerated degradation. Repeated treatment with one herbicide can sometimes enhance the rate of degradation of related herbicides, e.g. treatment with EPTC can increase the rate of degradation of butylate and vernolate. Cases in the UK have been reported where repeated treatment of soil with the fungicides iprodione and vinclozolin, and the insecticide carbofuran, increased the rate of degradation of subsequent applications. While enhanced degradation of herbicide does not appear, to date, to be a widespread problem in Europe the situation needs to be kept under view.

The pathways and rates of degradation are determined in the laboratory by incubating soil samples with ^{14}C-labelled herbicides under aerobic

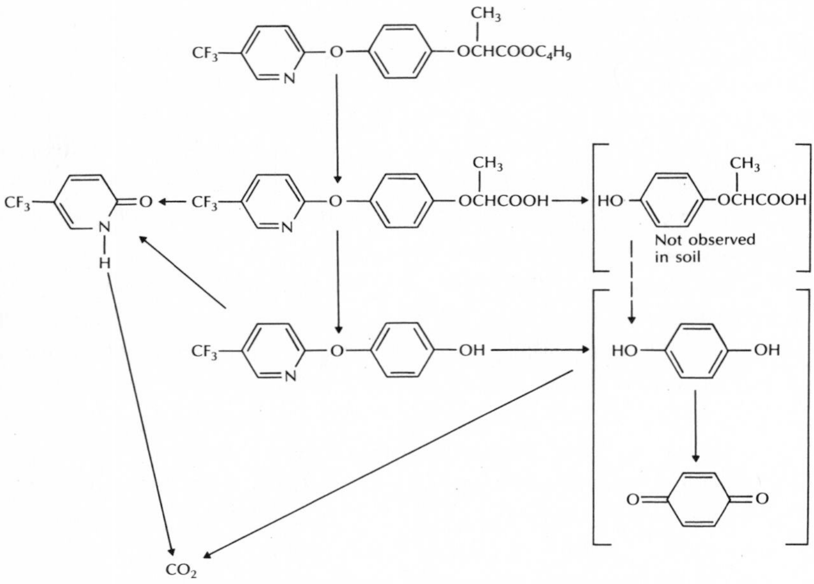

Fig. 9.3 Proposed pathway for the degradation of fluazifop-butyl in soil. * = postulated intermediates. From D.J. Arnold and D.W. Bewick, unpublished ICI data.

and anaerobic (flooded) conditions. The molecules are normally separately labelled in different positions so that the fate of the different parts of the molecule can be followed, e.g. the phenyl and pyridyl rings of fluazifop-butyl (Fig. 9.3). Volatile products, such as carbon dioxide are trapped and identified. At different times after treatment, samples of soil are then analyzed to determine the nature and amounts of ^{14}C-labelled chemicals using a wide range of analytical techniques such as thin-layer chromatography, gas–liquid chromatography, high-performance liquid chromatography and mass spectrometry. The degradation pathways are similar in all soil types, but rates can differ considerably.

Some herbicides contain an asymmetric carbon atom and therefore exist in two enantiomeric forms. For example, fluazifop-butyl contains an asymmetric carbon atom adjacent to the carboxyl group and consists of a 50:50 mixture of the R and S enantiomers. In soil they are rapidly (half-life <2 hours) hydrolyzed to fluazifop with retention of their optical configuration. However, the hydrolysis product with the S-configuration is inverted (50% inversion in 1–2 days) to the R-configuration, which is the more herbicidally-active form. There has been rapid progress in the technology for separating and synthesizing specific enantiomers and this has made it

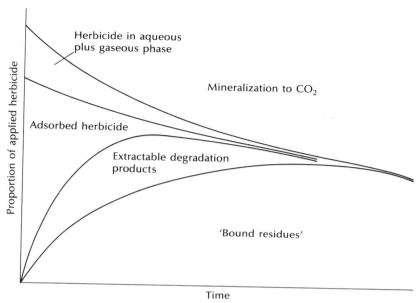

Fig. 9.4 Fate of herbicides in soil.

possible to market single enantiomers such as fluazifop-*P*-butyl, i.e. the *R* enantiomer.

The ultimate fate of all herbicides is mineralization, e.g. to carbon dioxide, or conversion to 'bound' residues (Fig. 9.4). These can only be released by extremely vigorous treatments, such as destroying the soil particles by refluxing with concentrated alkalis or acids. Thus they are difficult to identify. They probably consist of material incorporated into the soil organic matter by micro-organisms or parent herbicides, and/or degradation products which strongly bind, such as 3,4-dichloro-aniline released during the degradation of phenylamide herbicides. Frequently, at least half of the carbon present in the herbicide is converted to 'bound' residues but the amounts are very small compared to soil organic matter. An application of herbicide at $0.5\,kg\,ha^{-1}$ might give a 'bound' residue of about $0.1\,mg\,kg^{-1}$ soil in the top $20\,cm$ of soil compared with $10\,000-50\,000\,mg\,kg^{-1}$ of organic matter in most soils. They degrade slowly, which is not surprizing considering well-humified soil organic matter has a half-life of the order of 1000 years. Negligible amounts of 'bound' residues are absorbed by plants or other organisms such as earthworms; thus 'bound' residues are inactive.

Leaching

Leaching of a chemical depends on its concentration in water in the larger pores, through which water drains, and the amount of water moving through the pores.

Herbicides within soil aggregates or small pores ($<100\,\mu m$ diameter) through which water moves very slowly, are by-passed by water moving down the larger pores under gravity. Conversely, chemicals present in water in larger pores can be readily leached.

Movement of a herbicide is greatly affected by adsorption as only that fraction which is unadsorbed is free to move with the soil water. Solubility of the chemical also has some effect even though, in theory, 25 mm of rain is sufficient to dissolve $1\,kg\,ha^{-1}$ of a chemical such a simazine (with a solubility of only $5\,mg\,litre^{-1}$) and most herbicides are more soluble than this. Strong adsorption and limited solubility tend to retard the movement of a herbicide down the soil profile. Thus compounds that are relatively soluble and not significantly adsorbed, such as the phenoxyalkanoic acids, the sulphonylureas and the chlorinated aliphatic acids, are leached more readily than the ureas and triazines which are less soluble and more strongly adsorbed. Under UK conditions, evapo-transpiration normally exceeds rainfall during the growing season. Thus herbicides applied during the growing season can be degraded before leaching commences in late autumn.

In practice, even though some compounds have been detected deep in the soil, most residues stay in the top 1 m, and residues of the less-mobile materials are largely confined to the top few centimetres.

Volatilization

Under UK conditions, there is not a great deal of quantitative information about losses of herbicides from the soil by volatilization. In theory, even such involatile materials as the ureas could be lost significantly in this way but in practice movement of the herbicide into the soil and adsorption reduce such losses. Incorporation into the soil of the more volatile materials immediately after spraying is effective in minimizing vapour loss. For example, in the USA, when trifluralin was applied to the soil surface in the summer in an experiment in Maryland, 90% volatilized in 7 days, but in a similar experiment in Georgia, where it was incorporated to 2.5 cm, losses were only 22% in 120 days. Volatilization may also be reduced by using granular rather than liquid formulations. Vapour losses after spraying a wet soil are greater than from a dry soil because of the competition between water and herbicide for adsorption sites, so that as the water content increases there is a reduction in the amount of herbicide which is unable to volatilize owing to its adsorption. The same effect is produced by a little rain after spraying, but heavy rain reduces losses by washing the herbicide into the soil.

Evaporation usually increases with increasing temperature. However, if the rise in temperature produces a significant loss of soil water, more adsorption sites will be made available to the herbicide so that it is possible, under some circumstances, to demonstrate a decrease in the vapour loss of a herbicide with increasing temperature. Water has a second effect on vapour losses: as it evaporates at the soil surface, it is replaced by water from deeper in the soil and therefore solutes are concomitantly moved to the soil surface where they themselves may evaporate (the so-called 'wick' effect).

Uptake by plants

Only a small proportion of most soil applied herbicides is taken up by plants. Nearly all herbicides are retained near the surface so the majority of established crop roots are below the herbicide layer and uptake is minimal. Also the total amount taken up by weeds which are controlled by the herbicide is small so plant uptake is generally not a major route for herbicidal removal from the soil, i.e. probably <20% of that applied. However herbicides such as the sulphonylureas are very mobile in soils of high pH and when applied in the spring to an actively-growing cereal crop, the proportion taken up from the fraction reaching the soil might be significant.

Influence of soil texture on activity of soil-applied herbicides

The influence of rainfall and location of residues in the soil profile was discussed in Chapter 6. Considered here is the effect of soil texture on the adsorption and activity of soil-applied herbicides. Soil texture is determined by its physical constituents which include coarse particles (sand), very fine particles (clay), intermediate-sized particles (silt) and soil organic matter or humus. The Soil Texture (85) System, which is used for herbicides throughout the UK is fully described in ADAS Pamphlet 3001. It defines sand as particles between 0.06 and 2 mm in diameter, clay is <0.002 mm in diameter and silt is between 0.002 and 0.06 mm.

The main soil property determining its capacity to adsorb herbicides is its content of well-decayed organic matter or humus which is highly adsorptive. Organic matter contents can be >20% by weight in peaty soils from drained marshlands and most soil-acting herbicides are not recommended on high organic matter soils because the doses required for satisfactory weed control are too high to be economic. In arable areas of southern England, excluding drained fen or other marshland soils, organic matter contents range from little more than 1% in coarse sandy soils up to about 4% in clays. Levels are higher in the wetter and cooler parts of the country and range up to about 8%. Periods in grass increase soil organic matter because grass produces a greater amount of root residue which decays to give soil organic matter. Arable crops, particularly root crops, leave much less root residue and cultivation stimulates the microbial degradation of organic matter, so amounts gradually decline to a level which is fairly constant for the soil texture in mainly arable cropping systems within areas of similar climate. The activity of a particular dose of herbicide declines with increasing soil organic matter content.

The recommended dose for most soil-acting herbicides is varied for different soil textures to take account of these differences in soil organic matter. For example, the recommended dose of metribuzin is 50% higher on soils containing 2–3% of organic matter than that on soils containing 1–2%. The Soil Texture (85) system arranges textural classes into textural groups with similar ranges of organic matter content (see Table 9.1). Textural classes are quite readily recognized by 'feel' and it has generally been found unnecessary to have soil analyzed for organic matter content to make satisfactory recommendations on dose.

Influence of cultivations and cultural practices on herbicidal activity and selectivity

Cultivations for incorporation into the soil to minimize vapour loss are recommended for the volatile herbicides tri-allate (liquid formulation), and trifluralin applied for all crops except cereals. Cereal crops are much less tolerant of trifluralin so separation of herbicide and seed is imperative and incorporation is not safe. Incorporation is also recommended for napropamide which is readily degraded by sunlight. Shallow incorporation of root-adsorbed herbicides below the dry surface soil improves activity in dry years by placing herbicide in the root zone. This practice is only safe with herbicides which are highly selective for the crop such as the beet herbicides metamitron and chloridazon. It is not safe for the majority of herbicides which require separation between the herbicide and seed/roots.

Table 9.1 Classification of soils for adjustment of herbicide dose

Textural group for herbicides	Symbol	Soil texture (85 system)
Sands:	CS	Coarse sand
	S	Sand
	FS	Fine sand
	LCS	Loamy coarse sand
Very light soils:	LS	Loamy sand
	LFS	Loamy fine sand
	CSL	Coarse sandy loam
Light soils:	SL	Sandy loam
	FSL	Fine sandy loam
	SZL	Sandy silt loam
	ZL	Silt loam (85)
Medium soils:	SCL	Sandy clay loam
	CL	Clay loam
	ZCL	Silty clay loam
Heavy soils:	SC	Sandy clay
	C	Clay
	ZC	Silty clay

1 Adsorptive capacity for pesticides increases with soil organic matter content
2 Organic matter content tends to increase with clay content. For example in arable areas of the Midlands and Southern England clays generally contain 2.8–4.0% organic matter, compared with 1.0–1.5% in loamy sands
3 For each textural group the organic matter content within a particular area will be lowest under continuous arable cropping and highest in ley-arable systems. Soil organic matter generally increases with annual rainfall and distance north and west
4 The prefix 'organic' is applied to the above mineral texture classes if organic matter levels are relatively high. The 'sands' and loamy sands however appear organic when they only contain about 6% organic matter whereas 'clays' appear organic at about 10%. Organic matter levels between 1 and 10% cannot be detected by texture and must be determined by analysis
5 The prefix 'peaty' is applied when organic matter levels are between 20% and 35%
6 'Peat soils' are those containing more than 35% of organic matter
7 The organic matter percentages refer to the values determined by the chromic acid oxidation technique and not by the loss on ignition technique, which can produce considerably higher values
8 The risk of herbicide leaching on stony or gravelly soils is greater and textural grouping for such soils may need to be modified. In 'stony soils' stone (particles > 2 mm) content is 5–15% by volume. In 'very stony soils' stone content is more than 15%
9 The prefix 'calcareous' (Calc) is applied when soils contain more than 5% calcium carbonate

Safe and efficient use of soil-acting herbicides requires the preparation of good firm seedbeds free from large clods. Herbicide applied to a cloddy seedbed does not form a uniform layer of herbicide at the soil surface. This is partly because some of the herbicide is deposited in spaces beneath the surface and partly because herbicide-treated soil on the top of clods tends to weather down and fill the spaces between the clods. The result is that many weeds can emerge through soil which is no longer herbicide treated and so are not controlled. Crop damage can be caused by this weathering process when it occurs as a result of heavy rain soon after application and before the roots of germinating seeds have penetrated below the depth to which the herbicide moves.

For most herbicides, it is important to drill the seed below herbicide depth without placing it excessively deep. A firm level seedbed is necessary for good depth control. Shallow drilling can lead to crop damage from root-absorbed herbicides such as the substituted ureas or, in cereals, from trifluralin and pendimethalin. Deep drilling can lead to damage from shoot-acting herbicides such as terbutryne or prometryne.

Although sufficient cultivation to prepare a good seedbed is necessary for good weed control and crop safety it is important to avoid over-cultivation or cultivation under wet conditions which can give rise to a cultivation pan of compacted soil just below drilling or planting depth. Roots will not penetrate wet-compacted soil. Heavy rain, sufficient to leach herbicide into the root zone, may damage the crop when roots are prevented from penetrating below the herbicide layer. A well-known example is simazine injury to strawberries.

Limited movement of a moderately strongly-adsorbed herbicide down the soil can also lead to crop damage when there is sufficient rain to move a significant amount into the root zone. This is particularly likely to happen on readily-leached sandy soils.

Cultural practices can affect the adsorptive properties of a soil. The practice of direct drilling which does not involve ploughing or deep cultivation results in accumulation of organic matter near the soil surface and a gradual decrease in soil pH near the surface. Both of these changes increase the amount of herbicide adsorbed by the surface soil and decrease herbicide activity. Straw ash is highly adsorptive and the practice of straw burning with continuous direct drilling can eventually lead to poor performance of soil-acting herbicides, particularly chlorotoluron and isoproturon. When this happens ploughing is necessary to bury the highly-adsorptive surface soil.

Effects of herbicide residues on succeeding crops

The potential risk to a succeeding crop from herbicide residues depends on the persistence of the herbicide, the susceptibility of the following crop to the residues and the position of the residues in the soil in relation to the seed and root system. As residues tend to be near the surface, the risk of damage is greatest when seeds are sown without ploughing into the herbicide layer. Ploughing greatly reduces the risk of damage because it removes most of the herbicide from the vicinity of the seeds and young roots and also dilutes the residues.

The persistence of a herbicide in soil depends on its inherent stability, the physical and chemical properties of the soil, moisture content and the temperature. It is possible to assess the relative persistence of herbicides for a particular soil by determining their half-lives under controlled laboratory conditions of temperature and moisture. However, even under such controlled conditions, persistence can vary widely between soils. For example in a study of the persistence of simazine in 15 soils from eight countries, the half-life under similar conditions was found to vary by a factor of six. The variability was reduced by excluding soils from countries with a very different climate from the UK but was still by a factor of nearly three. In the field, conditions of moisture and temperature and hence rate of degradation vary greatly throughout the season. Summer rates of degradation are generally around three to four times faster than in winter but are much less than this when the surface soil dries out during extended dry periods so there is also variability between seasons. Because of all these variables, it is impossible to give a precise forecast for the persistence of active residues of a herbicide.

Table 9.2 gives a simple guide to the maximum persistence of active residues of herbicides used in the UK applied at normal doses and times. Foliage-acting herbicides with residual soil action are included as well as soil-acting materials. A very large range is quoted for atrazine because the recommended dose has varied widely from about 1 to 4 kg ha^{-1}. The range given for metsulfuron is very wide because persistence is exceptionally variable, being moderate in acid soils but long in alkaline soils. The range quoted for clopyralid is relatively large because of the great difference in susceptibility between crops which may follow a treated crop and there is evidence that residues in plant debris are more persistent than in the soil and these occasionally cause damage.

The type of risk from herbicide residues has changed in recent years. The susceptibility to residues of simazine, which used to be the herbicide most commonly-causing residue problems, does not vary greatly between crops. Recently-introduced herbicides have tended to be increasingly selective. Residues of propyzamide, which is widely used on oilseed rape, are safe to many crops but can seriously damage winter cereals unless sown after ploughing.

Residues of isoxaben applied to winter cereals are safe to most crops

Table 9.2 Guide to maximum persistence in the soil of active amounts of herbicides, based on practical experience in the UK (assuming full application rate and most sensitive plants)

Herbicide	Months	Herbicide	Months	Herbicide	Months
Alachlor	1.5–3	Diflufenican	5–10	Napropamide	6–12
Alloxydim	0.5	Diphenamid	6–12	Oxadiazon	9–18
Amitrole	1–2	Diuron	6–12	Pendimethalin	4–8
Asulam	1–2	EPTC	1–2	Phenmedipham	0.5
Atrazine	4–18	Ethofumesate	3–6	Prometryne	2–4
Aziprotryne	1–2	Fenuron	2–4	Propachlor	1–2
Bifenox	1–2	Fluazifop-butyl	0.5–2	Propham	1–2
Bromacil	9–18	Fluroxypyr	1–2	Propyzamide	6–12
Carbetamide	2–4	Hexazinone	4–12	Quizalofop	1–2
Chloridazon	2–4	Isoproturon	2–4	Sethoxydim	0.5
Chlorotoluron	3–6	Isoxaben	6–12	Simazine	4–8
Chloroxuron	3–6	Lenacil	3–6	2,3,6-TBA	6–12
Chlorpropham	2–4	Linuron	3–6	TCA	3–6
Chlorthal dimethyl	3–6	Mecoprop	0.5–1	Tebutam	3–6
Chlorthiamid	12–24	Metamitron	2–4	Terbacil	9–18
Clopyralid	2–6	Metazachlor	2–4	Terbuthylazine	3–6
Cyanazine	1–2	Methabenzthiazuron	3–6	Terbutryne	2–4
Dalapon	1–2	Methazole	4–8	Tri-allate	4–8
Desmetryne	0.5	Metoxuron	2–4	Triclopyr	2–4
Dicamba	2–4	Metribuzin	3–6	Trietazine	3–6
Dichlobenil	12–24	Metsulfuron	2–12	Trifluralin	5–10
Dichlorprop	0.5–1.5	Monolinuron	2–4		

but are highly damaging to brassicas and can cause complete failure of oilseed rape sown without ploughing. The sulphonylureas are unlike any other herbicide previously in use. Although only moderately persistent in acid soils, some can be highly persistent in alkaline soils. They have both foliar and soil activity and unlike most other soil-acting herbicides they are only slightly adsorbed. They are mobile in the soil and do not remain so near the surface as most other compounds, so the risk of damage may not be minimized by ploughing. Some crops, notably onions and beet, are extremely sensitive to residues of sulphonylureas and can be seriously damaged by <1% of the recommended dose.

Because of the risk of serious damage from residues of persistent herbicides, it is very important to follow label recommendations on safety periods before sowing a sensitive crop and to plough when advised. These precautions are particularly important in seasons when residues are higher than normal owing to slow degradation during an extended summer drought or an exceptionally cold winter. It is possible to predict using weather data and a computer program when the risk from residues is different from the norm. Computer prediction is used routinely each year by ADAS to advise whether residues are likely to be different from normal.

Analysis of the soil for herbicide residues can be helpful in assessing the risk to a crop. However, damage is generally worst in spray overlap areas so the maximum residue is likely to be at least twice the average level for a field. Analysis is difficult for herbicides active in very small amounts such as isoxaben and the sulphonylureas because there are great difficulties in detecting the minimum amounts that can cause damage.

Bio-assays of the soil to detect residues can be useful in some circumstances. Growing a susceptible crop such as tomato will detect the presence of active residues of hormone weedkillers such as 2,3,6-TBA, clopyralid and picloram. The characteristic distortions caused by these herbicides are readily recognized. Bio-assays are also useful for testing for residues of sulphonylureas by growing, for example, lentils which are extremely sensitive. However, the assay must be carefully done under controlled conditions with careful watering. In lentils, sulphonylurea residues cause stunting and root abnormalities which are clearly recognizable. Compaction of the soil, which can readily develop under glasshouse conditions may interfere with the test by restricting root growth. As sulphonylureas are mobile in the soil, it is necessary to test the sub-soil as well as the top-soil.

The activity of residues of herbicides such as the triazines, uracils and substituted ureas which act by inhibition of photosynthesis is greatly affected by growing conditions. Phytotoxic effects in the glasshouse may be greater than in the field because growth is more rapid, moisture conditions nearer to optimum and all the roots are in treated soil. Consequently, the risk from residues may be exaggerated. A negative response would be clear evidence of a lack of damaging residues but interpretation of a positive response is less straightforward, particularly as cultivations and especially ploughing can greatly reduce or eliminate risks from residues. Because of these difficulties and the time taken to do the bioassays, chemical analyses which are quick and reliable have generally been found to be more useful for herbicides in these groups.

Water

Herbicide residues can reach water by several routes:

Direct application for weed control
Spray drift
Run-off from treated fields
Leaching into ground water

Only a small number of herbicides are registered for controlling aquatic weeds. Initial concentrations applied to control floating and submerged weeds are typically of the order of 1 mg litre^{-1}. Most are applied as sprays but a few are formulated as granules, such as dichlobenil. After application the herbicides partition between the water, sediment and weeds, following the same principles as for soil. The major difference is the much higher ratio of water to solid phase. Consequently a much higher proportion of the residue is located in the water. For example, if a soil contains 50% water and a herbicide has an adsorption coefficient of 1, two-thirds of the herbicide will be adsorbed. In a pond 1 m deep, assuming the herbicide equilibrates with a layer of sediment 1 mm deep, only 0.1% would be adsorbed. Some herbicides, such as diquat, are rapidly adsorbed by weeds and suspended sediment and the concentration in water falls to <0.02 mg litre^{-1} in <10 days.

Degradation processes in the aquatic environment are the same as in soil except photochemical degradation of residues on plant surfaces and in the water is more important for herbicides which are decomposed by sunlight. The label on aquatic herbicides indicates the safety interval before treated water can be used for irrigating crops; this ranges from nil up to 4 or 5 weeks for residual herbicides such as dichlobenil.

Traces of herbicides can reach surface waters in run-off, from spray drift and sub-surface land drainage. In the UK, for most herbicides, $\leqslant$ 0.5% of the amount applied is carried off the field in run-off. When severe run-off events occur within 1 or 2 weeks of application, this can be up to 5%, particularly for wettable powder formulations. The maximum concentration of residues in run-off is highly variable but is typically in the range of 0.001 to 1 mg litre^{-1}. Most herbicides are lost mainly in the water phase, even when concentrations are 2–3 orders of magnitude higher in the sediment than the water, because of the large ratio of water to sediment. The traces of residues which reach surface waters are subject to the same adsorption and degradation processes as aquatic herbicides, although the relative importance of the different mechanisms and the rates are not fully understood. Dilution also plays a major part in reducing the concentrations. The recent development of analytical methods sensitive to residues down to 0.0001 mg litre^{-1}, or lower, has made it possible to monitor residues in water. For example, monitoring of surface waters in East Anglia, where herbicides are widely used, has shown residues of most herbicides to be <0.001 mg litre^{-1} and in most cases to be non-detectable (<0.0001 mg litre^{-1}). A small number of water samples contained traces of herbicides such as mecoprop, atrazine and simazine, but always <0.01 mg litre^{-1}. It is suspected that the traces of atrazine and simazine are due to use on industrial sites, railways and road edges rather than from agricultural use. The toxicological properties of herbicides are investigated in great detail to establish their safety to users and consumers of treated crops. Residues in water are well below levels which would cause any toxicological concern. They also do not present any threat to the environment.

Herbicides reaching soil are subject to both adsorption and degradation which minimizes the potential for leaching into groundwater. Extensive field studies have shown the maximum concentration of residues normally remains in the top-soil. Nevertheless, traces might be leached through the top-soil, particularly when applied in the autumn; leaching and persistence both being greater than when a compound is applied in the spring. Small amounts of residues might also be washed down channels and fissures in the sub-soil and underlying rocks. Monitoring of groundwater has shown, however, that residues are even lower than in surface waters, e.g. in East Anglia residues of the major herbicides in groundwater are <0.001 mg litre^{-1} and in most cases they are non-detectable, i.e. <0.0001 mg litre^{-1}

Further reading

The Soil Texture (85) System ADAS pamphlet No. 3001. MAFF Publications, Lion House, Willowburn Estate, Alnwick, Northumberland NE66 2PF.

10 / Evaluation of a new herbicide

L.G. COPPING, H.G. HEWITT AND R.R. ROWE

Dow Chemical Company Ltd, Letcombe Laboratory, Letcombe Regis, Wantage, Oxford OX12 9JT

Setting objectives
  Identification of targets
  Establishment of a screen
Synthesis strategies
  Random empirical screening
  Analogue synthesis
  Natural products
  Biochemical design
  Other considerations
Selection of candidate compounds for
    field testing
Formulation
Patents
Further synthesis
From glasshouse to field
Methods of field evaluation
  The objective
  Towards commercialization
Conducting the trial
  The trial site

Randomized complete block
Factorial design
Replication
Treatments
Techniques
Equipment
  Application
  Calibration
Assessment methods
Interpretation of data
Safety evaluation and regulatory
    requirements
  Toxicological studies
  Studies of effects on the environment
  Fate in the environment
  Ecotoxicity
  Residues in crops
  Supplementary studies
Time-scale and costs

The rationale, screening technique, targets and research philosophy of agro-chemical companies when searching for a new herbicide are often very different but the objective is always to identify a new molecule, preferably with a novel mode of action, which is patentable and marketable. The first part of this chapter considers different approaches to the discovery process.

Setting objectives

Identification of targets

Before work is begun sound markets must be identified. These may be in countries where the company has considerable expertise and hence a marketing/selling presence or in major crops. Traditionally, the major crops are maize, soybean, rice, small-grain cereals and possibly sugar-beet and cotton. Total vegetation control is also of interest. These markets are identified because of their size and the acceptance by the grower of the need for weed control to maintain productivity. Consequently, these markets are full of products and are very competitive. For this reason there is a need to identify the weaknesses of the existing products and establish screens which will select compounds that can exploit them. These may involve crop selectivity, varietal tolerance, spectrum of weeds controlled, timing of application (a greater 'window'), reduced (or increased) soil persistence, reduced soil mobility or compatibility with other products.

Hence, a typical set of herbicide discovery objectives would read as follows. To discover a compound, ideally with new chemistry and mode of action which can be:

1 Applied pre- or post-emergence for the control of annual and perennial grasses and broad-leaved weeds in field corn. The compound should be systemic.

2 Used pre- or post-emergence in soybeans (and possibly cotton) for the control of all annual and perennial weeds. It should be selective to determinate and indeterminate varieties and will ideally show selectivity in other leguminous crops.

3 Used to control the major weeds of wheat and barley, in particular blackgrass, wild oat *Viola* spp., *Veronica* spp. and cleavers. It should be active both pre- and post-emergence and selective up to Zadoks growth stage 39. All varieties must be tolerant.

4 Used in paddy rice. It should be selective to transplanted rice and control all important grass and broad-leaved weeds. It must be non-toxic to fish. Selectivity to direct-seeded rice would be an advantage.

It is clear that these objectives describe the ideal crop herbicide and that all the characteristics described will probably be unachievable. Nevertheless, all companies go through the process of identifying objectives so that they at least know what they are trying to discover.

The discerning reader will appreciate that the above objectives do not cover compounds such as paraquat or glyphosate and glufosinate and so an additional objective is always added to cover this type of activity:

5 Used to give rapid, long-term control of all plant species without residual activity. Perennial weed control is essential.

Establishment of a screen

Agro-chemical companies can be divided into those which use crops as indicator species in primary screens and those which use weeds. The argument for crops is that they can be selected from families that represent major weeds, the seed is readily available, genetically uniform, non-dormant and viable. In addition, primary screening is merely a process designed to reject as rapidly as possible inactive compounds and select those with interesting activity. The argument for weeds is simply that herbicides kill weeds and, therefore, should be evaluated, selected and rejected on their effects on weeds. Both schools have a good case and it is probably of little consequence which philosophy is adopted as the experimenter will quickly acquire a 'feel' for his screening process and will be able to distinguish between compounds of interest and those of no potential. Nevertheless, the authors lean towards weeds so it is the establishment of a weed screen that will be described.

Selection of species for a screen must make provision for several characteristics:

Temperate weeds   — grass
Temperate weeds   — broad leaved
Temperate weeds   — perennial
Sub-tropical weeds — grass
Sub-tropical weeds — broad-leaved
Sub-tropical weeds — sedge
Sub-tropical weeds — perennial

If the weeds are to be sown together in a tray then rate of growth of the species relative to each other has to be considered. Other important considerations are space available in glasshouse or growth room and quantity of chemical available to the investigator.

Perennial weeds always cause problems in primary screening systems and are usually only introduced at a later stage of the screening programme. The only exception is that of nutsedge (*Cyperus esculentus* or *C. rotundus*) which is incorporated in the screens of some companies. Its inclusion other than as a representative of the Cyperaceae is hard to justify since nutsedge is an agricultural problem because of its ability to produce 'nutlets' which may germinate immediately or which may lay dormant for many years. Consequently, pre-emergence control of glasshouse-grown nutsedge bears no relation to the field situation and may not reflect the potential for field control.

A good primary screen contains species which dominate major temperature or sub-tropical crop situations. A typical example is shown in Table 10.1.

Table 10.1 Typical primary screening species

Species	Biological name	Description
Temperate:		
Wild oats	*Avena fatua*	Large seeded grass weed
Blackgrass	*Alopecurus myosuroides*	Important cereal grass weed
Cleavers	*Galium aparine*	Large seeded important broad-leaved weed
Chickweed	*Stellaria media*	Small seeded abundant broad-leaved weed
Redshank	*Polygonum persicaria*	Representative of Polygonaceae
Mayweed	*Tripleurospermum inodorum*	Representative of Compositae
Sub-tropical:		
Cockspur	*Echinochloa crus-galli*	Important grass weed
Fingergrass	*Digitaria sanguinalis*	Important grass weed
Morning-glory	*Ipomoea purpurea*	Hard-to-kill broad-leaved weed
Cocklebur	*Xanthium pennsylvanicum*	Compositae — hard-to-kill broad-leaved weed
Velvetleaf	*Abutilon theophrasti*	Malvaceae — hard-to-kill broad-leaved weed
Common amaranth	*Amaranthus retroflexus*	Small-seeded broad-leaved weed

Growing conditions

A screen is only valuable if it is possible to compare biological effects week to week, month to month and even year to year. Glasshouses have usually been used for this type of work. Temperatures can be maintained during the winter and controlled (within limits) during the summer. Day length can be adjusted by the use of supplementary illumination during the winter. Nevertheless there is always variability in the glasshouse which cannot be controlled, thereby making comparisons between tests run in June and January difficult.

This can be eased by including an internal standard compound in every test. The variability of the weeds' response to this standard is used to modify the recorded response for each new test compound. This approach, however, increases the number of treatments in each screen and also demands additional calculations.

An alternative strategy is to use controlled environment (CE) rooms where growing conditions are uniform the year round. However, CE rooms provide unnatural environments and produce plants which may be atypical and respond differently to those grown in the field or glasshouse. Thus, you can have uniform-growing conditions and hence the ability to relate data generated at different times but with plants which may be more susceptible than natural plants, or you can have variable growing conditions and plants a little more robust but still 'softer' than those encountered in the field. Either situation is acceptable as long as the experimenter is aware of the potential problems.

Soil type is another important factor. Primary screening should give the compound the best possible chance so if it has no effect at the highest dose it can be rejected as inactive without worry. Hence a light sandy loam, which optimizes plant growth without significantly reducing the activity of most organic molecules is the usual choice.

It was once thought that soil type was of no importance in post-emergence herbicide evaluation and, as application is to plants which are at least two weeks old and assessments are made three weeks after spraying, a good potting compost was often used as the growing medium. However, we now know that many compounds applied post-emergence are taken up by the roots and hence could be bound in a soil containing a high level of organic matter. Hence, similar soil types are employed for both pre- and post-emergence applications.

Application rates

In the early days of screening for herbicidal effects, the rates of use in the field were high by modern standards and the chemistry was cheap. Field rates of $4\,kg\,ha^{-1}$ were cost effective. A primary screen is used to reject inactive compounds rather than to select active ones and so rates as high as $50\,kg\,ha^{-1}$ were often used. As expertise in herbicide design improved and as the complexity of chemical synthesis increased, cost-effective application rates under field conditions have reduced.

The introduction of the sulphonylureas has also gone a long way to changing the industry's attitude towards application rates. For this reason, current primary screening rates may be $2-4\,kg\,ha^{-1}$ or even lower if the compound under test is a member of a series known to be herbicidal from earlier analogues.

The secondary screen is designed to determine the activity of compounds not rejected at the high dose rate. This is usually accomplished by using the same indicator species and lowering the rate of application until activity is lost. This process should always include a relevant commercial standard and where appropriate the most active member of the chemical series under investigation. It is the data generated in the secondary screen which are used to direct further synthesis and to determine the biological potential of a compound. The weeds controlled in this screen also suggest commercial possibilities and all tertiary screening has to be crop directed, once again using the best commercial standard(s) and the most active chemically-related compound from the series under investigation. The crop screens will be those outlined earlier as the herbicide discovery objectives.

Synthesis strategies

Random empirical screening

Often herbicides (and other pesticides) were discovered by large chemical companies who tested compounds synthesized in other parts of their organization on weeds, insects or plant pathogens. This strategy was extremely successful in the period from 1945 to 1960 and those companies who managed to apply the most compounds discovered the most products.

As with all areas of research the task is not a simple matter of spray the compound, leave it for a few days and select those that kill the plant. Astute evaluation of the symptoms of injury, the ability or otherwise of a weed to recover from damage and some knowledge of the compound's potential for chemical modification are all essential to identifying a compound as a useful lead. The random screening that leads to the selection of a compound requires a degree of luck but subsequent improvement requires detailed observation and inspired synthesis.

As the number of patented compounds and uses has increased, areas of random chemistry available for screening have become more complex. The demands made of a pesticide are also increased in terms of level of activity, weed spectrum and crop selectivity. Hence, although many companies still screen several thousand new compounds each year, new strategies for discovery have been adopted.

Analogue synthesis

Analogue synthesis is often referred to as 'me-too' chemistry or 'patent busting'. Following the successful launch of a new compound, competitors use expertise or intermediates peculiar to themselves, to make compounds outside the patent, preferably with improved biological properties. This process is distinct from random synthesis/compound optimization, as the idea comes from outside. Examples of this type of synthesis are many and some are shown in Fig. 10.1.

Another strategy was to attempt to link together molecules known to be herbicidal, to produce a new compound which in theory would possess the desired properties of the parents. This strategy was generally unsuccessful and no good examples have been reported. The attempts to combine dichlobenil with diuron (Fig. 10.2) show that often a compound is produced with different activity for the wrong reasons. The hybrid compound produced (DU19111) showed no herbicidal effects but had a slow action on insects. Subsequent synthesis led to diflubenzuron and a new generation of insecticides. This is surely justification for wide screening of new chemicals!

Another approach is the cyclization of a known herbicidal structure to produce a new chemical which it is hoped will retain the activity of the original. Ureas have been particularly valuable in this area of chemistry. Attempts to form a cyclic variant of diuron led to the synthesis of methazole (Fig. 10.3) and inclusion of one of the urea nitrogens in the ring eventually led to the synthesis of epronaz.

Natural products

The pharmaceutical industry has long been aware of the value of natural chemistry as a source for new products and for compounds with new, unexploited modes of action. Since the discovery and development of β-lactams, most large drug companies are looking for new natural compounds from micro-organisms and from higher plants. The agro-chemical industry is now beginning to realize the potential of this source of products, leads or new modes of action. In the herbicide area this has been encouraged by the discovery in Japan of bilanafos (Fig. 10.4). Bilanafos is of particular interest because it is a relatively simple chemical structure by natural product standards and lends itself to chemical synthesis. Indeed two companies have evaluated the potential of close analogues (Fig. 10.5) for their herbicidal effects and one of these, glufosinate (phosphinothricin), is on sale as a total herbicide.

Unfortunately, not all natural herbicidally-active compounds are as simple as phosphinothricin (Fig. 10.6) thereby making analogue synthesis a much more difficult task. This means that the natural product must be easy to produce by fermentation or, alternatively, it should possess a unique mode of action which identifies a new biochemical target for directed synthesis.

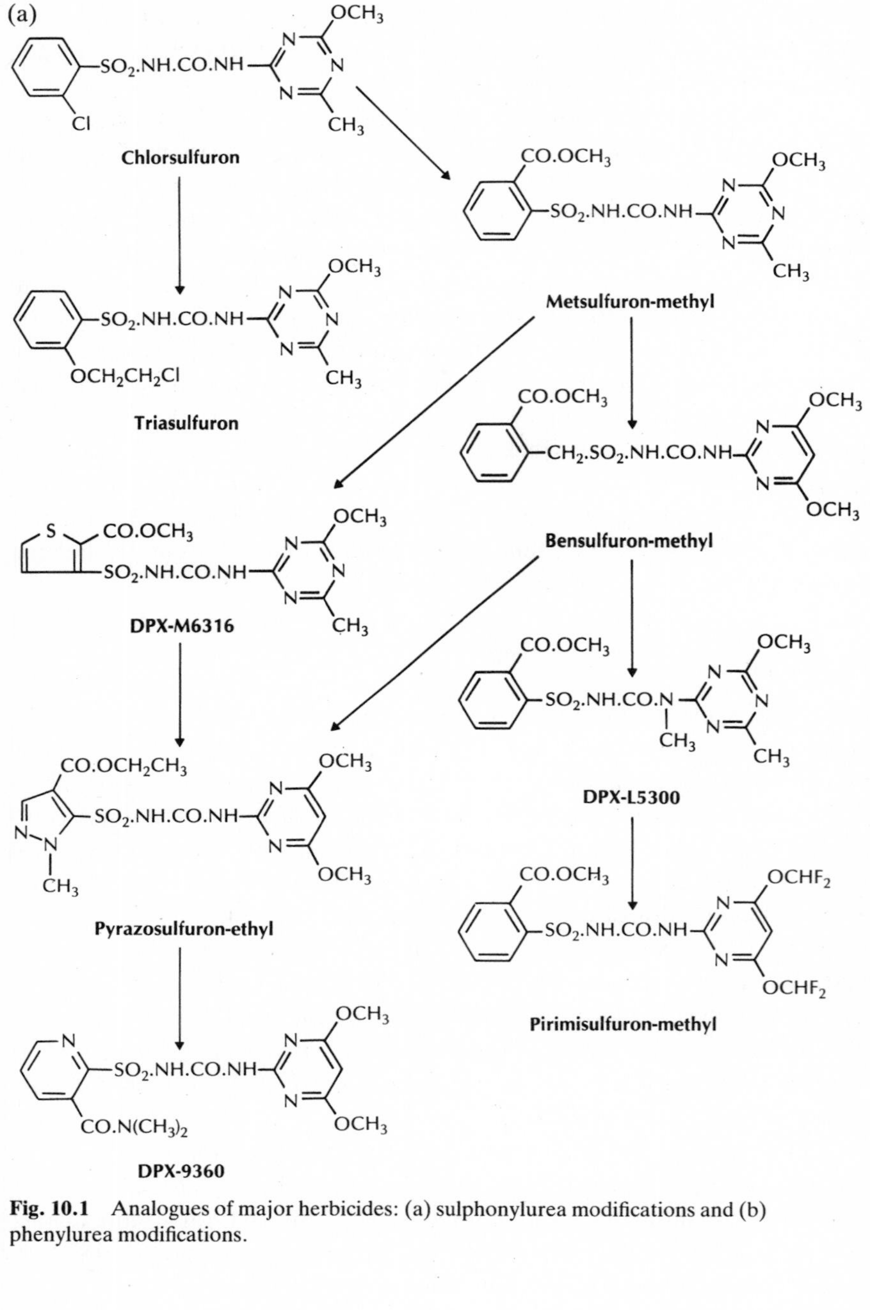

(a)

Fig. 10.1 Analogues of major herbicides: (a) sulphonylurea modifications and (b) phenylurea modifications.

(b)

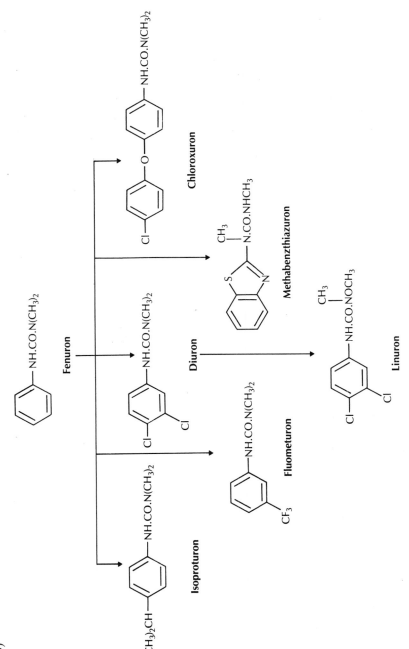

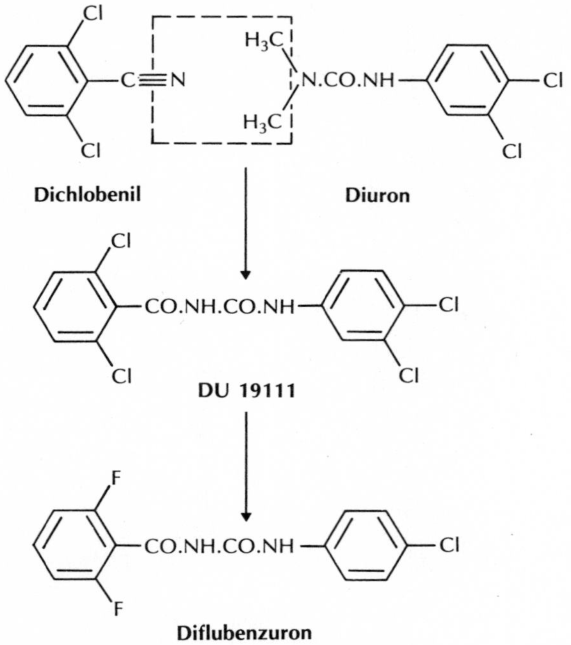

Fig. 10.2 Development of acyl urea insecticides.

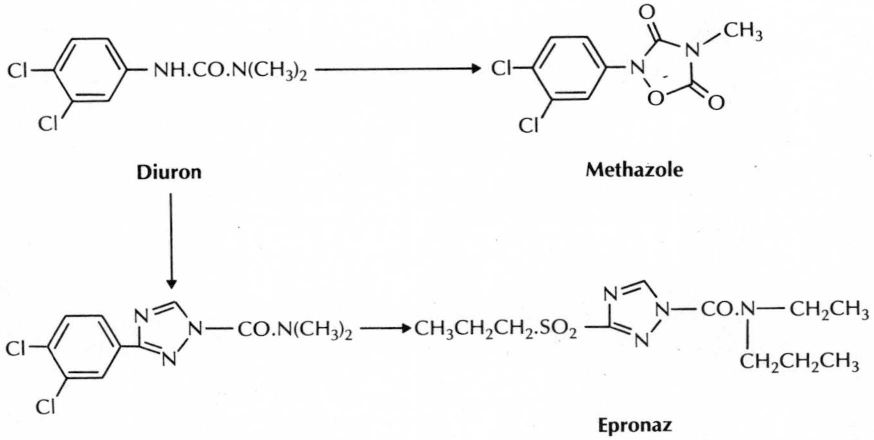

Fig. 10.3 Cyclization of diuron.

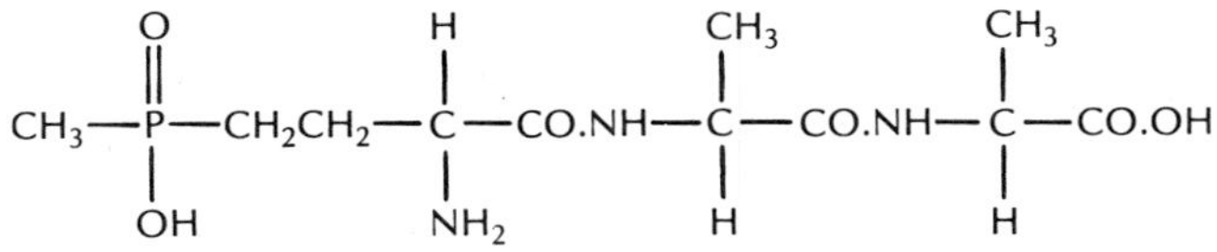

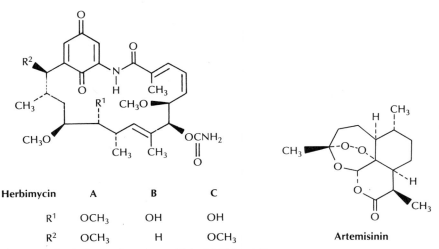

Fig. 10.4 Structure of bilanafos.

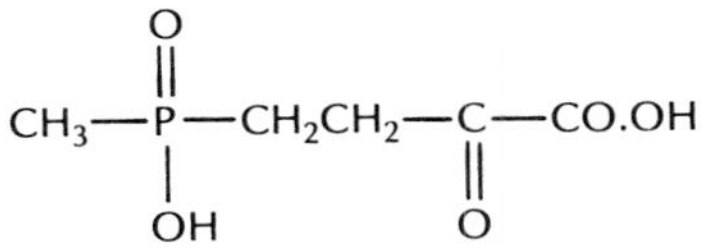

NC28260

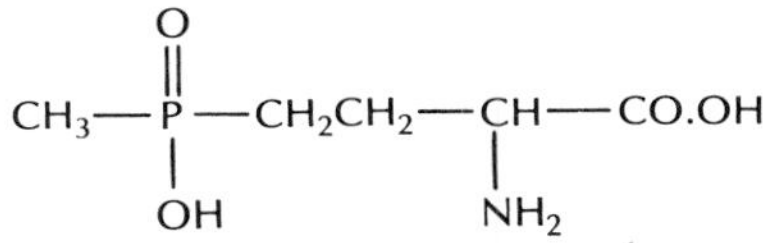

Glufosinate

Fig. 10.5 Structures of bilanafos analogues.

Herbimycin	A	B	C
R^1	OCH_3	OH	OH
R^2	OCH_3	H	OCH_3

Artemisinin

Fig. 10.6 Typical natural products with herbicidal activity.

The search for sources of biologically-active natural products requires some comment. The various approaches adopted by chemical companies include studies of the following.

New micro-organisms: Particular groups of micro-organisms produce a range of metabolites with potential value as pesticides and it is argued that the discovery of new species increases the chance of finding a new metabolite. For this reason streptomycetes, bacilli and pseudomonads are very popular targets for evaluation. Unfortunately this strategy often reveals compounds already described and identifying and characterizing a metabolite to confirm it is new, is extremely expensive and time consuming.

Micro-organisms that occupy particular ecological niches: The argument for this strategy comes from the fact that to be competitive, a micro-organism needs to have an advantage. This advantage could be the production of a biocide which inhibits the growth of competitors. This strategy has revealed producers of bacteriocides and fungicides but few herbicides.

Unusual micro-organisms: It is possible that previously undescribed micro-organisms will produce novel products with useful biological activity. In this case the expense of isolating and characterizing the product is more easily justified. Marine phycomycetes, bacteria from the guts of termites and micro-organisms that survive in deserts are all examples of potential new species which may produce active secondary products.

Necrotrophic plant pathogens: Plant pathogens can be broadly classified into biotrophs and necrotrophs. Biotrophs are parasites such as the powdery mildews which invade their host plant but do not cause the plant, or even the invaded cell, to die. Necrotrophs, on the other hand, are aggressive parasites such as the downy mildews (potato blight), which cause cell and tissue death during infection. These necrotrophs could act through the production of a chemical which causes the cell to die. Why not try to isolate it? It should be remembered that some pathogen-induced cell death may be caused by the host plant's hypersensitive reaction to invasion and is hence host, and not pathogen, mediated.

Plants which produce allelochemicals: Allelopathy has been variously described but for the purpose of this brief review the definition will be limited to the production by plants of chemicals which prevent seed germination or seedling establishment. The best known example is the production of

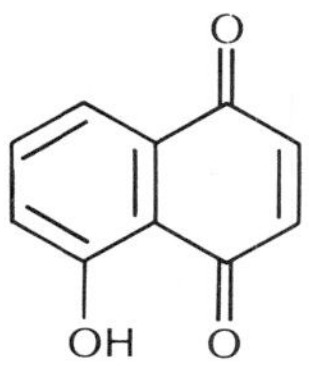

Juglone

Fig. 10.7 Structure of juglone.

juglone by the black walnut *Juglans excelsior* (Fig. 10.7). Not included in this definition are phenols or aliphatic acids which are produced in large quantities by some plants (such as sugar-beet) or which are produced when organic material is broken down in soil. Regrettably, many of these so-called allelopaths produce organic molecules which are general biocides and are consequently of little value as selective herbicides.

New plants or plants which feature in local tribal folklore: Remarkably, as the tropical rain forests of South America and the Far East are explored, new plants are still being discovered. Many feature in tribal folklore as medicines or 'magic' plants. The testing of these may reveal new chemistry.

Whatever the source of the natural product there are a number of possible outcomes:

1 A new herbicide may be produced by fermentation.

2 A new herbicidally-active structure can be synthesized or rendered more active or selective by additional synthesis.

3 A new biochemical mode of action identified.

Biochemical design

Throughout the above section on leads from natural products the possibility of finding a compound with a novel mode of action has been mentioned. One of the roles for a biochemical group within a pesticide discovery operation is to examine biochemical processes which occur in target organisms and identify those which, if inhibited, would lead to the death of the organism. A good example is photosynthesis. However, all plants photosynthesize and so there is a likelihood that an inhibitor would be non-selective and indeed such compounds exist (paraquat, bromacil). Selectivity can be introduced if there is differential metabolism of the compound between species. Here a good example is atrazine which can be detoxified in three different ways: firstly, via *N*-dealkylation, secondly by non-enzymic replacement of the chlorine with a hydroxyl group and thirdly via conjugation with glutathione. All plants dealkylate, albeit at different rates. Only maize and related species contain benzoxazinone which catalyzes the hydrolysis of the compound. All plants are able to conjugate with glutathione, a process catalyzed by the enzyme glutathione-S-transferase. Some plant species, notably corn and sorghum contain a very high titre of the enzyme and are thus able to conjugate and de-toxify atrazine before it reaches the chloroplast. Other species carry out this de-toxification more slowly and hence do not prevent accumulation in the chloroplast and subsequent plant death. It must be noted, however, that this understanding of the basis of selectivity was discovered retrospectively rather than being introduced rationally.

By understanding the efficiency with which conjugation, decarboxylation, β-oxidation, hydrolysis and dealkylation occur in crop and weed species, pro-herbicides can be synthesized which owe their selectivity to differences in specific metabolic processes. Here the selectivity of phenoxybutyrates, such as 2,4-DB, which require activation via β-oxidation is an example.

Another role of biochemistry is to help determine the mode of action and potency of a new herbicide lead which has been identified by empirical screening. If it is a poor herbicide but a potent inhibitor then clearly it is either a poor target site or the compound does not reach the target. Alternatively, if it is a poor inhibitor but a moderate herbicide, there is clearly scope for improving activity.

However detailed are the biochemical studies they must always be considered in conjunction with *in vivo* herbicide data.

A number of companies who have concentrated on cell-free assays to direct their herbicide synthesis programmes have discovered a large number of extremely potent enzyme inhibitors but no commercially viable product. Clearly, if used intelligently, good biochemical input can improve greatly the quality of a screening operation.

Other considerations

The physico-chemical characteristics of herbicides fall into a number of precise ranges of octanol/water partition coefficient, melting point, volatility and in the case of many phloem-mobile compounds, pK_a. Application of these relatively tight chemical requirements will frequently dictate which compounds of a relatively large group of candidates should be synthesized

1*H*-1,2,4-triazole   Imidazole

Fig. 10.8   Triazole and imidazole ring systems.

first, or alternatively which ester of an acid is most likely to enter a plant. It is unlikely that the use of physico-chemical or biophysical parameters will predict which member of a series will be the most active but rather their application will improve the chances of synthesizing an active compound. These parameters can be used in conjunction with the relative stability of different molecular substituents in the environment and in the plant to avoid making compounds which are too labile to be effective or too stable to be accepted by registration authorities. Comparison of the two five-member heterocyclic rings imidazole and triazole (Fig. 10.8) provides an example. The imidazole ring is generally more prone to photo-oxidation than the triazole and consequently under glasshouse conditions an imidazole compound may show the higher level of activity but in the field the triazole often performs the better owing to its improved stability.

Selection of candidate compounds for field testing

The purpose of the screen is to separate active from inactive compounds as rapidly as possible. It is the tertiary, crop-based screens which are used to select candidate compounds for field testing. Here two different criteria are used. If the compound is the member of a chemical series about which nothing is known and shows selectivity at what might be considered cost-effective rates, it should always be field tested. This is done to establish how well activity for that particular compound type is transferred from glasshouse to field. Often post-emergence herbicides transfer well with a reduction in activity in the order of 2–4 times.

Pre-emergence herbicides, however, can be very variable. Isoproturon pre-emergence activity is reduced by a factor of 2–4, like post-emergence compounds, whereas alachlor, which is highly active in the glasshouse, often shows an 8–16-fold loss of activity.

The second case is that of a compound from a well-tested series. Here intelligent evaluation involving comparisons with the most effective analogue should identify any possible advantage in selectivity, activity or persistence and only a compound which shows potential advantages should be field tested.

Formulation

All primary and secondary screening is usually undertaken using acetone dispersions in water of the compound under evaluation with added sur-factant to enhance plant wetting.

Compounds which are insoluble in acetone are frequently ground into a fine powder in the presence of a suspending agent and then dispersed in water with added wetter.This technique ensures that all compounds are compared under identical conditions. Some organizations, however, for-mulate each new compound prior to screening. These formulations may be wettable powders or emulsifiable concentrates in a complex solvent system. The arguments for these approaches are that the first produces uniformity of formulation and the second optimizes activity.

Regardless of the system used, a field candidate compound must be formulated in such a way that the biological activity is expressed with a physically stable system that approximates to a commercially viable formu-lation. Formulation chemists will confirm the physical stability of these experimental formulations but it is clear that their biological activity and crop selectivity must be tested to confirm that they have not been adversely affected by formulation. These tests are done under glasshouse conditions. Only formulations which retain the herbicidal efficacy of the experimental compound together with its crop selectivity will be field tested.

Patents

The filing of patents is always the objective of a chemical company to give it exclusive control of the manufacture, formulation, sale and use of the compound in all countries in which patents are filed. Unfortunately there is never a right time to file a claim. The aim of a patent is to cover as wide a group of compounds as possible but patent law requires that both chemical and biological examples of a representative selection of compounds must be presented as evidence that a discovery has been made. Clearly a claim such as that shown in Fig. 10.9 will cover several million compounds and the synthetic effort necessary to elucidate a representative selection of them would be enormous. Allowing the synthesis chemist time to make the necessary number of chemical examples is an essential feature of a patent application. Another, sometimes more important factor, is the area of

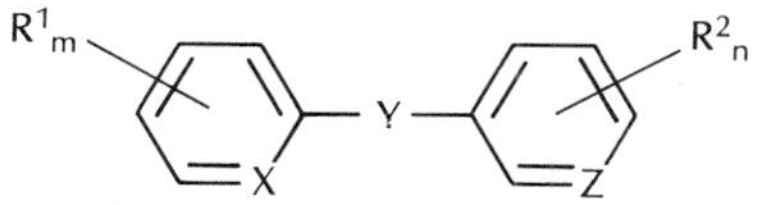

Fig. 10.9 General formula of a possible patent claim, where R^1 and R^2 can be independently straight or branched chain alkyl, alkoxy, halogen, nitrile, haloalkyl, aryl, hydroxy, hydrogen, variously substituted aryl. X and Z can be N or CH; Y can be NH, CH_2, O, S, SO, SO_2, CO; and m and n can be 0–4.

chemistry in which the activity lies. If the compounds, the activity or the mode of action are new, it is unlikely that another organization will be working in that area. This means that the chance of your patent claim being pre-dated by a competitor is small. This in turn allows more detailed examination of the chemical modifications that can be made whilst retaining good biological activity and consequently a broader, well-founded claim. If, however, the chemistry is in an area in which there is known to be considerable interest within the industry, a claim will often be filed after only a few compounds have been synthesized and shown to be biologically active. Patent law in Great Britain allows an inventor 12 months from the date of filing a provisional patent to complete that application. To complete an application, good chemical and biological data must be presented on a representative selection of compounds within the general formula claimed. If this is done then the date of filing the provisional claim is taken as that patent's priority date. The claim can then be extended to cover other territories which are usually major users of agro-chemicals and major manufacturing countries. The priority date of a patent is crucially important as this is the basis on which disputes between companies on the right to patent are resolved.

Further synthesis

When a compound has passed through the screening system and been evaluated in the field, there are three possible courses of further action. If a compound shows good levels of weed control, excellent selectivity to a major crop and preliminary costings suggest that the compound could be sold both competitively and profitably, it becomes a product candidate and progresses through the development system described later in this chapter. Secondly, a trial may give inconclusive results and consequently has to be repeated to enable a proper evaluation of the compound's potential to be made. Such a situation happens less frequently as organizations develop field research facilities in a number of geographical locations. A failure in the UK, France and USA in the same season is unlikely. Thirdly, the compound under evaluation may fail. If this is the case, it is important to find out why the compound failed in the field. Was it too labile, too volatile, too water soluble and washed off in rain, metabolized by the plant, unable to redistribute in the crop/weed canopy, unable to penetrate the cuticle of field-grown plants? The possible explanations for poor transfer from laboratory to field are many. It is important to determine which is the key feature affecting the performance of the field candidate compound and apply any knowledge and understanding gained to newly-synthesized compounds. Thus, the debate between organic chemist, biochemist and screening biologist is ongoing and continuous. The programme is driven by chemistry, advised by biochemistry but led by biology. The interactions of various disciplines are represented in Fig. 10.10.

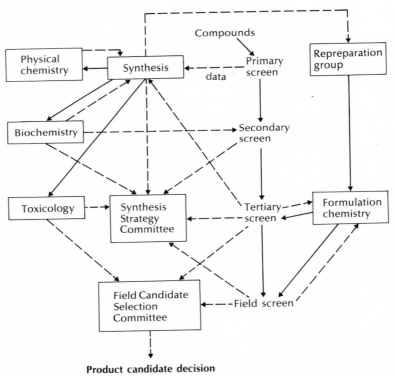

Fig. 10.10 Synthesis and screening diagram. ⟶ = compound transfer, – – – ⟶ = data transfer.

From glasshouse to field

Since the philosophy of screening is one which relies upon a sequential application of defined tests to identify herbicidal activity, the process of field evaluation can also be regarded as a screening test *en route* from newly-synthesized compound to product. Field research and development trials seek to optimize and extend activity under realistic grower conditions.

Environmental conditions early in screening are usually fixed to standardize the tests, enabling comparisons between compounds to be made. Commonly, under these conditions, crops and weeds are more sensitive to herbicides giving an optimistic view of efficacy but sometimes a pessimistic one for selectivity. In field work, environmental conditions cannot be controlled, can show extreme variation and can alter the basic performance of active molecules, especially when the desired effect relies upon physical factors such as light, temperature, humidity and/or biological features such as the integrity of the cuticle and leaf surface. Quite simply, field work accepts all the interactions within natural populations of weeds and crops and attempts to modify and refine herbicide performance to a guaranteed level. When specific interactions need to be studied more closely, field trials and glasshouse experiments may run concurrently. However, the precision needed to define the most active member of a chemically-related series of herbicides may only be achieved through a coalition of glasshouse and field studies.

Field testing is carried out over several seasons and many agro-chemical companies have trial capabilities in the northern and southern hemispheres so shortening development time. In all cases, however, field trials take two forms. These are field research and field development.

Field research identifies the field performance of compounds in terms of their economically important weed spectra and crop safety. Formulation types and environmental factors are investigated and their effects on herbicide performance or potential use are examined. At this stage the compounds are still experimental and frequently the first two years of field research are carried out on company-owned land in order to control trials and monitor progress closely. Other advantages of using company land are that weeds and crops can be planted to suit the trial's objective and that trials can be repeated, if necessary, within a single growing season. Also, trials carried out on company land do not require official approval or the identification of a third party, usually the farmer, to act as a company agent for the conduct of the trial.

Field research precedes field development, but once again the two aspects of field trials can run concurrently. For example, mixture work would usually take place in field development but, depending on the compounds and objective, may have begun at the earlier stage of field research. Field development is specific to a country to evaluate the commercial potential of active compounds under grower conditions and to generate data for registration.

Methods of field evaluation

The decision to develop an agro-chemical is based upon many factors. These include efficacy, selectivity, cost, markets, toxicology and reliability, some of which can be defined precisely, whereas others such as those describing field performance may not be as clear. In fact, the confidence which is given to field data is frequently questionable and often tempts the experimenter to present opinions rather than well-founded conclusions. However, in recent years much progress has been made in the acceptance of an objective approach to field trials.

At the outset it is accepted that under field conditions the performance of a herbicide will be variable and will not necessarily reflect earlier glasshouse results. Indeed, the bridge between herbicide activity under glass and the effects of that same herbicide in the field is a tenuous one. It can be argued strongly, that any correlation should not be attempted, the only assumption by the field researcher being that a compound elevated to field research has intrinsic activity for him to measure and exploit. Therefore, it is the task of the field researcher to design suitable trials to account for variability, to obtain reliable data and to form objective conclusions and recommendations.

The mechanics of field evaluation, i.e. the technology of cultivation, herbicide application, recognition of symptoms and their assessment, are well documented. For completeness they will be briefly described later in this chapter. At least as important, but often less appreciated, is the crucial nature of the trial's objective. The objective should be carefully considered since it will affect subsequent analysis of data and can also modify technique.

The objective

Dose–response

Herbicides can be compared in two distinct ways. Consider the comparison of several products at their recommended field rates. Here the objective is merely to distinguish one from another at their commercial-use rates. This type of trial is frequently used in field development to investigate the efficacy of competitor products. However, field research work generally seeks to distinguish between compounds which are new and for which few data, relevant to field usage, are available. In these cases, dose–response trials in which efficacy of compounds against specific weeds over defined rate ranges is evaluated, are the most suitable.

For the first case, straightforward analysis of variance techniques can be used to define differences. However, the second is a problem which must be solved by regression techniques and rate-for-rate comparisons should be avoided. Moreover, in a dose–response experiment, the optimum rate of herbicide will only rarely correspond to a treatment which has actually been used. It is the objective of the regression analysis to interpolate from the data to find that optimum rate. The use of extreme rates is to be encouraged since they stabilize the dose–response curve. However, it is worth noting that at the extremes, fewer replications are required to achieve precision.

If the dose–response is plotted arithmetically, a smooth curve may be achieved. Log transformation of the dosages and a transformation from percentage response produces a straight line. Such transformations, called 'probits', are commonly employed in dose–response studies to determine the line of best fit and to compare relative efficacies or selectivity. The comparisons may be done at any level but commonly the LC_{50} or LC_{90} (lethal concentration for 50% and 90% control, respectively) are used.

Probit analysis is designed for data which are based on measurements not subjective scores, e.g. mortality. Subjective assessment, e.g. percentage control, should be analysed by regression and comparison of regression methods.

A selectivity index between target weed and crop may also be derived from dose–response curves. It has been defined as the ratio between the dose giving acceptable crop effects and that achieving commercially-acceptable weed control. For levels of damage and control of 10% and 95% respectively, the selectivity index is:

$$\text{Selectivity index} = \frac{LC_{10} \text{ Crop}}{LC_{95} \text{ Weed}}$$

Comparisons between compounds can be made using their selectivity indices.

An essential step if registration requirements are to be met is to evaluate this calculated optimum dose in the field. Subsequent response curves containing the optimum treatment must comply with all the standard requirements of dose–response trials described previously.

Multi-factor trials — mixtures

The evaluation, the efficacy, spectrum and selectivity of combinations of herbicides is an important area of field study. Deficiency in spectrum, the need to increase speed of action or to extend residual control of specific compounds are the major reasons to initiate mixture trials with products of complementary action. Similarly, potentiation or synergism of one compound by another is a desirable and patentable feature, which may arise through the chance choice of mixture components or by rational design.

Irrespective of the type of activity sought, the broad aim of mixture trials is to define the interaction between a number of quantitative factors, such as concentration of herbicide, surfactant and oil, and determine a combination of levels, by interpolation, which gives the desired result. The technique involves regression rather than analysis of variance and in choosing the rates of the component herbicides it is an advantage to have some prior knowledge of their performance alone. However, if this is unavailable it is important that a wide range is chosen to include the two extremes of 'no effect' and 'complete kill'. Even if those extremes are totally unrealistic their inclusion is essential to stabilize the multi-dimensional dose–response (response surface) which will be achieved and which will describe the interaction between the mixture components. The choice of rates within that range should ideally involve collaboration between field researcher and statistician. Characteristically, the middle rates, and their combinations, must be in the region of the greatest rate of change in effect, remembering that whilst replication is essential, the addition of a treatment here will often be more useful in the definition of the surface than one at either extreme.

An analysis of the data from a trial with two herbicides A and B represented by dosages from 0 to $300 \, \text{g ai ha}^{-1}$ and 0 to $500 \, \text{g ai ha}^{-1}$ is shown in Fig. 10.11. An index of weed control is given on the Y-axis and the various interactions at the specific rates under test, have been linked in the form of a three-dimensional response surface.

The same surface can be represented as a contour plot (Fig. 10.12),

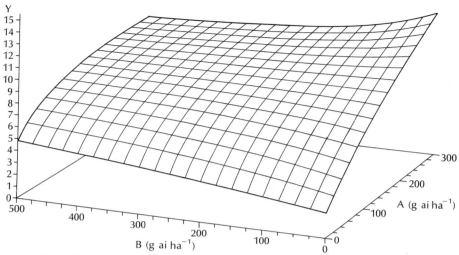

Fig. 10.11 Three-dimensional plot of fitted response surface.

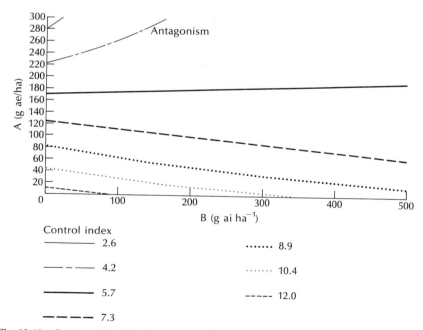

Control index

———— 2.6	⋯⋯ 8.9
— · — 4.2	⋯⋯ 10.4
———— 5.7	――― 12.0
— — — 7.3	

Fig. 10.12 Contour plot of fitted response surface.

in which each contour represents a particular level of weed control. Antagonistic and synergistic effects can be seen and 'ideal' mixtures, which in reality were not tested, can be determined.

Trials in mixed populations

Similar studies can be applied to crop tolerance in weed-free situations. However, the ultimate goal must be to evaluate the potential use of a compound in natural populations of crops and weeds. Such complex interactive systems require the use of multi-factorial designs and analysis, and one of the most useful methods to use is response surface technique. For a three-factor trial, e.g. dose, efficacy and selectivity, a true three-dimensional response surface is obtained. This is the situation encountered by a farmer and he would assess the usefulness of a herbicide only by its effects in this interactive situation. A convenient illustration is the control of a grass weed, such as blackgrass, in a cereal crop. Here, the herbicide or herbicide mixture elicits two responses of different types and associated with these are different and competing objectives. As control is exerted over the weed, the crop yield will increase. However, the use of an excessive amount of herbicide(s) while increasing weed control, may reduce crop yield.

Towards commercialization

The culmination of field trials programmes is the determination of the efficacy, selectivity and performance (in mixture with new or established products) of a new herbicide. At this point, it is essential to consider that herbicide research and development is expensive and the products of that work have to cover that expense and provide a return for the company. The advantage of such chemicals to the farmer is increased yield, assessed as bulk or quality, or merely as an aid to routine farming practice. The removal of weeds, whilst in some cases not affecting yield, may facilitate harvesting and hence reduce cost. Increased yield or decreased costs both lead to greater profitability. The task of industry is to balance cost of herbicide against the farmer's need for useful profit. Regression techniques and good experimental method allow the researcher to combine data which describe biological usefulness with costing and, using suitable comparisons with existing or future chemicals within that same market, make a judgement on further development. Such considerations are best made as early as possible in the research and development process but are usually well within the area of interest of field development.

Conducting the trial

To distinguish between treatment effects, the trial must be designed to be as precise as possible. Precision implies low variability and can be improved by increased replication, treatment selection, improved application and assessment techniques, increasing plot size, the use of appropriate statistical techniques and the choice of site.

The trial site

The field performance of a herbicide is influenced by a wide range of variables which may interact. These include the climate conditions, crop–weed competition, age of target, and morphology, agronomy, soil type, topography and genetic variation of the target weed.

Before proceeding with any trial, the objective must be clear because to address too many questions within a single trial frequently results in failure. Clearly it is important to decide upon those factors which will most profoundly affect the development potential of the herbicide and proceed stepwise to investigate them so that they can be eliminated or evaluated. The above variables imply that a large part of the total variation in any trial is due to its position and finding suitable sites for field trials is one of the most problematic aspects of herbicide evaluation, particularly on non-company land.

On company-owned land, trials are more straightforward in this respect. Field station sites are chosen to be flat, uniform areas to remove as many variables as possible and allow the researcher to carry out in-depth trials necessary in the early stage evaluation of chemicals.

The precision of any field trial can be further enhanced through the proper, but frequently unusual, management practices employed. For example, trial plots are commonly within a three-, preferably five-year, rotation with a cereal and grass crop. Perennial ryegrass is often chosen as a convenient rotational partner to trials because of its ease of management, but also because the indigenous weed population can be easily suppressed or controlled with established non-residual herbicides to leave a 'clean' area for herbicide trials. Cereals also have advantages related to the control of weeds and clearance of crop, but in addition offer the field researcher the opportunity to grade the uniformity of the area prior to the year of trials. Simply, this is done as a heterogeneity trial, using the crop area as a trial and harvesting it as a series of small (3 m × 15 m) plots. The yield figures are then plotted on a map and used to define areas of equal 'fertility'. These areas can be used to group subsequent trials and hence compensate for normally unseen variation.

The uniformity of on-station trials may be improved through the careful application of standard fertilizers and by choosing sites where animals have not been used in the rotation. Off-station trials may not be able to be sited with precision. The researcher then has to rely upon correct design to compensate for the variability within the trial.

Randomized complete block

Randomized complete block (RCB) design is especially suited to field experiments where the number of treatments is not large and the experimental area has a predictable gradient of inherent variability. Blocks or blocking is used to reduce the experimental error by accounting for that gradient. When the gradient or pattern of variability is unknown, then the blocks should be arranged as square as possible. However, if heterogeneity can be assessed beforehand, the blocks should be arranged accordingly. Consideration should be given to the restrictions inherent in the site such as slope and shape of field. The position of the trial slope and shape often determines the pattern of normal spraying operations; care should be taken to position the test site to avoid over-spraying by the farmer.

When treatments within an RCB design are unrelated, analysis of variance techniques should be used. Results which combine treatments, as in response work, should be analysed by regression.

Factorial design

Factorial designs are useful when more than one variable is being studied. Trial layout is basically as for RCB but split plots and other refinements can be added to accommodate objectives. At its most useful, the factorial design can allow for a complete picture of interaction between components, expressed as a response surface.

Replication

The precision of an experiment can always be increased by additional replications but the degree of improvement decreases rapidly as the number of replications increases. In planning an experiment, the field researcher should be reasonably confident that a true difference can be detected. If the probability of meeting that objective is low with the number of replications available and there is no other means of improving precision, then the experiment should not be carried out.

Treatments

Careful selection of treatments and the quality of subsequent observations can improve precision. The choice of treatment is very important in a dose–response trial where the objective is to determine how the herbicidal effects differ with increasing dose rather than to conclude the absolute value of different doses and if they differ significantly.

In dose–response trials the addition of extra treatments is often more advantageous than the inclusion of more replicates per treatment. However, in all cases, it is crucial to plan the dose ranges properly to allow the optimal analysis to be employed.

Techniques

The reduction of variability in the materials and methods associated with any trial will also lead to increased precision. For example, chemical application may be made more uniformly. In all cases, application is made at right angles to the drilled crop or weed. This reduces error, particularly in yield trials, which may arise from gaps in the drill or from variations in the field owing to previous cultivation or concurrent routine maintenance of the crop.

Trials conducted on company land can be planned according to requirements of the experiment. Typically, weed species are drilled alongside crop species and drilling is done to ensure the emergence of each at the correct time or their presentation at the correct growth stage at the time of spraying. Off-station trials, however, must be set out for the convenience of both experimenter and farmer, the essential consideration being to avoid over-spraying of the trial during routine herbicide applications to the field. The design of the trial may therefore involve a compromise.

Following application, variables such as rainfall, and growth-stage development, must be monitored. Above all, the effects of the treatments need to be assessed regularly and here it is important to record exactly what symptoms are apparent.

Equipment

Application

Application equipment used in field trials includes knapsack, microplot, logarithmic and tractor-mounted sprayers. Generally, the trial's objective will govern which is the most suitable but factors such as trial size and availability of chemicals also need to be considered.

Knapsack sprayers

Herbicides may be applied using knapsack sprayers, powered by compressed gases, usually carbon dioxide. These sprayers can be used to apply chemicals to small plots (no smaller than 5 m long) in a manner closely resembling commercial application.

The operator has to maintain a constant speed and ensure that pressure and nozzle height (boom height) above the target are correct and do not vary. Spray width can be altered from single nozzle to about 4 m. The size of the boom is restricted only by its weight and the volume and weight of the chemical needed to fill the system and meet the requirements of the treatment replicates.

This equipment is particularly suited to initial dose–response studies or spectrum trials. The area used can be small, approximately 500 m^2 for a twelve-treatment, four-replicate trial, with a corresponding economic use of chemical.

Micro-plot equipment

When plot size is limited by chemical availability, the use of micro-plot sprayers may be considered. Here again it is important to stress that the objective should be clear and the accuracy of the equipment and its ability to be used to meet that objective assessed critically.

The majority of micro-plot sprayers operate over areas of 1 m^2 and less. Most of them are hand powered and because of the difficulty in achieving a constant speed in a short distance, they tend to be highly inaccurate. However, static nozzle micro-plot sprayers have been used successfully.

Logarithmic sprayers

Logarithmic sprayers are used to apply chemicals over a dose range as the operator moves down the field plot. This is achieved by the continuous dilution of the test chemical with a suitable diluent. The system lends itself to 'ready-made' dose–response studies and has been used to investigate the interactions between chemicals, e.g. chemical A continuously diluted with a known amount of chemical B, or with a changing amount of chemical B. However, it must be stressed that this equipment can only give very crude ideas of effective ranges of herbicides or mixtures. It suffers from gross inaccuracy which arises mainly from the inability of the operator to maintain a constant speed along the length of the plot, together with the errors caused by the possible re-distribution of applied chemical through the soil and the sometimes variable doses applied to targets of different height.

Accurate calibration shows the variation as a change in applied dose. It must also be remembered that the plot is a dosage continuum and discrete doses theoretically occupy infinitely narrow bands across the plot.

Modifications have been made to overcome the inherent inaccuracy of log sprayers. Wheeled versions, complete with speedometers, single and multi-axle types and the use of governed motors have all been tested. None has succeeded, particularly on uneven ground. They can only be used with confidence on perfectly flat dry grassland. Generally, they should be avoided at all costs.

Tractor sprayers

Tractor-mounted commercial applications are used in later stage trials. The objective here is to assess the performance of a potential herbicide under grower conditions. The variability inherent in tractor spraying is justified and is an important feature of such trials.

Calibration

Standardization in application equipment is essential. Sprayers should be equipped with matched sets of standard nozzles. Their performance must be checked regularly, preferably using a patternator. Spray pattern uniformity can be evaluated colorimetrically using colour-sensitive paper or dyes.

Assessment methods

Chemical effects should be assessed in two ways — efficacy and crop safety. Generally, both require subjective measurements based upon a percentage scale (0 = no control; 100 = complete control). Certain assessment scales relate to commercial acceptability (e.g. 0–10 scale, 8 = acceptable level of control) but such methods, whilst of relevance to a particular trial, produce non-linear descriptive data which are difficult to analyse trial to trial. Using a percentage scale, correlations between trials can be achieved and provided percentage levels of commercial acceptability are indicated, the economic potential of treatments can be assessed. In other trials, more quantitative assessments may be required, e.g. plant number or head counts per unit area. The natural untreated population must also be assessed to obtain an estimate of zero control.

At the time of assessment the stage of weed and crop development must be recorded. Efficacy with respect to defined symptoms must be measured for each weed species but estimates of total control of a mixed population of weeds should be avoided.

Defined crop effects should be scored as percentage damage compared to untreated controls, with an estimate of commercial acceptability.

Interpretation of data

The successful completion of a trials programme allows the field officer, in collaboration with research, formulation, toxicology, residue, environment, registration and marketing groups, to make recommendations for further development. In making these decisions the cost/efficacy of the herbicide is assessed against the current and future market needs, the existing and developing competitors, the potential profitability as a mature product, toxicology and environmental safety. Often the herbicide is returned to the glasshouse or formulation chemist for further study prior to another field trials' programme. However, the ultimate objective is to be able to make a reliable product label recommendation for the herbicide which describes its use and usage to the farmer.

Safety evaluation and regulatory requirements

Once a compound has satisfactorily passed various biological laboratory screening trials, consideration is given to investigating other properties in order to evaluate the safety of the compound. Initially, acute toxicity studies are conducted so that relevant safety advice can be given to laboratory and field research people who start to work with the compound.

As the compound progresses through biological evaluation to a position of being a potential commercial compound, a full programme begins to generate basic data to support the development of the compound and ultimately to satisfy the national regulatory authorities. Such data are wide ranging and cover basic physical/chemical properties, toxicity testing, environmental fate, ecotoxicity, crop residues and various other ancillary studies. Generally the data are generated in stages usually in a stepwise progression in parallel with the biological development.

Before a new compound can be sold in any country, adequate data must be provided to the national registration authorities, that show the product has adequate safety for:

1 Operator handling, mixing and spraying the product.
2 Consumption of the treated crop.
3 Environment in the short and long term.
4 Crop that is being treated.

The fourth safety aspect is considered as part of the biological efficacy data generated to satisfy the need to know that the product works as designed and is commercially viable. These data requirements are considered elsewhere in this chapter and so discussion will deal with the first three criteria.

The data produced to satisfy national registration authorities are also used to assess continually the safety of the compound to research workers, manufacturing personnel and any other people who may come into contact with it during development, manufacture and ultimate introduction into commercial use. Today's emphasis on compounds for widespread use in major world crops or against major world pests means that safety data are developed on a global basis, and the requirements of all the world's major regulatory authorities must be considered. Countries such as Australia, Canada, Federal Republic of Germany, Japan, UK, USA and many others, produce detailed guidelines on data required for registration of a pesticide, as do several international organizations such as the Council of Europe (CEC), GIFAP and WHO/FAO.

Toxicological studies

In order to make adequate predictions regarding the effect of pesticides in relation to man, a wide range of toxicity studies are carried out in mammals, especially rats and mice, but also in dogs and primates. Whilst there is growing debate by numerous groups about the relevance of certain animal tests and the demand for the reduction in animal testing, to date there are few adequate, validated *in vitro* methods to substitute for direct animal testing, and even when available there is a reluctance by government regulatory authorities to accept them in preference to existing live-animal tests.

At the beginning of development when the pesticide is still in the research stage, acute toxicity studies are carried out, and on the basis of the derived LD_{50}, sub-acute or dose-ranging probe studies are begun in order to define more critically the nature of toxicity, such as target organs, clinical chemistry or haematology.

Once the pesticide has passed the research stage and a decision has been made to start general development, the range of additional toxicity studies is started. These include, 90-day feeding studies up to and including two-year chronic toxicity/oncogenicity studies, generally in two species. In addition to these, tests to investigate mutagenic potential, effects on reproduction and the pharmacokinetics and metabolism in animals, are also planned and initiated.

Often decisions in terms of the continued development of the product through research trials, pre-development trials and full country development trials, are linked to results of certain key studies in the toxicology programme. As indicated by Fig. 10.13, these studies are started at different times during the development of the pesticide as the continual safety assessment of the molecule is built up, based on the knowledge of the studies completed.

Studies of effects on the environment

Pesticides by their very nature are associated with the general environment. Before a new herbicide can be registered and sold, it is necessary to assess its environmental impact and to assure that its widespread use does not cause any environmental problems. To achieve this, many studies are conducted to investigate different parameters. Traditional application of a post-emergence herbicide to a major crop like cereals using conventional spraying techniques can lead to 30–60% of the applied herbicide landing on the surrounding soil. Once a herbicide has reached the soil, degradation starts either through chemical means or microbial action. Under certain extreme circumstances, the applied herbicide can move into neighbouring watercourses by direct run-off following rainfall and under certain circumstances, percolate down through the soil into the water-table below.

Similarly, any herbicide that lands on, and moves into, the soil could affect both macro- and micro-organisms. Although studies are carried out in order to determine such effects, they are often difficult to interpret particularly under field situations because of the influence of natural forces

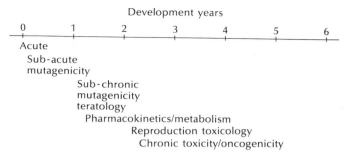

Fig. 10.13 Typical toxicology development schedule.

during any one season. Animals or birds can consume a newly-treated crop or treated seed and thus consume a level of pesticide that could adversely affect them and ultimately animals that prey on them higher in the food chain. Likewise, a herbicide entering a watercourse can adversely affect the various orders of aquatic life encountered, and which in turn can produce effects on the surrounding wildlife that live and breed close to water whether avian or mammalian.

Fate in the environment

As with the toxicology programme, initial assessments of the environmental fate of a herbicide are based on relatively simple studies in the laboratory to determine such characteristics as water solubility, the effect of acid and alkali, octanol–water partition coefficient, soil absorption/desorption properties and rate of hydrolysis. From these basic data, reasonable predictions can be made of the basic behaviour of the herbicide in soil and water and hence its potential environmental impact. As development progresses, laboratory studies are started to investigate fully the degradation, metabolism and persistence in soil and in water. If the indications are such that the herbicide may leach in soil, studies are carried out to confirm these results.

To study degradation and other transformation pathways the herbicide with a radioactive label in one or more specific sites is incorporated into uniform soil samples of various soil types and maintained in an enclosed environment. Samples are taken at regular time points and the breakdown of the herbicide investigated, initially by extraction and separation of the variously-labelled components followed by chemical analysis, usually chromatography, of the individual components. Volatile components evolved during the experiment, usually carbon dioxide, are measured after trapping in a suitable medium.

Concurrently, the effects of temperature, moisture content, concentration, soil type and time are investigated. The data from these detailed types of study are used to determine the metabolism of the herbicide and to calculate the half-life ($t_{1/2}$) and the time to 90% disappearance, DT_{90}, of the herbicide. Any significant metabolites produced by the breakdown of the herbicide can also be followed. In addition, anaerobic soil metabolism is usually studied as well as soil leaching and absorption/desorption studies in various soil types.

With increasing environmental concern from many directions, the use and interpretation of laboratory studies do not always define adequately the overall fate of a herbicide in soil. Ultimately, and particularly with herbicides, it may be necessary to carry out field studies, where the herbicide is applied in one or more typical agricultural areas and samples are taken to confirm the degradation and leaching characteristic of the herbicide.

Field studies play an important part in the evaluation of a herbicide; however, their main drawback is that weather conditions cannot be predicted over the duration of the study (often up to 18 months). In order to impart a degree of control over such work (in particular rainfall) and to maintain other environmental factors, lysimeter work is becoming more and more a preferred option to define critically the fate of a herbicide in soil.

Classical laboratory leaching studies invariably overestimate the potential of a herbicide to leach. Whilst useful as a comparative indicator against known standard compounds, practice has shown that field data and lysimeter data can give considerably different results. This area of investigation is becoming more and more important, particularly in light of concerns over the potential of herbicides to leach into groundwater and, ultimately, drinking water. The CEC Directive of 1980 on the quality of drinking water, sets levels of 0.1 ppb for individual pesticides and detailed studies are now necessary to investigate more fully the potential of a pesticide to leach. Analytical methods require refinement to meet this very rigorous demand. The Directive is the subject of considerable debate as there is no explicit reason for the level set but it is certainly not based on toxicological considerations.

Depending on the nature and use of the herbicide, other studies may be carried out to investigate the carryover of residues, following the harvesting of the crop or to investigate the fate to sequentially-planted crops following harvest or in the event of crop failure.

Studies are also carried out on the fate of the herbicide in water using either a radioactively-labelled compound or technical material. The intrinsic stability of the herbicide to acid and alkali is investigated as part of the early evaluation, so is already known, but detailed studies in natural waters including suspended matter and sediment are necessary. In addition to these basic studies, photo-degradation on soil surfaces and photolysis in water are usually investigated.

With all these data, a detailed picture can be drawn of the herbicide's likely environmental fate and a prediction of its potential risk can be made.

Ecotoxicity

The toxicity of the herbicide is also investigated in fish and other aquatic species such as Daphnia and algae. Because a pesticide could enter a watercourse by accident or through environmental factors such as erosion or run-off of surface water following its application in the field, a knowledge of any potential aquatic effects must be ascertained.

Whilst it was common practice some years ago, just to test the acute toxicity of the pesticide to fish in static test systems, nowadays pesticides are investigated in dynamic systems, for acute and chronic toxicity and potential bioaccumulation. Depending on the properties and use of the herbicide, tests may also be carried out on estuarine aquatic species, and reproduction studies may be performed.

Tests on birds are now routine. Herbicides may be applied in several different ways and onto an extensive variety of crops. Birds can consume crops at the seedling or early growth stage or when the crop is mature. Therefore studies are necessary to investigate the acute toxicity to various species of birds, together with feeding studies and investigation of the potential effect of the chemical on reproduction.

Tests on bees must be done particularly if the herbicide is applied at, or near, the flowering time of the crop or weed. Studies are also carried out in the laboratory and in the field to investigate the effects on soil macro- and micro-organisms, such as earthworms and bacteria. Similarly, although more particularly with insecticides, the effect on predator species must also be considered.

Residues in crops

Apart from people working in the manufacture of pesticides and those who apply them, the main potential exposure to a pesticide for the general population is by consumption of the treated crop. Pesticides applied to a crop or the soil in which it is growing, may be present in the harvested part of the crop so various studies are carried out to investigate the fate and nature of ultimate residue in the various food commodities treated and consumed. Residue trials are conducted in the field over at least two seasons, and from various locations.

Initially, residue samples will be taken from replicate trials where the herbicide has been applied at various dose levels, up to at least twice the anticipated maximum use rate. Depending on the nature and use of the herbicide, it can be important to investigate the residue in the harvested crop following application at different growth stages of the crop.

To complete the development of the basic crop residue picture, a further year of trials may be carried out in which the herbicide is applied to the target crop using conventional commercial application equipment under normal agricultural use. Many national regulatory authorities publish guidelines on the type and number of residue trials that should be carried out, as do GIFAP and WHO/FAO, and these should be consulted before trials are planned.

Should the use of the herbicide give rise to significant residues in the harvested part of the crop, it may be necessary to carry out further work on the crop commodity to investigate the fate of the residue following cooking, processing, etc. A significant residue in cereal grain may necessitate the investigation of the fate of the residue following milling and baking into flour and bread, respectively. Likewise, a residue in oilseed rape may require investigation into the partition of the residue in the oil that is produced for human consumption.

In addition to residue studies in crops, residue/feeding studies are also carried out in large animals, in particular cattle. If there is likely to be a significant residue in the crop, whether it goes for human or for animal consumption, it is important to investigate the potential for accumulation in meat or milk. These data, like those from crop residue studies, are important in the assessment of potential effects in the general public.

The appropriate residue data, together with toxicological data, allow an assessment to be made on potential effects to consumers.

Data from the long-term toxicological studies will be used to estimate a 'no observable effect level' (NOEL). This is the test level administered in the diet over the lifespan of test animals, that causes no observable toxicological effects. From this figure an 'acceptable daily intake' (ADI) for a person can be derived using an appropriate safety factor. Knowing the maximum residue level attained from the residue trials at the maximum commercial use rate, and the average daily consumption of the particular crop commodity, a comparison can be made between the derived ADI from the toxicological studies, and the maximum potential intake from the use of the product. As long as the latter does not exceed the former, it is considered there is no long-term risk to consumers eating treated crops.

In general, authorities have not set specific maximum residue levels for herbicides unless they have been set at the limit of determination of the analytical method. However, with the ever-increasing complex chemistry and the introduction of such chemistry into herbicide development, it is conceivable that many new herbicides will have specific maximum residue levels established that are reflected by their toxicological properties.

Supplementary studies

One of the prime objectives in obtaining the data described previously was to assess the potential risk to operators handling and using the products in the field. To achieve a satisfactory assessment of safety, it may be necessary to carry out operator exposure studies under field conditions.

Having investigated the fate of the herbicide, on or in the consumed portion of the crop in residue trials, there is often a need to assess the quality of the crop. This usually involves taint studies which are more often required following the use of insecticides and fungicides where application is often made much closer to harvest. However, it is feasible that the use of a herbicide could require such taint testing.

When the manufacturer is satisfied that the newly-developed pesticide is effective and does not pose a potential risk to users, the environment or the consumer, then the data are submitted for independent evaluation by the national regulatory authority. Only when the appropriate regulatory authority is also satisfied and has granted Approval or Registration for the herbicide product may the product be sold.

Time-scale and costs

Figure 10.14 shows a representation of the time-scale of screening and field testing of a compound. If the chemistry has come from an empirical or rational base, it will frequently take two years of synthesis, screening, debate and further synthesis to improve activity sufficient to justify a patent filing. This action precipitates significant synthesis effort as the filing of a provisional patent allows only one year for additional synthesis and biological evaluation before patent completion. This year also involves the selection of a field candidate, formulation studies and field screening.

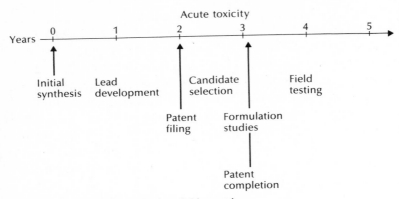

Fig. 10.14   Time-scale from synthesis to field screening.

Within this time-scale it is not unusual for a decision on a compound's future to be taken five years from the start of the project.

Calculations on the number of compounds which have to be synthesized for each successful product have been made by various organizations for many years. There probably is not a means of measuring this accurately but it is certain that the number has been increasing since 1960. The reasons for this are many. More products are now available and consequently broader-spectrum, flexibly applied, highly-selective herbicides are required. The hurdles each compound has to overcome are higher and therefore compounds which would have been products 20 years ago are now rejected. Quite correctly farmers and manufacturers are very concerned about the effect of compounds on non-target organisms and the environment. For this reason, detailed and extensive tests are carried out on each new chemical series to ensure there are no undesirable effects. This approach excludes certain compounds and whole series of chemistry. As the pesticide industry becomes more competitive it becomes increasingly important to provide a new discovery with adequate patent cover. This means making compounds around a claim which will not become products or even field candidates but which are synthesized to demonstrate the scope of the claim and the breadth of chemistry involved. An approximate calculation would show 20 000 compounds entering a primary screen, 2000 a secondary screen, 200 a tertiary screen and 20 being tested in the field for each product ultimately marketed.

Cash flow associated with product discovery is shown in Fig. 10.15. This shows that a company's expenditure may reach £20 million before any sales are achieved. Assuming a successful launch it will often take 14 years

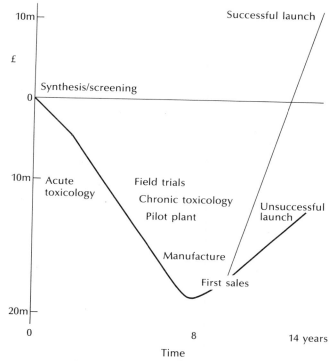

Fig. 10.15 Cash flow associated with product discovery.

from the initiation of the project before this money is recouped. There-after, of course, additional profit becomes cash in hand to be re-invested in the next pesticide discovery!

11 / Weed control in small grain cereals

D.R. TOTTMAN AND B.J. WILSON*

*ARFC IARC Broom's Barn Experimental Station, Higham, Bury St Edmunds, Suffolk IP28 6NP; and *AFRC IACR Long Ashton Research Station, Long Ashton, Bristol BS18 9AF*

Introduction

Cereals are grown on nearly four million hectares in the UK and virtually the whole of that area is treated with herbicides, much of it more than once a year. Many of the recent developments such as continuous cropping, more winter cereals, earlier drilling and reduced cultivations have been made possible only by efficient chemical weed control. Herbicides make a major contribution to profitable cereal growing.

Significant changes in cereal cropping, some not easily predictable, can be expected in the coming decades. Over-production in Europe may lead to a fall in grain prices. If so, herbicide use, in common with other inputs, must be considered for possible economies. Public concern about country-side conservation and a demand for organically-grown food is already generating pressure to minimize pesticide and fertiliser use. On the other hand, government payments for farmers to set-aside cereal land is likely to encourage high-input, intensive production on the remaining area. In the longer term, changes in weather patterns as a result of the greenhouse effect could have dramatic consequences for the growth of cereals in the UK and are likely to alter both the weed flora and its competitive ability.

Cost-effective decisions about cereal weed control will depend on a thorough understanding of the principles of crop and weed biology, competition and susceptibility to herbicides.

The weed flora

The species and abundance of weeds in a cereal crop depend on infestations in previous crops and are influenced by the system of cropping and cultivations. Traditional rotations were characterized by a wide diversity of species while a monoculture encourages a few species to dominate and sometimes to build up to high densities. The increase in the area of winter cereals in the 1970s favoured weed species that germinate in the autumn. Reduced cultivations encouraged plants whose seeds have little dormancy and can germinate quickly in the surface soil.

One of the more important factors affecting the distribution and abundance of weeds is the herbicide regime employed in previous crops. Before the advent of the 'hormone' herbicides, charlock and poppies were among the most widespread and competitive weeds. When these were effectively controlled, other weeds, such as cleavers and chickweed, resistant to the herbicides then available, flourished and became the new 'problem' weeds.

A successful annual weed germinates quickly after the seedbed cultivations, grows vigorously before the crop canopy becomes too dense and produces large numbers of seeds before cereal harvest. Seed dormancy may be short, which ensures quick germination in the following crop, or longer, which maintains a seedbank in the soil ready to germinate when conditions are favourable.

Perennial weeds generally emerge late but underground food reserves sustain their early growth in the competitive crop, allowing them to rise above the cereal canopy and then to make vigorous growth in late summer when the crop is ripening.

Although many weed species occur in cereals and can cause severe local problems, relatively few are common and widespread. Among the grass weeds, there are wild oat, blackgrass, couch and the meadow and brome grasses. The more important broad-leaved weeds in winter crops are common chickweed, cleavers and mayweeds. In recent years, speedwells and field pansy have increased. In spring crops, chickweed, knotgrass, redshank and pale persicaria are the main weeds and common hemp-nettle is a major problem in Scotland and the north of England.

Annual grass weeds

Wild oats: These compete strongly for light, nutrients and water and, at high densities, can severely reduce cereal yields. Green wild-oat plants interfere with harvesting and grain contaminated with wild-oat seed is reduced in value. Two species occur in cereal crops. The common wild oat is widespread throughout England and parts of Scotland, Wales and Ireland. It germinates through the winter and into spring. Although not completely frost-hardy, it can be a weed in both autumn and spring cereals. The winter wild oat is more localized, in the Midlands and eastern counties. Germination is restricted to the autumn and winter, it is frost-hardy and is, therefore, a weed of autumn-sown crops.

Wild-oat seeds can remain dormant in the soil for several years so a long-term strategy is required for control. A wild-oat population depends on the balance between seed production and mortality during the course of the weed's life cycle. Seed production is favoured by early germination, a thin and uncompetitive crop and adequate soil moisture in summer. Seed survival is reduced by delayed cultivations, by straw burning or in early harvested crops, when most of the seed will be harvested with the grain. The size of a wild-oat plant and the number of seeds it produces vary considerably, depending on its time of germination and competition from the crop. It is, therefore, difficult to define population thresholds at which herbicide treatment is economic. Between 5 and 20 plants m^{-2} are likely to reduce yield by a value equivalent to the cost of a herbicide treatment. However, the potential for rapid build-up warrants treatment at much lower populations and hand roguing can be used to prevent seed return if the population is below about 400–500 plants ha^{-1}.

Blackgrass: This is mainly a problem of winter cereals on heavy soils in central, southern and eastern England but has been extending its range to lighter soils, especially where drainage is impeded. It has been encouraged by the increasing frequency of winter cereal crops, earlier autumn drilling and the widespread use of reduced cultivations. About 80% of seed germinates in the autumn. Early cereal sowing means that fewer seedlings will emerge in time to be destroyed by the seedbed cultivations, leaving more to come up in the crop. Reduced cultivations leave most seed close to the soil surface in conditions ideal for germination and establishment.

Blackgrass can severely reduce yields of winter cereals. Losses average about 1% for every 5 plants m^{-2} at normal population levels. Very large populations of 800–1000 plants m^{-2} have been recorded. Heavy infestations produce vast numbers of seeds, with plants having three or four heads, each containing an average of 100 seeds. Seed production from smaller populations may still be substantial because, without intense competition from its neighbours, each plant is likely to bear many more ears. However, viability is relatively low and natural mortality high. Straw burning further reduces the seed stock and ploughing buries most of them below the 5-cm maximum depth from which seedlings can successfully emerge. Ploughing again for a subsequent crop can raise buried, but viable, seed to the surface. About 97% of the seeds in the soil die within three years. A small but sometimes significant proportion can survive for many years.

Because reduced cultivations favour blackgrass, a higher level of chemical weed control is needed to prevent the weed increasing than when ploughing is practised. However, the accumulation of burnt straw residues near the soil surface, associated with reduced cultivation systems, adsorbs soil-acting herbicides and can reduce their performance. A measurement of soil adsorptive capacity can help to predict herbicide performance.

The short-term economic threshold for control lies within the range of 10–50 plants m^{-2} depending on the cost of treatment and the efficacy of the herbicide. However, over the long term and allowing for variations in competitiveness and efficacy of the herbicides, spraying is advisable if only 2–5 plants m^{-2} are present. Such low numbers are difficult to determine in the seedling stage and it may be more practicable to base the spray decision on the presence of ≥ 2 heads m^{-2} in the previous crop during June.

The brome grasses: These are traditional hedgerow and field margin species but have recently increased dramatically as problem weeds. Although still most common on headlands, they sometimes spread through fields and are strongly competitive. Barren brome is the predominant species in cereals and is found throughout southern England, the Midlands and East Anglia and less frequently as far north as the east of Scotland.

The seeds of barren brome have little dormancy and germinate readily in moist soil. Germination is delayed by insufficient moisture and by exposure to light if the seeds remain on the soil surface. Dry autumns and very early drilling reduce the proportion of seed that will germinate and be killed by seedbed cultivations. The seeds subsequently germinate in the crop and are difficult to kill with herbicides. Nearly all seeds germinate in the autumn after shedding, so any break in the annual seed production cycle, for example, by a change to spring cropping, will dramatically reduce the infestation. Few seedlings are able to emerge from deeper than 13 cm so efficient ploughing is usually an effective treatment.

Meadow-grasses: These are prolific seed producers, and are prevalent in cereal/grass rotations in the high rainfall areas of the north and west of the country. Annual meadow-grass germinates throughout the year and is favoured by shallow cultivations. Rough meadow-grass, although potentially a perennial, behaves like an annual in winter cereal crops, growing from seed each year, with germination peaks in late spring and late autumn. Seed dormancy is induced by burial so stubble cultivations should be delayed as long as practical to encourage germination and mortality. Seedlings require a cold period (vernalization) if they are to produce seed. A succession of spring crops can act as a cleaning break, provided they follow efficient ploughing to avoid over-wintering transplants. In early-sown winter cereals, control depends on effective herbicide use.

Annual broad-leaved weeds

Common chickweed: This is the most common broad-leaved weed in arable crops. It thrives in cool, wet conditions and on fertile soils. It tolerates low temperatures and readily over-winters producing large competitive plants in early spring. Germination can occur throughout the year but with distinct peaks in the spring and autumn. Large numbers of seeds are produced and dormancy is very variable, some germinating immediately they are shed, others joining a long-term bank of viable seeds in the soil. The variable germination and the short life cycle, allowing several generations in a year, make this species a very successful weed.

Cleavers: This is the most aggressive weed of winter cereals. Its climbing and scrambling habit allows even small populations to form a dense canopy over the crop, causing severe lodging and loss of yield. Individual cleavers plants are more competitive than wild oats or blackgrass and 20–30 plants m^{-2} can halve wheat yield. Cleavers seeds are difficult to separate from wheat grain and are often spread by drilling contaminated seed. Seeds drilled with the crop are ideally placed for successful germination. Cleavers requires fairly low temperatures for germination and germinates mainly in the autumn and early winter. Dense infestations can produce over 20 000 viable seeds m^{-2} and individual plants, at low densities, over

1000 viable seeds per plant. Dormant seeds can persist in the soil for at least 18 months. Seeds are shed late in the season and the majority are still attached to the plants at harvest.

Mayweeds: These germinate in most months with peaks in the spring and autumn. They produce abundant seed which can survive and remain dormant in the soil for many years. Mayweeds tend to remain green and competitive throughout the life of the cereal crop in contrast to many other broad-leaved weeds which die away before the cereals ripen.

Speedwells: These have increased in recent years because they are favoured by autumn cropping and are not controlled by the widely-used, substituted-urea herbicides. Ivy-leaved speedwell germinates in the autumn and winter and is confined to winter cereals. Common field speedwell, which germinates throughout the year, is also a weed of spring cereals. Hydroxybenzonitrile herbicides are effective but ivy-leaved speedwell is generally less easily killed than common speedwell.

The polygonums: These are mainly weeds of spring cereals. Knotgrass requires low temperatures before germination and emergence usually begins in February, reaching a peak in March or April. Black bindweed germinates about a month later than knotgrass, starting in March with the main flush in April and May. It is a climbing weed and may cause crop lodging. Redshank and pale persicaria germinate over the same period as black bindweed. They are most abundant on wet soils.

Field pansy: This is becoming a severe problem in some winter cereal crops because it is resistant to many herbicides. Although individual plants are small and poorly competitive, very high populations can build up, particularly on light soils. They exploit the open canopy of a thin crop, reduce its yield and the wet bulk of their foliage at harvest can interfere with grain separation.

Perennial weeds

Couch grasses: These include a number of creeping grass weed species, the most common being common couch and black bent. Both have extensive rhizome systems which provide the main method of propagation. Shoot growth from the over-wintered rhizomes is slow, emergence occurring after the spring-sown cereals. New rhizome growth begins after the aerial shoots have started tillering, reaches a maximum in mid-summer and slows down during the autumn. Competition from the crop slows down aerial growth and delays rhizome formation but later in the season, when light begins to penetrate the cereal canopy, common couch makes rapid growth. It normally spreads its flag leaf above the crop and continues to grow while the crop is ripening. Seed production is greater in black bent than common couch but seedlings are not very vigorous and are susceptible to disturbance and most of the blackgrass herbicides.

Onion couch: This is a perennial form of false oat grass that propagates by strings of bulbs at the stem bases, although it also germinates readily from seed. Its distribution is localized but, where it exists, can be a difficult weed to control. Although ploughing buries the bulbils, it must often be supplemented with herbicides. Several wild-oat herbicides will control the aerial growth and kill some of the bulbils. Pre-harvest applications of glyphosate are effective if sufficient green leaf remains late enough in the season to be sprayed.

Creeping thistle: This is a persistent and troublesome weed of both grass and arable land and an increasing problem in winter cereals. It competes strongly with cereal crops. It spreads vegetatively from a deep underground root system which fragments easily during cultivation. Each piece produces new shoots. They emerge late in the spring, often after the normal time for herbicide spraying.

Field bindweed: This is a widespread weed of arable land that can be very persistent and difficult to eradicate. The climbing habit of the bindweed leads to lodging and so makes harvesting difficult as well as reducing yield. Bindweed has a perennial root system consisting of deeply penetrating vertical roots and shallower horizontal laterals, capable of regenerating new shoots. The main period of shoot growth is from June to September, too late to be controlled by normal spring applications of the 'hormone' herbicides. Cultivations after harvest fragment the shallow roots but the regenerated shoots usually perish. Successful re-growth comes from buds on the deeper roots below the depth of cultivation.

Effects of weeds on cereal crops

Competition

Cereal crops are most sensitive to weed competition during the early stages of stem elongation, when growth is most rapid. However, large weed populations may require earlier control if they are not to reduce crop yield.

Competition is likely to be most severe when large numbers of weeds emerge at the same time as the cereal crop. In autumn-sown cereals, high densities of the grass weeds, wild oats, blackgrass and sterile brome can be very competitive. With more than about 20 wild-oat or 50 blackgrass or brome seedlings m^{-2}, autumn control is advisable to avoid competition in early spring. Competition from lower populations will be less and control in the spring is likely to be early enough to avoid serious yield loss. However, autumn herbicide treatments may be preferred if they offer better or more cost-effective weed control.

Broad-leaved weed competition is generally less severe than from the grass weeds but varies widely with species. A number of weeds have been ranked in order of their mean dry weights compared to that of winter wheat (crop equivalents) in Table 11.1. The figures were obtained in a series of field experiments conducted over several years by staff at Long Ashton Research Station. While considerable variation was evident, the order of the species was fairly consistent. On this basis, cleavers emerges as a most competitive weed but even the least competitive, like the speedwells and field pansy will reduce yield if the population is very high. Current experiments are aimed at refining predictions of crop yield loss by weed competition, including the effects of populations made up of several different species.

Competition is for light, moisture and nutrients. Early competition from large numbers of small weeds will be mostly for nutrients. Moisture and light will be limiting later, when the weeds have produced abundant foliage. Of the later emerging weeds only those which rise above the crop

Table 11.1   Estimates of competitive ability of some common weeds based on crop equivalent measurements in experiments at Long Ashton Research Station

Weed	Crop equivalent
Cleavers	7.2
Wild oat	2.5
Poppy	0.6
Mayweed	0.6
Blackgrass	0.5
Common chickweed	0.5
Common speedwell	0.5
Red dead nettle	0.3
Ivy-leaved speedwell	0.3
Forget-me-not	0.2
Field pansy	0.1
Parsley piert	0.1

will interfere with the availability of light and are likely to reduce crop yields.

Competition is affected by the weather and the relative times of germination of the crop and weeds. A vigorous cereal crop suppresses weed growth well. The weeds emerging early in the life of the crop will remain larger and reduce yield more than those that emerge later and whose seedlings have to struggle for survival in competition with the rapidly-growing crop. Anything that reduces the crop's speed of establishment and vigour, like sowing too deeply, soil compaction, nutrient deficiencies or disease, will encourage weed growth. In weedy situations a high crop population will help to counter the effects of early weed competition. Extreme cold sometimes kills weeds; common wild oat and charlock are quite sensitive to frost. On the other hand, mild winters allow weeds like common chickweed to maintain growth and become very competitive in the spring.

Effect on harvesting and grain quality

The grain from a crop infested with weeds will cost more to harvest, clean and dry than grain from a clean crop. Weedy crops take longer to dry and hinder the harvesting programme. Weeds increase the total volume of material processed by the combine, reduce its efficiency and may increase grain losses unless the machine is driven very slowly.

Weeds with a climbing or clinging growth habit such as cleavers, field bindweed, black bindweed and barren brome, when present in abundance, can cause lodging. Where the crop has lodged, late weed growth can make harvesting very difficult. Grain harvested from such crops is likely to have a high moisture content, particularly if the weeds are green and immature at harvest. Grain from weedy crops, particularly those infested with weeds like cleavers that continue to grow and compete late in the season, may be small, shrivelled and fail to meet quality standards.

Weeds as grain contaminants

Grain, otherwise suitable for milling but contaminated with weed seeds, is unlikely to be accepted and will, therefore, lose its quality premium. Nor will grain be accepted for EC intervention purchase if it contains more than 3% total impurities, including weed seeds.

Seed regulations limit the amount of contamination in seed offered for sale. Home-saved seed also needs a high standard of cleanliness as even a small percentage of weed seeds can result in serious infestation. Sowing

188 kg seed ha^{-1} of seed contaminated with 5 g kg^{-1} cleavers will introduce 24 weed seeds m^{-2}.

Weeds as carriers of pests and diseases

Some weeds act as hosts to cereal pests and diseases. The weeds may build up disease levels in the crop. For example, blackgrass has been shown to increase the risk of ergot (*Claviceps purpurea*) infection in wheat. More commonly, weeds provide alternative hosts, enabling pests and diseases to survive between cereal crops, nullifying or reducing the effects of breaks in the rotation and providing a source of infection to newly sown crops. Examples include take-all disease (*Gaumannomyces graminis*), barley yellow dwarf virus, cereal cyst nematode (*Heterodera avenae*) and frit fly (*Oscinella frit*), all carried on various grass weeds.

Volunteer cereals have become troublesome weeds where cereals are grown intensively. They can act as a 'green bridge' from one season to the next, carrying foliage diseases such as yellow rust (*Puccinia striiformis*), brown rust (*Puccinia hordei*) and powdery mildew (*Erysiphe graminis*).

Weed control by cultural means

Weed prevention

Although weeds can be controlled with herbicides and appropriate cultivations, it is worth taking precautions to prevent the more aggressive weeds invading fields where they do not already occur. Some weeds invade from the field boundary. Herbicide sprays or extra cultivations on a barrier strip or on the headlands can prevent encroachment by perennial grasses. Fallowing a strip alongside a hedge with cultivations helps to break the seed cycle of the brome grasses. On the other hand, indiscriminate spraying of the perennial vegetation along field boundaries, such as hedge bottoms, is likely to kill harmless species and open the way to colonization by more aggressive weeds.

Sowing crop seed contaminated with weeds is a common source of new weed problems. Home-saved seed should be taken only from clean fields. Weed seeds are also carried from field to field, and to other farms, on vehicles and implements, in straw collected from weedy fields and in manure made from such straw.

Regular field inspections before harvest to pull and destroy any new weed before it seeds can prevent the establishment of a later problem.

Cultivations

The cultivation system has a profound effect on weed population dynamics and therefore, the herbicides subsequently needed in the cereal crop. Ploughing used to be the universal preparation for a cereal crop. The introduction of paraquat, a non-selective herbicide with almost no soil activity which controls most weeds in the stubble, allowed seedbeds to be prepared without ploughing. This alters the weed flora; annual broad-leaved weeds are generally reduced but the grasses and perennials flourish. Glyphosate is more effective than paraquat against perennial weeds and now makes an important contribution to the success of minimum culti-vation techniques. Continued weed and disease problems have, how-ever, in many areas, led to a reversal of the previous trend to minimal cultivation.

The burning of straw debris and stubble from the previous crop has an effect on weed populations. It can dramatically reduce the subsequent population of blackgrass seedlings (Fig. 11.1) and the quantity of freshly-shed viable wild-oat seed. It also doubles the autumn germination of freshly shed wild-oat seed, presumably by reducing the dormancy of seeds

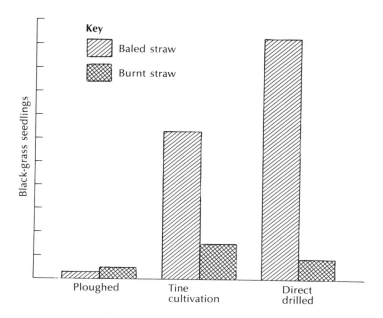

Fig. 11.1 The influence of cultivation and straw disposal on the density of blackgrass seedlings. Data from Moss (1981).

surviving the burn. Accumulations of ash in the seedbed reduce the efficacy of soil-acting herbicides. The chemical control of blackgrass can be enhanced by ploughing every 4–6 years to disperse ash and organic matter and to bury weed seeds.

Early stubble cultivations can be a key factor in the control of rhizomatous grass weeds such as common couch and black bent. Such cultivations also reduce some annual weeds by stimulating germination so that their seedlings can be killed by subsequent cultivations. On the other hand, early stubble cultivations may favour wild oat because burial prevents the natural mortality of freshly-shed seed on the soil surface.

Seedbed cultivations should be limited to those required to provide the necessary conditions for cereal germination and establishment. Deep cultivations may drag to the surface, seeds, roots and rhizomes that had been buried by earlier cultivations. A fine seedbed may be necessary for optimum activity of soil-acting herbicides. Clods can shield germinating weeds and allow them to grow without coming into contact with the herbicide layer.

Current methods of cereal husbandry are largely dependent on the use of chemical herbicides. However, public enthusiasm for food, free from pesticide residues, is reflected in premium prices for organically-grown grain. This may be sufficient to offset the extra costs of a change to traditional rotations and lower yields. Weed control in organically-grown cereals is essentially based on long rotations involving grass leys and mechanically-tilled root crops. Drilling dates and cultivations must be carefully timed to kill weeds and minimize seed return to the soil. Stale seedbeds can be used, in which an initial shallow cultivation stimulates weed germination and the seedlings are subsequently killed by a deeper cultivation before the crop is sown. It may be several years before such measures take effect and the transition from a pesticide-based husbandry to an organic one can be difficult. Weeds remain a severe constraint to growing cereals without agro-chemicals.

Weed control with herbicides

Annual broad-leaved weeds

The first herbicide to be used on a large scale in cereals was sulphuric acid but the advent of the dinitrophenol herbicides, DNOC and dinoseb, and the 'hormone' herbicides, first, 2, 4-D and MCPA, followed by mecoprop and dichlorprop, brought advantages of greater selectivity. Weeds, resistant to these chemicals, such as mayweeds and speedwells, prompted a continuing search for new herbicides. There followed the discovery of the benzoic acids, 2,3,6-TBA and dicamba, the hydroxybenzonitriles, ioxynil and bromoxynil and, more recently, isoxaben, diflufenican and the sulphonylureas, each with a slightly different spectrum of activity. Materials with activity against a more specific range of weeds have been introduced, like bentazone, benazolin, clopyralid and bifenox, each of them valuable in mixture with other herbicides. Fluroxypyr also controls a restricted range of weeds but is particularly active against cleavers.

Soil-acting herbicides are often used to control broad-leaved weeds and meadow grasses in the autumn. These include pendimethalin, trifluralin, linuron and low rates of the substituted urea herbicides.

Grass weeds

The first selective grass weed herbicides were barban and tri-allate, active against wild oats. More recent wild-oat herbicides include flamprop-methyl, flamprop-M-isopropyl, benzoylprop-ethyl, diclofop-methyl, difenzoquat and imazamethabenz-methyl. With the exception of tri-allate, they are post-emergence, mainly foliage-acting herbicides. In contrast, the herbicides developed for the control of blackgrass are mainly substituted ureas which act primarily through the soil. They also have activity against several other important grass and broad-leaved weeds. The broad spectrum of activity of chlorotoluron and isoproturon, which includes blackgrass, meadow-grasses, chickweed, several other broad-leaved weeds and some control of wild-oats, has made them very popular autumn treatments in winter cereals. Diclofop-methyl, chlorsulfuron and imazamethabenz-methyl all show activity against blackgrass but are often used in mixtures or sequences with other herbicides to achieve optimum control of the weed.

Perennial weeds

Glyphosate, although essentially a non-selective herbicide, can be used preharvest to control perennial weeds like couch, creeping thistle and field bindweed. It can be applied, in most cereal crops, once the moisture content of the grain has fallen below 30%.

New herbicides

Two groups of herbicides, the sulphonylureas and the imidazolinones interfere with the activity of an enzyme system unique to plants and are, therefore, very safe to animal life. The sulphonylureas, in particular, are highly active at very low rates, some at $<20\,g\,ai\,ha^{-1}$. Metsulfuron-methyl is already widely used in cereals in the UK. The imidazolinones are currently represented by imazamethabenz-methyl. Within each of these groups is a range of potential herbicides with different selectivities that encompass both grass and broad-leaved weeds and crops. They enter plants through both soil and foliage. They are readily mobile in the soil but some have persisted long enough to injure subsequent sensitive crops.

Herbicide mixtures and compatibility

No one herbicide can control all the weeds that might be encountered in a cereal crop. It is, therefore, often necessary to mix herbicides. There are proprietary mixtures containing up to four active ingredients and some recommendations permit herbicides to be mixed with each other and with other agro-chemicals in the spray tank. There are limitations due to the danger of incompatibility, either through physical or chemical changes in the mixture that might lead to precipitation and separation, or biological interactions which might result in poor weed control or crop injury. Care should be taken to follow manufacturers' recommendations for mixing their products with others and the time intervals sometimes necessary between one treatment and another. For example, the 'hormone' herbicides can interfere with the activity of several of the wild-oat herbicides if the treatments are applied within 7–10 days of each other.

Determination of weed and cereal growth stages

The response of a weed to a herbicide often varies with its size and stage of growth. To convey information and advice about weed control requires easily understood and standardized descriptions of weed growth stages. A suitable key is offered by Lutman and Tucker (*Annals of Applied Biology*, 1987, **110**, 683–687). The fifteen phrases used to describe the growth stage of annual dicotyledonous weeds are reproduced in Table 11.2 and examples are illustrated in Fig. 11.2.

Crop response to weed competition or a herbicide treatment often varies with its growth stage. The internationally-accepted standard for cereal growth stage descriptions is the decimal code proposed by Zadoks, Chang and Konzak (*Weed Research*, 1974, **14**, 415–421). This has been expanded and related to the specific problems of agro-chemical timing in

Table 11.2 Descriptive phrases for weed growth stages.
From Lutman and Tucker (1987)

Pre-emergence
Early cotyledons
Expanded cotyledons
One expanded true leaf
Two expanded true leaves
Four expanded true leaves
Six expanded true leaves
Plants up to 25 mm across/high
Plants up to 50 mm across/high
Plants up to 100 mm across/high
Plants up to 150 mm across/high
Plants up to 250 mm across/high
Flower buds visible
Plant flowering
Plant senescent

United Kingdom crops by Tottman and illustrated by Broad (*Annals of Applied Biology*, 1987, **110**, 441–454, reprinted as BCPC Occasional Publication No. 4). The stages of greatest relevance to herbicide applications are redrawn in Figs 11.3 and 11.4.

Sensitivity to the 'hormone' herbicides depends on the stage of development of the shoot apices. The salient stages of apical development are: apex elongation; double ridges, when about half the final number of spikelets have been initiated; terminal spikelet (awn primordia in barley) when the number of spikelet primordia has reached its maximum; and the green-anther stage, when the cells divide to form pollen and ovules (meiosis). Photomicrographs of some of these stages are reproduced in Fig. 11.5. However, these stages can only be identified after dissecting plants under a microscope and most spray decisions, in the field, have to be based on the external appearance of the plants. Correlation between the growth stages defined by the decimal code and apical development is sometimes poor but certain features offer a practical guide to safe spray timing.

In spring cereals the number of leaves unfolded on the main shoot indicates the approximate stage of ear development, although not all modern varieties have been checked and the relationship does not hold in other countries where temperatures and daylength are markedly different from the UK. A leaf is described as unfolded when its ligule has emerged from the sheath of the previous leaf and care is needed to avoid counting tillers and their leaves (Fig. 11.3).

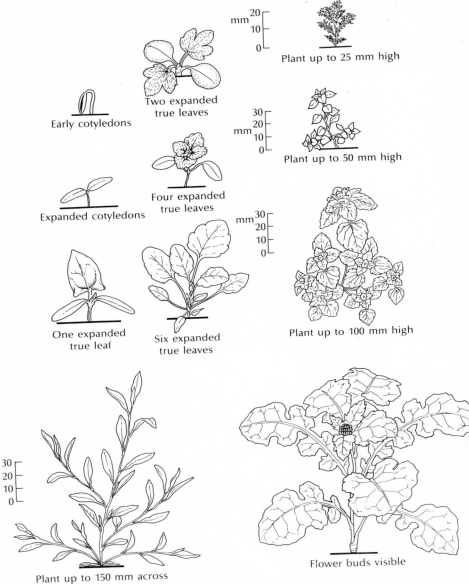

Fig. 11.2 Examples of weed growth stages. From Lutman and Tucker (1987).

For winter cereals the relationship between leaf number and ear development is less clear, and by the spring, leaves are difficult to count accurately because the lower ones often die and shrivel during the winter. A more reliable guide to spray timing is given by the length of the true

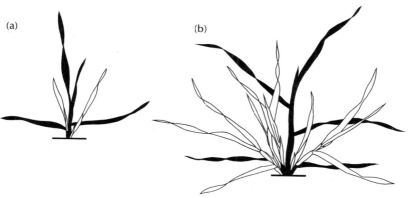

Fig. 11.3 Examples of cereal growth stages: (a) wheat, three leaves unfolded, main shoot and two tillers; and (b) spring barley, five leaves unfolded, main shoot and five tillers. From Tottman (1987).

stem. To see this, the plant must be dug from the ground and the main (largest) shoot split with a knife or finger-nail. The following stages can be identified and are illustrated in Fig. 11.4:

Ear at 1 cm (growth stage 30) = pseudostem erect: The stem, measured from where the lowest leaves are attached is 1 cm to the shoot apex. The first internode is <1 cm.

First node detectable (growth stage 31): An internode of ≥1 cm is present but the internode above it is <2 cm. Occasionally the node can be below ground level and may later bear roots.

Second (and subsequent) detectable nodes (growth stages 32, 33, etc.): These are counted when the internode below them is >2 cm.

When the ear is at 1 cm the apex will be beyond the double-ridge stage and floret initiation is likely to be in progress. Detection of the first node coincides approximately with maximum spikelet number. Meiosis occurs when the anthers change from colourless to green and the ear is 2–2.5 cm long. When this happens the third node is likely to be detectable and the flag leaf is usually beginning to emerge.

When sampling a crop to determine the stage of growth, it is essential to remove whole plants from the ground and samples should be dug rather than pulled, in order to avoid leaving leaves or tillers behind. A number of plants from several areas of a field should be checked and account taken of any differences in growth stage caused by uneven drought, waterlogging or disease.

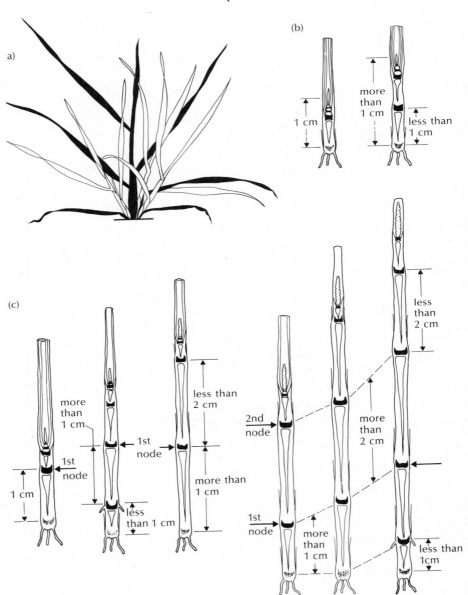

Fig. 11.4 Stages of cereal stem elongation: (a) wheat, ear at 1 cm, six leaves unfolded, main shoot and four tillers; (b) ear at 1 cm, main shoots split; (c) first node detectable; and (d) second node detectable. From Tottman (1987).

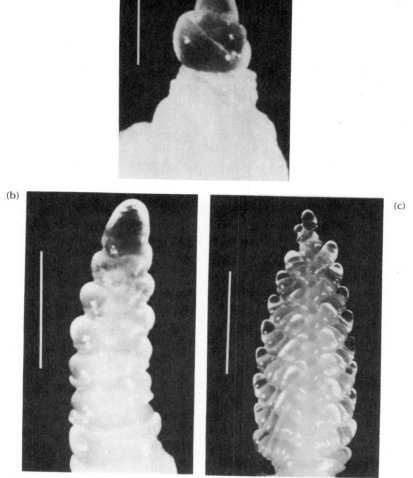

Fig. 11.5 Photomicrographs of wheat shoot apices: (a) vegetative stage — initiating leaf primordia, scale bar = 0.25 mm; (b) double ridge stage — initiating spikelet primordia, scale bar = 0.5 mm; and (c) terminal spikelet stage — initiating stamen primordia, scale bar = 1 mm. From Kirby and Appleyard (1984).

Cereal tolerance of herbicides

The incorrect use of herbicides can lead to serious crop damage. Great care is needed in the accuracy of application and its timing in relation to the crop growth stage, the weather and cultural operations. Crop reaction to a herbicide treatment is not always easy to predict and is often strongly influenced by weather conditions prior to, at and after spraying (see Chapter 6). Plants growing under stress whether due to drought, extreme temperatures or disease infection are likely to prove more sensitive to a herbicide.

The benefits of weed control usually outweigh any minor risk of crop damage but herbicides are often used as an insurance against weed problems or to prevent infestations in subsequent crops. In such circumstances, there may be no immediate yield benefit to offset the expense of the herbicide treatment and the extra cost of crop damage is unlikely to be acceptable. Where several effective treatments are available, the one that involves least risk to the crop is the obvious choice.

Pre-emergence treatments

Adequate soil moisture is necessary for weed seeds to germinate and to keep the herbicide in solution, so that it can be taken up by the developing seedlings. On the other hand, in wet autumns the herbicide may leach into the crop root zone and cause injury. A minimum drilling depth is often specified to ensure separation of the germinating crop seed from the layer of soil containing the herbicide.

Post-emergence treatments

Some weeds need to be controlled early in the life of the crop and the development of crop sprayers that can be used on wet soil now makes this possible in winter cereals. However, the 'hormone' herbicides, like mecoprop, are less active at low temperatures and sometimes give poor control. In frosty conditions several herbicides, notably mecoprop and the substituted ureas, chlorotoluron and isoproturon, can scorch crop leaves. This most often occurs when the frost is preceded by mild growing weather in the autumn and before the crop is 'frost-hardened'. Very rapid autumn growth can also render plants more than usually sensitive to herbicides, as happened in 1983 when instances of crop damage from the substituted urea herbicides were recorded. Usually quite severe symptoms of leaf scorch are outgrown with no effect on final yield but such a setback in growth is likely to reduce the crop's ability to compensate for any subsequent unfavourable conditions.

Other herbicides applied during early tillering can sometimes reduce the plant stand or interfere with the tillering pattern. Examples can be drawn from applications of high doses of herbicides to sensitive varieties. The substituted urea herbicides kill some plants but the survivors continue to grow vigorously. Early applications of the wild-oat herbicides, difenzoquat and diclofop-methyl, can injure the main shoots of sensitive varieties and the plants react by re-tillering prolifically. This results in many but small ears at harvest.

Applications of some 'hormone' herbicides during the early stages of ear initiation can lead to marked plant deformities. The most severe symptoms are caused by 2,4-D and MCPA (Fig. 11.6(a), (d) & (e)). Common symptoms are tubular or 'onion' leaves (which can trap and kink the emerging ear), enlarged and fused glumes or lemmas and derangement of the spikelets on the ear. Spikelets may occur opposite each other on the rachis instead of the normal alternate arrangement or several spikelets may be borne in a whorl with a length of bare rachis below them. In wheat, extra or 'supernumerary' spikelets may be produced by spraying just before the safe stage but they also occur naturally in some varieties.

Malformations do not necessarily result in a yield loss but grain quality is likely to be reduced and malting quality can be influenced by changes in nitrogen content brought about by early spray damage. Subsequent germination of the grain is rarely affected.

The tolerance of oats is less dependent on stage of growth. The oat panicle does not pass through the same distinct stages of development as the ears of other cereals. Spraying with 2,4-D can cause leaf deformities and bunched spikelets, due to abnormal branching of the panicles, over a wide range of growth stages. MCPA is less likely to do so.

There is evidence that several herbicides used in cereal crops can reduce yield if used when the nodes are detectable, but the effect is not easily predictable. The herbicides most likely to cause damage at this stage are the benzoic acids, such as 2,3,6-TBA and dicamba, although mecoprop, alone or in mixture with the hydroxybenzonitriles, has sometimes reduced yields. A crop affected by the benzoic acid herbicides produces thin ears with shrivelled grain and often turns grey owing to colonization by sooty moulds (Fig. 11.6(c)). Even later applications of 'hormone' herbicides like mecoprop, during flag leaf emergence when the cell divisions that give rise to pollen and egg cells are taking place, may lead to abortion of grain sites with a consequent drastic reduction in crop yield.

The growth-stage restrictions of the 'hormones' are not shared by the majority of the newer cereal herbicides and timing can be more readily tailored to the requirements for most effective weed control. For example, the recommendations for the use of fluroxpyr and metsulfuron-methyl in winter wheat allow treatment at any stage between two leaves unfolded and the appearance of the flag leaf ligule.

Pre-harvest treatments

When grain growth is complete and the crop is ripening, herbicides are unlikely to influence crop yield but effects on the biochemical constitution of the grain are still possible. Glyphosate can be safely used to control perennial weeds, pre-harvest, in most cereal crops. The mammalian toxicity of glyphosate is very low and once grain filling is complete its main effect on the crop is to hasten desiccation. It can be applied to crops intended for feed, milling or malting, but not for seed, when the grain moisture has fallen to <30%.

Varietal differences in crop tolerance

Differences in the tolerance of current cereal varieties to broad-leaved weed herbicides are generally considered to be of little practical significance. Apparent differences are often the result of slight variation in the growth stages of the varieties at the time of spraying. Sensitivity to the phenoxy-acid herbicides does exist. At least one potential cereal variety, the spring oat 'Margam', was rejected at a late stage in the breeding programme because tests showed that yields were reduced by commonly-used herbicides containing mecoprop. Continuing checks on the tolerance of all new varieties are therefore essential.

The tolerance of certain grass weed herbicides differs markedly from variety to variety. Recommendations for the use of chlorotoluron, metoxuron and difenzoquat include lists of varieties that can be sprayed safely. Other varieties may be damaged. Tolerance of the substituted urea herbicides is usually an all-or-nothing effect, simply inherited and amenable to manipulation in a breeding programme, if this were thought worthwhile. Where varieties show an intermediate response to other herbicides, many yield trials are needed to evaluate the risk of treating them on a commercial scale. The rapid turnover in new cereal varieties makes this procedure time-consuming and expensive. The possibility of introducing genes for tolerance to non-selective herbicides is currently being pursued with some commercial success, but not yet in the cereal crop.

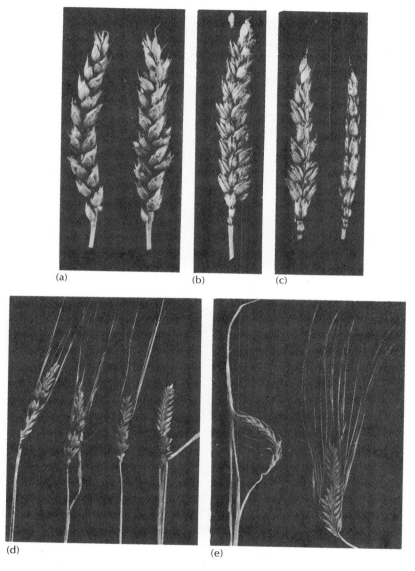

Fig. 11.6 Deformities of cereal ears caused by 'hormone' herbicides: (a) abnormal ears of wheat with opposite and supernumerary spikelets; (b) normal ear of wheat; (c) 'rat-tailed' wheat ears caused by late application of the benzoic acid herbicides; (d) abnormal ears of barley with whorled spikelets and extended internodes of the rachis; and (e) tubular leaf trapping the emerging ear and aborted spikelets.

Interaction with plant diseases

There is evidence of an interaction between herbicides and take-all disease
(*Gaumannomyces graminis*). Increased levels of infection on the cereal
roots and rat-tailed ears have been noted following over-doses of herbicide. Many diseases, particularly those causing cereal foot-rots, are likely
to flourish in crops weakened by a poorly-tolerated herbicide treatment.

Effects on foliar diseases are variable. While some herbicides have
been reported to increase infections, difenzoquat has given fairly good
control of mildew (*Erisyphe graminis*). Spraying diseased crops with
herbicides, other than those known to have a beneficial effect, involves a
risk of crop damage or increased disease infection.

Cultivations and herbicide damage

Physical damage to the crop in the 7–10 days before or after spraying can
allow increased entry of the herbicide into the bruised leaves and lead to
crop damage.

Cultivation of the soil after the application of soil-acting herbicides,
unless incorporation is specifically required, may result in crop damage by
mixing the herbicide into the crop root zone and/or poor weed control by
breaking up and diluting the herbicide layer at the soil surface.

Weed resistance to herbicides

There are several examples of weeds developing resistance to herbicides
(see Chapter 8), usually in maize or in perennial crops which may be
sprayed with the same or similar herbicides on a regular basis over many
years. The selection pressure on cereal weeds is generally less because
cereals are usually grown in rotation with other crops, cultivations encourage a dynamic seedbank of annual weeds in the soil and they have been
sprayed with a changing array of herbicides. However, some fields have
grown continuous cereals for many years and have been treated regularly
with the substituted urea herbicides to control blackgrass. On a few of
these fields there are now blackgrass plants resistant to several times the
normal dose of these and other herbicides. In another part of the country,
populations of chickweed resistant to mecoprop have been discovered. At
present, there is no evidence that these populations of resistant weeds

are spreading or that new ones are being induced elsewhere. Genetic resistance remains the least likely cause for any herbicide failure; unfavourable weather, incorrect spray timing or inaccurate application are more probable explanations. Nevertheless, the potential danger of weeds that cannot be controlled with the available herbicides is sufficient to justify vigilance by farmers and advisers and further research by the agrochemical industry.

Cereal weed management — a summary

The objective of weed control in cereals is to reduce weed populations to levels that do not affect the yield, quality or harvesting of the current or subsequent crops. To achieve this economically generally requires a strategy based on the intelligent use of herbicides coupled with appropriate cultural practice. It is often easier and cheaper to control broad-leaved weeds in cereals than it is in other crops and this should sometimes influence decisions about cereal weed control.

It may not always be necessary to kill every weed in the crop to achieve the best economic return. Some of the common broad-leaved weeds are not very competitive and large yield increases cannot be expected from the removal of low populations.

A few weeds, mainly the annual grasses and cleavers, are fiercely competitive and high levels of control are needed to reduce the seed burden in the soil. Cultural measures such as ploughing, stale seedbeds and straw burning may be necessary to supplement the available herbicides. Sometimes the measures required to reduce different weeds conflict. For example, early post-harvest cultivations may benefit the control of barren brome and common couch but encourage the seed survival of wild oats. Priority should be given to the control of the weed most difficult to control with herbicides, in this case, barren brome.

Spraying thresholds

Spraying just a few weeds in a cereal field is unlikely to be economic but defining the threshold population that justifies treatment is difficult and will vary with the relative times of germination of the crop and weeds. The patchy distribution of many weeds and the difficulty of counting low populations further detract from the simple application of spraying thresholds. The short-term economic threshold for wild oats in winter cereals lies between 5 and 50 plants m^{-2} and for blackgrass between 20 and 50 plants m^{-2}. Ten-year economic thresholds, allowing for potential population increases, may be one-quarter to one-seventh of these densities. Cleavers are so competitive that even very low populations need to be sprayed. In any case, populations need to be kept well below the notional economic thresholds in case herbicides in future years fail or cannot be applied as intended.

Although eradication of a weed species from a field or farm has often been attempted it has proved very difficult in practice. Some success has been achieved with wild oats which, at low populations, can be hand-rogued.

Timing of weed control

Cereals are most sensitive to competition in the early stages of stem elongation, which, in autumn-sown crops, usually occurs during early April. However, weeds that emerge with the crop can be 10 times more competitive than those that come up when the crop is tillering and some species, like blackgrass, chickweed and speedwells, make vigorous growth in a mild autumn. There are, therefore, circumstances in which autumn control may result in a significant yield advantage over a spring treatment. Other reasons for spraying weeds in the autumn include the ease with which some weed species can be killed at that time of year, either because they are more sensitive to pre- or early post-emergence herbicides or because they become resistant to herbicides as they grow larger. Autumn weed control relieves the work load in the spring when spraying may be delayed by the weather and conflict with other crop husbandry operations.

Early post-emergence applications of herbicides like isoproturon and chlorotoluron often give better control of annual grass weeds than pre-emergence treatments. Later treatments also have the advantages of persisting longer into the season and allowing assessment of the weed population before spraying. However, soil and weather conditions can make late autumn treatment difficult.

After autumn herbicide treatments, more weeds may germinate in the spring and lead to significant seed return to the soil. Sequences of autumn and spring treatment, perhaps at reduced herbicide rates, may give optimum yield response and prevent seed return.

Spray opportunities

Opportunities to spray winter cereals may be quite restricted and require appropriate equipment and careful planning to achieve optimum timing. In autumn and winter, rain, wind and low temperatures reduce the number of days on which spraying is possible. Wet soil may also prevent spraying unless a low ground pressure vehicle is available. The crop passes through the 'safe stage' for the application of the 'hormone' herbicides in as little as 10 days in April and, even at that time of year, unsuitable weather conditions can restrict spraying to one day in four, or even less where adjacent crops or gardens present a drift hazard.

Economic considerations

Husbandry systems such as continuous winter cereals established with minimum cultivations encourage annual grass weeds and expensive herbicides are needed to control them. On fertile soils where high yields and good financial returns are possible, the expenditure is justified and the intensive use of herbicides is likely to continue even if cereal prices fall. If cereal growing is to continue on the poorer soils, where yields are restricted by low fertility or inadequate soil moisture, cheaper methods of weed control must be sought. A greater emphasis on crop rotations and cultivations to minimize problems from aggressive grass weeds can help to avoid the need for expensive herbicides.

Only the careful integration of cultural control measures and rational herbicide use, judged on the basis of a full understanding of weed population dynamics, competition and the factors that influence herbicide performance, will enable farmers to grow cereals economically and without detriment to wildlife and the countryside.

Acknowledgements

The authors gratefully acknowledge the assistance of Mr Jim Orson (MAFF/ADAS, Cambridge), Mr Dick Makepeace (Oxford Agricultural Consultants Ltd) and Mr Allan Lock (Integrated Crop Protection Systems, Milton Keynes) who made significant contributions to the previous edition of this chapter and who have helped with its revision. Thanks are also due to Mr Steven Moss (Long Ashton Research Station) for his contribution to the sections on blackgrass and cultivations.

Further reading

Kirby, E.J.M. & Appleyard, M. (1984) *Cereal Development Guide*, 2nd edn. Arable Unit, National Agricultural Centre, Stoneleigh. 96pp.

Lutman, P.J.W. & Tucker, G.G. (1987) Standard descriptions of growth stages of annual dicotyledonous weeds. *Annals of Applied Biology*, **110**, 683–687.

Moss, S.R. (1981) The response of *Alopecurus myosuroides* during a four year period to different cultivation and straw disposal systems. *Conference on Grass Cereals*, UK (University of Reading) pp. 15–21. Association of Applied Biologists.

Tottman, D.R. (1987) The decimal code for the growth stages of cereals, with illustrations. *Annals of Applied Biology*, **110**, 441–454.

Zadoks, J.C., Chang, T.T. & Konzak, C.F. (1974) A decimal code for the growth stages of cereals. *Weed Research*, **14**, 415–421.

12 / Weed control in other arable and field vegetable crops

C.M. KNOTT

Processors and Growers Research Organization, The Research Station, Great North Road, Thornhaugh, Peterborough PE8 6HJ

Introduction

This chapter deals with crops other than cereals, including oilseed and protein crops, sugar-beet, annual vegetable and forage crops. The crop, its method of culture, place in the rotation, and the market for the end product all affect weed-control strategy, together with herbicide availability, timing, trends in usage and crop tolerance.

The crop and growing systems

The speed with which the crop covers the ground is a major factor in suppressing weeds and can be encouraged by providing the best possible conditions for crop growth, as for example with stubble turnips. However,

329

some crops, such as onions, with slow initial growth and erect habit are naturally poor competitors.

As mechanization of crop production has increased, crops which were labour intensive and grown on a small scale, such as onions and green beans, are now grown on a field scale with almost total dependence on herbicides. Farms which specialize in high-value salad and vegetable crops, may have to carry a labour force for harvesting and preparation of produce for market which is available for handwork and inter-row cultivations so reduced herbicide use is still possible. The standard of weed control sought becomes higher as production methods increase in sophistication. For example, the weed interference tolerated in hand-harvested over-wintered spring cabbage is unacceptable in a short-season, mechanically harvested crop of calabrese grown for quick-freezing.

In most crops, as weed control with herbicides has removed the need for inter-row cultivations, row widths can be reduced with consequent changes in plant populations. However, some husbandry changes have recently taken place regardless of potential weed-control problems. The development of methods for mass producing modular or block-raised transplants and new transplanting techniques has changed weed control in crops such as brassicas, celery and onions. These transplants can be more sensitive to herbicide damage but existing herbicide recommendations were developed for bare-root transplants. There has been rapid uptake of the use of floating plastic film; over 6000 ha of outdoor vegetables were covered in 1987–1988 in the UK. The objectives are earliness and a longer cropping season, but there are also improvements in emergence, yield and quality. At present very few herbicide labels refer to the new techniques and the situation regarding Approvals is unclear. Problems range from residual herbicides having no effect because the soil surface dries out, restricted leaching causing herbicide residues to remain near the soil surface, and because of warmer conditions some weeds germinate early and grow too big for control by herbicides when the cover is removed. Possibly soil sterilants may prove to be the best means of weed control. There is therefore a need for further development work and re-assessment of herbicides for such special conditions.

Market requirements

The market outlet, whether for processing, fresh market, compounding for animal feed, crushing, seed or 'organically-grown' produce also plays an important role in determining the methods used for weed control.

The Food & Environment Protection Act has introduced legal constraints on the use of pesticides and there are set limits for maximum pesticide residue levels (MRL) in food crops (Chapter 21). In addition, herbicides used in a crop for processing must leave no taint or 'off flavour'. Processors require that pesticides are tested and cleared by Campden Food & Drink Research Association before use is permitted. This is not a legal requirement, but the processor has a duty to the consumer to provide food which is of wholesome quality and free from taint.

Markets for horticultural crops demand uniformity in size, quality, maturity and continuity of supply. Any treatment which causes blemishes or malformation of produce, uneven or delayed maturity, or a wide size range distribution is unacceptable and these aspects have to be considered during herbicide development.

Seed may not be marketed unless it has been officially certified. To obtain certification the seed crop must pass official field inspection and the harvested seed must, after cleaning, meet the minimum prescribed standards. Some weed species are scheduled in the Seeds Regulations and the details of those statutory requirements are available from the Seed Production Branch, National Institute of Agricultural Botany, Huntingdon Road, Cambridge. In all crop certification schemes there is a general requirement that crops shall not be so weedy that a proper inspection for trueness to variety cannot be carried out. In addition, any herbicide treatment used must not cause damage effects which mask symptoms of disease and thus prevent thorough crop inspection.

There is a small but increasing interest in 'organically-grown' produce and the potential has been forecast for 10% of fresh market sales. Guidelines for achieving weed control while attaining organic quality standard are laid down by the Soil Association Standards for Organic Agriculture and a National Standard is currently being considered. The use of 'chemical and hormone herbicides' is prohibited, and weed suppression rather than elimination is the aim using rotations, cultivations, stale seedbed techniques and biodegradable mulches. Weeders with thin flexible tines or revolving brushes, as well as gas burners are under evaluation. In crops grown in wide rows, such as cabbage, it is possible to achieve a reasonable level of weed control by planting on the square and hoeing both ways, but in crops grown on close intra-row spacing, for example green beans, weeds remaining within the row cause serious competition. In addition, difficulties with machine harvesting and the risk of poisonous contaminants in produce may prove insurmountable.

The weeds and crop rotations

Annual weeds are usually the major problem in annual crops and previous cropping and soil type influence the number and species of weeds occurring. Broad-leaved crops are at present grown in rotation with cereals and the intensive winter cereal production in the UK, with earlier sowing and minimum cultivation techniques results in weed problems such as blackgrass and volunteer cereals, and also perennial weeds such as creeping thistle. In the future, changes in cropping sequences are likely, as alternatives to cereals are sought and it is anticipated that these weed problems may become less. The consequences of set-aside schemes where arable land is taken out of production are difficult to predict and some weed problems, particularly perennial species, may increase. Particular difficulties in control arise when the prevalent weeds are botanically related to the crop, where Cruciferae occur in brassica crops or Compositae in lettuce. There are often few, or no, herbicides selective in such circumstances.

Reasons for controlling weeds

Competition

Weeds that emerge before or with the crop usually cause the greatest yield reduction by competing for moisture and nutrients and tall species which shade the crop are particularly damaging.

Crop quality

The value of some crops is determined by quality of the produce. Size and uniformity are the main criteria in vegetables such as onions and these can be reduced by weeds. Weedy contaminants which are difficult and costly or impossible to remove at the factory or packing-shed downgrade the crop and poisonous contaminants such as berries of black nightshade in vining peas may cause crop rejection. In crops grown for seed, complete removal of weed seeds which ripen with the crop and are of similar shape, weight and size is not practicable. Some species are also scheduled under the Seed Regulations and if limits are exceeded, the crop may be rejected for certification for seed. For example, in rape seed the content of charlock shall not exceed 0.3% and there are standards for dock in kale and swede, blackgrass in linseed and in most crops for wild oats and dodders although the latter occur infrequently.

Harvesting

Crops which are mechanically harvested must be free from weeds that may interfere with the operation and woody, or climbing species such as cleavers and black bindweed are a problem. These climbing weeds sometimes cause crops to lodge, thus adding to harvesting difficulties. Use of a desiccant as a harvesting aid to kill weed growth adds to production costs. Some species such as small nettle and creeping thistle in a hand-harvested crop are objectionable to pickers, or obstruct access and slow down picking — a major cost.

Pests and diseases

Weeds are hosts to a wide range of pests and diseases which also affect crop. Shepherd's purse is a host to *Sclerotinia sclerotiorum*, a disease of oilseed rape and peas, perennial fleshy rooted species of weeds are hosts to a fungal disease of carrots *Helicobasidium purpureum*, and Cruciferae are hosts to club-root caused by *Plasmodiophora brassicae* a disease affecting brassicae. Common couch and volunteer cereals are hosts to cereal diseases and if these weeds are not controlled in the 'break' crop they act as a 'green bridge' carrying over disease.

Maturity

Most horticultural crops are sown and planted over long growing seasons to provide a continuous supply to processors and supermarkets of uniformly mature produce. Weed problems which cause delayed or uneven maturity must therefore be avoided. Weeds will also delay senescence and drying out of a crop harvested at the dry seed stage, for example beans.

Effect on other crops in the rotation

A weed species which is of no consequence in one crop, can well prove a problem in another; weed-beet for example is not a nuisance in cereals but serious in sugar-beet. Recognition and avoidance of specific weed problems in future crops in the rotation is an increasingly important aspect of management.

Herbicides

The extent of herbicide development in crops other than cereals is a reflection of crop area (Table 12.1) and there is thus a wide range of pre-

Table 12.1 Cropped area, estimated 5-year average yield and crop value of some agricultural crops and field vegetables in the UK, 1977 and 1987. Data from MAFF Statistics Branch, Guildford, and Nix (1986)

Crops	Thousand hectares		Estimated 5-year average yield t ha^{-1} 1983–87	Crop value £/t 1987
	1977	1994		
Agricultural:				
Wheat	1076		6.7	93 (feed)
				103 (milling)
Oilseed rape	55	388	3.1	270
Sugar-beet	201	200	40.6	27
Potatoes (early & maincrop)	232	179	38.2 (ware 4 yr av.)	70 (ware)
Peas harvested dry	37*	117*	3.2	169 (feed)
Field beans	37	91	3.5	163
Maize (incl maize for threshing)	34	23	37.0	na
Turnips & swedes	97	56	59.3	na
Fodders beet & mangolds	7	12	64.9	na
Kale, rape & other brassica- for stock feeding	55	30	43.3	na
Linseed§	0	8	1.8	175†
Horticultural:				
Roots & onions				
Beetroot	3.3	2.6	37.8	102
Carrots	19.5	14.7	41.5	93
Parsnips	3.4	3.1	21.9	153
Turnips & swedes (vegetable)	6.0	4.6	33.4	82
Onions, dry bulb	8.7	7.8	32.3	123
Onions, green	1.8	2.2	14.6	741

Brassicas				
Brussels sprouts	14.8	10.9	14.3	187
Cabbage (& broccoli)	26.8	22.4	a 16.5; b 37.7; c 35.1	a 196; b 71; c 97
Cauliflower	15.4	18.1	20.2	158
Legumes				
Beans, broad	14.4	2.3	4.9‡	262
Beans, runner & dwarf green	13.4	6.7	9.1	273
Peas (fresh market)	4.0	2.3	8.5	311
Peas (vining for processing)	55.3	43.9	4.9‡	196
Others				
Celery	1.6	1.3	52.6	269
Leeks	1.8	3.4	22.6	307
Lettuce	7.1	5.4	22.2	313

* 1977 all for human consumption; 1987 mainly for animal feed but includes 16 000 ha for human consumption
† Plus a very variable EEC subsidy which can be over £300/ha; seed producers get an additional subsidy
‡ Shelled weight not available
§ All figures estimated, area increased to 14 000 ha in 1988
a spring cabbage; b summer and autumn cabbage; c winter cabbage
na = not available

and post-emergence herbicides for oil-seed rape and sugar-beet. In the past, growers of minor crops have had few herbicides at their disposal since development cost is high, sales small and a damage claim in a high value crop could be considerable. Reliance was placed on materials approved for other crops. Non-approved use of pesticides is now an offence under the Control of Pesticide Regulations 1986 (Chapter 21), although there is a provision for application for approval for off-label uses by bodies other than the chemical manufacturer. Resources for independent evaluation of herbicides in minor crops have also been reduced. The pesticide legislation could also affect the continued availability of existing products as they come under review.

Residual herbicides, applied at a dose often dependent on soil type, are used in most crops. Disadvantages are that activity is usually reduced under dry soil conditions; several are not safe for use on very light soils, or are ineffective on highly organic soils. The residues of some materials persist in the soil restricting the choice of following crop and this is a particular problem if the treated crop fails. The choice of a residual herbicide to suit the anticipated weed flora is not always easy.

Foliar applied herbicides often have a restricted weed range and development of more broad-spectrum, selective herbicides would do much to improve weed control in some horticultural crops. Sequential treatments are therefore often used, which may control complementary weed spectra or extend the period of control. There is also increased use of mixtures of active ingredients, either in a formulation or as a tank-mix, to broaden the weed spectrum.

Selective application of a non-selective herbicide, usually glyphosate, applied with a rope-wick or other applicator which utilizes the height differential between crop and weed has proved useful in a few situations.

Brassicas for forage, stockfeed, and oilseed

Kale

Kale is sown from April to July, but mainly in June following a grass silage cut. Direct drilling is used as a means of establishing kale quickly after grass and to conserve moisture and thus improve germination. Row widths vary from 110 mm to 450 mm and within these limits appear to have little effect on yield.

Although kale is very sensitive to weed competition during the first few weeks, once the leaves have met across the rows the crop effectively suppresses weeds. The main problem weeds in kale are fat-hen, Polygonums, charlock, mayweeds and chickweed. Common couch can be serious since it persists throughout the life of the crop and the underground rhizomes are not destroyed by the treading of stock during feeding. Shepherd's purse is rapidly overgrown by the crop, but it competes with small kale seedlings and produces substantial numbers of seeds.

There are recommendations for pre-sowing, pre- and post-emergence herbicide treatments, and sequential use widens the weed spectrum. Clopyralid is useful alone or in tank-mix to control creeping thistle and other Compositae. Some problems still remain on farms where kale is grown intensively and there is a build-up of species such as charlock. The pre-emergence herbicides used in horticultural brassicas could be useful but are generally considered too costly for use in kale.

In kale crops direct-drilled into grass swards destroyed by glyphosate, weed control presents little difficulty. Where paraquat is used, re-growth of any perennial docks or creeping thistle may be a nuisance.

Forage rape and quick-growing turnips

These catch crops are of short duration, low value, are broadcast or drilled on narrow rows and establish quickly so little is spent on weed control. These crops are commonly sown from June to August for use from September to December. They often follow cereals and, as in oilseed rape, volunteer cereals are the most frequent weed problem. Volunteer cereals are controlled with TCA pre-sowing or a post-emergence material, but no graminicides are currently registered for forage rape.

The possible effects of herbicide residues on succeeding crops must be considered since rape and stubble turnips are short season crops.

Swedes

Swedes are grown for stock feed and also for culinary use. Swedes are precision drilled in 450–660 mm rows, and in Scotland and the north of England they are frequently grown on ridges. A small area is grown on the flat. They are sown from April to mid-June and, particularly for later drillings, weeds which have already germinated can be destroyed during seedbed preparation.

The most troublesome weeds are the same as those listed for kale. Swedes are less effective at covering the ground than kale or fodder rape and they are more susceptible to weed competition. Weeds also interfere with mechanical harvesting. There is a greater emphasis on effective weed control, particularly in the higher value human consumption swede crop.

In the fodder crop most growers use only one herbicide, followed by mechanical cultivation, whereas in culinary swedes sequential programmes are often used with trifluralin incorporated pre-sowing followed by a residual treatment. Foliar-acting materials for broad-leaved weeds are currently limited to clopyralid which is used to control Compositae. Blemishes on, or malformation of, roots caused by herbicides are unacceptable in culinary swedes.

Oilseed rape

At present, most oilseed rape is sown in the second half of August or in early September on close rows 115–200 mm apart and spring-sown crops occupy a small area. However, there is now more emphasis in quality and in the early 1990s only 'double low' (low in erucic acid and glucosinolates) varieties of oilseed rape are likely to be grown.

Winter oilseed rape competes strongly with weeds during the period of rapid growth in the spring, but crops of low vigour or areas where establishment is poor or delayed are vulnerable to weed competition (Fig. 12.1). Winter oilseed rape is frequently drilled into cereal stubbles with insufficient time to control volunteer cereals by cultivations or with sprays before sowing. Volunteer cereals are the most serious problem, particularly barley which grows vigorously in autumn and will rapidly smother a sparse crop. Where competition is severe the oilseed rape plants remain small and are more susceptible to frost and bird damage. Blackgrass is also very competitive if large numbers emerge at the same time as the crop. Cleavers, an increasing problem, is the most yield damaging broad-leaved weed. It causes harvesting difficulties and the seeds are costly to remove from those of rape. Control in rape is difficult since herbicides currently available are not reliable and priority should therefore be given to controlling the weed in previous cereal crops. Common chickweed and speedwells are troublesome broad-leaved species because of their vigorous growth during winter. Populations of the latter, common poppy and red deadnettle are increasing on many farms. Mayweeds can have a significant effect on open backward crops in the spring and the flower-heads can be a contaminant in the harvested rape seed. Field pansy, although not competitive, cannot yet be controlled reliably in oilseed rape and may carry over to infest the following cereal crop.

Seed of some cruciferous weed species, including charlock and wild radish are difficult or impossible to separate from the oilseed rape sample. There is a price deduction where admixture exceeds 2% and in some cases

Fig. 12.1 Volunteer cereals in oil seed rape; control on left, herbicide treated on right. Photograph courtesy of ADAS, MAFF.

the produce is rejected by the crusher since weed seeds adversely affect oil content and some can cause unacceptable taint or discoloration of the oil. These weed seeds also contain high levels of erucic acid and glucosinolates. Since the area of 'double low' varieties of oilseed rape is increasing, such weed problems are likely to become even more serious. In Canada, triazine herbicide-resistant varieties of canola ('double low') varieties have been bred to try to overcome this problem, but so far yield and quality is lower than for standard canola varieties. Fortunately most charlock does not survive the winter frosts and may also shed its seeds before winter-sown rape is harvested. Charlock thus poses more of a threat in spring-sown oilseed rape crops.

In crops grown for seed production, some cruciferous weed species including charlock and wild radish are scheduled in the Seeds Regulations and can only be rogued with extreme difficulty. Fields known to have a high population should therefore be avoided.

In winter oilseed rape, weed control is mostly with herbicides and pre-sowing, pre- and post-emergence herbicides are available (although ploughing before sowing will avoid a volunteer cereal problem). Not all crops are treated however, and while control of grass weeds, cleavers, mayweeds and possibly common chickweed may be cost-effective, there is little evidence to show that use of expensive materials is justified for other broad-leaved weeds. The necessity for weed control in the absence of heavy infestations or species likely to contaminate seed is the subject of current research.

Experiments to identify the optimum time for weed removal have shown that early removal of volunteer cereals is not needed unless the infestation is severe (as found in old combine swaths), or in a poor, low vigour crop. Therefore less emphasis is now placed on pre-sowing treatment with TCA and there is more reliance on post-emergence graminicides. In considering herbicide timing, factors other than early weed kill are also important. Pre-sowing treatments involving soil incorporation may delay sowing, or emergence if the soil dries out.

Activity of residual pre-emergence herbicides is reduced by dry or cloddy seedbeds (although so are weed numbers), and where some materials are absorbed onto burnt straw residues and organic matter. Selectivity of metazachlor pre-emergence is dependent on drilling depth of the crop, and as it can be leached by heavy rainfall it is not safe for use on very light soils, and here a split treatment using pre- and post-emergence applications is useful for cleavers and common poppy. Several species including cleavers, common poppy, shepherd's purse and field pansy are more effectively controlled pre- than post-emergence of the weed.

Post-emergence applications have to be timed according to growth stage of weed and crop. A crop key for stages of development of oilseed rape was proposed by Sylvester-Bradley, Makepeace and Broad (1984). Foliar-acting herbicides are applied in autumn if the oilseed rape is at a sufficiently advanced stage, or sometimes in spring in warm conditions while there is still an open crop canopy to allow spray penetration. In a dry autumn, oilseed rape can emerge over a long period and this can lead to problems of timing treatments in relation to crop safety. Delay in application may mean soil conditions become too wet to allow tractor passage, and efficacy of some herbicides is reduced under cool conditions.

Propyzamide is a very widely used winter oilseed rape herbicide. It is applied post-emergence and has residual activity. It controls the most important grasses and volunteer cereals, but not large chickweed or cleavers. Some control of the latter is achieved with the contact material, pyridate. Benazolin/clopyralid is also used post-emergence. Mayweeds can be killed pre-emergence, with metazachlor (early post-emergence), or in spring; or left until as late as possible in autumn, but when weed growth is active and sprayed with clopyralid alone or in mixtures. The recent introduction of a cyanazine recommendation for post-emergence use in winter oilseed rape has improved the prospect for control of cruciferous weeds such as charlock. However selectivity of cyanazine is very dependent on adequate leaf wax cover. Leaf wax can be affected by some oilseed rape herbicides, for example TCA, and this can reduce safety of all post-emergence treatments.

Tank-mixes are sometimes used to increase weed spectrum or improve control of difficult species. Sequential applications are frequently necessary, particularly for cleavers and grasses. Blackgrass germinates over a long period especially in a dry autumn, and at the time of application of a residual herbicide some may be too large to be controlled adequately so that a post-emergence graminicide is needed subsequently. Conversely foliar-acting graminicides timed for emerged volunteer cereals are often used in sequence with residual herbicides applied to control late emerging blackgrass.

It should be borne in mind that oilseed rape is susceptible to some herbicides used in other crops and spray drift from growth-regulator herbicides used in cereals on to a flowering rape crop can cause damage.

A rape crop presents a useful opportunity to eradicate common couch with a pre-harvest application of glyphosate.

Some oilseed rape herbicides are highly persistent in the soil and where, for example, propyzamide is used, mouldboard ploughing is recommended before subsequent cropping. Limitations are also imposed regarding drilling an alternative crop in the event of oilseed rape failure. Since the latter frequently occurs as a result of poor establishment in dry soil conditions, the interval before sowing susceptible crops should be extended. The use of a foliar-acting herbicide has an advantage where the seedbed is dry and failure seems a possibility. Many failed winter oilseed rape crops are followed by spring-sown rape to overcome persistence problems.

Spring-sown oilseed rape establishes and grows rapidly, but during establishment in April and early May it is sensitive to weed competition. Few crops receive herbicides except where control of wild oats is required. Propyzamide has no label recommendation for the spring-sown crop, and there are few herbicide recommendations for broad-leaved weed control; those available have a limited weed spectrum.

The rape crop is either swathed before combine harvesting or desiccated with diquat and combined direct. A desiccant is not normally used specifically to kill green weedy material present in the crop at harvest.

Volunteer oilseed rape

Oilseed rape volunteers are becoming an increasing and persistent problem in other broad-leaved crops. In the future, volunteers in the seedbank of 'single low' oilseed rape cultivars with high levels of erucic acid and glucosinolates could cause contamination and rejection of the crop from a 'double low' cultivar. Harvesting techniques which avoid seed return and allow lost seed to germinate with subsequent cultivation (as opposed to seed burial which would aid survival) will prevent the problem to some extent.

Horticultural brassicas

Cabbage, cauliflower, Brussels sprouts and a smaller area of sprouting broccoli and calabrese are grown throughout the country on a wide range of soils. Trifluralin pre-sowing and propachlor pre-emergence have been widely used in horticultural brassicas for many years because of their complementary weed spectra. The need for fewer mechanical cultivations has meant that narrow rows can be used with consequent changes in plant population. Crops grown for processing (Brussels sprouts), or where produce of small size is required for the fresh market, are grown closer so there is an early formation of crop canopy. Calabrese for small spear production are grown at high density on a bed system. Most brassica crops are transplanted but some are drilled. The cost of F_1 hybrid seed and less available hand labour has encouraged the practice of precision sowing to a stand. Transplants establish quickly and crop canopy cover which suppresses weeds, develops earlier than in field-drilled crops. In recent years there has been rapid adoption of new techniques of crop establishment using modular or block-raised transplants. Plastic-film cover used to advance harvest date has become popular and is in significant use in calabrese and cabbage.

Horticultural brassicas are increasingly grown in arable rotations, often on rented land, and thus inherit weed problems of previous arable crops. Volunteer potatoes and oilseed rape can also be a nuisance. In market gardens, on the other hand, where brassicas are intensively grown, shepherd's purse can be a serious spring weed. Weeds which emerge in large numbers in autumn such as annual meadow-grass and common chickweed, and also mayweeds, are the main problems in spring cabbage which are drilled or transplanted from July to September.

In Brussels sprouts, tall growing weeds such as fat-hen interfere with mechanical harvesting. Where cauliflowers or calabrese are hand-harvested the presence of small nettle is unpleasant for the pickers. Weed seeds sometimes contaminate curds and calabrese spears when the crop is wet with rain and dew. Weed control is achieved with herbicides. In the case of cauliflower and cabbage crops, sometimes cultivations are used as well.

It is important to avoid crop damage in horticultural brassicas. In a precision-sown crop, such as Brussels sprouts, a full plant stand is essential and herbicide damage which impairs a uniform and full crop emergence cannot be tolerated. Herbicides must not affect uniformity, quality or maturity of crop produce, or cause defects such as abnormal curds on cauliflower.

Fortunately, several brassica herbicides are safe on all the major crop species. Examples include residual herbicides trifluralin, napropamide/trifluralin and tri-allate pre-sowing and chlorthal-dimethyl, propachlor and tebutam pre-emergence. Some post-emergence materials are less selective and only have recommendations for cabbage and Brussels sprouts. Experience suggests that these crops appear more herbicide tolerant than those with a less well-developed leaf wax layer such as broccoli, calabrese and cauliflower. The grower relies mainly on residual programmes. Selective foliar-acting herbicides currently available for use in brassicas, for example clopyralid, generally have a restricted weed spectrum; they are often used for weeds remaining after residual treatments.

In transplanted crops, weeds are removed by contact herbicides or cultivations prior to transplanting and rapid crop establishment allows the early use of post-planting treatments. Most existing recommendations for brassicas are for bare-root transplants. Modular or block transplants on the other hand can be particularly sensitive to herbicide damage and some labels have been modified to include warnings, while others forbid use on modular or block transplants. At the time of writing there is only one recommendation, for chlorthal-dimethyl plus propachlor, either as a formulated product or as a tank-mix, for use soon after transplanting but before weed emergence, and any weeds present must be removed before transplanting. At present, no brassica herbicide label makes a specific reference to a floating plastic cover except propachlor which is excluded from use on 'protected crops'.

The more persistent materials, for example trifluralin, may be unsuitable in the short-term brassica crops, since herbicide residues in the soil might adversely affect the crop following. Maintaining adequate harvest intervals after application of foliar-acting herbicides may also be difficult. Both factors restrict herbicide choice.

New brassica herbicides are likely to appear as long as oilseed rape remains a major crop, but it may not follow that these will be necessarily safe for use in all horticultural brassicas.

Sugar-beet and related crops

This group of crops includes fodder-beet and mangolds and also the vegetable crop red-beet (beetroot) mentioned later. Sugar-beet occupies the greatest area and most are grown in eastern England. It is in this crop that the technical developments in weed control have initially been made.

Sugar-beet, fodder-beet and mangolds

Traditionally, sugar-beet seeds were sown thickly and then the seedlings were singled by hand to obtain the required plant population. Technical developments since 1950 have allowed the crop to become increasingly mechanized, from drilling right through to harvesting. Introduction of monogerm seed has allowed the universal practice of precision seeding and most crops are drilled at a spacing that is left to produce the optimum final stand of 75 000 plants ha^{-1} on a 500-mm row width. Since the early 1960s, there has been a progressive increase in the proportion of the crop in which weed control is achieved with herbicides and now the normal practice is to employ a sequence of chemicals to obtain broad-spectrum control for as long as is required, although some non-chemical methods may be used for low weed populations or difficult weeds such as weed-beet.

Annual broad-leaved weeds are the main problem in sugar-beet, particularly black bindweed, knotgrass and other Polygonaceae and fat-hen which is troublesome because of its extended germination period, rapid growth and competitive ability. If not controlled, they seriously reduce yield and their tough stems also make mechanical harvesting more difficult and costly. Problems can also occur in handling the beet after harvest; if the crop is stored in clamps, weeds can reduce root quality and if trash is taken into the factories, processing can be affected.

Weed-beet is an increasing problem. In the late 1960s and early 1970s various forms of annual beet were introduced as impurities in sugar-beet seed. These set and shed seeds in the growing beet crops and annual beet may also hybridize with bolters which can themselves be a source of weed-beet. Weed-beet seeds can remain viable in the soil for many years, longevity and dormancy being influenced by depth of burial. There is often only a two-year break between sugar-beet crops and this aggravates the problem.

Fodder-beet and mangolds tend to be grown in wetter, more westerly areas than sugar-beet. Here these fodder crops frequently suffer problems from late germinating weeds such as fat-hen and they are also more likely to be affected by bents, meadow-grasses and common couch which have been encouraged by the previous cropping sequence. The crops are often weedier than sugar-beet for other reasons also. The soils may have a higher organic content, so affecting activity of soil-applied herbicides, the fields are often small and sloping so that accurate application is difficult, while specialized application equipment and experience is less than in intensive sugar-beet areas.

Until the late 1970s, chemical and non-chemical methods of control were used in close combination. Band spraying, in which a restricted area over the crop rows was treated, was widely used and weeds between the rows were then killed by tractor hoeing. Typically, three band sprays were used, one pre-emergence followed by two post-emergence. The area treated with herbicide by each application was approximately one-third of the cropped field and spray volumes were usually $240\,l\,ha^{-1}$. Two passes of a tractor hoe were normally required. The main reason for adopting band spraying was the high cost of the herbicides used.

In sugar-beet, and to slightly lesser extent in fodder crops, this system has now been replaced by the low-volume, low-dose technique. In this, the whole of the field is treated, but the dose rate of herbicide and volume applied to that sprayed area are reduced by 66% compared with the traditional band-spray applications. For successful weed control, low-volume, low-dose sprays must be applied to small, cotyledon weeds. The success of low-volume, low-dose post-emergence sprays has meant that doses of pre-emergence herbicides can also be 33% lower than previously used, and an increasing number of growers are able to rely solely on post-emergence treatments for their weed control. Typically, two post-emergence sprays following a pre-emergence herbicide or three post-emergence sprays without a pre-emergence, are required. The new technique does not offer savings in the total amount of herbicide used in the crop but its success lies in the ability to achieve sufficient weed control to prevent loss of yield by weed competition, to achieve it reliably over a wide range of conditions and seasons (more so than the traditional band-spray system) and to do so at an economical cost.

Inter-row cultivation is still employed to remove unwanted beet from overlapping rows on the headlands and to control low populations of weeds or for weed-beet and other difficult weeds. New high-speed, self-steered band-spray and tractor-hoe systems have been developed that allow the

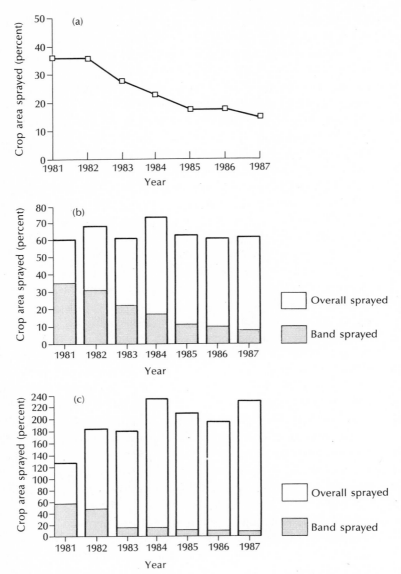

Fig. 12.2 Herbicide usage in sugar-beet between 1981 and 1987: (a) predrilling; (b) pre-emergence; and (c) post-emergence. Data from British Sugar Beet Survey.

use of low-volume, low-dose band sprays. However while these save on herbicide and total weed-control costs, growers are reluctant to adopt such systems because they take twice as many man-hours as the overall low-volume, low-dose system.

Selective herbicides can be applied pre-drilling, pre-emergence or post-emergence. The use of pre-drilling selective herbicides, particularly those for grass weed control, has declined greatly (see Fig. 12.2). The reason for this is the growing concern by growers and advisers about the effect that wheelings can have on the establishment of sugar-beet. Cultivation techniques for establishment of the crop are also changing and where plough and press or similar techniques that result in reduced cultivations are employed, then non-selective contact herbicides are used pre-emergence of the crop. The majority of growers use paraquat plus diquat, but paraquat alone and glyphosate are also employed. On mineral soils, the normal weed-control programme is a pre-emergence soil-acting herbicide followed by low-volume, low-dose post-emergence sprays. Chloridazon with or without ethofumesate, or metamitron alone, are the most popular pre-emergence herbicide choices. Metamitron is one of the safest pre-emergence herbicides and for this reason is often used on light soils unless black bindweed is likely to be a problem. Chloridazon plus ethofumesate is used particularly on heavier soils, where cleavers or high weed populations are expected.

The main factor influencing herbicide choice is efficacy, both alone and as part of a mixture, or sequential spray programme. A deficiency in the spectrum of weeds controlled by a pre-emergence herbicide may be acceptable provided that a follow-up treatment or treatments can deal with it. The use of a residual soil-acting component in post-emergence sprays has provided growers with a high degree of weed control and is especially useful against late germinating weeds such as fat-hen. Because sugar-beet is grown to a stand, crop safety is a prime consideration for all treatments that may be used. For this reason metamitron plus adjuvant oil is especially useful as the first post-emergence spray applied when beet are emerging or in the cotyledon stage. Subsequent sprays will depend on weed species present, but phenmedipham plus metamitron is a popular mix; phenmedipham alone or in mixtures with chloridazon, ethofumesate, lenacil, tri-allate or with both chloridazon plus ethofumesate are also widely used. Most annual and perennial grass weeds are usually controlled post-emergence with selective graminicides.

Weed-beet are a particular problem in sugar-beet and are dealt with by one or more of the following methods: hand hoeing, inter-row cultivations, cutting and the use of height selective rope-wick applicators later in the season.

Long-term control of weeds should consider the rotation and cropping, cultivation techniques and herbicide selection. Some of the herbicides used in cereals and other crops are particularly effective against weed-beet.

Beetroot

Beetroot are grown for the fresh market or for processing including
pickling. They are sown from April onwards for harvest beginning in
August, on row widths of 380–500 mm. Hoeing is seldom used and weeds
are controlled by herbicides. Weed infestations have been shown to affect
root size and therefore marketability, as well as yield and harvestability
and so a high level of weed control is sought. The weed problems, with the
exception of weed-beet, are similar to those encountered in sugar-beet.

Some sugar-beet herbicides also have label recommendations for beetroot and similar sequential spray programmes, including low repeat-dose
post-emergence spray systems are used.

Potatoes

Potatoes were traditionally regarded as a cleaning crop. They are grown in
ridges on wide rows and, in the past, weeds were controlled with several
post-planting cultivations which inevitably caused some root damage. Research with selective herbicides has clearly shown that excessive cultivations can cause serious loss of yield. Now fewer cultivations are done
and weed control is mainly based on herbicides, although it should be
noted that sprayer wheelings at a late stage also cause yield loss.

Potatoes are grown mainly for human consumption (ware), as first
earlies, second earlies or maincrop. A proportion of crops, especially
in Scotland, are grown for seed. The weed-control requirements and
limitations differ for each category. Potato seed crops are of high value and
any factors which delay tuber production, affecting seed size and yield, are
of particular concern. Thus inter-row cultivations should be avoided, as
should the use of herbicides which distort or discolour crop foliage and
mask symptoms of disease, as this may result in crop rejection for seed
certification. Weed control in ware crops is less demanding. Early potatoes
require weed control for a shorter period than maincrop varieties but
early crops must not suffer a growth check or root disturbance and so
weed control is mainly dependent on the use of herbicides rather than
cultivations.

Crop husbandry may also influence methods of weed control. The use
of stone and clod separators produces a finer tilth within the ridge and thus
the efficacy of soil-acting herbicides is increased, but soil incorporated
treatments cannot be used. Post-emergence herbicides are the only means
of control of perennial weeds, for example common couch, growing in the

rows of collected stones and clods. The increased reliance on herbicides for weed control has meant that the final ridge is often created at planting. The increasing production of early potatoes under the protection of clear floating film poses its own weed problems as traditional cultivations or late pre-emergence herbicide treatments cannot be used.

Although the crop offers good weed suppression, potato yields can be severely reduced by heavy weed competition. Yield loss is related to time of weed emergence, the weeds which emerge early, if not controlled, being the most damaging. Weeds also influence tuber size and affect rate and ease of harvesting, particularly late-germinating weeds which form a strong stem such as fat-hen and redshank.

Perennial grass and broad-leaved species including creeping thistle, perennial sow-thistle, dock and occasionally field bindweed and coltsfoot and cleavers are reported as problems mainly in maincrop potatoes. The normal spectrum of annual weeds occurs in potatoes.

Cultivations are still used for weed control, particularly on organic soils where chemical control is difficult. The development and use of herbicides in potatoes has been recent in comparison with many crops.

Screening of potential herbicides and desiccants takes account of: (1) effects on quality, such as poor skin set, scab, cracking, discoloration, malformation, internal vascular browning; (2) specific gravity and effect on sugars (of particular interest to the processor); and (3) for seed potatoes, effects on foliage and progeny tubers. Spray drift or sprayer equipment contaminated with herbicides, in particular glyphosate or clopyralid, can affect the growth of progeny tubers without necessarily showing symptoms in the contaminated mother crop. Where this is suspected the progeny need to be tested for abnormal growth before they are used for seed.

Contact pre-emergence treatments of paraquat, or paraquat/diquat are widely used to kill weed seedlings which emerge before the crop. Damage to potatoes which has already emerged at the time of application is usually only temporary. This treatment does not eliminate the need for subsequent inter-row cultivations or herbicide treatments for controlling weeds that germinate later.

Soil residual herbicides for potatoes also have some contact action and most produce temporary damage symptoms on crop leaves. Linuron is more effective pre-weed emergence, and is often used in mixtures. Metribuzin is the most persistent of the currently recommended residual herbicides and is versatile but there are varietal restrictions and it is soil-moisture dependent. Increased activity of metribuzin on soils of >10% organic matter is achieved by incorporation into the ridge before planting. Low-dose sequential application programmes are often used.

There is a need for a foliar-acting herbicide for post-emergence control of weeds which escape control with residual materials in dry conditions and for varieties that are susceptible to post-emergence application of metribuzin. There is now a recommendation for bentazone post-emergence, with or without adjuvant oil, but there are some restrictions on its use.

While residual materials are desirable for control of late germinating weeds, the choice of crops that can be sown, particularly after early potatoes or in the event of a failure, is restricted. For example, lettuce is very sensitive to linuron, monolinuron and metribuzin residues. Winter cereals may also be affected, though mouldboard ploughing will reduce this risk. In a dry summer, the likelihood of damage of susceptible crops increases.

Potato haulm destruction

The destruction of potato haulm is carried out for the following reasons: (1) to restrict tuber size; (2) to encourage tuber skins to set before the planned harvest date which is essential before potatoes are put into store; (3) to reduce the spread of *Phytophthora infestans* from foliage to tubers; (4) to prevent virus spread by aphids; and (5) to kill weeds in order to facilitate mechanical harvesting. Mechanical means, or more usually, chemical desiccants are used.

For prevention of tuber blight it is essential that at least 14 days should elapse between haulm death and lifting since blight spores can remain viable for some days on the soil surface. This 14-day period is particularly important in the case of mechanical haulm destruction as the blight fungus might continue to produce spores until the chopped haulm dies. Haulm destruction in seed crops by chemical rather than mechanical means will avoid the spread of *Erwinia* spp.

Particular care is needed in crops for seed or canning. The premature destruction of haulm by cutting, or by chemical desiccation when potatoes are too immature, will result in yield loss and affect tuber quality in the form of thin skin, poor skin set and a reduction in specific gravity. Skin setting becomes firmer as tubers remain in the ground even if they are detached from the parent plant and haulm destruction at the time when natural senescence is approaching, under normal weather conditions, has no effect on tuber yield or quality. If destroyed when the soil is dry, the vascular tissue of the tuber may become necrotic causing an internal brown staining of the vascular ring and in extreme cases, jelly end rot. Diquat is most likely to cause this effect. Sulphuric acid is relatively free from this problem, but special spraying equipment is needed and its use normally depends on availability of contractor services.

Volunteer potatoes

In other crops, volunteers often present a major problem. They can be aggressive and damaging weeds and potato 'berries' or stem can contaminate peas and green beans for example. The persistence of potatoes as volunteers is a special problem for the seed-potato producer and can lead to crop rejection since there are only limited tolerances in the certification regulations for non-crop cultivars of potato. Plants must be hand rogued if possible. The volunteers act as reservoirs for potato pests and diseases, posing a threat to the health of subsequent crops and virus diseases are also covered by the regulations.

The majority of volunteers originate from tubers, usually small, which are not removed from the field by potato harvesters, but some are derived from true seed. Experience shows that volunteer potatoes can survive through five-year rotations at least. Herbicide-resistant varieties of potato have now been developed; if these were introduced, the consequences of herbicide-resistant volunteers could be serious.

Control of volunteer tubers: This has been the topic of considerable research and there is still no easy way of eliminating the problem. Sowing competitive crops such as winter cereals immediately after potatoes minimizes daughter tuber production. Tuber numbers are reduced by the effects of frost, small mammals and birds, and therefore ploughing aids tuber survival. However, a non-ploughing regime may conflict with a requirement to plough after certain potato herbicides have been used to avoid risk of damage to succeeding crops.

Treatment of the growing potato crop with maleic hydrazide may sometimes reduce the multiplication potential of tuber-derived volunteers but this option is not available to the seed producer.

A system of similar row widths for potato planter and harvester reduces losses. Modified harvesting machines were developed that could collect or crush small tubers, but they were not adopted because of high cost and unsuitability for use on stony soils.

Herbicides do not always give high levels of control. The most effective herbicide for the control of volunteer potatoes is glyphosate and 100% control of daughter tubers has been achieved when application is made to plants with well-developed foliage. Glyphosate used pre-harvest of cereals is thus successful if potatoes have not senesced at this time. Selective application of a non-selective herbicide, usually glyphosate, with roller or wick applicators is only suitable where there is sufficient height differential between target weed and crop. It is possible in carrots and *Phaseolus* beans but sometimes unacceptable levels of crop damage are reported. The method is seldom used in sugar-beet because the height differential is usually insufficient.

Although some selective herbicides damage potato leaves, rapid regrowth is not prevented. However a number of post-emergence herbicides used in sugar-beet suppress potato volunteers, particularly where sequential applications or tank-mixes of different products are used, and repeat doses of dendritic salt plus non-ionic wetter are also effective. Some success has been achieved with metoxuron in carrots. Fluroxypyr in cereals can be applied at a late cereal growth stage and has achieved 65% reduction in daughter tuber viability, although effectiveness may be related to potato growth stage at the time of application.

Control of volunteer potato seedlings: Varieties that produce many berries such as 'Desiree' and 'Maris Piper' lead to the occurrence of volunteer potatoes derived from true seed. True seedlings occur in large numbers in sugar-beet and in many vegetable crops everywhere potatoes are grown. They are difficult to distinguish from small plants produced from small tubers or pieces of tuber, and they may occur together.

Mechanized haulm removal and chemical desiccation has no effect on number or viability of seeds returned to the soil. Seed can remain viable in soil for at least 7–8 years and there are indications that there is a residual seedbank of 8–12 million ha^{-1} in some fields. Seedlings will germinate in May and June in favourable conditions in any crop in the rotation, and if not controlled or suppressed, will produce small tubers by late summer, thereby contributing to the groundkeeper problem. Those emerging in July or August are unlikely to develop tubers. The late emergence results in poor control by less persistent residual herbicides and some vegetable herbicides for example chlorthal-dimethyl and trifluralin, are ineffective. Early removal of other weed species prevents inter-weed competition.

Leguminous crops

Since the introduction of an EC subsidy scheme for protein crops the area of peas and field beans grown for animal feed has increased markedly in the UK. These crops offer a profitable alternative to cereals, which are in surplus. The expansion has resulted in increased herbicide development for peas and field beans.

Peas

Peas harvested as dry seeds are grown mainly for animal feed and some for canning or dry-packet sale for human consumption. Peas are also grown as a vegetable harvested at the green immature stage for quick-freezing, canning (vining peas) or for the fresh market. Peas have slow initial growth, and an open sprawling habit and do not form a dense canopy able to smother weeds. Many varieties lodge at an early stage and in the dry-harvest crop weeds grow through the canopy before combine harvesting. Semi-leafless varieties are becoming widely grown and the reduced foliage and more open-plant habit have a reduced weed-smothering capacity.

Peas are sown on a row width ≤200 mm and this precludes the use of inter-row cultivations, so the grower relies entirely on herbicides for weed control. Peas for dry harvest are sown as early as possible in spring, usually in March and harvested from the beginning of August. Sowing programmes are used for vining peas to ensure continuity of supply to the processing factory, from mid-February finishing sometimes as late as the first week in June.

Weeds can seriously reduce pea yields and tall weeds which shade the crop such as wild oats and fat-hen are particularly damaging. Black bindweed germinates from depth and hence is sometimes not controlled by residual herbicides, and rapidly overruns a lodged dry-harvest pea crop. Other species also interfere with harvesting including common couch and volunteer oilseed rape. The latter is becoming a widespread and persistent problem and it is not controlled by most pre-emergence herbicides. If weeds are not controlled in the dry-harvest crop, a desiccant is needed and this adds to production costs.

In the vining pea crop, weed contaminants such as flower or seedheads of creeping thistle, mayweeds and common poppy are difficult to separate from produce and some may cause taints. Vining pea crops are sometimes rejected to avoid risk of contamination with poisonous berries of black nightshade or volunteer potatoes (Fig. 12.3). A high standard of weed control is therefore necessary in vining peas for processing to avoid problems in the factory.

One herbicide application for broad-leaved weeds is normally sufficient for peas, which are a short-season crop. Pre-emergence residual treatments are the most widely used means of control where soil type and seedbed

Fig. 12.3   Weed contaminants in vining peas for freezing — a potential problem. Photograph courtesy of Processors and Growers Research Organization.

conditions are suitable. They are ineffective at an economic rate on organic soils, although sometimes used to weaken weeds or delay their emergence before following with a foliar-acting treatment. On sandy soils, pre-emergence choice is limited to pendimethalin (for dry-harvest peas only) or aziprotryne, a relatively expensive material.

Post-emergence treatments are used for late sowings where dry conditions would reduce activity of residual herbicides. They are foliar-acting and rely on adequate epicuticular wax on pea leaves for selectivity. Peas are assessed for leaf wax by retention of crystal (methyl) violet dye. Dinoseb products were widely used but approval has now been revoked. Alternatives are less flexible in timing and 'hormone' materials are applied earlier to avoid pea flower-bud abortion. Some varieties are more sensitive to these materials than to dinoseb-amine. A key for stages of development of peas to aid timing of application has been described by Knott (1987).

Post-emergence graminicides are applied if and where necessary to control wild oats. There has been a reduction in use of 'blanket' pre-sowing treatments, which may often have been unnecessary and the soil incorporation required is detrimental to seedbeds when soil conditions are wet. Common couch is usually best eradicated with autumn application of glyphosate before sowing peas, but it can be suppressed in the growing crop with a post-emergence graminicide. In the dry-harvest crop, there is a further opportunity using glyphosate pre-harvest of peas.

Peas are unusual in their varietal reaction to herbicides; some varieties are tolerant whereas others are sensitive. The principles involved remain unclear, although small-seeded varieties, including forage types, tend to be more sensitive to some residual materials, and plants with pale, soft leafy growth are often damaged by foliar-acting herbicides.

Desiccation of dry-harvest peas with diquat before combining is a useful harvesting aid, to kill off green weeds which might cause difficulties.

Beans (*Vicia faba*) — field and broad

Field beans are sown in autumn (winter beans) on heavier soils, where it is difficult to achieve a reasonable seedbed in spring. Spring-sown field beans (spring beans) are sown as early as possible, usually on a 200 mm row width; yields are reduced where row width is >400 mm. The majority of broad beans are grown for processing and are sown in spring from mid-April onwards for harvesting in August, following peas. Broad beans (*Vicia faba* var. *major*) usually have large seeds and since they are also expensive, are sown with special precision drills on wider rows of 450 mm. Beans grow rapidly in spring and the tall plants achieve better weed suppression than many crops. Newer cultivars of spring beans are less prone to vegetative growth and broad beans are also shorter than field bean cultivars.

Weed problems in bean crops are related to the germination periods of various weed species. Cleavers and blackgrass are most troublesome in autumn-sown field beans and control measures are best sought in the cereal crop in the rotation. Black bindweed, which can cause lodging of the crop and severe harvesting problems, and wild oats are predominant in the spring-sown crop. Common couch is often associated with beans autumn sown on heavy soils. However, the widespread use in the rotation of glyphosate, is reducing the severity of perennial grass weed problems.

Inter-row cultivations are sometimes used in field beans where row widths allow, but this method does not control weeds within the row and herbicides are the main means of control.

Field beans sown in the autumn are often ploughed in, and this method gives adequate depth protection for the safe use of simazine, a cheap and widely-used surface-applied residual herbicide. There are several other residual herbicide ˙options. However, the effectiveness of residual materials is reduced on the cloddy seedbeds which are prevalent on the heavier soils. Drills used for spring-sown field or broad beans cannot usually sow deep enough to allow the use of simazine and it is then preferable to apply an alternative which is less dependent on depth protection for crop safety. Simazine is a persistent material and residues can affect the cereal crop, which usually follows, if the minimum time interval is not observed. However, growers must rely on residual herbicides where soil type allows.

There are no recommendations for residual herbicides on sands or highly-organic soils. On these soil types, or if a follow-up is needed where species escape control, or if activity of residual herbicides is reduced in a dry spring, a selective foliar-acting herbicide for post-emergence application is needed. However, beans are very susceptible to most of these materials, since leaf wax is often insufficient to avoid scorch and translocated 'hormone' herbicides cause severe epinasty. Bentazone is now the only option and although it is useful for cleavers control, it has a limited spectrum.

There are several herbicide recommendations for controlling annual grasses. If common couch becomes aggressive in the crop, a graminicide offers good suppression, or an application of glyphosate pre-harvest of field beans is used to eradicate the weed which presents an actively-growing target at this timing.

Where weeds are likely to interfere with harvesting, diquat is used to burn off green weed growth.

Pulse-seed crops

Much of the seed for dry-harvest pea and bean (*Vicia faba*) crops is produced in the UK. The climate is less suitable for vining pea-seed production and all seed of *Phaseolus* beans is produced elsewhere. Peas and beans are easily separated from smaller weed seeds during combine harvesting. The only occasional problem is where the seeds of wild oats become lodged in holes in the pea seed made by the larvae of *Cydia nigricana* (pea moth).

Fig. 12.4 Weed competition in green beans. Photograph courtesy of Processors and Growers Research Organisation.

Green beans (*Phaseolus vulgaris*)

Green beans (sometimes called 'dwarf' or 'French beans') grown for quick-freezing or canning, and much of the fresh-market crop are harvested by machines capable of harvesting a crop on any row width. Beans on narrow rows are higher yielding, but many are still sown on 400-mm rows since a limitation is imposed by precision drills which operate on a minimum row width of 300 mm. Sowing programmes, beginning in mid-May, are used to achieve continuity of supply to the factory.

The crop does not produce much foliage until fairly late in the season so competition against weeds is not very effective in the early stage of growth, and tall weed species can grow above the crop canopy (Fig. 12.4). For efficient machine harvesting the crop must be free from bushy or woody-stemmed weeds such as redshank and fat-hen. As with peas the presence of poisonous weedy contaminants can result in rejection of the crop, and volunteer potatoes and black nightshade are a frequent problem.

Inter-row cultivation is unsatisfactory and little used now because weeds within the rows are not controlled; beans are shallow rooting and may be damaged; and soil build up round the bean stems may interfere with mechanical harvesting.

Soil conditions are often dry during the sowing period of May and early June and a stale-seedbed technique is used with contact-acting herbicides to kill emerged weeds before sowing, or a contact material included in a pre-emergence residual spray if weeds emerge before the crop.

Few herbicides are registered for the crop. A herbicide programme is used for broad-leaved weed control, and may include a pre-sowing incorporated treatment with trifluralin where moisture loss is reduced by sowing immediately and rolling. This, followed by pre-emergence residual monolinuron and later a post-emergence application of bentazone, are all needed to ensure freedom from weeds. Trifluralin and bentazone control complementary weed spectra.

Height-selective application of glyphosate is sometimes used to control volunteer potatoes, otherwise hand pulling is necessary.

Land infested with grass weeds is avoided for dwarf bean cropping, and although annual grasses can be controlled with pre-sowing or post-emergence, no post-emergence graminicide is currently registered for perennial species.

Runner beans (*Phaseolus coccineus*)

The climbing (runner) bean is usually grown on sheltered sites to avoid wind damage since the early crops fetch a premium and sometimes there is little or no rotation. There is also an area being grown under low-level plastic cover to advance maturity; either seeds are sown through black polythene mulch, or covered after sowing with clear polythene.

In market garden crops grown on quadrapods of canes, weeds are controlled by inter-row cultivations with a two-wheeled rotary cultivator supplemented by hand hoeing. On a large-field scale, the crop is grown up strings supported by a semi-permanent system of strained wires and posts at a spacing wide enough to allow the use of small tractors. Here a herbicide programme is used, with trifluralin applied pre-sowing and incorporated, followed by a post-emergence treatment with bentazone. Approval for use of dinoseb has been revoked and, at the time of writing, the only pre-emergence materials registered for runner beans are diphenamid, and chlorthaldimethyl often used as a tank-mix to widen the weed spectrum. The latter is expensive but very safe and also controls black nightshade, a major weed. In the case of runner beans protected with polythene a more effective pre-emergence herbicide is sought since weeds not controlled may become too large before a contact-acting post-emergence bentazone application is possible.

Carrots and related crops

Carrots

The total area of carrots grown in the UK declined in the early 1980s, whilst canning-carrot production had also decreased owing to the continuity of good-quality fresh produce. There has been a slight increase in the area grown for quick-freezing. A small area is grown for soups. Carrots are mainly grown for fresh market, with out-grades or defective roots used for stockfeed. There is also an increasing interest in growing carrots for early production for the fresh market under low-level clear polythene and it is estimated that in the UK there were over 1000 ha grown in this way in 1988.

The sowing arrangement depends on market outlet, time of harvesting and type of harvester used. Carrots are grown on 420 mm rows where top-lifting harvesters are used up till late October. As the carrot tops die back, share-lifting harvesters are necessary and here carrots are normally grown on double rows at 760 mm centres. Baby carrots for canning or quick-freezing whole are sown on a mini- or full-bed system and are share-lifted from the end of August onwards.

The carrot crop is grown mainly on two contrasting soil types — very light sands containing little organic matter, and on highly organic soils. Both are deep, easily worked and the organic soils are free from stones. Soil blowing can occur on both sands and peats. Where this is likely, early weed growth is not removed and is left to protect the soil until the crop is large enough to provide ground cover, and post-emergence herbicides are applied later.

Carrots were the first vegetable crop in which effective chemical weed control was achieved with the introduction of mineral oils in the late 1940s and since then virtually all carrots have been treated with herbicides. The requirements depend on the market for the crop, production systems and method of harvesting.

In carrots, a high standard of weed control is needed to avoid yield loss and to control weeds likely to affect size and quality, particularly in baby carrots for processing, or species such as knotgrass, annual meadow-grass or common couch, which impair mechanical harvesting. Tall weeds such as fat-hen and mayweeds are a nuisance where top-lifting harvesters are employed. No effective selective control has been found for wild migno-nette which remains a problem for a few growers on sandy soils.

A range of herbicides for broad-leaved and grass weeds and which can be applied at different timings pre-sowing, pre- and post-emergence of the crop, give good control. In some cases, there may be a slight growth check, but recovery is likely without significant effect on yield. It is essential that crops grown in the open or under plastic cover for early market do not suffer any damage which results in delayed maturity, therefore post-emergence herbicides are usually avoided if possible.

Linuron pre- or post-emergence has been widely used on carrots for many years. Control of mayweeds, once a problem, has been overcome by post-emergence treatment with metoxuron or pentanachlor. A tank-mix of metoxuron with linuron improves control of large knotgrass and annual meadow-grass. Tri-allate, or post-emergence graminicides control wild oat but, in common with most vegetable crops there is less choice for post-emergence control of common couch.

Plastic-covered carrots are difficult to keep weed-free, as these con-ditions favour emergence and growth of weeds as well as crop. Residual herbicides are normally used between drilling and covering the crop and usually provide control at least until uncovering time. However, the use of plastic cover may enhance or reduce the effect of pre-sowing or pre-emergence herbicides on weed or crop. There have been attempts at lifting plastic, and replacing after application of post-emergence herbicides but this does not appear to be a viable proposition. The cover is usually removed at about the seven true-leaf stage and contact herbicides applied before the growth hardens off can be damaging. Except for linuron, there are no specific recommendations for carrots grown under cover. There is a need for more research in this area.

Celery

Early and late 'self-blanching' celery, harvested from July to November, is grown from transplants mainly on organic but some on mineral soils, and few crops now are drilled. The area of trenched ('blanched') celery grown, harvested from October to late December, is very small.

The self-blanching crop is grown in beds and is dependent on irrigation. Weeds are controlled with sequential applications of residual/contact-acting herbicides combined with some hand weeding. Trenched celery is grown on wide rows, relying on residual/contact-acting herbicides applied soon after transplanting followed by inter-row cultivation and some hand weeding. Annual grasses are the main weed problem remaining.

Weed control in drilled celery is difficult because of the very slow emergence and development of the crop and thus the long interval between residual herbicide application and the stage at which the celery seedlings can safely be treated post-emergence. Application of a contact herbicide to kill weed seedlings before the crop emerges is useful and essential on organic soils. This technique is also used in parsley, a biennial crop which is also slow to emerge.

Some of the materials used in carrots, for example linuron and pentanochlor, are also suitable for celery. The only special instructions for the transplanted crop are 'apply after transplants are established'.

The national area of celery grown is small, and it is difficult to relate celery to other crops in terms of weed control. Owing to the morphology of the plant, larger amounts of pesticides may be retained resulting in higher levels of pesticide residues than in many crops. Therefore the number of herbicides with label recommendations is likely to remain small.

Parsnips

Parsnips are grown mainly on sandy or organic soils with an open texture. They are drilled to a stand and there is an increased use of cultivations and a bed system. Parsnips are slow to emerge, their growing season is long and suppression of weeds is poor. Weeds hinder harvesting where a top-lifting harvester is used. Inter-row cultivations, if any, must be carried out with care since damage to parsnip crowns increases risk of *Itersonilia pastinacae*.

The weed problems are very similar to those of carrots. In parsnip-seed crops, wild parsnip can be a field contaminant in some locations and can not only hybridize with the cultivated crop, but may also contaminate the harvested seed. Weeds are controlled with herbicides developed for the carrot crop. Parsnips however, are less tolerant of post-emergence herbicides than carrots, as the larger parsnip leaf retains more herbicide. Thus there are fewer recommendations, and for crop safety, herbicides are applied at a more advanced growth stage than is allowed for carrots.

Onions and related crops

Onions

Most onions are grown for bulb production and are sown in early spring. Much of the crop is direct drilled but field losses are high and there is an increasing area grown using transplants of multi-seeded (about seven seeds) modules or peat blocks. The onions are transplanted at the two-leaf stage in late March or early April at a density of about 100 000 blocks per hectare. The transplanting system reduces seed cost and achieves earlier maturing crops with improved quality and bulb size compared to field drilling; however the labour requirement is high and there is therefore an interest in growing onions from sets, which also obviates the necessity for specialist equipment. Over-wintered onions are sown in August on a smaller but still significant area; this is declining, but improved varieties could reverse this trend. Salad onions are drilled from early spring to August for continuity of supply.

As the habit of onions is upright and they are slow to form a significant leaf canopy, weeds are not suppressed early in the life of the crop. Heavy weed infestations affect bulb size and quality. Bulb onions are harvested mechanically and so must be free of knotgrass, common couch and other species likely to interfere with this operation. Volunteer cereals are the main weed in the over-wintered onion crop, which is usually sown after cereals. Mayweeds are often a problem in spring-sown and over-wintered crops and there is still no effective post-emergence control for annual meadow-grass which is becoming an increasing problem.

Onions are very susceptible to weed competition and weeds must be controlled throughout the life of the crop. This is done by several applications of herbicides.

Because initial growth of spring-drilled onions is slow, control of early germinating species of weeds is essential. It is achieved with an initial pre-emergence herbicide application of a contact/residual treatment. Sometimes there is a follow-up residual treatment 4–6 weeks later, and sequential applications of contact or contact/residual treatments are usually necessary. Split low-dose treatments are also used. The choice of the initial pre-emergence herbicide is determined by soil type, efficacy and crop safety. Later sprays are chosen on the basis of the weed problem present and crop growth stage. The safety of all post-emergence contact herbicides such as ioxynil or cyanazine, which has a recommendation for peat soils, depends on adequate onion leaf wax formation. If weather conditions have been unfavourable for wax development or there has been an abrasion of the foliage, damage can result.

On some sands or peats, soil blowing can cause injury and crop loss and here shelter rows of barley or other cover may be sown. These are later killed with a post-emergence graminicide.

The August-sown crop is drilled when the soil is warm and irrigation is often used, so that emergence occurs in as little as 5 days compared with

3–6 weeks for the spring-sown crop and weeds also emerge more quickly. A pre-emergence treatment is still essential but the crop stage at which a post-emergence herbicide can be safely used is soon reached, although a follow-up residual treatment 4 weeks after the pre-emergence treatment is usually desirable. If emergence is delayed until mid-September, the seedlings may be too small and leaf wax insufficient for safe use of a contact-acting herbicide in autumn.

Transplants (and onion sets) have the advantage over drilled crops of eliminating the seedling stage during which weed control with crop safety is difficult. Weed control should be cheaper, with cultivations or application of contact-acting herbicide pre-transplanting, the application of a residual herbicide soon after planting, and appropriate post-emergence treatment later. However, herbicide recommendations are different from the drilled crop and sometimes applications '14 days after transplanting' or 'when transplants are established' are specified. Some treatments are excluded since there is a risk of damage from residual herbicides particularly where shallow planting occurs and roots may be exposed.

Salad onions occupy the ground for a shorter period and the higher crop plant densities are more competitive than the bulb crop, so weed control may be less of a problem, but care is required in the use of post-emergence contact herbicides since onion leaf damage renders the produce unmarketable.

Leeks

Most leek crops are drilled to a stand, but where produce with a longer white shank is sought, the crop may be transplanted. The method used also depends on the time of year. Multi-seeded block-transplanting techniques for leeks have produced some problems of bent shanks and difficulties with timing so far. Leeks are usually earthed up in ridges and this may reduce the number of herbicide applications required.

The drilled crop is slow to emerge so it is essential to achieve the best possible pre-emergence weed control. Weeds are difficult to control post-emergence if they become too large and the problem is being overcome by development of low-dose treatments applied at earlier growth stages. Onion herbicides are used, but since leeks have a larger 'funnel' type of leaf and there is less run-off they are more sensitive to post-emergence herbicides than onions. Leeks may also collect more pesticide residues than onions, which limits the use of some chemicals.

Lettuce

Lettuce is grown throughout the UK on a wide range of soils and is both drilled and transplanted. There has been a rapid uptake of the use of cover and about 400 ha of outdoor lettuce are grown under clear polythene or non-woven film cover. These techniques advance crop maturity, and improve emergence, yield and quality of produce.

Planned continuity in lettuce production is essential. Lettuce, particularly butterhead varieties, grow quickly and must be harvested at the target date so any crop check from herbicides or delay in maturity caused by weed competition must be avoided. Weed seed contaminants also reduce produce quality and hinder hand-harvesting.

Block transplants of early-maturing varieties do not usually suffer from severe weed infestation but in later-maturing, and all drilled crops, the problems can be acute. In specialist market garden areas, lettuce is often grown repeatedly with several crops on the same land in a single season. Thus while other species are controlled there is a build-up of Compositae; under such conditions the application of soil partial sterilant applied each autumn can be economic.

Although most weed species can be controlled with herbicides there is a limited choice and none achieves control of mayweeds or groundsel. In the past, propachlor pre-planting sometimes followed by propyzamide, or as a post-planting tank-mix, was used to overcome the problem, but propachlor is only currently approved for use pre-planting on lettuce, and it is not recommended for use on any protected crop.

Although a non-woven material may offer the possibility of post-emergence applications through the cover, this has been unsuccessful in practice. At the moment, research has suggested the best means of weed control in outdoor-covered lettuce is with a soil sterilant.

Several varieties of lettuce are grown of crisp, butterhead and novel types and there may be differences in tolerance to herbicides.

Linseed and flax

Linum usitatissimum is grown for oilseed (linseed) for industrial oil and cake for animal feed, or for fibre (flax) production and is an alternative to cereals. Linseed occupies the larger area, but there is some flax grown mainly in Northern Ireland and both crops are suited to the UK climate. Linseed varieties are larger seeded and shorter strawed, more resistant to lodging than the traditional flax grown for fibre. The crops are shallow rooting and perform best on moisture-retentive soils and are drilled in March or April in narrow rows. The plant is fine-leaved, initial growth is slow and it has poor ground cover. It is a poor competitor to weeds at early stages, although a crop with an adequate population density effectively smothers weeds later on. Weed growth seriously reduces yield, hinders harvesting, weed seeds affect throughput of linseed crop driers and tall weeds contaminate harvested flax fibre.

A range of spring-germinating weed species including common chickweed, charlock, knotgrass, fat-hen and annual meadow-grass affect linseed and flax crops, and black bindweed and wild oats cause harvesting problems. Crops grown for seed production should be free from the latter, and from blackgrass and dodder.

Weeds are controlled with herbicides but there are few approved for linseed and flax in the UK. The same herbicides are not necessarily recommended for linseed and flax. For example, MCPA which is only recommended for linseed, causes unacceptable distortion of flax fibre, hence loss of quality, and sometimes yield.

While control of weeds during early crop establishment is important, many pre-emergence materials for broad-leaved species are not highly selective in linseed so reliance has now to be placed on post-emergence treatments such as bentazone, MCPA, or bromoxynil/MCPA. Recent evaluation has shown that early post-emergence treatment with metsulfuron-methyl is selective in linseed. There are recommendations for control of some grass weeds with tri-allate pre-sowing, and diclofop-methyl post-emergence.

The linseed crop is usually desiccated with diquat or sulphuric acid before direct combining and any weeds present will also be killed. In flax, the use of glyphosate is being developed for pre-harvest retting.

Maize and sweetcorn

Maize is grown mainly for ensilage as cattle feed, and there is also a small area of sweetcorn for human consumption. The forage varieties are considerably more vigorous and competitive with weeds than the vegetable varieties. Maize is drilled in late April and May for harvest in late August and September. Both crops are sown in very wide 750–800-mm rows to suit forage (or sweetcorn) harvesters or hand picking. The crops seldom form a complete canopy before the end of July, and even when mature, the canopy is still sufficiently open to allow light penetration and weed growth beneath it.

Most spring-emerging annual weed species compete against maize, often completely smothering it at an early stage if no herbicide is used. Perennial weeds, especially common couch can also suppress young maize crops. On some farms, maize or sweetcorn is grown continuously on the same field and this may lead to a build-up of black nightshade following repeated use of atrazine.

Weeds are controlled by herbicides, mainly atrazine which is both very safe and highly effective against almost all the annual weed species that occur in the UK. It can be applied at any time during the growing season to control germinating weeds, but is usually sprayed pre-emergence. Atrazine is also used for controlling common couch where it is applied at a higher rate pre-sowing and incorporated. It has limitations in that strains of annual meadow-grass or groundsel have become resistant to triazines. Lower rates of atrazine in an atrazine/cyanazine formulation, or cyanazine alone, which is a less persistent alternative, can be used pre-emergence for annual broad-leaved weeds and annual meadow-grass.

Atrazine residues in the soil may affect subsequent crops if the interval is too short or the soil remains dry so that degradation is delayed. Where higher doses are used for perennial weed control, an interval of 18 months should be left or only maize or sweetcorn grown within that period.

Acknowledgements

The authors of the section on sugar-beet, fodder-beet and mangolds were: M. May (Norfolk Agricultural Station, Morley) and T. Breay (British Sugar Corporation, Kidderminster).

The following contributed to the remaining sections: A. Greenfield (MAFF, ADAS, Oxford), H. M. Lawson (Scottish Crops Research Institute, Dundee), J. H. Orson (ADAS, Cambridge), P. C. Rickard (ADAS, Arthur Rickwood Experimental Husbandry Farm, Mepal), P. Bowerman (ADAS, Boxworth Experimental Husbandry Farm, Boxworth), M. Askew (MAFF, ADAS, Wolverhampton), P. Lutman (Weed Research Division, Long Ashton Research Station, Bristol) and S. Perkins (ADAS, Institute of Horticultural Research, Wellesbourne).

Further reading

Knott, C.M. (1987) A key for stages of development of the pea (*Pisum sativum*). *Annals of Applied Biology*, **111**, 233–245.

Nix, J. (1986) *Farm Management Pocketbook*. Wye College, University of London.

Sylvester-Bradley, R. Makepeace, R.J., & Broad, H. (1984) A code for stages of development in oilseed rape (*Brassica napus* L.). In *Aspects of Applied Biology*, Vol. 6, *Agronomy, Physiology, Plant-breeding and Crop Protection*, pp. 399–419. Association of Applied Biologists, Wallesbourne, Warwick.

13 / Weed control in fruit and other perennial crops

D.V. CLAY, H.M. LAWSON* AND A.J. GREENFIELD†

*AFRC IACR Long Ashton Research Station, Long Ashton, Bristol BS18 9AF; *Scottish Crop Research Institute, Invergowrie, Dundee DD2 5DA; and †MAFF ADAS, Government Buildings, Marston Road, Oxford OX3 0TP*

Introduction

Perennial crops include all the fruits, many herbaceous and woody ornamentals and a few vegetables such as asparagus and rhubarb. By definition, they are plants that have a life cycle lasting more than a year, although many are grown for a shorter period before they are transplanted. Growth habits vary from trees with a permanent woody framework through the various bush and cane fruits to herbaceous perennials, some of which die down completely during winter. Such diversity makes generalization on weed control difficult.

Many perennial crops are grown on specialist holdings which are small and labour-intensive. Since other aspects of production require detailed attention, weed control may be neglected. Before herbicides became available, repeated inter-row cultivation was supplemented with hand hoeing in the rows. In the wider-spaced crops the area between the rows was ploughed or rotary cultivated during the winter. Now that herbicides are used extensively there is no post-planting cultivation in many plantations. Most orchards still have grassed alleys, but in some orchards they have given way to overall bare soil.

Weeds of perennial crops

Nearly all soils contain an abundance of weed seeds and this is the main source of infestations. But there are others; for example, seedlings from wind- and water-borne seeds can establish rapidly in the composts used for pot plants, while birds are responsible for the introduction of brambles into orchards. Self-seeding of crops such as asparagus and blackcurrants creates new weed problems, while on many soils perennial weeds are also present. With the absence of competition from other weeds and the generally poor competitiveness of most perennial crops, weeds that are not controlled develop into large plants. This can lead to rapid increases in the numbers of weed seeds and a build-up of vegetative propagules of perennial weeds.

Reasons for controlling weeds

The general principles are covered in Chapter 1, but the relative importance of particular factors is often different for perennial as compared with annual crops. Efficient weed control needs to be based on an objective assessment of weed effects and crop requirements, not merely on the cost and convenience of the treatments available. Competitive effects are of prime importance but the neglect of other aspects can have serious consequences.

Competition

The main reason for controlling weeds is to avoid reductions in crop growth and yield caused by competition for moisture. Newly-planted crops are particularly sensitive to competition and weeds that develop early in the season are more competitive than those developing later (Fig. 13.1). In apples and blackcurrants for instance, failure to control weeds during the first year can reduce extension growth by as much as 60% while competition up to mid-June reduces growth 20–40%. Crops that are checked in this way may never recover, even if they are subsequently kept weed-free.

Fig. 13.1 Effect of increasing (left to right) severity of weed competition during the year of planting out in the nursery on the size of 2-year-old plants of *Chamaecyparis lawsoniana*; all plants were kept weed-free in their second year. Photograph courtesy of ARC Weed Research Organization.

The growth of older crops can also be affected by the presence of weeds, and in orchards, grassed alleys may be another source of competition.

Crop quality

The commercial value of the produce from perennial crops is determined by several characteristics. Size is one of the main criteria and is reduced by weeds in many crops including apples, flower bulbs and nursery stock, with a consequent reduction in value. Nursery stock plants with thin trunks or poorly-developed lower branches, are downgraded with a consequent reduction in saleability and value. The forcing performance of flower bulbs can be reduced by weed competition before lifting.

Harvesting

Weeds can impede picking, and this is particularly important for soft fruit, where harvesting is one of the major production costs. They can also cause uneven ripening or result in part of the crop remaining unpicked simply because it is not seen. Pickers are deterred by physically-objectionable weeds such as creeping thistle and common nettle. Climbing weeds such as field bindweed interfere with both hand and mechanical harvesting. Cider apples are a special case; they are harvested from the orchard floor and are difficult to pick up in the presence of tall grasses or weeds.

Convenience

Other field operations become more difficult or less efficient in the presence of weeds; for instance, climbing weeds such as field bindweed hinder the pruning of blackcurrants and gooseberries. Weeds also slow down the lifting of nursery stock, especially when weed roots or rhizomes have to be removed.

Appearance

On many holdings to which the public has access, e.g. retail nurseries and self-pick fruit farms, the visible presence of weeds may discourage the potential customer. Weedy crops may also reduce the morale of workers.

Pests and diseases

Weeds are hosts to a wide range of crop pests and diseases of fruit crops and may contribute to their spread. By modifying the micro-climate they may also predispose crops to attack. The incidence of *Phytophthora cactorum* (collar rot) and trunk damage by voles has declined since tree base vegetation has been controlled with herbicides. In hops, the elimination of cultivation has reduced the spread of *Verticillium albo-atrum* (verticillium wilt).

Weed increase

The widespread long-term use of herbicides has proved generally effective, but a failure to control various annual and perennial weeds which were hitherto of only minor importance has resulted in their rapid increase in some plantations. Examples of annuals include mayweeds in bulbs, knotgrass, speedwells and pansies in strawberries. Perennials too have flourished, the most widespread being common couch and creeping thistle, while others such as field horsetail and field bindweed are less common but more difficult to control. Species not previously considered to be weeds of cropped areas have spread to become problems in some plantations. They include hogweed, St John's worts and willowherbs.

Methods of control

Cultivation

The individual plants of many perennial crops are large and widely spaced, enabling a high standard of weed control to be achieved by mechanical means. Rotary cultivators have been used extensively to chop and bury weeds, but cultivations of this kind favour the germination of yet more annual weeds, spread perennial weeds and also damage the roots of the crop. In prostrate and spreading crops, mechanical weeding is difficult and not very effective, and even in orchards it is not possible under low branches. A major limitation of mechanical weed control is the need to repeat it several times during the growing season. The loose surface soil leads to impaired traction, rutting and increased erosion. Mechanical methods, other than hand-hoeing, are not very effective in dealing with weeds growing close to the crop.

Mulching

Annual weeds can be suppressed by mulching with pulverized bark, straw, peat or other natural materials, or with plastic film. Straw is widely used in newly-planted tree crops, mainly to conserve moisture. Plastic film has been employed less widely in the UK than in many other countries though it is now being used in dessert strawberry production. In experiments, mulching with black polythene has often increased the growth of newly planted woody plants by 30–40%, but not on poorly-drained soils. Benefits are greatest on light soils in dry seasons. The cost of the film, and of the necessary laying/planting machinery is high; as a result, the availability of relatively cheap herbicides has restricted its use. Other drawbacks have been the build-up of perennial weeds under the film, damage from vermin and occasional crop damage after mulch removal from herbicides which reach roots developed at or near the soil surface.

Chemical

Herbicides have many advantages over mechanical weeding. They control weeds close to the crop plants and can give season-long weed control. They require less labour than mechanical methods and can often be applied at times when soil is too wet for cultivation.

Herbicides do not kill weed seeds, they only control germinating or established weeds. Therefore, application must coincide with a susceptible stage of weed growth or the herbicide must persist in the soil until the weeds germinate. Some partial sterilants kill weed seeds, but they cannot be used selectively. As they are relatively expensive they are used mainly as pre-planting treatments for very high value crops, e.g. propagation beds for which there may be no selective treatment. They also control certain soil-borne diseases, and when used for this purpose should also provide a degree of weed control as a bonus. However, the rates of application and depth of penetration required for control of pests and diseases may not be optimal for weed control.

Development of herbicide use

Since their introduction in the late 1950s, herbicides have gradually replaced soil cultivation in fruit and other perennial crops. The early soil-acting herbicides, chlorpropham and 2,4-DES, were soon superseded by the more persistent simazine. Initially, inadequacies in chemicals were overcome by combining chemical and mechanical methods of weed control. However, the introduction of diquat and then paraquat in the early 1960s enabled established annuals to be controlled chemically. At that time, common couch was a major problem and dalapon was the only material that could be used to control it selectively. MCPA, MCPB and 2,4-D were available for the control of perennial broad-leaved weeds.

After nearly 30 years, simazine is still the most widely-used herbicide in tolerant crops because it controls most annual weed species and is cheap. In soft fruit, it is usually part of a programme involving other soil-acting herbicides such as lenacil, napropamide, propachlor and trifluralin either because they are safer to the crop, control simazine-resistant weeds or are less dependent on soil moisture. Other soil-acting herbicides have been introduced for specific crops, including dichlobenil, bromacil, terbacil and propyzamide. All of these herbicides control perennial weeds, including common couch, in addition to some annuals not controlled by simazine.

Chloroxuron was the only residual herbicide used for many years on container-grown perennials; because of foliar injury in some species, it had to be washed off leaves by irrigation after spraying. More recently, oxadiazon and oryzalin have been introduced for these crops.

Contact herbicides are still used to supplement the soil-applied herbicides. Non-selective materials such as paraquat are applied as directed sprays to control annuals but in apples and other tree fruits, translocated herbicides are also used, especially where there are perennial weeds. Phenmedipham, which is used to control seedling weeds in strawberries, is the only contact herbicide that is applied overall in an actively growing perennial crop.

Translocated herbicides employed as directed treatments for control of perennial weeds in fruit include alloxydim-sodium, amitrole, clopyralid, glyphosate, MCPA, MCPB and other phenoxyalkanoic herbicides. Alloxy-dim-sodium and clopyralid are also used as overall sprays in strawberries. Alloxydim-sodium permits the selective control of common couch in the growing crop, while clopyralid effectively controls creeping thistle.

Factors influencing the effectiveness of herbicides

Weed growth

Many weeds have clearly defined periods of emergence, so that herbicide treatments can be timed for maximum effect. Any weeds that survive, however, usually develop freely in the absence of serious competition from the crop or from other weeds. Annuals may thus produce large numbers of seeds and perennials become more established making subsequent control more difficult. Repeated use of the same herbicide for several years may select for a weed flora of more resistant species. A change to a different herbicide may overcome the initial problem, but other weeds are likely to develop; it is normal, therefore, to apply herbicide mixtures to overcome such problems.

Unsprayed areas are potential reservoirs of weeds. For instance, weeds such as creeping bent, creeping buttercup and silverweed can spread from the grassed alleys of orchards into the herbicide-treated strips. These may necessitate special treatments, or it may be possible to control them in the grassed alley with treatments that will not harm the grass. Species with wind-dispersed seeds spread much greater distances.

Herbicide-resistant weeds

Herbicide-resistant biotypes of weed species previously uniformly suscept-ible to a given herbicide have become a major problem in perennial crops in the UK since the early 1980s (see Chapter 8). Triazine-resistant groundsel and American willowherb occur in most fruit-growing areas and Canadian fleabane in Essex and Kent (Fig. 13.2). There are widespread reports of triazine-resistant annual meadow-grass and a few of triazine-

Fig. 13.2   Simazine-resistant Canadian fleabane (*Erigeron canadensis*) infesting a fruit tree nursery.

resistant pineappleweed. The triazine-resistant types appear to be suscept-ible to soil-acting herbicides of other chemical groups except uracils (lenacil, bromacil), so alternative residual herbicides can be used in fruit and ornamental crops, e.g. diuron, napropamide or oxadiazon, but such treatments are often more expensive. The only other fruit herbicide for which resistance is proven is paraquat to which resistant biotypes of American willowherb and annual meadow-grass occur. These biotypes are susceptible to the foliar-acting herbicides of other chemical groups recom-mended in fruit. Resistant biotypes have generally occurred where a single herbicide has been used for many years as the basis of the herbicide programme. Development of resistance can be deferred by using differing types of herbicides in rotation. Once resistance has developed other herbicides will need to be used to control resistant biotypes either instead of, or in addition to, the original herbicide.

Undisturbed soil

The compacted surface soil of plantations which are not cultivated remains moist for longer than does recently cultivated soil, and this is favourable for the activity of soil-applied herbicides. A firm surface also offers more support to men and equipment, so that the opportunities to apply herbicides are more frequent. Herbicide activity is reduced, however, if there is an increase in the organic matter and acidity of the surface layers. Under some conditions moss will develop, and growers may try to encourage this by using only those herbicides which do not affect moss, e.g. simazine, propyzamide, napropamide, glyphosate and paraquat. The lack of soil disturbance favours perennial weeds which were previously checked by the cultivations undertaken to control annual weeds. Firm surfaces also favour efficient harvesting by hand or by machine.

Timing

Most soil-acting herbicides are effective at any time of year provided that there is adequate soil moisture. This means that there is a long period during which application is possible in established crops. However, applications are normally made in late winter or early spring to ensure that the maximum herbicide concentration is present in the soil when germination of most weeds begins. Recommended rates of many soil-acting herbicides have little effect against weeds which have already emerged. They should be applied pre-emergence or used in conjunction with a post-emergence herbicide.

Annual weeds are usually most susceptible to contact herbicides when they are small. Phenmedipham, which is used in strawberries, kills only seedling weeds, so that weed size is the main factor determining when the spray should be applied. Paraquat and diquat control most annuals at any stage, and it is factors such as crop development and the timing of soil-acting treatments which determine when they are applied.

Translocated herbicides are usually most effective against perennial weeds when applied to well-developed shoots. Earlier application, or a reduced dose, may give good initial control but are less effective in preventing re-growth than a full recommended dose applied at the correct time. Because perennial weeds vary greatly in their growth patterns, it is seldom possible to choose a single application date which gives maximum effect against all species in a mixed population.

Seasonal patterns of crop growth affect the level of crop tolerance to herbicides; timing must often be a compromise between effectiveness against weeds and the tolerance of the crop. It is for this reason that the application of simazine to strawberry is restricted to the period between picking and the end of December. This avoids the risk of high concentrations remaining in the soil in late spring when there is maximum root activity and uptake from the surface layers of the soil. Application of foliage-acting herbicides may also be restricted to particular periods, e.g. glyphosate in orchards during the winter. Crops such as asparagus and rhubarb that die down completely can be treated safely with contact herbicides such as paraquat once senescence has occurred. With bulbs, application should only be made after the dead foliage has been removed. Premature application to dying foliage can cause serious damage because it coincides with maximum translocation.

MCPB may be applied over blackcurrants to control perennial broad-leaved weeds once extension growth has stopped. Earlier application would provide more effective weed control, but would damage the crop.

Dose

Commercial recommendations are based on the amount of herbicide needed to control specific weeds, the amount tolerated by the crop and economic factors including the cost and alternatives. This invariably results in a compromise which is capable of improvement in specific situations. In general, increasing the dose improves the level of weed control but it also increases the risk of crop damage. Exceeding the maximum dose shown on the product label is now prohibited under the UK Control of Pesticides Regulations 1986.

With most soil-acting herbicides, the dose is adjusted to the soil type but the amount entering the plant is largely determined by the weather conditions after treatment. Increasing the dose prolongs the period of weed control but this can be achieved more effectively with repeated applications, usually at lower rates. This technique reduces the maximum concentration in the soil and hence the risk to the crop, but increases the cost of application and requires suitable spraying conditions at regular intervals.

Soil type

Soil type is important in determining the availability and movement of herbicides and hence the extent to which weeds are controlled and crops are damaged. The activity of most soil-applied herbicides is greatly reduced in organic soils as a result of adsorption, and many recommendations exclude these soils since weed control would not be adequate. On sandy soils with low adsorptive capacity, excessive or unpredictable activity may occur, with consequent crop injury, and such soils may also be excluded from recommendations. The response to certain foliage-applied herbicides may also be related to soil type where they have some soil activity, e.g. dalapon.

Weather and soil moisture

The weather before, during and after application can have a profound effect on herbicide performance (see Chapter 6). The main effect of weather before application is on the condition of both crop and weed leaves, which may affect the retention and entry of foliage-applied herbicides. Most herbicides require a rain-free period for entry into the leaf after application; it can be very short for paraquat, but is stated to be 2 hours for alloxydim-sodium and 6 hours for glyphosate. The time interval needed between application and subsequent rain for optimum effect will vary with temperature, humidity and light intensity.

Soil moisture is the most important factor influencing the effectiveness of soil-acting herbicides. Ideally, they should be applied to a moist soil and there should be sufficient rain to ensure adequate penetration before the weeds germinate. However, excessive rain may move the herbicide deeply enough to damage the crop or too deep to control the weeds. In practice, growers often have to decide whether to apply a herbicide to a dry soil in the hope that rain will come in time, or to delay application until the soil is moist, thereby running the risk of being unable to get on the land before the weeds germinate. For most herbicides it is better to make the application to the dry soil in anticipation of rain, but this decision should be influenced by the material to be applied.

Irrigation can be used to 'activate' soil-acting herbicides but the requirements are generally ill-defined. They are also difficult to attain because uniform distribution cannot be achieved with most sprinkler systems. Nevertheless, it is used successfully with lenacil on summer-planted strawberries. When irrigation is applied for other reasons, it is important that there should be no conflict with the herbicide requirements. For instance, repeated irrigation of blackcurrants for frost protection in the spring on a soil already at field capacity can lead to damage from simazine, which is normally safe in this crop.

Application

It is usually important to ensure that the correct dose is applied uniformly, but with foliage-applied herbicides there are occasions when it is more important to avoid contact with the crop foliage. Although spray volume, pressure and drop size can be important, they are generally less critical than in other crops. This is because the herbicides are applied to relatively large weeds and there is no crop foliage to penetrate. The size, growth habit and spacing of many perennial crops enables sprays to be directed beneath the leaf canopy. This technique is most advanced in the tree fruits in which amitrole and other translocated herbicides are used extensively. Application is mainly with conventional hydraulic nozzles including wide-angle and off-centre types with swathes of 2–4 m. They tend to be used at lower than recommended pressures. The resulting uneven distribution is accepted because the materials used are cheap and application is simple, but weeds are often poorly controlled and this leads to problems later in the growing season. In addition, there can often be serious damage to low branches weighed down by fruit from these sprays in summer. Difficulties become even greater with the two- or three-row orchard beds now being planted.

The soil-acting herbicides used in perennial crops can be applied over a wide range of volumes. Many can also be safely applied over the crop foliage and since simazine, the most widely-used material, has a wide margin of tolerance in most fruit crops other than strawberry, the method of application is usually not critical.

Alternatives to hydraulic nozzles include spinning disc applicators which apply the herbicide at very low volume rates and uniform drop size. The large drops and the closeness of the disc to the ground, which permit application beneath the leaf canopy, make this type of equipment well suited for orchard use and in bush and cane crops where contact with the basal part of the plant is acceptable. Spinning discs are widely used overseas in plantation crops but there are currently no recommendations for use in the UK. Applicators which place the herbicide by physical transfer from a rope-wick or roller are suitable for the control of tall weeds in low-growing crops and between the rows of taller crops. This permits drift-free spot applications of foliage-applied herbicides that cannot be applied with conventional equipment without risk of much greater crop damage.

Factors influencing the tolerance of crops to herbicides

Innate tolerance

The tolerance of the crop to the herbicide is the most important single factor determining suitability for selective weed control. The mechanisms conferring tolerance are described in Chapters 6 and 7. Nearly all the herbicides used in perennial crops, including simazine, can be damaging under certain conditions because the crops are not innately tolerant. For soil-acting herbicides, field tolerance is determined by the inter-play of factors affecting the relative distribution of herbicides and roots, the uptake of herbicide by the plant and the growing conditions.

Age, size and variety

Large and long-established plants are usually more tolerant of soil-acting herbicides than small or newly-planted ones, mainly because they have a greater proportion of their roots beneath the zone penetrated by the herbicide. However, depth of rooting is not the only factor which determines tolerance. In flower-bulb crops there are few, if any, roots in the depth to which soil-applied herbicides penetrate, and damage indicates entry through the leaf bases.

Many recommendations for soil-acting herbicides exclude newly-planted crops because of their presumed greater susceptibility. There is, however, experimental evidence that they can be more tolerant than crops that are a year older; this is attributable to the small amount of roots near the surface in the year of planting.

Varietal differences in tolerance to herbicides exist in fruit crops, but it is impracticable to carry out full-scale field tests with all varieties. New recommendations for strawberries tend to be confined to one or two of the most important varieties for which experimental evidence is available; others are added later when there is more information from experiments and commercial experience. With apples and other crops that are grafted or budded, both stock and scion varieties contribute towards the overall tolerance. Differences attributable to the rootstock have been observed in experiments — M27 apple rootstock, for example, appears to be relatively sensitive — but they are usually not of practical significance.

Certain varietal differences in susceptibility are associated with a differential ability to metabolize the herbicide. It has been known for many years, for instance, that the apple 'Cox's Orange Pippin' can degrade 2,4-D whereas 'Bramley Seedling' cannot and is more susceptible to damage; seedlings from crosses between these two varieties are intermediate in response.

Crop vigour

There is no evidence of growth or yield reduction from repeated use of herbicides on vigorous crops. Where damage does occur it is often associated with poor plant growth. Strawberries suffering from verticillium wilt may be more susceptible to damage from soil-acting herbicides. Poor root growth on compacted, poorly-drained soil can lead to greater uptake of residual herbicides and consequent damage. Many recommendations restrict applications to vigorous crops and also require an interval between previous and/or subsequent treatments.

Area treated

Most recommendations are made on the assumption that there is adequate crop tolerance when the entire crop is treated and reductions in growth or yield are normally unacceptable. Weed problems sometimes occur which cannot be overcome with these recommended treatments, and it may be necessary to use a treatment not normally considered suitable either because it is known to be damaging or there is insufficient information on crop safety. Such treatments may be justified, provided they are safe to the operator, consumer and wildlife, but the decision rests with the individual grower having regard to local circumstances and approval of the use under the Control of Pesticides Regulations 1986. Often, the risk of crop damage can be minimized by treating only those parts of the plantation where the weed problem occurs or spot-treating individual weed patches. The recommendation for ethofumesate for controlling cleavers in strawberries, for example, is restricted to 'Cambridge Favourite' because yields of some other varieties can be reduced by 20–30%.

Impact of herbicides on crop production

Herbicides have eliminated the need for cultivation. Therefore it is no longer necessary to use spacings that allow access for cultivation and mowing. Initially this led to the introduction of bed systems for crops such as blackcurrants, but these have been superseded by spacings that are more easy to manage and are suitable for mechanical harvesting. In apples, the ability to use herbicides to control weeds under the branches has permitted the development of trees that crop much closer to the ground and which are easier to prune and spray. Multi-row beds are also only feasible with an effective herbicide programme.

The spatial arrangement of the plants may also have a bearing on subsequent herbicide use; with very close planting, for example, directed applications of foliage-acting herbicides may not be possible. Any new technique likely to change the ultimate plant size or affect root distribution could influence crop tolerance to soil-acting herbicides, and such consequences must be considered before new systems are adopted commercially.

The improved weed control and elimination of cultivation which are now possible with the aid of herbicides allow greater utilization of the upper layers of soil by the crop roots. In some plantations, surface acidity has increased and this in turn has led to manganese toxicity, while as already mentioned, a build-up of surface trash may reduce the activity of soil-applied herbicides. Undoubtedly, however, herbicides have allowed improved crop production methods to be adopted with a much reduced labour force.

The ability to control all vegetation under fruit trees has led to the introduction of complete bare-soil systems with no grassed alleyways. It has been shown experimentally that absence of grass and other competing vegetation can lead to quicker establishment of trees and higher yields in the early years of the orchard, as well as saving the cost of mowing alleys. On level, well-drained sites where a moss cover develops this system may be acceptable. Increase in surface acidity can be corrected by liming and the decline in organic matter levels associated with a long-term bare soil system can be corrected by mulching with straw or chopped prunings. There is evidence however that crop quality may be reduced in some orchards under bare soil and the lowered organic matter level can result in poor soil structure and difficulties in establishing subsequent crops.

On most sloping sites, grassed alleys are necessary to reduce soil erosion and aid traction. Grassed orchards may be necessary for other reasons, e.g. in cider orchards to keep fallen fruit clean before gathering and for better access by pickers on 'pick-your-own' enterprizes. Many bush and cane fruit plantations are successfully operated with complete bare-soil management.

Economic circumstances are making growers increasingly aware of the need to reduce costs. In most crops weed control and herbicides account for a small proportion of the total production cost. There is ample evidence that both newly planted and established perennial crops are very sensitive to competition from weeds, and that it can take, at best, several years to outgrow any check they cause. Besides the measurable losses, any shortcomings can make subsequent weed control more difficult and expensive. Both short- and long-term consequences must therefore be considered.

When costing weed control it is important to distinguish between the cost of the herbicides and the cost of application. Economies in weed control should also be considered in relation to other possible economies. A relatively large saving in weed control may be needed to give the same financial benefit as a small saving in some other cost such as harvesting, storage or marketing.

Weed-control programmes

It is not possible to prescribe a programme without defining its objective and considering the various options available. Most growers can base their programmes on their experience with herbicides in previous seasons and their knowledge of the requirements for machinery and labour. In general, year-round control of all weeds on treated areas is the objective.

The simplest programme is an annual application of the same soil-acting herbicide. If it is effective and there are no problems with residues, there is no need to change. In many situations, however, this simple programme is inadequate even when a persistent, broad-spectrum herbicide like simazine is used. By late summer or autumn some weeds become established, and these over-winter. They can be killed in spring by using a contact herbicide in conjunction with next season's soil-acting treatment, and a programme of this kind based on simazine and paraquat has worked well in many fruit plantations. Increased confidence in simazine, coupled with a reduction in its price compared with that of paraquat, has resulted in wider use of a second application of simazine in late summer or autumn. This prevents the establishment of over-wintering annual weeds which would have to be treated with a foliage-acting herbicide in spring. The residual effect of an autumn application reduces the dependence of the spring treatment on favourable soil conditions and, if necessary, allows it to be delayed.

Programmes have to be modified if any weeds are found to be tolerant. There have been many instances of weeds such as knotgrass and cleavers greatly increasing in numbers where control has been only partial. Incomplete control has also led to increases of perennial weeds, a phenomenon which is now well understood. This may necessitate a change of herbicide; knotgrass, which is often not controlled by simazine is susceptible to diuron. However, a change to diuron can result in plantains becoming a problem. A second approach is to supplement the original herbicide with one known to be effective against the tolerant weed species; pendimethalin or terbacil, for example, can be added to simazine in order to control knotgrass in apple orchards.

The possibility that repeated use of the same herbicide will result in weed species developing resistance has been discussed earlier. Where herbicide resistance is confirmed, an effective herbicide needs to be included in the programme. This may be a supplementary treatment to a broad-spectrum herbicide such as simazine, which will continue to control most species and is cheap to use. The timely use of mechanical methods of weed control may also be an effective means of preventing selective seeding by weeds escaping initial herbicide treatments.

Programmes should be realistic in terms of the amount of labour required and its seasonal availability, as well as in the cost of materials and their effectiveness in controlling the weeds. The crop, the variety and the management system all play a part in determining which herbicides can be used and when they can be applied. For instance, in intensive orchards, it is inadvisable to apply foliage-acting herbicides after the branches have become weighed down with fruit — usually July — but weeds must be kept under control until the crop is harvested in September to October. An early summer application of a translocated herbicide will control perennial weeds during this period, whereas a spring-applied soil-acting herbicide may control them only until the summer. In the latter case, it would be difficult to control the re-growth before harvest. Gooseberries are harvested in June or July and there are opportunities for post-harvest treatments. Raspberries are harvested slightly later, but in many plantations post-harvest applications are impracticable because access is prevented by new canes; pre-harvest treatments are therefore preferred.

With all programmes involving residual herbicides, an adequate time interval must be allowed between application and grubbing to avoid toxicity problems to following crops (see p. 386).

Special situations

Pre-planting treatments

Pre-planting treatments enable weed populations to be reduced by means of herbicides and techniques that cannot be used once the crop has been planted. The term normally refers to a specific measure, usually against perennial weeds for which there is no selective control in the young perennial crop. A wider interpretation includes any control measure that has had a significant effect on the numbers of weed seeds or vegetative propagules in the soil.

The traditional fallow, based on cultivation, can be very effective in the right conditions against those perennial weeds which are relatively shallow-rooted, such as common couch; it is less so against deep-rooted broad-leaved weeds such as creeping thistle and field bindweed. These latter species are better controlled with translocated herbicides, which are most effective if applied as the weeds approach flowering. In a fallow situation, the perennials can often only reach this stage in the absence of annual weeds, which would otherwise reduce their growth and make patches difficult to locate for spot treatment. To maximize growth of perennial weeds under these conditions, a suitable soil-acting herbicide (without effect on perennials or the subsequent crop) should be applied beforehand to suppress annual weeds. Alternatively, the need for a fallow can be avoided in cereal rotations by applying glyphosate before harvest. Both techniques can be very effective, but they rely on there being sufficient shoot development to ensure adequate translocation to the root systems of the weeds, and this is not always so. Treatment of deep-rooted perennial weeds in a fallow rarely eradicates, but may suppress, re-growth for up to two years. Where effective translocated herbicides can be used early in the life of the next crop, e.g. among trees, a fallow period may not be worthwhile.

There may be opportunities for pre-planting control in the previous fruit crop where spraying of patches of perennial weeds with a translocated herbicide after final harvest but before grubbing should be more effective than spraying re-growth after grubbing and land preparation.

Chemical sterilization of soil prior to planting can also be an effective means of controlling annual weeds in high value crops. Although the main purpose and economic justification is generally disease and pest control, weed control is a further reason for using the technique, particularly if it means that potentially inhibiting herbicide treatments can be avoided on young crops. However, control of weeds is seldom complete and re-invasion by means of wind- or bird-borne seeds will occur fairly quickly.

Newly-planted crops

Many herbicide recommendations are restricted to crops which have been established for 1–4 years. However, uncontrolled weeds have their severest effects on crops during the first two years. If there is not an effective re-commended herbicide, there may be a suitable treatment available which has off-label approval under Control of Pesticides Regulations 1986. Off-label treatments are used at the grower's own risk. Where severe weed growth threatens to smother a young crop, some crop damage (e.g. from a contact herbicide applied as a directed treatment) may be considered acceptable. Similarly, low rates of residual herbicides may be safe and effective.

Grass swards

In many orchards, grass swards are used to regulate tree growth by means of varying the frequency and height at which the swards are cut. The possibility of controlling grass growth by chemical retardants rather than cutting has been investigated, mainly using maleic hydrazide, but this technique has so far met with only limited acceptance. Future development will depend on obtaining consistent results at an acceptable cost. The composition of existing swards, of course, was not determined with chemical growth regulation in mind; success is likely to be greater with swards composed of species selected deliberately on the basis of suitability for management using growth retardants.

Control of unwanted crop vegetation

Herbicides, desiccants or growth regulators may be used to control unwanted vegetative growth in a number of perennial crops, such as suckers in top fruit and raspberry, runners in strawberry, to defoliate the basal areas of hop bines and kill unwanted lateral shoots, and to remove the first flush of young canes in vigorous raspberry plantations. Some of these were previously controlled by the cultivations designed for weed control. In many cases the new techniques involve interactions with other aspects of crop protection, crop husbandry or choice of variety and should be applied only as part of a management package. For example, a raspberry plantation which produces a few canes, is suffering from virus infection or is non-vigorous for other reasons, is not a suitable candidate for annual cane desiccation. Manufacturers' labels, of necessity, tend to give only the basic recommendations for chemical treatment. Growers should seek advice on the management aspects from their horticultural adviser.

Protected crops

Special care and caution are required in the use of herbicides under glass and polythene because of the exceptionally high value of most protected crops. It is also essential to ensure favourable conditions for soil-acting herbicides. In structures without internal irrigation systems, e.g. strawberries growing in low polythene tunnels, this means adequate moisture between applying the herbicide and covering with polythene. Trickle irrigation provides insufficient surface moisture to activate soil-applied herbicides. Where heavy hose watering is practised, there may be some

risk of crop damage on sandy soils. Damage may also be caused by volatilization of herbicides that are normally safe under field conditions. In
general, manufacturers are unwilling to make recommendations for the
use of herbicides under glass or polythene.

Container-grown plants

Container-grown ornamentals pose special problems. Weeds should not
occur when sterile rooting media are used and there is no ingress of wind-
and water-borne seeds but most nurseries have problems with hairy bittercress, willowherbs and liverworts. The shape of individual plants and
their density precludes directed applications of non-selective treatments
and the diversity of species and the relatively small area of each on most
holdings require treatments that can be used on a wide range of species.
Several herbicides are recommended for this use including diphenamid,
oryzalin and oxadiazon, applied as sprays or granules. The range of weeds
controlled, persistence of weed control and the crop species on which they
are recommended, differ for each herbicide.

Herbicide residues

The principles affecting herbicide degradation in soil and the occurrence
of residue problem are discussed in Chapter 9. There is no evidence of
repeated use of soil-acting herbicides for 10 years or more in perennial
crops leading to build-up in the soil and adverse effects on growth. However, persistent residues can affect subsequent sensitive crops. For this
reason recommendations for persistent soil-acting herbicides used in perennial crops indicate appropriate intervals between final treatment and
grubbing and may suggest which crops should follow. Where there is doubt
soil analysis can help. The risk may be accentuated in some crops where
difficulties in application lead to repeated over-dosing of strips along or
between rows.

Growers may not encounter problems on their own land but planting
crops on newly-rented land can pose problems where previous herbicide
usage is unknown. This particularly affects crops for propagation of stock
necessarily grown away from production areas. There have been instances
of damage from isoxaben and sulphonylurea herbicides used in preceding
cereal crops. The previous history of herbicide use on all land to be used
for growing perennial crops needs to be carefully checked.

14 / Weed control in agricultural grassland

R.J. HAGGAR, D. SOPER* AND W.F. CORMACK⁺

*AFRC IGAP Welsh Plant Breeding Station, Plas Gogerddan, Aberystwyth, Dyfed SY23 3EB; *Rhône-Poulenc Ltd, Fyfield Road, Ongar, Essex CM5 0HW; and ⁺MAFF ADAS Trawscoed, Dyfed EHF*

Nature and importance of weeds

Definition of weeds

An agricultural weed can be defined as a plant species which, when present in sufficient numbers, reduces the profitability of an enterprise. Weeds may do this by reducing output, by reducing the value of the product, or by increasing costs.

It is fairly easy to identify arable weeds by these criteria, since relationships between weed densities and crop yield can be determined directly. Weeds also reduce the value of grain so reducing profitability, while the control of arable weeds is a major cost item in crop production. It is more difficult to define which plants are weeds in grassland, since grass is rarely a traded commodity, being an intermediate product in livestock production. Poisonous plants, such as common ragwort, are obviously weeds but many other species, such as meadow-grasses, seem to have little detectable effect on economic returns — at least in the short term — and so are not obviously weeds.

Some people define weeds more broadly, to include plants which are unsightly but have no detectable effect on profitability. Such a wide defini-

387

tion is difficult to justify in modern agriculture, where the cost of control measures must be set against profits.

The problem of defining weeds in grassland is made more difficult by the fact that a species may be truly a weed, as defined above, under certain conditions but not under others. For example, Italian ryegrass would certainly not be regarded as a weed in a pasture, but can become a severe weed in a following arable crop if the pasture forms part of a rotation. Similarly, there are good reasons for considering fine-leaved fescues as weeds in lowland pastures because of their low quality herbage, though they have considerable value as high-yielding constituents in upland pastures.

Types of weeds

Weeds of grassland can be classified into four main categories:

Poisonous plants

A plant which is acutely or chronically poisonous when eaten by stock is clearly a weed. Acutely poisonous plants include common ragwort and water dropwort. Chronically poisonous plants include foxglove, field horsetail, bracken and rhododendron. Buttercups, too are poisonous plants but are often unrecognized as such because of the insidiousness of the effects.

Harmful plants

Plants which cause physical injury to stock include grasses with sharp, barbed awns, e.g. wall barley, which penetrate the mouth and eyes causing great discomfort and loss of production. Other plants may reduce the value of animal products by becoming entangled in wool, or by imparting undesirable colour or taint, e.g. wild onion.

Unpalatable plants

Plants which are not eaten freely by grazing animals and which restrict the abundance and productivity of more palatable species include nettles, rushes, thistles and tufted hairgrass. Palatability is a relative term, varying according to the availability of other plants, stage of growth, etc.

Unproductive plants

The productivity of some grasses, e.g. annual meadow-grass is low in comparison with other plants which could be grown. These species, although unproductive alone, may not necessarily affect yield in the short term when present in mixtures of other species growing at high density. In the long term sward productivity is likely to suffer, especially if the distribution and abundance of crop species (e.g. white clover) is adversely affected. Some broad-leaved plants, e.g. docks will reduce pasture utilization without having other desirable properties.

Weed occurrence

Newly sown leys

Young leys in a mixed rotation are commonly infested with arable weeds including annual species such as fat-hen, cleavers, groundsel, common chickweed, annual meadow-grass and redshank. The actual species which occur vary depending on previous cropping, time and method of sowing. Perennial weeds are less frequent, especially if weed control in the previous crop has been satisfactory.

Some species are more damaging than others in depressing the density, speed of establishment and yield of the sown species. Of particular concern is chickweed, especially when allowed to form large, smothering clumps; this can occur with infestations as low as 15 plants m^{-2}. On the other hand, annual meadow-grass tends to occur at much higher densities, often 3000 plants m^{-2}, and causes substantial reductions in the tiller density of sown species.

Where new leys are established after old grassland, a different type of weed flora can be expected, including buttercups, daisy and various grasses. However, it is not always possible to predict from observations of the original sward what will be the main weed problem. Arable weeds, such as charlock and poppies can establish in large numbers from viable buried seeds, even after 5–10 years. Similarly, the seeds of grassland weeds such as rushes can remain viable for many years and, when disturbed by ploughing, may produce enormous populations of seedlings. Areas on which weed-infested hay has been previously, or where manure has been spread, can also produce unexpected weed floras in the subsequent grass crop.

Weeds of established pastures

Surveys of weeds in established pastures have shown that creeping thistle is the most common weed on beef- and sheep-producing farms, especially on older pastures. On dairy farms, docks are most abundant, especially in

Fig. 14.1 Sheep grazing thistle-infested pasture. Photograph courtesy of Welsh Plant Breeding Station.

young swards, while buttercups are also very widely distributed. Other evident weed species include bracken, tufted hairgrass, rushes, nettles and common ragwort.

Some grasses found in permanent swards which have previously been regarded as weeds, e.g. Yorkshire fog, are of questionable weed status since there is no firm evidence that they reduce the output or profitability of grassland enterprises. The status of some broad-leaved species is also in doubt though there is evidence that thistles can reduce animal output. In the case of docks, although they may be grazed by cattle and will certainly contribute to hay and silage bulk, there is evidence that removal of broad-leaved dock populations giving over 20% ground cover can result in increased grass production. In practice excessive amounts of this weed leads only to poor quality silages.

Factors affecting weed content

Newly sown leys

The greatest opportunity for weeds to invade swards is at the time of sowing and during establishment. Large numbers of weed seeds are stimulated to germinate during seedbed preparations, especially when rotary cultivation is carried out. Rapidly establishing broad-leaved weeds, such as fat-hen and corn spurrey, often outgrow and overshadow the sown grasses and legumes, suppressing leaf growth, tillering or stolon production. Even quite prostrate species, such as common chickweed and annual meadow-grass, can be extremely competitive in poorly establishing crops, especially in mild, wet autumns.

The extent to which weeds invade newly sown leys is often a reflection of the density of the sown species. When crop density falls much below 250 plants m^{-2}, substantial invasions of weeds will occur, especially if the crop plants are not uniformly distributed. Crop vigour, also, plays a key role in determining weed ingress. Thus, rapidly establishing grasses, such as Italian ryegrass, tend to be relatively weed-free, especially when crop growth is stimulated by fertilizer nitrogen. Conversely, slower establishing grass, such as timothy and tall-fescue, and crops stunted by poor growing conditions, or by pest and disease attack, are more likely to be invaded by weeds.

Established swards

As swards age, the density of sown species tends to decline, with fewer but larger plants remaining, often in a clumped distribution. The sward then takes on a mosaic appearance. Surveys have recorded the rate at which sown species decline and the extent to which weed species invade older swards. Factors which influence sward vigour — and hence weed ingress — are considered in the next section.

Drainage

Impeded drainage affects sward composition directly, by creating unfav-ourable conditions for crop grasses (such as, poor aeration for root devel-opment and slow release of nitrogen from organic matter). It also has indirect effects by imposing limitations on grazing practice (e.g. by reduc-ing stocking rates at key times of the year, notably in late autumn). Surveys have shown that both perennial ryegrass and white clover disappear rapidly from swards on soils with poor drainage while improved drainage leads to increases in the proportion of sown species.

Grasses which tolerate waterlogged soils include Yorkshire fog and tufted hairgrass. Weeds like creeping buttercup and field horsetail are particularly associated with unsatisfactory drainage.

Soil pH

Perennial ryegrass, white clover and red clover do not thrive at pH values below 5.5. When the pH is between 4.5 and 5.5 grasses such as common bent, sweet vernal-grass and Yorkshire fog, often with common sorrel, usually predominate. On markedly acid soils (pH below 4.0), grasses such as mat-grass and purple moor-grass become dominant.

Liming raises soil pH and makes major nutrients available to plants. Liming to a pH of at least 6.0 is commonly advised for soils under permanent grassland, especially where white clover is important.

Soil nutrients

Phosphorus: Species associated with low available soil phosphate include common bent, sweet vernal-grass, and sheep's fescue. As the phosphate index increases so the content of perennial ryegrass and white clover usually increases while the content of dicotyledonous weeds tends to decline. Phosphorus is especially important in increasing the content of white clover. However, phosphate does not move readily in the soil and a sharp gradient in the soil profile may build up with time.

Potassium: Grasses occurring on potassium-deficient soils include red fescue, common bent and smooth meadow-grass, as well as several broad-leaved weeds, e.g. docks. The effect of potassium in promoting clover growth is well known, although applying potassium can also increase some broad-leaved weeds, for example, dandelion.

Nitrogen: Grasses such as common bent, red fescue and Yorkshire fog can tolerate soils which are relatively deficient in nitrogen. These species, however, are not inherently low yielders and they should be treated as symptoms of nitrogen deficiency rather than causes of low productivity.

Increasing the nitrogen supply usually increases the ryegrass component of permanent pasture; other grasses reported to be encouraged by nitrogen include meadow-grasses, cocksfoot, meadow fescue and common couch.

Increasing use of nitrogen fertilizer, which is usually accompanied by intensification of grazing management, decreases the content of certain broad-leaved weeds, e.g. creeping buttercup, but can increase the content of others such as docks, common chickweed and common nettle. Another detrimental effect of increasing nitrogen fertilizer is that legumes are severely suppressed. Also, nitrogen applied to conservation fields results in more open swards, which allows weeds (particularly docks) to establish.

Physical damage

Pasture species vary in their resistance to animal treading; perennial ryegrass and white clover are among the most resistant and Yorkshire fog least. Heavy treading, especially when the soil surface is dry, can increase the proportion of ryegrass. However, treading when the soil is wet breaks the turf and compacts the soil; this poaching can have a very adverse effect on herbage production and sward composition. Many undesirable plants are better adapted to poor soil structure and waterlogging than are the more useful grasses and clover.

Molehills, and to some extent dung pats, by obliterating sown grasses and exposing friable soil, provide an ideal medium for weed seeds to germinate, and for other weeds to spread by vegetative propagation; rushes, annual meadow-grass and bents are frequent colonizers of such areas.

Dung, urine and slurry

The nutrients in dung stimulate herbage growth but dung also causes the herbage to be rejected by grazing animals and can affect botanical composition. Sheep dung initially increases clover growth but this is followed by grass dominance.

Urine, being rich in nitrogen and potassium, stimulates grass growth and depresses clover. Occasionally, urine scorch results in bare patches which are rapidly colonized by broad-leaved weeds and annual meadow-grass.

Slurry and farmyard manure when spread on grassland often lead to heavy infestations of docks, especially on fields cut for silage.

Defoliation

It is well known that different patterns of grazing and cutting change sward composition. Hard spring grazing usually encourages white clover; over-grazing in winter, followed by under-grazing in early summer, is most likely to cause an increase in weed species. Hard grazing, especially in spring, encourages white clover but can also encourage prostrate weed species, such as buttercups.

Infrequent cutting or lax grazing encourages tall and stemmy plants, e.g. cocksfoot, false oatgrass and unpalatable species, whilst continuous grazing, or set-stocking, usually increases the clover content compared with rotational grazing, provided the stocking rate is carefully matched to herbage growth so as to avoid over- or under-grazing.

Sheep graze more selectively than cattle but horses are even more fastidious. Consequently, weeds are a conspicious feature of horse paddocks. Whenever possible, horses should graze with cattle or sheep, which will eat areas rejected by horses.

Mowing in late autumn to remove herbage neglected during grazing, may benefit botanical composition. Also mowing at key times of the year, especially when plants are making maximum growth, helps to control certain weeds. For instance, creeping thistle is most vulnerable to mowing at the early flower-bud stage. Mowing must be used with care, however, since close cutting after a long rest period opens up the sward and favours the establishment of seedlings of some weeds, e.g. soft brome, rough meadow-grass and docks.

Mowing once or twice each year for several years will, like lax grazing, encourage tall species, many of which yield low-quality herbage. This practice also results in sward thinning and weed ingress. However, in old hay meadows it is difficult to separate the effects of cutting from those of continual removal of plant nutrients. Many undesirable plants may occur in hay meadows because they tolerate lower nutrient status than more desirable and productive species. The alternation of grazing and mowing in successive seasons is a useful way of reducing nutrient depletion and weediness.

Winter kill

Low temperatures, often interacting with nutrient deficiency, pests and diseases, affect sward composition. For example, Italian ryegrass is generally much less winter hardy than perennial ryegrass and this effect is increased by pests, leading to weed ingress later. Frost heaving can force young plants out of the ground and so reduce stand density.

Pests and diseases

There is a wide range of pests afflicting grass swards; and the influence on botanical composition is now beginning to be understood. Damage is usually more apparent in newly-sown swards, but older swards also suffer damage from pests. Frit fly can reduce the productive life of swards, especially those containing Italian ryegrass. Slugs, stem eelworm and pea and bean weevil attack clovers, leading to a clover-deficient sward.

Fungal diseases such as *Fusarium*, often cause problems in newly-sown leys, while clovers are commonly infected with viruses. These infections can affect sward composition, to the detriment of yield and seasonal production.

Losses can be reduced by treatments at seeding (either seed coatings or granules) and there is scope for the development of resistant grass and clover cultivars.

Weed control during establishment

Preventing weed ingress

Rectifying site limitations

Before sowing grass seed, every opportunity should be taken to rectify inadequate drainage and excessive soil acidity; deficiencies of any of the major elements should also be corrected. Phosphorus, which is relatively immobile, should be applied during seedbed cultivations.

Choice of species to sow

Of the grasses commonly sown, the ryegrasses (including Westerwolds and Italian) establish most rapidly and are therefore more likely to suppress weeds during the first year. Next in order of aggressiveness comes cocksfoot, followed by timothy and meadow fescue, with tall fescue being the least aggressive during establishment.

If nurse crops such as rape or cereals are used they should be sown at moderate rates and harvested as soon as they have served their purpose of initial weed control.

For long-term swards, high-tillering, persistent varieties of grasses and disease-tolerant clovers are needed.

Seedbed preparation

Every care should be taken to control previous weed infestations, either chemically or by appropriate cultivations. A fine, firm and moist seedbed in needed, free of trash and weeds.

Seed rates

A target density of up to 400 plants m^{-2} is usual; to achieve this, grass seed rates of 15–$25\,kg\,ha^{-1}$ are adequate. Italian and tetraploid ryegrasses, being relatively large seeded, should be sown at slightly higher rates than diploid ryegrasses. There is little scope for preventing weed ingress by increasing seed rates much above these amounts. Clovers have small seeds, and sowing rates of up to $3\,kg\,ha^{-1}$ are adequate.

Method of sowing

Sowing in widely-spaced rows is more likely to lead to a weedy sward than is sowing in narrow rows, and crops, established by broadcasting seed are often freer from weeds than drilled crops. In neither case, however, should seed be buried deeper than 2 cm.

Undersowing: It is claimed that a cereal cover crop helps to suppress weeds. Although this is often the case, the cereal crop also suppresses the sown sward, while the undersown species restrict the range, and timing, of herbicides that can be used, particularly if legumes are included. Success with this technique is best achieved by early removal of the cover crop.

Direct sowing: Direct sowing can be done at any time from March to mid-August, or later in the case of ryegrasses. With direct sowing, the main weed problem is the flush of arable weeds that germinate soon after sowing. It occurs, regardless of sowing date, because many important annual weeds, especially common chickweed, germinate at any time.

Minimum cultivation: It is possible to resow swards without the need for expensive ploughing or cultivation. The sward should be closely grazed or cut, and then allowed to regrow before applying herbicide (e.g. glyphosphate or paraquat) to kill the old sward. Heavily matted swards should be sprayed in autumn, left overwinter to allow breakdown of the crop, then sprayed again in spring before sowing. A suitable drill is essential and the seed should be protected against slugs, insects and pathogens. It is especially important to provide adequate nitrogen in the seedbed. This technique is not suitable for fields which are acid, compacted, badly drained or otherwise unsuitable for productive grasses.

Post-sowing management

The first few weeks of a sward's life are crucial. Some seedlings will inevitably die but it is important to minimize losses due to weed competition, pests and pathogens. Cover crops should be removed as early as possible. Newly-sown swards are especially prone to winter kill if they enter winter with too much lush foliage.

Killing weeds in newly-sown leys

Cultural methods

Some control of erect annual dicotyledonous weeds may be achieved by mowing when their apical parts are above the cutting height. Most prostrate or rosette species, such as shepherd's purse, common chickweed, knotgrass and common orache, may actually be encouraged by mowing.

Grazing, preferably by sheep, encourages rapid tillering of sown grasses, and helps to control some weed species, e.g. common chickweed.

Chemical control

Choice of herbicide: Having identified the problem weeds, suitable herbicides can be selected. Most herbicides are sold as mixtures, to give a wide weed spectrum, and care is needed to ensure the safety of sown species, especially legumes. The choice of herbicide will also depend on its cost, availability, and relative ease of application.

When to spray: Spraying should usually be carried out as early as possible, to kill weeds before they become too competitive, and while they are most susceptible. However, when using post-emergence treatments care should be taken to ensure that the young grass and legume seedlings have developed sufficiently to withstand any adverse effects of the herbicides.

Perennial weeds are often more difficult to kill than annual weeds and spraying may need to be delayed until flower-bud formation has started. In practice, weeds are often present over a wide range of growth stages and may need repeat treatments; pre-treatment mowing, so that the weeds regrow evenly, can improve the effectiveness of spraying. Spraying should take place when crops and weeds are growing actively (when soils are warm and moist) and not in high winds or if rain is likely within 6 hours.

Application of herbicides: The product label should be read carefully before spraying and the herbicide applied at the correct pressure and volume, avoiding misses or overlaps, and ensuring safety standards (to operator, surrounding crops and wildlife), both during use and in the disposal of any excess herbicide. It is an offence under the Food and Environment Protection Act 1985 to use herbicides outside the manufacturer's label recommendations. Most herbicides can be applied overall by conventional sprayers; those that rely on contact action should be applied in a high volume of water. Accuracy of application is more important with these herbicides than with growth-regulator herbicides.

Weed control in established pastures

Invading weeds are useful indicators or symptoms of sub-optimal growing conditions and/or mismanagement. There is little point in treating the symptoms without rectifying the conditions which caused them.

Preventing weed ingress

The main factors which favour weeds in pastures have already been considered; these include inadequate drainage, deficiencies of phosphorus, potassium and nitrogen, physical damage brought about by poaching, molehills and dung, grazing mismanagement, and the attacks of pests and diseases. If the ingress of weeds is to be prevented, then these deficiencies must be diagnosed and corrected.

Killing weeds in established swards

Pulling or digging out weeds is not practicable, except for light infestations of potentially damaging weeds that cannot be dealt with economically in any other way.

Cutting flowering stems will prevent seed production but even with regular and frequent topping it will take at least two years to eradicate creeping thistle. In some cases, such as with common ragwort, cutting may do more harm than good, since it encourages prostrate growth.

Serious weed problems should be tackled by spraying with selective herbicides. There is a wide range of proprietary herbicides to choose from. Most include MCPA and/or 2,4-D, which control many of the common broad-leaved weeds without damage to grass. More resistant weeds require mixtures of herbicides including mecoprop, dicamba and triclopyr: even then complete control may not be achieved with a single application. Where the content of clovers is to be maintained, mixtures containing asulam, benazolin, bentazone, 2,4-DB and/or MCPB may be used.

For the best results, swards should be sprayed during the period of active growth but before flowering. The most suitable time for most weeds in the UK is May or early June but there are exceptions; common chickweed is best sprayed in autumn; bracken is best sprayed in early August.

Times of spraying may need to be modified to suit grassland management. For instance, fields containing common ragwort should be treated in early May if grazed but if hay is required to be taken in the following year, spraying needs to be carried out in the autumn. Nettles and docks can be sprayed after June or July only if there is adequate regrowth of leaves.

Clumps of difficult weeds, e.g. nettles, are best tackled with a knapsack sprayer. Where tall-growing weeds are scattered throughout the sward, 'weed-wiping' applicators can be used to smear the weed with translocated herbicides, without damaging the underlying grasses and clovers. However, overall spraying may then be needed to kill smaller plants and germinating seedlings.

Post-spraying management

Following spraying, fields should be left for about 14 days (but consult label) before being grazed or cut so as to allow for herbicide translocation. If poisonous plants are present, such as common ragwort, they may become more palatable to livestock after being sprayed. In this case, grazing must be delayed until the dead plants have been removed or have disappeared.

The death of weeds leaves gaps which other weeds can invade. Most sown grasses, because of their tufted habit, are slow to invade large gaps and rapidly spreading species, such as creeping bent and rough meadow-grass usually invade the space. Spaces may also be invaded by seedlings produced from the buried seed populations, such as annual meadow-grass. White clover, however, is one of the few sown species that is well equipped to colonize bare spaces quickly.

When large clumps of weeds have been killed, e.g. nettles, the large bare patch can be filled by sowing grass seed, but additional spraying will probably be needed to stop further invasion by weeds.

Unless the deficiencies that lead to weed invasion (inadequate drainage, nutrient deficiency, grazing mismanagement) are corrected or ameliorated, weeds will reappear after treatment, so that chemical weed control will have only transient effects. For example, buttercups are best tackled by the combined use of herbicides and nitrogen fertilizers, and bracken will re-encroach in the absence of treading.

Manipulating botanical composition with herbicides

There are certain circumstances in which a specific change in botanical composition may be required. An example of this is the need to increase the legume content of swards. Many beef and some dairy enterprizes, are run with low nitrogen fertilizer inputs ($<50\,kg$ nitrogen ha^{-1}). Under these conditions, the proportion of clover is of paramount importance, yet most of these swards contain less than 5% clover.

White clover can be changed from a minor to a dominant component within 4 or 5 months by using grass-suppressing herbicides such as carbetamide, propyzamide and paraquat in autumn or late winter. Although the concomitant loss of early grass growth is considerable, this is largely offset by a substantial increase in production from mid-summer.

There are some circumstances in which it is desirable to increase the proportion of perennial ryegrass in a sward. This can be done by using a low dose of dalapon in early summer which suppresses other grasses, such as bents and rough meadow-grass, and is especially effective where a herbicide-tolerant variety of perennial ryegrass has been sown.

Using selective herbicides to manipulate botanical composition can only be effective if there is an adequate presence of desired species. If there is not, then they may be introduced by scattering seed and treading it in with livestock (mob-stocking), by drilling seed into the sward, usually after temporarily suppressing the other species with paraquat, or by sowing seed into slots.

Once again, in all these cases when the sward composition has been changed rapidly, the new composition can only be maintained if the factors which caused the original composition have been rectified. If nothing is changed, the sward will soon revert to its original composition.

Weed control in rough grazing

Rough and hill grazings occupy nearly a half of the total agricultural area of Scotland, one-third of Wales, one-ninth of England and one-sixth of Northern Ireland; this makes a total of about 7 million ha. These grasslands occur typically in areas of high rainfall, on acidic, poorly-drained and nutrient-deficient soils.

Vegetation types

Some rough grazings occur in the lowlands in very wet habitats: bogs, fens and riverside meadows. The vegetation in these areas is characterized by sedges, tufted hairgrass, rushes and reeds which are practically worthless as forage, though other more palatable species may also be present. Chalk downland is a different type of rough grazing usually dominated by fine-leaved fescues or, if undergrazed, by false oat grass and tor-grass, often invaded by shrubs such as hawthorn or blackthorn.

On very acid, peaty soils, the predominant vegetation consists of either mat-grass or purple moor-grass, with variable quantities of heathers, cotton-grass, rushes, tormentil and bilberry.

These vegetation types may be divided into three categories: (1) species with long growing seasons which include most of the acid grassland species; (2) species with restricted growing seasons but remaining winter green, such as heather; and (3) species which die back in autumn, e.g. purple moorgrass. In the last two categories herbage quality rather than quantity is the major factor limiting livestock output.

Changing the vegetation

If the existing vegetation is to be changed to increase both the quantity and quality of herbage available, then the first essential is to improve soil conditions and management. In almost every case, lime or fertilizers must be applied and fences erected so that grazing management can be improved. In many cases the land will also need draining. Since many of the existing species are of little value, yet persistent, the existing vegetation must often be destroyed and replaced by more desirable species. Ploughing and reseeding is extremely costly and does involve a risk of failure during establishment. The risk is increased in wet areas, where buried seeds of rushes will be encouraged to germinate.

Another approach is to destroy undesirable species with selective herbicides, or with surface cultivation and then broadcast seed to introduce desirable species, especially white clover. Dalapon has been used to control selectively mat-grass and purple moor-grass; and paraquat and glyphosate are now also widely used. The best technique is to apply these herbicides in September so that the mat decomposes over winter; re-seeding can then be carried out in the following spring. Troublesome species, like tormentil and bilberry need special treatment.

On certain moorland areas, where it would be uneconomic to modify the existing vegetation by cultivation or chemicals, herbage quality can be improved by controlled burning. For instance, burning molinia moors at intervals of about 7 years helps to prevent further deterioration in herbage quality. Similarly, burning at such intervals rejuvenates heather moors.

Controlling specific weeds of rough grazing

The principles of controlling problem weeds like thistles, bracken and nettles have been considered earlier, while the control of sedges, rushes and reeds is dealt with in Chapter 18.

Fig. 14.2   Bracken threatening agro-forestry plantings. Photograph courtesy of Welsh Plant Breeding Station.

Scrub

Because the stocking rate of rough grazing is normally low, and because grazing is less controlled than on lowland pastures, rough grazings are often invaded by shrubs, such as hawthorn, gorse or sometimes by brambles. Where clearance by mechanical means or by hand is not possible, spraying with triclopyr provides effective control. It can be applied to the foliage in the summer or to cut stumps in the winter.

Gorse

Although gorse can be checked by spraying triclopyr onto growing bushes, it is a prolific seeder and produces hard seeds which persist in the ground for many years. As a result, gorse seedlings remain a continuing problem, even after initial clearance. One of the most effective control measures is frequent hard grazing.

Agro-forestry

This refers to a mixed farming practice in which livestock is grazed amongst deliberately planted widely-spaced trees. Agro-forestry is now being adapted to the hill and upland farm to meet changed economic conditions.

Neglected areas invaded by undesirable bracken and scrub and no longer required or used purely for extensive cattle or sheep grazing can be utilized in this manner, providing an acceptable and pleasing environment. Broad-leaved trees such as ash and sycamore, provided from the nursery at 2–3 years old, are planted at 5–10 m wide spacings. The shade and light patterns eventually encourage a diversity of fauna, particularly insect species, and flora. The native grass flora accompanied by natural re-seeding may provide sufficient keep, but some deliberate re-seeding with grass and clover species (the latter to obviate or reduce use of nitrogen fertilizer) may be required.

Such novel approaches have their own problems. Experiments are currently in progress on land preparation for planting, nutrient requirements, livestock density, and the most suitable type of livestock (e.g. beef cattle or sheep breed).

In general, it appears that herbicides such as asulam or glyphosate, and triclopyr can be used, e.g. for bracken and gorse clearance respectively, prior to planting, but much work needs to be done on the tolerance of young broad-leaved tree species to these herbicides and others. Tree guards might assist selectivity, but there could be root uptake. Legume-safe herbicides are generally desirable. Existing spray equipment may need some adaptation. Above all, safety of use and absence of any environmental problems, e.g. water pollution, are of paramount importance.

Economics of weed control

The control of a weed in grassland can only be justified on economic grounds if the benefits of control, in the medium or long term, are greater than the costs. Some farmers may wish to control weeds for aesthetic reasons; however, they should be aware of the direct costs (chemicals and spraying) and the indirect costs (opportunity costs and possible reductions in output) which they will incur.

A major difficulty in measuring the economic benefits of controlling weeds in grassland is to attribute a value to grass; any increased herbage production has to be utilized by ruminant livestock before it can give a financial return. The values inputed will depend on whether the herbage is used: (1) to increase stocking density; (2) to reduce concentrate feeding; or (3) to buy less hay.

Selective herbicides used to control invading weeds in newly-sown leys usually increase the contribution of the sown component and bioeconomic models have shown that the control of common chickweed can be worthwhile. However, few experiments have continued long enough to determine the long term effect on total yield and pasture production. There have been relatively few studies of the effect of treating establishing pastures with herbicides on animal production.

Rather more information is available on the effects of weed control in established pasture. The short term effect of herbicides is usually to reduce the total herbage yield of the sward but increase the content of sown species. In the longer term, herbicides may have little substantial effect on total yield, though the effect on botanical composition can persist for some while, with associated improvement in forage quality.

The effects on animal production have been variable. Applications of dalapon, which reduce the content of invading species such as Yorkshire fog, meadow-grasses and bents, often have only marginal effects on animal output. However, applications of paraquat have resulted in substantial live weight gains, apparently because of an increase in the clover content of the sward. In New Zealand the control of certain harmful weeds is known to increase animal output; control of wall barley by a mixture of herbicides usually leads to an increase in lamb production the value of which greatly exceeds the cost of treatment. An economic advantage from controlling creeping thistle has also been reported. In Australia, estimates have been made of the economic impact of controlling serrated tussock, and a number of bioeconomic models of weeds in pastures have been constructed.

The literature on weed control in UK grassland reveals that information on the economic threshold levels of weed infestation, above which economic benefits of improved grass yield outweighs the cost of spraying, is scarce. However, a study of the effects of controlling dock has shown that if only the first year following spraying with asulam were to be considered, then ground cover by the weed needed initially to exceed 15% before spraying was economic. On the other hand, if a 10-year time horizon was considered, the economic threshold was lowered to a dock ground cover of 2.5%.

In conclusion, the economics of weed control in grassland have been inadequately considered and insufficient information is usually available to predict the likely long-term outcome of weed control. However, there can be little doubt that infestations of poisonous species, e.g. common ragwort, and harmful species, e.g. wall barley, or unpalatable species, e.g. rushes, are likely to justify the use of control measures.

In all cases, it is essential to recognize that weeds only occur because of the particular environmental conditions and pasture management. Unless these conditions are changed, weed control will only be temporary and reinvasion will occur; it is therefore essential to correct deficiencies in environmental conditions and management if long term control of weeds is to be achieved.

Further reading

Brockman, J.S. (1985) *Weeds, Pests and Diseases of Grassland and Herbage Legumes.* Monograph No. 29. BCPC Publications, Bracknell.

Charles, A.H. & Haggar, R.J. (1978) *Changes in Sward Composition and Productivity.* Occasional Symposium No. 10., British Grassland Society, Hurley.

Doyle, C.J., Oswald, A.K., Haggar, R.J. & Kirkham, F.W. (1984) A mathematical modelling approach to the study of the economics of controlling *Rumex obtusifolius* in grassland. *Weed Research*, **24**, 183–193.

Peel, S. & Hopkins, A. (1980) The incidence of weeds in grassland. *Proceedings of British Crop Protection Conference—Weeds*, **3**, 877–890.

Williams, R.D. (1984) *Crop Protection Handbook: Grass and Clover Swards*. BCPC Publications, Bracknell.

15 / Weed control in sports turf and intensively managed amenity grassland

J.P. SHILDRICK

National Turfgrass Council, 3 Ferrands Park Way, Harden, Bingley, West Yorks. BD16 1HZ

Introduction

Amenity grassland is 'all grass with recreational, functional or aesthetic value, and of which agricultural productivity is not the primary aim'. On this basis, it was estimated in 1977 that in the United Kingdom about 3.5% of the total area (*c*. 24 million ha) could be classified as amenity grassland, in four broad divisions, lettered A–D in Table 15.1.

Nearly all Category A is sports turf, but it omits Armed Services 'outfield' areas (19 000 ha) and golf roughs (49 700 ha). Another estimate in 1987, on a slightly different basis, gave a total of 134 301 ha for the same kinds of area as Category A, including in this case 6 000 ha of high-quality lawns. Thus the best estimate of intensively-managed sports turf seems to be *c*. 125 000 ha, with an additional 50 000 ha of golf rough. About 5 000 ha of lawns might receive an intensity of management comparable to the intensively-managed sports turf.

The remainder of UK domestic lawns (probably about 145 000 ha) makes up, together with urban parks and road verges (*c*. 150 000 ha), the

Table 15.1 Amenity grassland in the UK

Category		Ha
A	Intensively managed areas	109 650
B	Trampled open spaces: man-made	268 500
C	Trampled open spaces: semi-natural	406 357
D	Untrampled open spaces	62 780
	Total	847 287

Category B amenity grassland, totalling about 300 000 ha. The maintenance of this category would usually only consist of mowing (or sometimes the use of growth retardants) with very few other operations, if any, and hardly ever any weed control.

Approximately 480 000 ha is thus accounted for. The remainder, totalling *c.* 400 000 ha (i.e. Categories C and D, with golf roughs extracted) are discussed in Chapter 16.

Turf for sport or in traditional ornamental lawns in the UK ideally consists of only one, two or at the most perhaps half a dozen grass species. Uniformity of sward is desirable for appearance and consistency of playing surface. In such swards all other plants are weeds, i.e. unwanted grasses, all broad-leaved species, and mosses.

The main turfgrass species in the UK are as follows:

Perennial ryegrass
Red fescue (Chewings fescue and different types of creeping red fescue)
Bents (browntop — including Highland — bent, creeping bent and velvet bent)
Meadow-grasses (smooth-stalked and rough-stalked meadow-grasses, and — in some circumstances — annual meadow-grass)

Minor UK turfgrasses include:

Very fine fescues (hard fescue, sheep's fescue, fine-leaved sheep's fescue)
Tall fescue
Timothy
Crested dogstail

Weed control in swards of these species entails the elimination of broad-leaved species, rushes and mosses, and also of undesirable grasses if possible. The last may include: cocksfoot or Yorkshire fog, which are seldom considered acceptable in any turf; annual meadow-grass, which has a uniquely ambivalent status in turf management, sometimes considered a pernicious weed and sometimes an essential turfgrass; and perennial ryegrass, which is a weed in fine turf.

In some lawns, broad-leaved species are tolerated, or even encouraged, for their flowers and the diversity they create. There are even broad-leaved species grown as mono-cultures for turf, very rarely for lawns in the UK, e.g. chamomile, and sometimes for sports in other countries, e.g. *Cotula* spp. for bowling greens in New Zealand.

Chapter 16 deals with the management of amenity grassland in which it is desired to keep a balance between grasses and broad-leaved species, whether in domestic 'meadow gardening' or in larger areas with low intensity of maintenance and use. In the latter, white clover or other legumes might be sown as part of the original seeds mixture.

With the range of herbicides now available, most broad-leaved weeds in grass swards are fairly easily controlled, but the use of chemicals is only one method of weed control in turf. Indeed, hand weeding at an early stage is often the cheapest and simplest method of control, and the only one for grass weeds (though it must be done efficiently, cutting weeds out at root level rather than just pulling up leaves). All weed-control measures are, in turn, part of good general management, which can maintain a first-class turf and so keep weed problems to a minimum. The discussions below explain this more fully.

The influence of general management on weeds in turf

Land preparation

The value of a turf area for sport or recreation depends principally on sward quality, which in turn depends largely on the underlying soil and its drainage and fertility. The initial construction methods, the quality of seedbed preparation, the amount of seedbed fertilizer and the choice of good cultivars will determine the vigour of the sward in the important establishment phase and influence its subsequent ability to keep out weeds.

When preparing areas for seeding or sodding (turfing) it is important to have weed-free soil. It is particularly necessary to eradicate weeds which will be troublesome in a newly-established sward. The usual weed-control measures at this stage are cultivation or herbicide application, or both. The aim is to kill weeds already present when an area is taken in hand and any weeds that appear subsequently, and they should be dealt with before they set seed. Seedlings can be controlled either by cultivations or by herbicides. Perennial weeds sprouting from well-established rootstocks (e.g. docks and thistles) or rhizomes (e.g. common couch) are best treated chemically. It may be advisable to make special treatments, overall or as spot application, against such perennials, perhaps using translocated herbicides such as asulam or triclopyr: but with these it would not be possible to sow until at least 6 weeks after application, and therefore careful advance planning is needed. The ideal herbicides for use on a seedbed are those with short or negligible persistence. Glyphosate is particularly useful as it deals with perennials and also kills all annual and perennial grass and broad-leaved weeds. Paraquat, possibly combined with diquat, will kill annuals, and also check perennials or hasten their desiccation if applied 7 days after appropriate translocated herbicides.

With glyphosate, the weed foliage should be treated during a period of good growing conditions and must then remain undisturbed for at least 3 or 4 days to allow absorption and translocation of the herbicide to occur. Paraquat and diquat may be applied at any time of the year, provided green leaf is present. With diquat, paraquat or glyphosate, sowing can take place, if desired, the day after treatment.

Alternatively, the soil may be sterilized with methyl bromide or dazomet if the cost is justified and the conditions are suitable. Such treatments can be applied either to the seedbed, whether made of natural soil or of special soil/sand/peat mixes, or as necessary to the components of such mixes while still in heaps prior to mixing and laying.

Sowing into a well-prepared seedbed may be done at any time between spring and early autumn. Soils should be warm and moist, to give quick establishment and so overcome competition from any further weeds and attacks from soil pests and fungi. For the same reason, the seedbed should receive complete fertilizer, with adequate phosphorus.

Young swards

In spite of pre-sowing precautions a newly-sown sward may be infested with weeds, especially after sowing in the spring. Nevertheless, many of the weeds present in a new sward will be broad-leaved annuals. These weeds will disappear once regular mowing is started, without any other treatment being necessary. If, however, they are abundant they may smother the grass, so that special treatment would be necessary.

It is possible to use some selective herbicides on young grass, with due caution and mostly at special low rates, but such treatments should generally be avoided unless conditions make them absolutely necessary, as for example if the grass is being smothered and reasonable mowing will not improve the situation, or if the weeds are clearly likely to persist under regular mowing. Young grass plants are best left to grow vigorously without any risk of a check from herbicide or from mechanical damage by feet or wheels. When control of potentially persistent weeds is needed, careful pulling or cutting out by hand may be the best method if numbers are limited; for grass weeds this is the only method of control.

Effects of soil type and moisture on established turf

The type of soil can influence appreciably the botanical composition of a sward. Fine-textured or badly-drained soils present problems for grass growth in any type of turf. Wet conditions weaken the grass, encourage shallow rooting and a soft spongy sward, and particularly favour certain weeds such as mosses and small rushes. The turf can be improved, and the weeds probably eliminated, by correcting the conditions by efficient drainage. This may require the installation or improvement of a pipe drainage system, coupled with measures to maintain water channels from the surface to the drains. Such measures may simply consist of regular mechanical operations for aeration or surface slitting. More elaborate measures would be: a by-pass system of sand slits, overlaid with a shallow top layer of sand; a thicker sand carpet about 100 mm deep, laid over slit drains; or, at the extreme, complete reconstruction in which the unsuitable existing soil is replaced by a sandy (or pure sand) root-zone mixture about 250-mm deep.

Linked to the problem of naturally wet and anaerobic soils, is that of compaction under use. For intensively-used sports turf, such as golf greens or football pitches, the provision of a suitable soil, by initial construction or by subsequent soil amelioration, requires a compromise between good drainage and the adequate supply of water and nutrients for plant growth. In such a situation, weed problems may be relatively unimportant compared with other problems. Nevertheless, annual meadow-grass is a grass of major importance in heavily-worn turf, whose status as weed or useful species is discussed below. Although broad-leaved weeds do not generally persist under treading or heavy wear, some species such as greater plantain or white clover may present problems under such treatment. Knotgrass is a frequent weed of muddy goal mouths, and herbicide treatment at the end of the season has to be carefully co-ordinated with re-seeding to avoid damage to the grass seedlings (see p. 425).

At the other extreme, excessively dry conditions can also adversely affect turf quality and encourage weeds. Not only does drought weaken the grass and let in weeds: the weeds will be particularly encouraged if excessive watering then takes place in an effort to correct drought effects, perhaps when water restrictions are relaxed. At all times, over-watering leads to shallow rooting, disease and an increase in thatch (see p. 416), and if these weaken the turf or cause bare spots, moss and other weeds can develop. On golf greens and similar fine turf, annual meadow-grass is particularly likely to be encouraged in this way.

Effects of wear

The wear tolerance of a turfgrass is usually expressed as the percentage cover remaining immediately after a period of wear (games, footpath traffic, etc.). The main turfgrasses can be grouped as follows:

More tolerant: Perennial ryegrass
                     Annual meadow-grass
                     Smooth-stalked meadow-grass
Less tolerant: Red fescues
                     Bents
                     Rough-stalked meadow-grass

The wear tolerance, and particularly the recovery from wear, of the finer-leaved species (fescues and bents) is appreciably greater in mature swards than newly-sown ones, because thatch of moderate thickness protects plant crowns and a mature sward will have developed rhizomes from which new shoots develop. But if a sward of fescue and bent is seriously damaged by wear, down to the root-zone material, then it takes a long time to recover, and faster-growing annual meadow-grass is likely to colonize the gaps. Perennial ryegrass survives wear because of its relatively large tough leaves, and recovers quickly, but when the above-ground plant cover and the crowns are destroyed there can be no regeneration, and annual meadow-grass colonizes the bare areas unless there is a deliberate dense re-sowing of perennial ryegrass, as in end-of-season football-pitch renovation.

Annual meadow-grass plants in turf are perennial, and the fast growth rate and dense tillering make them outstandingly persistent under treading (i.e. under vertical wear forces), although in conditions which compel shallow rooting, e.g. wet or compacted soils, the density of plants makes them more likely to be kicked out (i.e. displaced by sudden horizontal wear forces). Annual meadow-grass is dealt with in more detail below.

Fertilizer and lime treatments

The correct use of fertilizer, and particularly nitrogen, is important for a hard-wearing, weed-free turf. Nitrogen stimulates grass growth, affects the colour of turf and its ability to withstand wear and tear, and influences the botanical composition of the sward.

The standard form of nitrogen for fine turf remains the fairly fast-acting ammonium sulphate, which has value for weed control by nudging turf towards a low pH and, sometimes, by contact scorching of broad-leaved weeds. The traditional organic sources of nitrogen, such as dried blood or the slower-acting hoof and horn meal, are still used to some extent, in spite of scarcity and expense, but tend to encourage broad-leaved weeds. Some forms of synthetic slow-release nitrogen fertilizers have also encouraged weeds. Recently, there has been increased use of more satisfactory slow-release nitrogen fertilizers, particularly those based on IBDU or formulated with coatings or with nitrification inhibitors. These level out the growth stimulus, with effects visible even into winter and the following spring, and do not have any marked effect on pH or weed populations.

Parallel with these changes in recent years has been renewed recognition of the value of ferrous sulphate to improve the appearance of grass without unwanted growth, to restrain disease and to act against broad-leaved weeds by scorching.

By contrast to this basic reliance, for fine turf, upon nitrogen and the judicious use of ferrous sulphate, has been the recognition that amounts of phosphorus in fine turf are often excessive and encourage weeds, particularly annual meadow-grass. It is therefore now recommended that fine turf should only be given phosphorus when the soil analysis requires it, and the same applies to potassium — although it is not implicated so positively in the encouragement of annual meadow-grass.

Coarser turf, based on perennial ryegrass, routinely receives potassium — especially on sand constructions — and some phosphorus, within the broad framework of the ratio 4:1:3 for $N:P_2O_5:K_2O$, conforming approximately to the nutrient ratio in plant tissues and clippings. This is usually applied as granular fertilizer.

The finer turfgrasses (red and fine fescues, and browntop bents) grow well at a fairly low pH (4.5–5.5) whereas most weeds of turf grow best in soils ranging from slightly acid to slightly alkaline, i.e. within a pH range of approximately 6.0 to 7.5. Hence the benefit for close-mown fine turf of using ammonium sulphate which tends to maintain acid conditions, and so helps to keep turf free of broad-leaved weeds and some grass weeds.

Grasses for heavy wear such as perennial ryegrass and smooth-stalked meadow-grass grow well at a slightly higher pH, about 6.0 to 6.5. Sports turf based on these grasses is therefore generally kept at a slightly higher pH than fine turf.

Over-acidity is sometimes a problem. The growth of grass is retarded and certain acid-tolerant weeds become established, e.g. sheep's sorrel, field wood-rush and mosses such as *Polytrichum juniperinum* or *Ceratodon purpureus*. The judicious use of small doses of ground carbonate of lime will correct over-acidity, but great care is needed to avoid creating surface conditions which will favour disease.

Top dressing

In fine turf management, 'top dressing' implies applications of sandy soil, compost or other material to smooth out surface irregularities, to improve the physical condition of the complex surface layer of live and dead grass and top soil, and — in some cases — to improve fertility. Any soil or compost used in the top dressing should be sterilized both to kill weed seeds and to obtain the stimulus to plant growth imparted by some sterilization methods, particularly heat treatment. Top dressings of sand alone (provided it is of the appropriate grade) can do something to improve the texture of compacted muddy playing areas, but this procedure is less useful than a by-pass system of sand slits covered with a complete surface layer of sand (see p. 411).

Mowing

The quality of turf is largely determined by the height and frequency of mowing. There are approximate guideline figures for the minimum height tolerated by various species, from 18 to 25 mm for perennial ryegrass to 5 mm for bents. Nevertheless, good cultivars, carefully managed and mown regularly, will adapt to and persist at even lower heights of cut, whereas with indifferent management or infrequent drastic mowing even the best cultivars will thin out and allow broad-leaved weeds or moss to develop. These points, and particularly the damaging effects of infrequent mowing, apply to all turf.

Table 15.2 gives a broad classification of intensively-managed turf areas by height of cut. The very fine sports turf, mown at ≤ 5 mm, would be cut several times per week in the growing season and perhaps even daily. Fine and medium-fine sports turf or ornamental amenity areas would be mown

Table 15.2 Broad classification of turf areas by height of cut

Type of turf	Height of cut during growth season	Typical sports use	Height of cut	
			summer (mm)	winter (mm)
Very fine sports turf	5 mm or less	Bowling greens — flat	3–5	8
		Bowling greens — crown	5	8
		Golf greens	5	8
		Cricket tables*	5–6	8–13
		Tennis courts	5–6	8–13
Fine and medium-fine sports turf, & close-mown ornamental turf	6–20 mm	Golf tees	6–13	13–20
		Cricket outfields†	6–20	10–25
		Ornamental lawns	10–15	15–20
		Golf fairways	13–20	20
		Hockey pitches	13–25	13–25
Heavy-duty sports turf, & general-purpose lawns & amenity areas	20 mm or more	General lawns, etc.	15–25	20–50
		Recreation areas	20–50	25–50
		Soccer pitches	25–38	25–38
		Rugby pitches	25–50	50–75
		Racecourses	75–100	75–100

* Lower in final pitch preparation
† When outfields are used for winter games the height of cut must be related to the type of sward and its uses

2–3 times a week at heights up to 20 mm. Longer-cut areas would be mown once or twice a week for sports use, and every 7, 10 or 14 days in amenity situations, depending on the standards adopted.

Newly-sown grass is particularly vulnerable to the removal of a high proportion of leaf at once, by infrequent mowing or by abrupt reduction in height of cut, and a new sward of red fescue and bent needs the greatest care of all, to create a dense weed-excluding sward.

Grass clippings are collected and removed from fine turf such as golf greens and bowling greens for reasons of play and appearance, and to control diseases, worms and thatch build-up, as well as for weed seed removal. On some lawns and on many amenity areas in parks, etc. the clippings are not collected, for simplicity and economy in mowing, and to maintain fertility cheaply. On most sports turf, practice varies according to the type of turf and its use. Obviously, where clippings are returned, weeds such as annual meadow-grass are encouraged, but annual meadow-grass will quite easily set seed below mowing height and the removal of clippings will make little difference if conditions suit the species. Conversely, the occasional return of clippings will not matter on a well-managed sward.

Growth retardants

There are three growth retardants which can be used on amenity grass in the UK to suppress grass growth and reduce the need for mowing — maleic hydrazide, mefluidide and paclobutrazol. The first two are leaf-absorbed and somewhat faster-acting and shorter in duration of effect than paclo-butrazol, which is soil-absorbed. They are used by themselves, or — subject to current approvals — in mixtures, and can be supplemented by herbicides against broad-leaved species. With all three retardants, the margin between effectiveness and unsightly damage to the turf has been too narrow for any to be recommended for fine turf or any intensively managed turf, although, as Chapter 16 shows, they can be useful on banks, road verges, cemeteries and other areas which are difficult or laborious to cut.

Scarification and vertical cutting

All turfgrass swards accumulate organic matter at the soil surface, but this is normally only a problem with fine turf, because of the high density of shoots, the lack of physical disturbance and the relatively low pH (favoured for reasons already discussed). The term 'thatch' is used for the more or less tightly intermingled organic layer of dead and living shoots, stems and roots which develop between the zone of green vegetation and the soil surface. 'Litter', 'mat' or 'fibre' are other names used in various contexts for the same material.

Various treatments are made on turf to remove dead leaf and fibre, to reduce the thatch layer and to limit the horizontal spread of stoloniferous turfgrasses. These treatments are mostly made on fine turf, and represent a vital part of good management. Obviously, the nature and frequency of the work strongly influences the quality of the turf and contributes substantially to the costs of intensive maintenance. According to the nature of the turf and the size of the area, anything from tractor-drawn harrows to wire hand-rakes may be used but for fine turf there are numerous special scarifiers and vertical cutters. Some of these treatments will help to control creeping broad-leaved and grass weeds and all the treatments should benefit the turfgrasses and so increase their competitiveness. If, however, a treatment is too drastic, the turf will be thinned out and weeds allowed to establish.

Scarification can also be an important element in moss control, dead moss being raked out to allow grass, stimulated by nitrogen, to replace it. There is, however, a risk to guard against, that scarification may spread moss to new areas.

Aeration

To function properly, plant roots require sufficient oxygen. Treading, mowing with heavy roller machines and almost every kind of use or routine maintenance practice on turf tends to create compacted soils with restricted permeability to air and moisture. There is a wide range of special aeration equipment to correct this, by slitting, spiking or hollow-tine coring. The proper use of it, and the interrelationship with soil-type factors, is a complex subject, but clearly whatever improves the general vigour of the turf will also increase its competitive ability and reduce the tendency for weeds to invade the sward. Hollow-tine coring may sometimes allow weeds to establish in the small bare patch of sandy soil or compost used to fill each hole, before the turf closes over it, but normally the benefit from the treatment greatly outweighs this minor drawback.

Worm control

Turf with numerous worm casts soon becomes infested with weeds. Dormant weed seeds may be brought to the surface by worms, and weed seeds from any source can easily establish in the casts. Efficient worm-killers may therefore minimize weed invasion, although management problems arise in worm-free turf and all factors need to be considered carefully before using a worm-killer.

Disease control

Several fungi attack turfgrasses and sometimes severely disfigure fine turf. Dead patches may ultimately appear unless remedial treatment with a fungicide is given as soon as the first symptoms are seen. If diseased areas are left to heal naturally they may be filled by broad-leaved or grass weeds. Re-seeding or re-turfing may be needed to avoid a severe weed infestation.

Selective organic herbicides for broad-leaved weeds in turf

Choice of herbicide

Most selective weed control in turf depends on 'broad-spectrum' mixtures of herbicides, and the main selective herbicides for professional use are still the long-established translocated herbicides MCPA, 2,4-D, mecoprop and dicamba. Ioxynil, which acts mainly by contact, is useful to professionals against some weeds, though no longer approved for any amateur use, or for application by professionals through hand-held equipment. The residual herbicide chlorthal-dimethyl has one specialized approval for turf, for control of slender speedwell.

On the less intensively managed amenity grass areas, dealt with mainly in Chapter 16, the translocated herbicides picloram and triclopyr are used — alone or in mixture with other translocated herbicides — against difficult perennial and woody weeds.

This volume does not give details of the response of common turf weeds to herbicides, but it is important to note that weeds in turf do not always respond as they do in agricultural or other situations. In general, higher rates are needed in turf. Physically, herbicide spray cannot reach the whole plant surface, especially undersides of leaves, as effectively as in taller and more open plant communities, and plant litter and thatch may also give some protection to surface stolons and growing points close to the ground. Physiologically, plants tend to become more perennial under close and frequent mowing; for example, reproductive cycles are extended, and the spread and frequent rooting of stoloniferous species encouraged.

Method and volume of application

Herbicides are still usually applied to turf and amenity grassland by conventional hydraulic spraying at volumes of about 200–1100 litre ha^{-1}. The spray volume actually selected will depend on the label recommendation and the available spraying equipment. For large areas, tractor-mounted or towed sprayers would be used, generally applying a total volume of liquid of about 250–600 litre ha^{-1}. For medium-sized areas, knapsack sprayers are normally used, and a higher volume rate, in the range 500–1000 litre ha^{-1}, is usually recommended. To minimize spray drift, pressures should not exceed 2 bar (200 kPa).

On small areas it might be appropriate to use a watering-can fitted with a fine rose or a dribble bar. A fine rose, by giving small droplets, provides — within the limitations of watering can application — the best possible foliage cover and consequently the greatest chance of adequate herbicide absorption. Even with the finest rose available, high volumes of water are necessary, e.g. up to 2500 litre ha^{-1} or more. Because of this, some manufacturers recommend increased doses of chemical to compensate for possible run-off from the weed foliage. A dribble bar gives less drift, and a lower volume of application, than a rose.

A wetting agent is frequently included in commercial formulations for control of pubescent weeds such as speedwells, and can also help to prevent run-off from waxy-leaved weeds such as clovers.

Drift can be reduced by keeping pressure at <2 bar (200 kPa), and by choosing appropriate low-pressure nozzles. In confined areas or near susceptible plants it may be preferable to use a dribble bar or one of the various sizes of no-drift application machines which transfer liquid from the tank to the turf by means of a fluted roller.

Hand-held 'rope-wick' applicators which are gravity-fed for wiping weeds are mainly suitable for non-selective pre-sowing treatments, but there have been some attempts to develop similar applicators for turf, taking advantage of the fact that some weeds grow faster than the desired turfgrasses, and are therefore taller a few days after mowing. Nevertheless, the fine turf in which weed control is usually considered most important has the shortest mowing intervals and therefore the least height differences: this technique therefore seems unlikely to prove useful in the UK.

Various spot-weeder sticks, aerosol cans and other devices are available for selective weed control in turf, mainly as amateur products. Spot treatment is generally not advisable without such special equipment or special herbicide formulations: if ordinary herbicide formulations are applied, whether with a watering-can, a knapsack sprayer or a paint brush, even a professional user may have difficulty in ensuring safe but effective treatment.

Recently, there has been increasing interest in reduced-volume application with rotary atomizers. In these, spinning discs (or spinning cages on certain equipment not used for turf) throw out spray droplets of very uniform size; hence the term 'controlled droplet application' or CDA. A single spinning disc may be mounted at the end of a hand-held lance, or several can be mounted on a boom. The advantage of these sprayers is the accurate pattern of droplets of approximately uniform size (typically *c.* 250 μm), without large droplets which run off foliage or very small ones liable to drift. The disadvantages are that spray application and cover are difficult to observe on vegetation, and there is a likelihood of turf scorch owing to the high concentration of active ingredient when a droplet shrinks because of evaporation. An adjuvant oil or wetter will decrease this shrinkage and make the relatively small droplets spread, with greater effect, on leaves. Rotary atomizer application on turf or adjoining areas with CDA products approved for use with droplets of at least 200 μm diameter would entail application rates in the range 10–50 litre ha^{-1}. Reduced-volume application can be achieved in other ways, e.g. by producing a very fine spray mist for bracken control, with droplets of 50–70 μm diameter, but this technique would not be used on turf or in normal amenity situations.

The Code of Practice for the use of approved pesticides in amenity areas contains guidance on reduced-volume application of pesticides, which

— in the absence of more specific guidance on the label — indicates when and how reduced-volume application can be made, and in particular how fine a spray quality is permissible at various volume rates.

To minimize the effects of misses or overlaps, particularly on turf where appearance is important, it is advisable to apply herbicides in two half-doses, one at right angles to the other.

There are some proprietary herbicides and fertilizer/herbicide mixtures formulated as granules. They can normally be applied with a suitably-calibrated fertilizer distributor or by hand. Herbicide cover will be less thorough than with liquid application, and efficiency therefore less.

It is sometimes possible to treat turf with a herbicide combined with a fungicide, if the timing recommendations coincide. Special care may be needed with the order and thoroughness of mixing and of course only approved tank-mixes should be made.

Season of application

Herbicides against broad-leaved weeds can be applied whenever weeds and grass are in active growth, i.e. usually from April to September inclusive. Best results are generally obtained in April and May. Good results may be obtained in autumn but at this season turf fills in very slowly and consequently the removal of weeds may leave a scarred surface which could be re-colonized by weeds the following spring.

Weather conditions

Fine, warm weather, moist soil and active growth lead to the best results. Light showers of rain soon after application are not likely to cause appreciable loss in efficiency but heavy rain may nullify treatment.

Application during drought is inadvisable unless regular judicious watering can be given, to maintain a steady growth of both grass and weeds. Although the effects on weeds may be satisfactory, especially if the drought ends quite soon after application, there is risk of serious damage to the grass by applications during drought, particularly when temperatures are high.

Linked fertilizer treatment

Because the best results are obtained from growth-regulator herbicides when both the grass and weeds are growing vigorously, an application of nitrogenous fertilizer one or two weeks before spraying improves the uptake and translocation of herbicide by the weeds and encourages the grasses to fill in after the weeds have been killed. Compound mixtures of fertilizer and selective herbicide are available for greater convenience and economy of labour. Unlike the separate treatments first of fertilizer and then of herbicide, the compound mixtures do not give any stimulus to the growth of grass or weeds before application of herbicide, but grass growth is increased afterwards. Rain falling soon after treatment may make the compound mixtures less effective than separate applications.

Mowing before and after herbicide treatment

For maximum absorption, weeds should present a reasonably large total leaf surface to the herbicide. Therefore, when turf is only mown infrequently, e.g. every 7–10 days, treatment should not be made until a few days before the next mowing is due, to give weeds time to produce new leaves. In more frequently-mown turf, there is no time for weeds to develop much new leaf between mowings, and most weed leaves are probably below mowing height anyway. Therefore, in turf mown close and frequently, the timing of treatment depends not on weed recovery but on the theoretical possibility of turfgrass damage if herbicide is taken in more readily by leaves when freshly cut than after they have healed over. As there is no definite information on this point, it is prudent to defer treatment until the day after mowing if possible, although in practice many greenkeepers mow fine turf and spray the same day with no apparent damage to the grass.

Turf should not be mown for at least 1 day, and preferably 2 or 3 days, after treatment, to allow time for the herbicide to be translocated through the plants from the sites of absorption. This point is again most important in turf mown infrequently, where much of the weed is likely to be removed by mowing: it is much less important for prostrate weeds in fine turf.

Precautions

Particular care is needed with herbicide treatment of newly-sown grass (see the discussion on young swards above).

After spraying with growth-regulator herbicides, any grass clippings removed from the turf at the first four mowings should not be used directly as a mulch round broad-leaved plants or shrubs but may be incorporated into compost heaps provided that they remain there for at least 6 months before use as compost for broad-leaved plants or shrubs. After the first four mowings no special precautions are needed.

Normal care is needed in all matters relating to drift onto adjacent susceptible plants, cleaning sprayers, safe storage of herbicides and operator safety as described elsewhere in this volume (Chapter 5) and in the Code of Practice for the use of approved pesticides in amenity areas.

Where valued plants surround a small, confined lawn, liquids should be applied with a watering can, dribble bar, special no-drift application machine, or aerosol spot weeder. Alternatively, granular materials could be used with a drop spreader (a distributor in which material from a hopper falls uniformly along the length of the hopper during travel). Although it is also theoretically possible to use barriers of hardboard or plastic bags to protect sensitive plants from spray applications, it is normally more sensible to find safer application methods. Moreover, even if growth-regulator herbicides are applied accurately, volatilization can take place for some time afterwards, and much the best course is to avoid the use of such herbicides in any doubtful situation, relying instead on hand weeding or lawn sand.

Great care should be taken when spraying near glasshouses, even though the doors and ventilators are closed. Vaporization of herbicides can cause havoc among sensitive plants. Water supplies for glasshouses and other garden use must not be contaminated by herbicides from watering cans and other equipment used for the application of herbicides.

Users of professional products working with the Code of Practice should be well aware of the principles of safe storage and disposal of containers and unwanted pesticides. Even for the generally safer amateur products, similar procedures should be followed unless specific guidance is given on the product label.

Ferrous sulphate and lawn sand

Calcined ferrous sulphate has been used for many years as a partially selective herbicide on all types of turf, mainly in the form of lawn sand, i.e. mixed with ammonium sulphate and sand, for example in the proportions of 1 part ferrous sulphate : 3 parts ammonium sulphate : 10 parts sand, applied at about $140\,g\,m^{-2}$. Repeated applications can control many turf weeds, but the use of lawn sand was for some time largely limited to the control of moss and a few problem broad-leaved weeds. Nowadays, ferrous sulphate has become more widely used again, with the current concern about the use of more complicated herbicides, and the realization of its wider value in turf management.

The advantages of ferrous sulphate are that it:

1 Scorches weeds and moss.
2 Improves turf colour, without stimulating growth.
3 Discourages some diseases (e.g. fusarium patch) and algae.
4 Acidifies soil slightly at each application (thus discouraging the weeds and diseases which may be favoured by high pH).
5 Is claimed to harden off growth in autumn, although this effect may be simply the result of improved colour without new growth that is susceptible to disease.

The disadvantages are that ferrous sulphate may:

1 Cause over-acidity.
2 Scorch turf.
3 React with or tie up other elements.

Ferrous sulphate might be applied with nitrogen fertilizer once or twice in summer, as lawn sand, at rates up to $10 \, \mathrm{g \, m^{-2}}$ with the main aim of weed or moss control. Alternatively, with less direct attention to weeds, the ferrous sulphate might simply be included, at about $5 \, \mathrm{g \, m^{-2}}$, in the fertilizer mixture applied to fine turf two or three times a year. A third use might be simply as a foliar spray in winter, at about $1 \, \mathrm{g \, m^{-2}}$.

In applications of lawn sand specifically directed against weeds, the best results are obtained if it is applied on a dewy morning and fine weather follows. The material adheres well to the large leaves of weeds and scorches them more severely than the grass. The turf usually assumes a very dark, almost blackened, appearance for a few days but subsequently recovers to a very good green colour. If rain does not fall within 48 hours of application, watering is advisable.

Special broad-leaved weed problems

Legumes (clovers and trefoils)

Species occurring commonly in turf are the annual lesser trefoil, the annual or short-lived perennial black medick, and the perennials white clover and common bird's-foot trefoil. These weeds occur under a wide range of soil and management conditions. Apart from common bird's-foot trefoil, they are capable of withstanding close mowing and can become a serious problem in fine turf. Their growth is favoured by alkaline conditions and the excessive use of some phosphatic fertilizers but discouraged by regular use of certain nitrogenous fertilizers, particularly ammonium sulphate. The annuals produce seed freely even under close mowing, and worm casts may assist new plants to establish. Wire raking before mowing enables these prostrate weeds to be cut by the mower more effectively and may assist in the reduction of the size of the patches. Scarifying and 'vertical cutting' can also be useful. On small areas and young sown turf, hand-weeding is recommended.

Control of clovers can be achieved with various herbicides, but the timing of treatment can influence the results. Applications in early summer when the weeds are in full growth give the best results, in conjunction with a programme of adequate nitrogen fertilizer. Repeated treatment may be needed for well-established perennials or if there is re-establishment from seed.

Speedwells

Several perennial speedwells occur in turf, and spread by creeping stems which root at the nodes. Slender speedwell is probably the most serious. It normally occurs in damp situations and propagates readily from pieces of cut or broken stem. For that reason it is usually more troublesome on fairways or winter games pitches where clippings are returned, than on fine turf where the clippings are removed. It is advisable to attack the weed as soon as it appears, by spraying or perhaps by hand weeding on areas where only a few plants exist, by improving drainage if possible, and by collecting and removing clippings.

Slender speedwell (and some other speedwells) can be controlled reasonably well with an ioxynil/mecoprop mixture applied in early spring before the flower heads have formed, followed (usually) by a second treatment 4–6 weeks later (depending on weed growth) before the weed has completely recovered. Application in early autumn will suppress the weed but is less effective than in early spring. Treatment is most effective if, prior to spraying, the turf is mown as close as is reasonable, and the time of spraying arranged when there is maximum weed exposure. It is important to wet the weed's hairy leaves thoroughly; a wetting agent, included in the formulation or added to the herbicide if permissible, will help. It may also be possible to use a sprayer with relatively high pressure.

Chlorthal-dimethyl can also be used, and lawn sand containing ferrous sulphate also has some effect against slender speedwell.

Knotgrass

Knotgrass is particularly troublesome on heavily-worn and compacted areas such as football pitches. The renovation of such areas at the end of a season requires careful co-ordination of the herbicide application and the sowing of grass seed (normally perennial ryegrass alone).

If knotgrass seedlings are evident before re-seeding, 2,4-D (alone) can be applied to kill or check them. Mecoprop and dicamba are liable to persist too long in the soil for safe grass establishment and should not be used. The knotgrass seedlings may not, however, appear sufficiently early for this preliminary spraying to be done before re-seeding, which obviously has to be undertaken as soon as play finishes. Even 2,4-D persists for a short period on the soil surface. If it is not possible to arrange an interval of 2–3 weeks between spraying and re-seeding, the seed rate should be increased.

When perennial ryegrass seedlings have two or three expanded leaves, ioxynil alone can be used (at the rate recommended for young turf) if a second treatment of the young knotgrass plants is needed, or if they have just germinated with the ryegrass. Ioxynil, at the rate safe for the young grass is, by itself, only likely to kill seedlings and very young plants of knotgrass, but if established plants of the weed have been weakened by earlier spraying or cultivations there may be some useful effect on them. Alternatively, half-normal rates of formulations containing MCPA, meco-prop or 2,4-D with dicamba could safely be used at the two- or three-leaf stage.

Later in the summer when the ryegrass is well established, the full normal rate of weed-killers containing dicamba can be used, except in periods of drought. Such treatment may be repeated after an interval of about 1 month if any of the weeds are left and provided growth is good. Removal of cuttings will help to reduce the seedbank for next year.

Rushes

Two species of rush can be important in turf — toad rush and field wood-rush. Both are good examples of the importance of cultural practices. For field wood-rush, indeed, cultural treatment gives much better long-term control than chemical treatment, and the latter is seldom worth considering.

Toad rush is an annual, found in wet areas of fine turf. It is moderately susceptible to some herbicide mixtures (e.g. mecoprop with 2,4-D or ioxynil, or 2,4-D with dicamba); chemical control can therefore be achieved by repeated monthly treatments. Nevertheless, long-term exclusion will depend on the success of surface drainage measures, including reduction of the thickness and impermeability of the thatch layer.

Field wood-rush is a perennial, spreading by underground stolons as well as by seed. It is characteristic of very acid conditions, and found in areas such as over-acid fairways and cricket outfields rather than in more intensively-managed fine turf areas. In fairways and outfields it is seldom likely to be objectionable enough to merit special control measures, although these may be desired in domestic lawns. It is most obvious in spring when the brown flower spikes are produced. When the soil pH is so low (i.e. conditions so acid) as to favour field wood-rush, a lime dressing to raise the pH slightly is probably desirable for the sake of the grass and this is likely to reduce or eliminate the weed without special attention being given to it. If, however, field wood-rush is the particular object of control efforts, then the same strategy applies, of raising the pH slightly and reducing the amount of the thatch which accumulates in acid conditions and in turn tends to increase acidity — see the discussions on fertilizer and lime treatments (p. 413) and on scarification and vertical cutting (p. 416). Herbicide treatments are relatively ineffective; mecoprop alone or in mixture may check the weed, but even repeated applications are unlikely to kill it completely.

Weed grasses

There are as yet no recommended selective herbicide treatments for weed grasses in turfgrass swards, although the minor-use approval of ethofumesate and the potentialities of residual herbicides such as bensulide are mentioned in the discussion on annual meadow-grass below. A detailed discussion is given on annual meadow-grass because of its unique status in UK turf; it is regarded — according to circumstances and viewpoint — either as a useful turfgrass in its own right or as a very successful and indestructible weed. Points are also made on general management which, appropriately modified, apply to other weed grasses also.

Annual meadow-grass

This single species name covers a wide range of plant types, ranging from true annuals, through various biennial and short-lived perennial forms (perhaps with creeping stolons that root at the nodes), to longer-lived perennials (small-leaved, and forming very low dense rosettes which spread only slowly). The perennial forms are hard-wearing and tolerant of close mowing, able to replace themselves by seeding at all heights of cut and to grow nearly all the year round even on compacted soils.

In turf subject to heavy wear, e.g. football pitches, annual meadow-grass often gives more grass cover at the end of the season than the sown species, although it tends — especially on the wings of soccer pitches — to form a thick, spongy, water-retentive turf which has shallow roots and can be kicked out easily during play, especially on compacted soils where only shallow rooting is possible. Annual meadow-grass is not necessarily a shallow-rooting species but it is better able than other species to tolerate compacted soil. Unless something better can be found to take its place, there will be only limited demand for elimination of annual meadow-grass from football pitches or similar heavy-duty turf, although this is now possible with sequential summer applications of ethofumesate on turf consisting solely or mainly of perennial ryegrass. Two or three applications at monthly intervals should be made when soil is moist in May–August (continuing later if pitches remain unused). This treatment will eliminate most, or all, of the existing annual meadow-grass, and prevent establishment of new plants from seed. It will not damage mature perennial rye-grass, but if seed of perennial ryegrass is sown to renovate the sward no treatment should be made until the new seedlings have reached the two- to three-leaf stage and are growing vigorously in the presence of adequate moisture (from rain or irrigation). Nevertheless, total grass cover on a renovated pitch area may be less after successful control of annual meadow-grass than if no treatment is made. Careful management is needed to create satisfactory grass cover before the playing season, and then to maintain it. In a mature perennial ryegrass sward there is more likelihood than on a renovation sowing that the perennial ryegrass will fill in as the annual meadow-grass is killed.

In fine turf, annual meadow-grass may occur as isolated patches in the sward, that look unsightly and spoil the playing surface's uniformity (in texture and height of growth), or as a more or less complete cover which has supplanted the sown species. Its main drawbacks are susceptibility to fusarium patch disease, poor colour in drought and in winter, and the seed heads which are unsightly and may affect playing surfaces on fine turf. It is often associated with thatch problems, partly because it helps to produce thatch and partly because it is then well adapted to survive in thatchy conditions.

In established fine turf areas where annual meadow-grass has become dominant because it is better adapted to prevailing conditions than anything else, the groundsman or greenkeeper would not want to lose the grass he has learnt to live with, whatever its faults, without assurance of an equally robust replacement, that could be introduced gradually into the turf without disastrous treatment effects.

In new fine turf areas, however, a careful programme of seedbed preparation and subsequent management may succeed in keeping out annual meadow-grass. The first step is to provide a really clean seedbed, in which clean seed of good turfgrass cultivars can produce a dense weed-excluding sward before seeds of annual meadow-grass arrive. Soil sterilization, e.g. with methyl bromide or dazomet, is the only effective method at present.

Once a turf is established free of annual meadow-grass it may one day be possible to prevent establishment of adventitious seedlings by selective herbicide treatments. Various herbicides such as bensulide are used regularly in this way in countries where annual meadow-grass mainly behaves as a true annual, going through a vulnerable seedling stage each year. Such treatments under UK conditions have been disappointing, because of more persistent ecotypes of annual meadow-grass and some risk of turfgrass damage from repeated doses. Much, however, can be done by turfgrass management to avoid invasion by annual meadow-grass, or at least limit its spread:

1 The use of densely-growing cultivars.

2 Sensible irrigation (annual meadow-grass thrives in moist conditions).

3 Correct fertilizing. In particular, avoid excess phosphate, especially organic and alkaline types (bone-meal is both). Phosphate assists (1) seedling establishment, (2) rooting and (3) seed production. These are three essential stages in annual meadow-grass take-over.

4 Maintenance of fairly low pH for inland fescue/bent turf (pH 4.5–5.5).

5 Adequate aeration (to get rid of compaction and surface water).

6 Scarification frequent enough that it need never be severe (thus avoiding deep scars and bare areas where annual meadow-grass can establish).

7 Light top dressings (which will not create a seedbed for annual meadow-grass).

8 Sterilization of compost used in top dressing.

9 Avoidance of worm casts.

10 Removal of clippings.

11 Cleanliness of equipment.

Yorkshire fog and creeping soft-grass

Yorkshire fog is a persistent perennial, coarse in texture, which forms unsightly patches in closely-mown turf. There are no chemical treatments for selective control. It is best to pull out or cut out young plants as soon as they are noticed. Well-established plants may be weakened and perhaps eliminated by slashing with a knife across the patch, coupled with raking before mowing. Scarification or vertical cutting may have the same effect. The only certain treatment, however, is to cut plants out at the roots and, if necessary, re-turf. Creeping soft-grass is similar to Yorkshire fog in the treatment it requires. It is less common and much less likely to come as a seed contaminant, but it has underground rhizomes and, once established, is harder to destroy.

Mosses

Many different mosses occur in turf. The small cushion-forming mosses (e.g. *Bryum* spp.) are mainly a problem in thin acid turf mown closely, whereas the feathery or trailing species (e.g. *Hypnum* spp.) are common in swards mown less closely but allowed to become moist and spongy. A third type (*Polytrichum* spp.) is found in dry acid conditions, e.g. the surrounds of golf greens. To reproduce themselves, most mosses require wet conditions; all turf may sometimes be wet, but well-drained soils usually dry out sufficiently quickly to prevent extensive moss development. Moss thrives where compaction or excessive thatch reduces the vigour and density of the grass and at the same time holds water at the surface, but anything that weakens and thins the grass — even drought — will encourage moss.

Complete elimination depends on removing the predisposing factors and creating a vigorous, healthy turf. Otherwise, moss-killers will only give temporary benefits. The main cultural treatments to consider, according to the situation and the type of moss, are:

1 Improvement of drainage (surface and sub-soil).
2 Avoidance of unnecessary compaction.
3 Correct irrigation practice.
4 Correct fertilizer treatment.
5 Liming over-acid soils.
6 Aeration.
7 Not mowing too closely or scalping on uneven surfaces.
8 Reduction of shade.

There are many proprietary moss-killers based on ferrous sulphate. These are mostly lawn sands combining ferrous sulphate with nitrogenous fertilizer in the form of ammonium sulphate. See p. 422 for more details of this treatment. The main alternative is dichlorophen, which is applied as a spray: it is essential to wet the moss thoroughly, during a period of active growth, i.e. any time when conditions are moist but not freezing or frosty. Dichlorophen rapidly turns moss dark brown; it is not very persistent but it is highly selective, leaving turfgrasses unharmed, and it is also fungicidal.

There is some, mainly amateur, use of moss-killers based on chlorox-uron, which is slow-acting and very persistent. It may be used alone, or in conjunction with other moss-killers. Various coal tar derivatives (tar oils, phenols and cresylic acid) are effective against moss, but some discolora-tion of grass may occur.

Algae and lichens

Bare ground in turf areas is often quickly covered by a green scum which is a mixture of algae and moss. Algae such as *Nostoc* spp. sometimes form black or very dark green jelly-like masses in turf during autumn or spring. Algal slime is not damaging to grass, but the slippery surface can be dangerous to players and groundstaff. Cultural control should always be attempted by improving grass growth and taking steps to dry the surface. Spiking and sanding are often useful. Chemical control involves the use of ferrous sulphate, tar oils or dichlorophen.

Lichens may occur on damp turf areas or in dry acid situations where grass growth is weak, though they are rarely as troublesome as moss. Cultural control consists of improving grass growth, e.g. by fertilizer treatment and careful liming. Ferrous sulphate, tar oils or dichlorophen are effective if chemical control proves necessary.

16 / Weed control in amenity areas and other non-agricultural land

E.S. CARTER

Farming and Wildlife Trust Ltd, National Agricultural Centre, Stoneleigh, Kenilworth, Warwicks. CV8 2RX

Introduction — types of amenity land and problems

This chapter is concerned with the management of areas of herbaceous vegetation, usually non-agricultural grasslands in various forms. Some of this land may be 'un-used' whilst some other will have an amenity value. These areas are found on roadside verges and railway land, as grassed areas on river banks, golf course roughs, in cemeteries, country or urban parks and on land (whether landscaped or not) associated with housing, public works and defence installations. (Weed control on industrial sites is dealt with in Chapter 19.)

Amenity land may be broadly defined as areas to which the public has access together with areas not accessible to the public that screen or enhance the appearance of public sites, railways and motorways. In the UK, these lands are administered by a wide range of public and private bodies, which include city, district and county councils, universities, health authorities, water authorities, British Rail, British Coal, the Central Electricity Generating Board and the National Trust. Amenity land fulfills varied requirements from the intensive use of limited areas of recreational open space in city centres to extensive countryside parks. The enterprize is not (usually) run for a profit, although, with the emphasis on diversification in the countryside, there is more interest in getting a return from such land. It is the visual appearance that is of prime importance and levels of weediness, which might be ignored under other circumstances, may prove unacceptable in amenity areas. The concept of certain plants being considered weeds depends on the context — any sort of plant out of place in a formal city bedding display is a weed, whereas the term has an entirely different meaning in the diverse flora of a countryside park. All concerned share a common interest in preventing weed competition with newly-planted subjects, especially woody plants.

Less formal areas of recreation and amenity such as country parks, rural picnic areas, footpaths, bridleways and green lanes come under the general category of rural open space. The management of nature reserves, whether 'official' or privately managed, is of increasing importance and nature conservation as an objective of management, is of concern to nature conservation bodies and to individual land managers who wish to encourage wildlife. Habitat managed for game conservation also requires special consideration and some aspects of this may impinge on cropped areas of the farm. In an attempt to control agricultural production and to reduce the impact of modern farming on the countryside, the Ministry of Agriculture, Fisheries and Food has introduced the Environmentally Sensitive Area Scheme, with a number of localities so designated where farmers, through special grant aid, will agree to practise less intensive agricultural systems. Weed control in these areas will require special consideration and the use of herbicides may be restricted. Set-aside is a voluntary scheme designed to reduce surpluses of arable crops with annual compensation payments of up to £200 per hectare. The use of herbicides on set-aside land is only permitted in certain limited circumstances with the authorization of the Minister. Such authorization covers spot treatments to control specified weeds. Herbicides used on fallow land must be absorbed primarily through the leaves and stem with little or no persistence in soil or water.

In the UK, some types of non-agricultural land are extensive (*c.* 180 000 ha of rural grass road verges; 30 000 ha of grass, scrub and woodland on operational railway land) but others are very small, inaccessible or inconvenient. Such areas often have no particular function, except as an incidental part of primary land use and in many instances they are regarded as wasteland. Even so, they fulfill needs for amenity and recreation and serve as refuges for wild plants and animals, whose other habitats may be endangered on neighbouring land. The limiting factor in management is usually cost of labour, suitable machines and chemicals if they are available. As the land concerned usually shows no financial return, managers aim to achieve an acceptable standard, at least expense, which usually means non-intensive management systems. The definition of an acceptable standard will depend on the location and is a subjective judgement, so that solutions differ widely.

Grasses provide surfaces for many types of activity in amenity areas. These vary from intensively-managed sports turf to semi-wild grassland in countryside parks. On sports turf, all plants, other than the carefully-chosen grass cultivars, are weeds because they reduce the playing quality or performance of the turf for games. A diversity of grasses and non-grass species in non-playing areas and in informal recreation sites is generally tolerated, unless they produce excessively tall flowering stalks. Plants such as *Plantago* spp. that die back leaving bare patches are also unacceptable. The species that cause pain or discomfort, such as bramble, common nettle or which are poisonous may need to be controlled when they occur in places used by the public. Local herbicide applications using a selective chemical can be used and will do little permanent harm to the grass sward. White clover occurs spontaneously but patchily in nitrogen-starved turf, but an even stand can be ensured when it is sown along with the grass seed. Clovers can be particularly valuable for establishing grass on the nutrient-poor substrates, common to many disturbed planting sites. Clovers are not suitable for sports turf subject to hard wear, but in other less-demanding situations, clover will impart a green colour to the turf even under dry conditions and provides the grasses with a source of nitrogen.

The largest urban areas under grass are usually playing fields and park land, but urban-land managers must also maintain many small parcels of turf in cemeteries, on roadside verges and around housing estates. Although individually small in area these may add up to a considerable total. Maintenance costs of such small, sometimes isolated patches, are high and the current demand for greater economy has forced managers to examine cheaper chemical methods using herbicide and growth regulators such as maleic hydrazide. Roadside verges often accommodate street furniture supported on posts for signs and utility services. Despite regular cutting of the surrounding sward, tall or unsightly weeds, such as wall barley grow vigorously against the posts particularly where high nitrogen levels are maintained by the urine of visiting dogs.

Maintenance costs must be kept to a minimum, but not to the extent that the desired effect is lost. Chemical control of large established weeds in which unsightly dead plants remain in place after the treatment is generally unacceptable on amenity sites, but may be of less concern to those responsible for industrial areas or countryside parks.

The areas under consideration are mainly grassland, quite often rough and containing a wide range of plants, both narrow and broad leaved, Neither the productivity nor sward composition is important in the agricultural sense. The principal requirement is to maintain a healthy sward of appropriate composition and height which involves some form of management at least once a year. Although a number of plants occurring in these situations are commonly referred to as weeds, there are strong arguments for considering them and the majority of non-weedy plants, as part of the desirable flora of non-productive land. The dangers of the spread of agricultural weeds from roadsides and such places to cause economic infestations of neighbouring land are often over-emphasized. Vegetation management is a better term for describing what is involved, rather than 'weed control' which often implies eradication of an unwanted species. There may, of course, be some circumstances where weedy herbaceous plants may endanger the establishment of newly-sown grass and then methods to eradicate them may be necessary.

Many of the wide range of grassland flowers are plants of short sward and are often suppressed by coarse grasses and scrub. Once scrub has become established, the build-up of fertility, especially under such nitrogen-fixing species such as broom and gorse, may make it impossible to maintain short, diverse swards even if the scrub is physically removed. It is important to be aware of this problem at an early stage as it is easier to dig out or cut-off young woody plants than to tackle fully-grown bushes and young trees.

Bramble and thorn are usually regarded as troublesome invaders on amenity areas and on heaths and moorlands bracken and broom can quickly eliminate heathers and other characteristic species.

Thistles, docks and some mayweed and chickweed may rapidly invade newly-seeded areas and smother emerging seedlings. The site should be as free as possible from these weeds before sowing; control at a later stage may be difficult and will involve careful management.

Aims of management

The most important objectives in management of non-productive land for most purposes are usually the control of height and species composition. Invasion of scrub or the early stages of secondary woodland may, or may not be tolerated. The public especially those with little experience of land management often do not appreciate, that left alone, most areas of the UK would revert through scrub to the climax vegetation. This is the final, stable community which results after a series of changes in the vegetation in a particular area. Depending on the area, birch, aspen and sallow will be followed by pine and hazel giving way to alder and oak and later holly, ash, beech, hornbeam and maple. Such a process gives rise in time to 'natural' wildwood. This can be seen 'in action' in the corner of Broadbalk

Field at Rothamsted which has been left untouched since the last wheat crop in 1882 and is now a well-established oak wood. It is the aim of management to arrest this process at a predetermined stage.

In many areas, additional objectives of management are to preserve stability of banks and cuttings, to prevent erosion, and to provide an acceptable appearance. Where motorways or reservoirs have been constructed, there is a need to heal the scars and blend the new works into the surrounding areas. Management policies must take into account factors affecting visibility, wear, the effects of vegetation on structures, fire risk, snow drift, litter control, access for emergency and maintenance and drainage.

Other important considerations are as follows:

Good neighbourliness. This includes adherence to statutory provisions for the control of weeds and pests. There may be occasions when weed control is necessary in order to maintain goodwill although there may be no statutory obligation.

Amenity. The standards to be achieved may differ widely, but amenity is an important reason for management, especially by local authorities and public bodies and those with a tourist or recreation interest. Under the Food and Environment Protection Act 1985, a voluntary Code of Practice covers the users of pesticides in areas of public access.

Wildlife conservation. There are statutory obligations (Countryside Act 1968, Section 11; Wildlife and Countryside Act 1981, Sections 41 and 48; Food and Environment Protection Act 1985) and special considerations apply to land scheduled by the Nature Conservancy Council as a Site of Special Scientific Interest (SSSI). Care must be taken in the way in which land adjoining an SSSI is managed, as some operations may affect the scheduled area. Public bodies have statutory obligations regarding amenity and wildlife conservation on land under their control and other land managers will wish to have such regard on a voluntary basis. Positive planning is essential. It by no means follows that the best interests of wildlife conservation will be served by leaving vegetation entirely unmanaged. In general terms, it may be said that what is good for wildlife will usually be good for the landscape, but what is desirable in landscape terms will not always benefit wildlife.

Public responsibility. Amenity and safety considerations, including the provisions of the Health and Safety at Work Act 1974, and the Food and

Environment Protection Act 1985, must be incorporated into management plans. The Control of Pesticides Regulations (1986) state that users 'shall take all reasonable precautions to protect the health of human beings, creatures and plants, to safeguard the environment and in particular to avoid pollution of water'. Recognition of public comment and opinion is clearly an important factor. Where management practices may be controversial, it is as well to discuss and explain these in advance with local leaders and opinion formers. A better understanding of the reasons for management and the methods to be used will reduce the likelihood of controversy through misunderstanding.

Methods of weed control

Control of soil fertility

Many of the most attractive species of rich grasslands in the country occur on soils of low fertility where, because of the reduced vigour of growth, the vegetation requires little management. Herb-rich, chalk grasslands have usually been created by the practice in the past of grazing with sheep by day, which were then folded over low-ground arable at night. Over many years fertility was transferred from the grass downs to the lowland arable. This system is no longer practised.

If possible, sites should be maintained at a low nitrogen status by nutrient-extractive techniques of management. Removal of cuttings or winter burning of standing vegetation prevents nutrient recycling. Like unenclosed grazing and hay cropping, fire removes large amounts of nitrogen and phosphorus and contributes to the impoverishment of the ecosystem which is important in maintaining amenity grasslands and heathlands. Repeated removal of vegetation from fertile sites leads to a reduction of soil fertility and more diverse and easily-maintained swards. However, removal of cuttings is not economic for sites such as rural roadsides. Building-up soil nutrients is relatively easy but reducing nutrients is difficult, certainly in the short term, especially the phosphate status, as phosphates are firmly held in the soil matrix.

On reclaimed industrial spoil heaps and similar areas of disturbed ground, sward failure associated with nitrogen deficiency may occur and this may lead to erosion. Under these circumstances the use of nitrogenous fertilizer and the establishment of nitrogen-fixing plants, such as clover and other legumes, is recommended.

Cultivations

In some situations it may be desirable to eliminate competition from all other plants, for example round trees and shrubs, but cultivation has disadvantages. Cultivation for weed control carries with it the risk of damage to roots and the loss of soil moisture. Soils left undisturbed for a number of years will carry fewer annual weeds as the number of weeds that emerge and set seed declines. The situation will be different with some perennials where bare ground, but not necessarily disturbance, may be all that is needed to encourage growth.

In the past, the maintenance of amenity areas and other open spaces depended on the availability of a large labour force. Economic pressures and changed attitudes towards physical work, force the use of cheaper alternatives. In commercial production soil cultivation may no longer be necessary where herbicides are employed for weed control. Farm crops may be drilled with the minimum of soil preparation and perennial crops, such as orchards, rely on herbicides for weed control rather than cultivations. Long-term trials of these methods have shown no disadvantages. Reduced cultivation under trees and ground-cover planting should encourage the accumulation of a protective mulch of leaf litter, which will also discourage weed seed germination.

Mulching

Mulches of organic material, such as peat, sawdust and bark chips used alone, or in combination with a herbicide, can be effective in conserving water and suppressing weed growth round individual trees and shrubs. A depth of at least 5 cm is required making organic mulches expensive to transport and apply. Experiments and practical experience have shown that an alternative such as black polythene mulch is at least as good as chemical weed control in terms of the growth of the trees.

Black polythene used for mulching must be thick enough to withstand being punctured by perennial weeds and must fit closely around stems of trees and shrubs to prevent weeds emerging. On amenity sites, a covering of stones or bark chips improves the appearance and protects the mulch from deterioration by ultra-violet radiation. On farms, a turf turned upside down will hold the polythene in position. A cheap alternative is to use old plastic fertilizer sacks held down with a couple of inverted turfs. Although initially expensive in labour and materials, black polythene mulch requires little or no maintenance and plant growth is superior to that with other weed-control treatments.

Grazing

In some non-agricultural areas, management by grazing is possible and economic, the grass usually being let out. This form of management has considerable attractions for conservation and amenity. It is important as a management tool for country parks, recreation areas and other sites, such as reservoir embankments where the animals can be fenced in. The choice of grazing animal, or combinations of animals, significantly affects the composition and growth of the sward, but usually depends on the nature of the site and the availability of stock. It will be in the land manager's interest to control the grazing periods in order to prevent impoverishment of the sward. Agricultural weed control may, or may not be required, depending upon the value of the grazing and other uses to which the area may be put.

For maintaining a short sward, sheep are best. Cattle tend to break up the sward, do not graze so closely and tear-up rather than bite the grass. Grazing by horses is usually best avoided, unless very large areas are available. Horses graze closer but produce a more uneven sward than sheep and they also tend to dung in restricted areas. It is, of course, important to ensure that any arrangements include provision for removing stock from the area when they may do damage through poaching or over-grazing at times when growth is slow. There is an increasing interest in keeping rare breeds of grazing livestock, quite often by people with only small areas of land available, who may only have a few animals of the breed in which they are interested. Such individuals may be keen to help graze amenity areas and will be particularly interested in helping to retain their conservation interest. The Rare Breed Survival Trust (National Agricultural Centre, Kenilworth, Warwickshire, CV8 2LG) is a useful point of contact.

Mowing

The tall sward communities of meadow grasslands are the product of mowing for hay. The herbage is usually cut in June or July and sometimes the aftermath is grazed later in the year. Just as pasture grasslands are best maintained for amenity by grazing, so amenity management of mowing meadows should simulate traditional practice as closely as possible. This may mean quite ruthless cutting of plants in flower. Many of the interesting component species will have set seed by cutting time or will flower again after defoliation. Large amounts of nutrients are lost in a hay crop. Traditionally, these were replaced by farmyard manure and other slow-acting manures. Today, artificial fertilizers are used which reduce plant diversity by encouraging growth of the more vigorous grasses.

Mowing does not favour or reject particular species as does the grazing animal. It also levels the ground, flattens heaps, molehills and tussocks and does not recycle nutrients in dung and urine. It does enable a good deal of control to be exercised over the season, frequency and height of mowing so that different species can be favoured and swards produced in which they are abundant. It is a useful management technique where spraying is difficult to arrange. The swards of golf courses, large estates and parks often show how mowing can maintain grassland free from coarse plants and scrub and of considerable wildlife and amenity value.

The frequency of cutting depends on the kind of sward desired and the inherent productivity of the site. Nature Conservancy Council trials have shown that on infertile hills, pastures cut once a year will maintain a short species-rich sward, a winter or early spring cut being just as effective as summer cuts. Cuts in summer may also interfere with ground nesting birds and reduce the display of wild flowers.

On better soils, more frequent cutting will be necessary with a shorter cut of 5–10 cm. Many species cannot survive a short cutting regime which will result in a sward with only those herbs which are able to persist and flower very close to the ground.

To achieve a species-rich sward, it is important to remove the cut material otherwise there will be an accumulation of nutrients leading to a sward with less diversity but greater productivity. Collecting and disposing of the cuttings presents problems. Silage harvesters and trailers can be used but this is time-consuming. Dumping the cuttings will result in nettle beds; suitable landfill sites can be used or the material may be burned after allowing it to dry.

Fire

Controlled burning at the right time of year (November to March) can be a valuable management technique and has been traditionally practised on railway embankments, with the prime purpose of reducing scrub and bramble. Winter burning benefits herbaceous swards by removing old plant litter and trash before the new season's growth begins. It also disposes of the dried-up material that provides fuel for casual, summer fires which affect wildlife severely. Burnt areas can be re-colonized rapidly, but the speed and success of re-colonization depend upon the size of the area and the proximity of populations of plants and animals available to re-invade. Wherever possible, burning should be done in patches and care taken to avoid burning off the whole of the site in any one season.

Much amenity grassland is on steep slopes; indeed it is likely that if tractors and machinery could have worked the ground, the land would have long since been ploughed or otherwise improved for agriculture. Burning or 'swaling' of rough grasslands to remove coarse grasses and stimulate fresh growth in the spring has always been a part of pastoral management in many parts of the world; it may be a key factor in creating and maintaining some grasslands. Many conservationists feel that burning threatens the survival of invertebrates and other animals and may encourage the spread of undesirable coarse species, particularly tor grass in limestone grasslands and purple moorgrass, bracken and birch in heathlands and moorlands. Once these species are established in a sward, they are rarely tackled effectively by stock unless weakened by the use of mowing or perhaps burning. Burning of tor grass-dominated swards has proved a useful technique and burns in alternate years with hardly any grazing proved effective in controlling the vigour of the tor grass and sustaining a diverse, species-rich, chalk grassland sward. Burning is still widely used for maintaining heather swards on deer and grouse moors, where the cycle is commonly 10–15 years.

There are regulations governing grass burning, which is not permitted in the UK between 31st March and 1st November. Great care is needed to ensure that the area to be burned is confined and that there is no risk to adjacent property, especially woodland.

Chemical treatments

Weed and vegetation control with herbicides and growth regulators is generally cheaper, more reliable and requires a smaller labour force than traditional methods. Herbicides can be applied easily and rapidly, so that the problems associated with cultural control of weeds under prolonged wet conditions will not arise. Several soil-acting herbicides can be most effectively applied during the winter months, when there is usually less demand for labour.

There are few ornamental plants that are completely resistant to herbicides and damage can occur when non-selective foliar-acting herbicides such as paraquat or glyphosate are carelessly sprayed onto foliage. Damage can also occur if herbicide vapour or droplets are allowed to drift onto susceptible plants from adjacent sprayed areas. Some soil-acting herbicides may produce transient leaf symptoms in the more susceptible species, but the effect on growth is usually insignificant compared with the damage caused by hand hoeing and forking around the stems, or by uncontrolled weed growth.

It is impossible to test the response of the many thousands of ornamental species and cultivars to herbicides, but sufficient is known to make it possible to plan mixed plantings, so that species and cultivars with similar herbicide susceptibilities are grouped together, so simplifying maintenance.

Selective herbicides

In general, the use of selective herbicides on non-productive land should be confined to the control of agricultural weeds where there is a real danger of infestation of neighbouring crops, or to those plants which may be classed as local weeds because of specific circumstances. There are also particular amenity situations where selective herbicides are useful; these include such weed problems as common nettle and brambles which can obstruct footpaths or interfere with recreational areas. In these cases, local application using hand-held equipment to individual plants or clumps of plants is all that is necessary.

Chemicals for use on roadside and motorway verges, waterway embankments, cemeteries and similar areas must be of low volatility to avoid damage by vapour drifting onto nearby farm crops or gardens. Non-volatile formulations of MCPA or 2,4-D can be used, often with localized application. Some of the more specific and selective herbicides can be used for the control of particular herbaceous problems. Mixtures of 2,4-D, dicamba and mecoprop can be used for the control of woody weeds. Other chemicals used for the control of woody plants include fosamine, glyphosate or triclopyr. Selective materials such as asulam may be used on bracken. In all cases, correct timing of application is critical for success and it is essential to follow-up the treatment with pasture improvement measures, otherwise the bracken will recover after a few years.

Selective weed control on a wider scale may sometimes be required in newly-seeded areas, where broad-leaved weeds seem likely to compete seriously with the growth of the new sward or prove a nuisance. Growth regulator-type herbicides must not be used on sites where wild-flower seed mixtures containing broad-leaved plants have been sown. If broad-leaved weeds present a problem under these circumstances, then a 'weed wiper' may be employed to control the tall growing plants without resort to overall spraying.

On less-coarse swards, such as amenity grassland and housing sites, cemeteries and recreation areas, 2,4-D, MCPA and mecoprop are suitable selective herbicides for broad-leaved weed control.

Annual weeds are less difficult to control chemically than established perennial weeds, especially when they can be exposed to herbicides as they germinate. Perennial weeds such as common couch, ground elder, creeping thistle and field bindweed can be difficult to control and selective chemical control may be impossible when they occur among amenity plantings. A substantial degree of control, using appropriate herbicides, is possible before, rather than after, planting-up. Regeneration from deep underground root systems may prevent elimination. Lasting control of perennial weeds using translocated herbicides, such as glyphosate, is best achieved when the herbicide is applied to an actively growing and well-developed shoot system. This presents problems on amenity sites where it may not be possible to allow weeds to develop to this unsightly stage before spraying. Where trees and ground cover are planted on new housing or recreational sites before the public have access, then the most effective form of weed control can be used regardless of appearance.

Several herbicides that act primarily through the soil, such as diuron, lenacil and simazine, are sufficiently selective to be applied over the aerial growth of many kinds of established ornamental plants. They control germinating and small seedling weeds only. Whereas a light cover of weeds after harvest and into autumn will do little harm to commercial plantings, beyond producing a crop of weed seeds, such a weed cover will be unacceptable in amenity plantings. In these cases two or more applications per year of a soil-acting herbicide may be necessary, depending on the dose applied and the persistence of the herbicide. Larger weed seedlings can be controlled by local applications of a non-selective herbicide such as glyphosate or paraquat. The action of alloxydim-sodium and fluazifop-butyl is largely restricted to grasses, so they can be applied selectively to certain grass species growing among non-grass amenity species. Some annual grasses are resistant and the growth of long-established infestations of creeping perennial grasses may only be checked. Alloxydim-sodium and fluazifop-butyl are applied to grass foliage during the growing season; propyzamide, a soil-acting herbicide, will also control perennial grasses selectively among non-grass plants, but must be applied during the winter. Dichlobenil destroys weed seedlings as they germinate and also has pre- and post-emergence activity against many kinds of annual and perennial weeds. Propyzamide and dichlobenil are formulated as granules and are especially convenient for local application around walls, gravestones, poles and pylons to prevent perennials developing to an unsightly stage. They are tolerated by many woody species and have been used to control herbaceous weeds growing under trees and hedges, but care must be taken to prevent an accumulation around stem bases of granules which can damage or kill woody plants. These materials should be used on farm hedges where the maintenance of undergrowth should be encouraged (see later).

Most conifers are more resistant to several foliar-applied translocated herbicides than are the majority of broad-leaved species and treatments recommended for overall applications to conifers must only be applied to broad-leaved trees and shrubs when directed away from the foliage and bark using a suitable guard.

Growth inhibitors

Maleic hydrazide can be used to inhibit the growth of perennial swards and is especially useful on slopes and inaccessible areas where cutting may present problems. It effects mainly grasses and is often combined with a herbicide effective against broad-leaved weeds, which might otherwise grow away and dominate the swards. Wherever possible, growth-regulator herbicides should be omitted in the interests of conserving the desirable broad-leaved plants. Other growth inhibitors such as paclobutrazol and mefluidide may also be used as growth suppressants.

Maleic hydrazide on its own can control, but not necessarily kill, the growth of some common tall-growing weeds, such as cow parsley, docks and hogweed. It has little effect on other broad-leaved plants except those at a sensitive stage of growth at the time of application. Its use over a number of years will encourage finer and shorter grasses such as red fescue and smooth-stalked meadow-grass at the expense of taller, coarser species such as false oatgrass and cocksfoot. If the use of maleic hydrazide is discontinued after a number of seasons, the coarser species may re-establish themselves. If the full benefit of this material is to be obtained, its use needs to be continued over a number of years.

Uptake of maleic hydrazide by plants is affected by the weather following application and depending on formulation, 8 hours dry weather is necessary for effective results. Combining maleic hydrazide with dicamba and MCPA makes it less dependent on weather.

Application of maleic hydrazide during March to early May after growth has started will normally retard grasses for 10–14 weeks and is effective in preventing the production of seed heads. The success of the treatment depends on application at the right time of year and the period of retardation will be affected by subsequent growing conditions. Some discoloration of herbage may occur in the weeks following application, especially if the weather is dry.

Non-selective herbicides

These can be used to control herbaceous vegetation around the base of trees and shrubs in the establishment phase after planting. Competition from vegetation significantly affects the establishment and growth of trees and shrubs and control is essential. Total herbicides provide a convenient means, especially in areas such as motorway and roadside plantings and for other areas where it is inconvenient or uneconomic to clear by other methods. Clearing of vegetation by carefully-directed spot treatment around the base of trees and shrubs can also facilitate mowing in these areas and save damage to trees from the careless use of machines. Suitable guards are available to protect trees from sprays when treating surrounding vegetation.

Paraquat, propachlor and simazine will control annual dicotyledons and annual grasses but bindweed and cleavers will need treatment with oxadiazon and perennial weeds and grasses with glyphosate and simazine or propyzamide.

Application

The methods of herbicide application described in Chapter 5 are also suitable for amenity areas. However, some modifications may be necessary because individual areas to be sprayed are often small and irregularly shaped and there may be species sensitive to the herbicide being used.

On amenity sites, it is seldom possible to achieve the degree of accuracy in application that is possible with crops spaced regularly in rows and it is safer to aim at applying relatively low rates of herbicide more frequently. It is particularly difficult to ensure accuracy when making local applications by hand with knapsack sprayers to small, ill-defined areas, such as those around the bases of trees. Specialized sprayers delivering a known volume of spray have been devised for this purpose.

Normal agricultural sprayers are unsuitable for the application of sprays to roadside verges, embankments and similar areas. It is necessary to use equipment designed or modified to deal with special conditions which will include variations in verge width, obstructions such as signposts, guiderails and reflector posts, the varying height and density of vegetation and the special dangers of spray drift. The most effective and safe method of applying herbicides is to spray at high volumes of water (700–1100 litres ha^{-1}) at low pressures using special jets for the purpose. This ensures a uniform coverage with large droplets that are not liable to serious drift. Spray drift into hedges and neighbouring crops and gardens must be avoided completely. The safest condition in which to spray is a steady force 2 light breeze blowing away from susceptible crops, open water or neighbouring land.

Other methods of application, including wet ropes and wicks and spray-droplet formation by spinning discs have advantages for special purposes. Using ropes or wicks wet with herbicide which are moved forward at a fixed height, it is possible to apply non-selective herbicide, such as glyphosate in a selective manner killing only those plants that grow above the height of the machine. Besides the uses in amenity plantings, this method has wider application. Control of weeds such as ragwort, docks and thistles may be achieved by using a 'weed wiper', so avoiding damaging low-growing desirable plants. This method is particularly useful in managing sites where conservation is important, such as SSSIs.

Hand-held lightweight spinning-disc sprayers, capable of applying herbicide at reduced volumes, have important advantages in difficult terrain in which a normal sprayer cannot be carried. Although using a low-volume rate of application, some of these sprayers can produce sprays relatively free of small droplets, which might otherwise drift and damage adjacent sensitive species. Specified formulations may be needed for this type of sprayer. Gel or invert emulsion formulations have also been used to reduce drift risk.

Special situations

Non-cropped areas on farms

On every farm there are areas where for some reason or another cropping is not possible and vegetation grows rankly. Farm road and track-way verges may need attention from time to time, but are usually kept open by the passage of vehicles and very little else will be required. Tidiness on farms is sought after and is a matter of pride, but a distinction should be made between the control of weeds on non-cropped areas, because they are actually likely to cause an infestation in the crop, or because they harbour crop pests and diseases, as against control purely for appearances. Excessive tidiness should be avoided. Some weeds are actually attractive plants, or like common nettle are a valuable food plant and a habitat for desirable insects and other animals. Herbicide sprays most certainly give the farmer a chance to clean-up non-cropped areas on the farm, but thought must be given to the real need to do so.

Field boundaries

Boundaries between fields have existed for centuries in the UK and their agricultural function was originally for stock impoundment. The commonest boundary structure is the hedge, but there are many others including ditches, dykes, stone walls, woodland edges, fences and grass banks. It is estimated that there are 800 000 km of hedgerow in the UK at present, occupying 160 000 ha. Whilst many hedges still maintain their important function of stock impoundment the role of the boundary has altered and field boundaries are now regarded as of interest as a reservoir of inoffensive and beneficial wildlife, including some game species.

In hedge bottoms where there is a need to control herbaceous growth, a range of selective herbicides and maleic hydrazide have been used, both on young and mature hedges. Total herbicides with some residual action like dichlobenil can be used to help in the establishment of newly set and young hedges. Care must be taken in these situations as destruction of a particular weed flora will often encourage the development of another which is more resistant and aggressive. The use of paraquat without cultivation encourages the development of common couch. It should also be remembered that the flora at the bottom of the hedge and around field margins provides a habitat for wildlife, including game, which will be absent from cultivated or intensively-managed fields, so that the timing of mechanical or chemical weed-control operations may be very important for conservation.

A diversity of boundary structures increases the landscape interest and the role of field boundaries in wildlife conservation is of vital importance in areas of intensive arable cropping. The grey partridge relies on field boundaries for nesting and shelter.

It is often assumed that weed species spread from boundaries into the crop and whilst this is the case for a small number of species, the majority of plants found in field boundaries do not successfully colonize the crop and are, therefore, inoffensive. Many species recorded in field boundaries may disperse up to 2.5 m from the hedge bottom, but few survive annual cultivations. The hedge bottom may also support some common annual species, typical of the main crop area. On average, only 25% of the species recorded in the boundary are also present in the adjacent crop. A few common weeds, notably couch and barren brome, do spread from the field edge, but these weed species represent only a small part of the flora of field boundaries. Cultivations close to the hedge can reduce the habitat for ground flora species and affect shrub root growth. Inadvertent or deliberate applications of fertilizer and herbicides may selectively affect species composition. This may modify the ground flora by encouraging annual weed species at the expense of perennials. A high nutrient status will favour dominance by a small number of species which are capable of exploiting the conditions at the expense of slower growing species. Field rates of many herbicides will affect hedgerow species with broad-spectrum chemicals likely to cause the most damage.

Management of field boundaries should aim to enhance beneficial features of the margin, whilst limiting adverse characteristics. Those species capable of spreading into arable crops, such as barren brome and cleavers, cannot be controlled selectively in the boundary with overall applications of herbicides. The aim should be to maintain populations at as low a level as possible. Such problem species are usually annuals which are favoured by disturbance and management should aim to maintain an ecologically-stable situation which is disturbed as little as possible. In this respect, the requirements of agriculture and the wildlife of the field margin may be complementary in that the maintenance of a diverse, perennial ground flora will also discourage weed populations in the boundary.

In general, the overall application of herbicides to field boundaries is not recommended as compounds do not show sufficient selectivity between species. Selective control of species may be achieved with selective application techniques, such as weed wipers, but long-term control should address the underlying reasons for particular species assemblages. Under conditions of severe weed infestation where barren brome and cleavers are the major constituents of the flora, then chemical intervention is justified. The use of plant growth regulators to inhibit flowering in annual grass weeds might also be justified. Present information indicates that barren brome would not be inhibited and species diversity may not be maintained.

The creation of barrier strips between the hedge bottom and the crop may be useful for the reduction of disturbance in the field boundary and for the prevention of weed ingress from infested margins. The effectiveness of such barriers has not yet been established. Bare ground strips may be created by repeated rotavating or by the application of approved residual herbicides in winter or spring. The practice of spraying out the crop edge in early summer with broad-spectrum herbicides may facilitate the harvest, but unless applied under optimum conditions and with accuracy, poses the risk of destroying the perennial ground flora.

Another option for the barrier strip is the establishment and management of grass strips between the hedge bottom and the crop. These grass strips may act as weed barriers and may be used for access. Management of grass strips will be by cutting or grazing. Experimental work indicates that plant growth retardants may be useful for strip management, but may not encourage diversity within the created sward. Ideally cutting treatments should take place after wild flowers have set seed and should not interfere with nesting birds.

Once a field boundary has become infested with annual weeds, usually following severe disturbance, the ground flora may need to be recreated. Prevention of disturbance will allow successional change in the flora although this may take several years to occur. The alternative is to destroy the existing weed flora, create a seedbed and sow a mixture of desirable non-invasive perennial species.

Footpaths often follow field boundaries and the landowner is under a statutory obligation to keep these open and free from obstruction. Trimming the sides of the hedge is an obvious measure but it may be necessary to use chemical control measures to keep plants such as briars and brambles in check. Diquat + paraquat + simazine may be used and 2,4-D + dicamba and triclopyr can be used for perennial and woody weed control. Elder can be controlled in hedgerows by painting glyphosate into a cut in the main stem at the base of each bush during the period October to April.

Farmers and other commercial users of pesticides should advise those who may come close to the intended area of operations. They can take action to prevent pets straying into treated areas. The obligation is stated in the *Code of Practice of Agricultural and Horticultural Use of Pesticides* issued under the Control of Pesticides Regulations 1986. It is important to make sure that the public are fully informed about work in progress, and the materials applied and when it is safe to enter.

Weed control on rural roadside verges

Roadside verges form a network of semi-natural vegetation covering approximately 180 000 ha of the UK. Most of the area of roadside verges is grassland but there are also substantial areas of scrub and woodland. Responsibility for the management of roadside verges lies with Highway Authorities. The Department of Transport is responsible for major trunk roads and motorways, although management is delegated to local authorities. Local authorities are themselves responsible for management on all other roads.

Aims of management

The principal aim of roadside verge management is to ensure the safety of road users. It has been recommended that the maximum height of vegetation should not exceed 30 cm and this has been used as the maximum for sight lines, but keeping the vegetation short also prevents obstruction of signs, helps keep drainage channels operational and visible to motorists, reduces fire risk and allows the verge to function as a footpath all of which contribute to road safety.

Verges are occasionally managed, in compliance with the statutory requirement of the 1959 Weeds Act, to prevent the spread of injurious weeds namely docks, thistles and ragwort but in practice the control of agricultural weeds rarely features as a main reason for management. In contrast, management of verges in the interests of amenity and conservation is becoming increasingly important. Precise objectives may vary from keeping the verges tidy and free of litter to managing them in the wider interests of the countryside and wildlife. In particular, road verges support a diverse flora and fauna and many represent examples of species-rich grasslands which are becoming increasingly rare elsewhere. For this reason, some 4% of all road verges are designated as roadside nature reserves.

Tall herbs growing close to the road may be visually attractive and harbour a diverse community of insects making them of considerable wildlife value, but in terms of road safety they would be 'weeds' needing control of weeds in roadsides. Weeds thrive as a result of poor manage-Tall grasses such as false oat and some umbellifers such as hogweed and cow parsley are particularly dangerous. Even in rare cases where verges are managed primarily in the interests of wildlife, some weed control may be necessary. An increase in some species such as nettle and thistles is frequently indicative of a deteriorating verge, and may be associated with the loss of other, more desirable, species.

Standards of management

Current policy originates from 1975 when the Department of the Environment issued a Technical Memorandum instructing its Agent Authorities that, with a few exceptions mainly concerned with road safety, there was to be no further mowing of roadside grass; the Department of Transport's *Code of Practice for Routine Maintenance (1985)* reiterates this policy. In practice it means that on most verges only a single swathe width of verge is cut, and it results in large areas of unmanaged grassland and scrub. The code also recommends that selective herbicides are likely to be more appropriate and economical than cutting for the control of injurious weeds, but that herbicides and growth retardants should only be used in exceptional circumstances, not as a general rule.

Methods of verge management and weed control

Traditional methods of vegetation management on verges, namely grazing, burning and cutting for hay are no longer appropriate on most verges. There are now three distinct management options:
1 Mechanical cutting.
2 Grass growth retardants.
3 Herbicides.

Most cutting is now done with mechanical flail mowers and the only factor which highway engineers have any real operational control over is the frequency of cutting. Cutting once per annum or every other year is adequate to control scrub growth but is unlikely to give control of tall weed species on any but the poorest soils. Cutting twice per annum (in May and September) gives good control of weed species, reduces populations of hogweed and cow parsley and encourages the development of a diverse flora. Cutting twice should also, in most cases, keep the vegetation <40 cm throughout the year. Where a lower height is necessary an additional cut in June or July may be required. In general, it is best to avoid mid-summer cuts because control of weeds is likely to be of short-term cosmetic benefit only, and cutting at this time is likely to be most damaging to wildlife. Cutting strategies on roadside nature reserves should also take account of the special needs of the species they contain. For example, to allow cowslips to flower and seed, cutting may have to be at a greater height or delayed until July.

Grass growth retardants may be used as an alternative to cutting to control vegetation height. Maleic hydrazide has been available since the 1940s but has not been extensively used on roadside verges, because of its relatively high cost, unreliable performance in wet periods, and because of concern over its environmental effects. Two new products, mefluidide and paclobutrazol, are now being marketed.

Significant savings by using retardants are only likely on verges which are regularly cut three times per annum. On these verges the use of a retardant combined with a cut late in the year is likely to give more effective and cheaper control. The main advantage from retardants is the ease and speed with which they can be used. Furthermore, spraying can be done earlier in the year than cutting (particularly when using paclobutrazol) and in contrast to cutting, it is preventative rather than curative, never allowing the vegetation to get excessively high.

Grass growth retardants also reduce the growth of most broad-leaved species. This may result in some beneficial weed control (e.g. maleic hydrazide will practically eliminate hogweed and cow parsley within 2–3 years), but it also raises concern that these chemicals may be damaging to the environment. Information on the effects of maleic hydrazide indicate that although in the short term it may lead to increased plant species richness, eventually (after 4 years or more) it leads to a decline in the number of broad-leaved species. There is no long-term information on the effects of mefluidide and paclobutrazol, but relatively short-term results show that although paclobutrazol is likely to lead to coarse, species-poor grass sward, mefluidide may lead to an increases in plant diversity.

Although retardants may have a damaging effect on the vegetation to which they are applied, their effect on plant species is by no means as indiscriminate as the herbicides used for weed control on roadsides. Furthermore, if their use is confined to the narrow strips adjacent to the road and visibility splays, influence on the remainder of the verge as a whole should be small. The main environmental problem associated with the use of retardants is that, in practice, they are often used in a mixture with broad-leaf herbicides. Wherever possible this practice should be avoided.

The main target species for herbicides are those listed in the Weeds Act, but increasingly herbicides are being used to control tall umbellifers such as hogweed and cow parsley, and also nettles. The two most appropriate products for roadside use are 2,4-D and MCPA, repeated applications of which will control most roadside weed problems. Unfortunately, many harmless broad-leaved species are likely to be affected, resulting in very species-poor verges. For this reason herbicides should only be used on limited areas and should not be used indiscriminately. In order to reduce the visual impact of dying plants, these herbicides should be used as soon as is practicable in the spring. The rate of disappearance of 2,4-D and MCPA from the environment is rapid, ranging from a few weeks to a few months. However, the use of more persistent herbicides such as picloram, which may be more effective in controlling deep-rooted perennials, should be avoided wherever possible because of possible drifting or leaching of the chemical into non-target areas.

Unless used repeatedly, herbicides are unlikely to provide long-term

control of weeds in roadsides. Weeds thrive as a result of poor management, localized disturbances by road widening and dumping of spoil, and
even selection of less susceptible species by spraying. Without a return to
more intensive management, involving more frequent cutting, weeds are
likely to be a permanent problem on many verges.

Weed control on nature reserves

The main reason that weed control is needed in nature-reserve management is where the objective for conservation is to maintain a sub-climax
community, for example a heathland, a grassland or other early successional habitat. These communities depend on management for their existence and if the management is relaxed or eliminated, then invasion by a
later successional species will occur. Examples of this problem include the
invasion by bracken, birch, gorse and Scots pine on lowland heaths and by
hawthorn in chalk grassland. Without intervention, succession will proceed and the habitat of interest will be lost. Over the last 40 years, some
50% of lowland heaths and grasslands have been lost in this way. This is
a classic vegetation management problem involving the invasion of unwanted species, in other words, weeds. As with any weed problem there
are several potential solutions. Weeds may be tackled either mechanically
or by using herbicides, but the use of the latter on nature reserves is beset
with problems. In farming or forestry there is usually only one species that
must be safe from damage and all other species can be considered targets,
but with conservation the opposite is usually the case. Only one species is
the target, that is the plant invading the desired community, and all the
other species should remain unaffected by the herbicide treatment. Before
using herbicides to manage vegetation on nature reserves it is essential to
consider the effects of the materials on the non-target vegetation. Whilst
some information is available from the manufacturers, it is strongly suggested that before any herbicides are used in specific conservation tasks,
screening trials should be done to test the effects of the herbicide on the
community on which the herbicide is to be used. One such screening trial
has been carried out on a *Calluna*-dominated heath and a grass heath in
Suffolk. The results suggest that asulam and fosamine-ammonium did not
damage *Calluna* and asulam, and fosamine-ammonium and triclopyr did
not damage the grass heath. On the basis of this trial it may be suggested
that these should be the herbicides of choice for weed control on heathland
nature reserves. There are two major factors concerning the development
of an optimal herbicide strategy for weed control in nature reserves:

1 *Choice of herbicide*: Selectivity is necessary if it is to be applied by spraying where it may become deposited on underlying non-target species if present.

2 *Application methods*: Conventional sprayers are the most commonly used. It is usual in conservation work to spray the foliage of the plants to 'run-off' where a balance is achieved between maximum coverage of the leaves and minimum loss of liquid running off the surface.

Application through direct contact with either rope-wicks or rollers, continually replenished with fresh herbicide solution from a reservoir, is another option. There are tractor-mounted and hand-held versions of such applicators available. This equipment eliminates the hazard from spray drift and selectivity can be achieved if there is a height differential between target and non-target species.

Herbicides may also be applied through the cut surfaces of scrub after cutting. The herbicide may be sprayed onto the stump or a drench or spot sprayer used, or in some cases the herbicide can be applied with a paintbrush.

The exact choice of herbicide-based strategy would depend to a large extent on the problems of individual sites, the equipment and manpower available, and the sensitivity of the underlying vegetation.

Scrub control

There are basically two methods for the control of scrub. Mechanical treatment on its own is usually sufficient for the control of species such as Scots pine, which do not regenerate after cutting. Deciduous scrub will regenerate from cut stumps and mechanical control by cutting with no follow-up herbicide treatment is ineffective, as there will be substantial regeneration after 2–3 years. At best, cutting will give a brief respite, but at worst the large number of suckers produced may make subsequent mechanical treatment more difficult. Herbicide treatments on their own are often unsuccessful because they leave a large number of dead trees standing, which in most cases will be aesthetically unacceptable. In most conservation work a combination of both herbicide and mechanical cutting is needed. Two herbicides are recommended for the control of scrub as foliar sprays, fosamine-ammonium and triclopyr. Fosamine-ammonium is recommended for use in late summer from August until just before leaf fall. The species will die back as usual and appear unaffected, but leaf development is prevented in the next season. Triclopyr can be used throughout the summer from the period of maximum leaf expansion onwards, with the plants dying in the season of application.

Another technique is to cut the scrub and allow it to regenerate from the stump and then treat the re-growth with a selective herbicide such as 2,4-D + dicamba + triclopyr. This has the advantage in that cutting can be done at any time throughout the year allowing a flexible approach and spraying small plants is often easier than tall ones, so that less herbicide will be used and costs and effects on possible non-target species will be lower.

Applying herbicides to cut stumps is a very useful method, because there is no spray drift and it can be carried out at any time of the year. The trees should be cut at, or near, ground level and the stumps sprayed or painted with herbicide. This treatment must be carried out immediately after cutting, otherwise stumps will be missed especially in dense under-growth and will consequently regenerate. Herbicides approved for cut-stump application by paintbrush include triclopyr and glyphosate, whilst ammonium sulphamate can be applied by spraying and granule application.

The main problem as far as conservation management is concerned is that weed control is also part of the management required to reverse and arrest succession. In many cases where a dense weed is present, much of the interesting vegetation will be lost or reduced in the area. When this problem occurs, restoration methods may need to be developed to ensure regeneration of the desired community. Weed control is only a small part of the conservation management of sub-climax communities.

The final choice in any particular situation would depend on the site wardens and the resources which are available.

Public relations

Commercial growers usually apply chemical sprays on their own property without directly involving the public, but amenity managers must learn to control weeds without upsetting people who use the amenity. All ac-cidental exposure of people or pets to pesticides must be avoided and posi-tive action taken to enhance public confidence. The visiting public should, whenever possible, be informed about the Food and Environment Protec-tion Act and the methods whereby manufacturers test herbicides thoroughly to ensure safety before they are marketed.

Chemicals during mixing are also subject to low-level exposure, some-times for long periods. Because staff may have to wear appropriate protec-tive clothing and face shields to comply with the Food and Environment

Protection Act, this may cause alarm unless the public is properly informed and, if necessary, kept at a safe distance during spraying.

Acknowledgements

The help of Dr E.J.P. Marshall (Long Ashton Research Station) and Dr R.H. Marrs (Monks Wood Experimental Station) in contributing material to this chapter is gratefully acknowledged.

17 / Weed control in forest nurseries and forests

D.R. WILLIAMSON AND W.L. MASON*

*Forestry Commission, Alice Holt Research Station, Wrecclesham, Farnham, Surrey GU10 4LH; and *Forestry Commission, Northern Research Station, Roslin, Midlothian EH25 9SY*

Introduction

Forest nursery production has many similarities with ornamental horticulture since a perennial crop is grown which is easily checked by weed competition. The financial value of the crop is considerable since plants ready for dispatch to the forest may be worth between £5000 and £50000 per hectare depending upon species. Every effort is made to maintain weed-free conditions and herbicides are not normally excluded upon grounds of cost.

There are some 600–700 ha of forest nursery area in the UK. Some 90% of this area is devoted to raising conifers and most weed-control techniques have been developed for conifer species. Most nurseries are sited on acid soils (pH 4.5–5.5) of light texture (sands to sandy loams). The microflora of such soils is very different from that found in most arable or horticultural sites so that the breakdown of herbicides cannot be safely extrapolated using data for agricultural soils. The weed flora also tends to be different, the most important being annual meadow-grass, sheep's sorrel, common chickweed, corn spurrey, groundsel and horsetail.

The normal crop rotation in forest nurseries is a 3–4 year cycle of seedbed, transplant line and fallow ground. The height of the final crop is normally 30–60 cm depending upon species. The fallow period provides an excellent opportunity to get rid of perennial weeds. In some nurseries, the bare fallow is replaced by a green crop of lupins, oats or rye.

The crop, whether seedlings or transplants, seldom covers the ground completely until late August or September. Until complete crop cover is achieved, weeds have to be controlled by herbicide, machine or by hand. The aim is to prevent weeds over-topping the crop and/or setting seed. In practice, this means preventing weeds germinating or removing weeds within 3–4 weeks of germination. As in other areas of horticulture, irrigation is generally essential to ensure adequate incorporation of herbicides.

Seedbeds

Pre-sowing treatments

Soil sterilization with either dazomet or methyl bromide in the autumn of the year before sowing is a common practice. The latter is subject to the 1982 Poisons Rules and the 1972 Poisons Act so it can only be applied by certified operators. It is important to aerate the soil and release any sterilant residues before sowing takes place. These sterilants will usually improve the growth of the crop as well as providing weed control. An alternative approach is the stale seedbed technique using paraquat to remove weeds before sowing the crop.

Pre-emergence treatments

Seeds of conifers and of small-seeded broadleaves are sown onto the surface of a raised seedbed and covered with coarse sand or fine grit. The seeds and roots of young seedlings of such species have little protection from direct contact with applied herbicides. Diphenamid, applied within 7 days of sowing, is used safely on a wide range of species though not birch and alder species. The larger seeds of species such as oak, beech and sweet chestnut are normally drilled into seedbeds and covered with at least 25 mm of soil. This allows simazine to be safely used provided it is applied immediately after drilling. Paraquat can be used up to 3–4 days before crop emergence to clean-up germinated weeds. Spraying should not be done if seeds have radicles longer than 5 mm or if sporadic germination has occurred.

Post-emergence treatments

Diphenamid can be applied from 2–4 weeks after crop emergence once the first true leaves are apparent. Thereafter, repeat applications can be made at intervals of 6 weeks. Propyzamide applied in October to January is used to control annual grasses and some broad-leaved weeds, and is particularly useful in stand-over seedbeds. Simazine is sometimes used, but can only be applied at low rates to a dormant crop. The lack of herbicides for controlling germinated weeds during the growing season means that hand weeding can often be necessary. This can frequently damage small seedlings. Mechanical weed control using inter-row powered hoes or tines is an option in crops that have been drill-sown.

Transplant lines

Transplants are set out in lines 15–20 cm apart with 3–8 cm between plants within rows. Residual herbicides, such as simazine, have greatly reduced the need to hand-weed in transplant lines. However, in the last decade, populations of triazine-resistant weeds (e.g. groundsel) have become established in a number of nurseries. Control of these resistant weeds has required extensive hand weeding and the use of alternative herbicides.

The standard practice is still to apply simazine as an overall spray to moist, clean soil immediately after transplanting. Lower rates are used on sensitive species like larches and Serbian spruce. Atrazine is a reliable alternative to simazine when applied in spring although some conifers are sensitive to it. However, it cannot be applied overall to crops in active growth and heavy rain can cause the chemical to be washed into water-holding areas resulting in local over-dosing. Diphenamid can be used as an alternative to simazine or atrazine and is particularly useful for application to sensitive species. Propyzamide can be applied from October to January to control germinating and established weeds. However, up to 6 months should elapse between application of a triazine herbicide and the use of propyzamide.

Directed sprays can be used in the growing season to control germinated weeds. Paraquat or glyphosate can be applied through special shielded applicators to prevent any chemical contact with the crop foliage. Best control is achieved when weeds are growing actively. Repeat applications may be necessary to achieve optimum control.

If triazine-resistant weeds are present, then oxadiazon or oryzalin may be used as alternatives. Both are applied to clean soil after transplanting although oxadiazon may control weeds up to the first true leaf stage. Oryzalin can be applied as an overall spray in either spring or summer, but oxadiazon should only be used in late winter or spring before the buds begin to swell.

The fact that transplants are set in lines means that mechanical weed control is a viable option. It can be particularly useful in the growing season to remove germinated weeds that may spoil the appearance of the crop or seed into nearby seedbeds.

Fallow and other land

The fallow period in the nursery rotation provides a chance to control perennial weeds using a combination of herbicides and cultivation. Repeat applications will generally be necessary to achieve satisfactory control. Glyphosate and paraquat are the most widely-used herbicides for this purpose. In extreme cases, sodium chlorate may be used to control persistent weeds. At least 6 months should elapse between treatment and growing a crop on treated ground. A cress test should be carried out before any crop is planted on the treated area.

Weeds on uncultivated land, notably paths, fencelines and areas round buildings are potential sources of weed seeds in forest nurseries. Controlling such weeds by cutting or cultivation is frequently difficult. Non-selective, persistent herbicides can be used in early spring to control most annual weeds and keep sites weed-free for up to 12 months. Such chemicals can be applied close to existing crops as there will be little lateral movement of chemical in the soil provided the crop is screened against drift during spraying. Techniques are similar to those used in industrial weed control (see Chapter 19).

Weed control in the forest

Forestry differs from most agricultural crops in that the period between planting and final harvest is many years. But the management of the forest area prior to planting and during the early establishment phase is critical if plantations are to be established successfully at least cost. It is during the establishment period when most costs are incurred. Once established, trees will suppress most competing herbaceous vegetation but woody weeds can compete for the whole rotation unless removed, or killed with herbicides. The amount of weed control needed is reduced if the establishment phase, when the trees are vulnerable to weed competition, is shortened. Rapid establishment can be achieved by the use of planting stock of high quality which has been handled with care and well planted at the correct time of year, using cultivation to inhibit weed growth and promote early tree growth whenever appropriate and economic.

Weeds reduce the survival and early growth of newly planted trees by competing in the following ways.

Competition for light

Tall weeds may compete for light, physically damage the trees when they collapse in the autumn and harbour bark-gnawing rodents such as voles. Tall weeds may protect young trees from desiccation but if their roots are close to the trees they will also compete for moisture and nutrients, and on grassy sites subject to water deficits, this can be more important than competition for light. Mowing or cutting these weeds reduces competition for light but not root competition. On sites with a summer moisture deficit, mowing often increases the swards' transpiration and thus the moisture stress suffered by the tree.

Competition for soil moisture

Relatively little moisture evaporates from bare soil before a layer of dry soil forms reducing further evaporational losses; vegetation transpires moisture faster and for longer before moisture availability limits further transpiration. Therefore soil-moisture deficits are greater under weeds than bare soils. Weed-induced moisture stress kills trees or reduces their growth. This factor is particularly significant in the southern lowlands and on freely-drained soils with a southerly aspect in the uplands.

Competition for nutrients

Moisture and nutrient competition are interrelated; weeds may compete directly for nutrients or, by drying the soil, render them unavailable to the trees. Heather is one of the best examples of direct competition, inhibiting the uptake of nitrogen by the roots of certain conifer species, especially spruces and firs. The competition is relieved by a complete application of herbicides in some instances, but on sites with little available nitrogen, the heather can be difficult to control, and there may be no alternative to nitrogen fertilization. Nowadays, the preventative action of planting pines or larches in mixture with sensitive species would be employed.

On grassy areas, fertilizing alone rarely relieves competition — it often invigorates weeds at the expense of the trees.

Tree growth is related to the area weeded around a tree. A 1 m diameter spot appears ideal in upland conifer plantations with little benefit from more intensive weeding. Larger planting stock should receive a larger weed-free area, and on droughty sites, e.g. in southern England, a larger weed-free area generally results in better early tree growth.

On sites where modest weed growth is expected, weed control should start before the trees are planted when there is no risk of damaging them with contact herbicides or cultivation and weeding implements. If serious weed problems are expected, it may be better to delay planting for one year and first tackle the weeds with a combination of cultivation and herbicide application.

Until recently, afforestation sites have usually been poor-quality grazing land acquired from agriculture. Under these circumstances, heavy grazing pressure is best kept up until as late as possible before tree planting thus reducing colonization by perennial and woody weeds. On many impoverished upland sites, where new planting is undertaken with vigorous species such as Sitka spruce, re-invasion after ploughing or scarification is sufficiently slow to allow satisfactory crop establishment with no further weed control. Control by herbicide application will be needed on rip-cultivated areas, the more fertile afforestation sites and on many weedy re-afforestation areas. Most lowland sites, even when they are cultivated, will require chemical weed control.

Control of weeds before planting trees

Where problem weeds such as bracken, gorse, heather, bramble or woody growth are present on afforestation sites or where such weeds are anticipated to invade within one or two seasons after clear felling on a re-stock site, it may be better to control such weeds over the whole site before planting. Heather, for example, is generally burned before afforestation as this delays re-invasion and facilitates ploughing.

The use of herbicides prior to planting allows a wide choice since trees do not have to tolerate direct application. Because the trees will be planted fairly soon after herbicide application to obtain maximum benefit from the weed-free conditions, the residual effects of any chemical must be considered.

Unless the herbicide being used pre-plant has very rapid and visible action, or is applied along a cultivated strip, overall applications are used because of the difficulty of making sure that the trees are planted in the chemically-treated area.

Where undesirable woody weeds, e.g. rhododendron or vigorous *Rubus* spp. (brambles) are present, the use of a herbicide alone will not facilitate access to the site for cultivation, planting or subsequent operations. In these situations, it may be best to cut down the offending weed and then treat the cut stumps with herbicide or spray the young re-growth. Mature woody weeds are so expensive to remove that a long-term containment programme with eradication as the overall objective is the only solution. Rhododendron, for example, can colonize roads, rides and wind-blown areas and even develop as an understorey when trees are thinned. After felling, this nucleus of weeds can spread rapidly and compete with newly-planted trees. The most efficient method of control may therefore be to kill the rhododendron prior to felling.

Control of weeds after planting

Small trees (15–40 cm tall) are generally used in large-scale plantings because such trees are less costly, cheaper to plant than taller trees and are more likely to survive owing to the balanced nature of the root and shoot systems. However, small trees require more protection from the effect of weeds. A reduction in weed competition can be achieved by cultivation and the use of herbicides. But on weedy sites, some form of weed control is usually required for several years after planting.

The period during which most trees are in active shoot growth is short, usually late April to late August, although there is substantial variation between tree species, site type, latitude, elevation and season. The greatest shoot extension takes place in the early part of the season (late April to June) and it is during this period that the effects from weed competion are most damaging. Leaving weed control until the effects of weed competition are obvious means that it is too late to give benefit to the trees in the current season. It is therefore important to inspect plantations in the autumn and take steps to relieve trees from weed competition in the early part of the growing season. Once the trees are tall enough to stand above the non-woody weeds, the competition effect for moisture and nutrients from weeds is probably not enough to justify further weed-control measures. Hand weeding has long been the traditional method of weed control and although this does stop the trees being smothered or physically damaged, the cutting of a grass/broad-leaved weed sward makes it more competitive on sites subject to drought, as a cut sward transpires faster than an uncut sward. Hand weeding, where it cannot be mechanized, is slow, labour-intensive and expensive and frequently fails to achieve the prompt release of the crop.

The optimum benefit from weed control is obtained where a residual herbicide is used in anticipation of weed competition and is applied during the winter before growth starts. In this respect, propyzamide is particularly useful as it can be applied in liquid or granular form from November to January and has long-lasting residual action, particularly on soft grasses. Residual herbicides have the added advantage of being less demanding in terms of weather conditions and date of application.

The choice of which herbicide to use after planting is determined not only by the weed species to be controlled but also by the species of tree which is present. Very few herbicides are tolerated by any tree species when in active growth. Spruces and pines are the most tolerant of overall herbicide application outside the period of active growth. In contrast larches, firs and broad-leaved tree species are much less tolerant.

Paraquat used in spring and summer will give some measure of release from weeds for that season but can only be used if trees are carefully protected from spray. Glyphosate can be used as an overall spray with spruces and pines provided they have ceased active growth but all other tree species must be protected from the spray even when dormant. Glyphosate acts more slowly than paraquat and although weed control may be effective the main benefit from summer applications will be in the following season.

Two contact and translocated herbicides are recommended for the control of bracken. Asulam is effective and can be used post-planting with the majority of conifers and some broad-leaved species. Glyphosate, although slightly less effective than asulam on bracken, is useful when there is a mixture of weed species.

Heather can cause trees several times taller than itself to be severely checked in growth through competition for nitrogen. Sitka and Norway spruce and Douglas fir are particularly susceptible to check by heather, while pines are relatively insensitive. Spruces may be brought out of check by applications of fertilizer containing nitrogen (or in special cases, phosphate) or by killing the heather with herbicides such as 2,4-D or glyphosate or by a combination of fertilizer and herbicide application. Hand cutting to form a mulch is not cost-effective.

Woody weed species tend to grow up with the crop trees on fertile sites and especially on old woodland or lowland clays of high pH. The sites can sometimes require periodic control up to 10 years after planting. Now that 2,4,5-T is unavailable, alternatives such as glyphosate and triclopyr are used for foliar application for the control of woody weeds, although rhododendron is only susceptible to high rates of glyphosate and triclopyr. Where woody regrowth is a problem in a broad-leaved crop, stem incision

Table 17.1 Main herbicides recommended for use in the forest or in areas to be planted with forest tree species

Herbicide	Weed type controlled
*Pre-planting:**	
Asulam	Bracken
2,4-D Ester	Heather
Glyphosate	Bracken
	Bramble
	Broad-leaved weeds
	Grass
	Heather
	Rhododendron
	Woody weeds
Paraquat	All green vegetation is burnt off
Triclopyr	Bramble
	Gorse and broom
	Rhododendron
	Woody weeds
Triclopyr + dicamba + 2,4-D	Bramble
	Gorse and broom
	Rhododendron
	Woody weeds
Post-planting:†	
Asulam	Bracken
Atrazine	Grass
Atrazine + dalapon	Grass
Cyanazine + atrazine	Grass and broad-leaved weeds
Dicamba	Broad-leaved weeds
Dichlobenil	Grass and broad-leaved weeds
Dichlobenil + dalapon	Grass and broad-leaved weeds
2,4-D Ester	Heather
Glyphosate	Bracken
	Bramble
	Broad-leaved weeds
	Grass
	Heather
	Rhododendron
	Woody weeds
Hexazinone	Grass and broad-leaved weeds
Paraquat	All green vegetation is burnt off

Table 17.1 Cont.

Herbicide	Weed type controlled
Propyzamide	Grass and broad-leaved weeds
Terbuthylazine + atrazine	Grass and broad-leaved weeds
Triclopyr	Bramble Gorse and broom Rhododendron Woody weeds
Triclopyr + dicamba + 2,4-D	Bramble Gorse and broom Rhododendron Woody weeds

* These herbicides are the main chemicals used to deal with problem weeds prior to planting
† Although these herbicides can be used post-planting, use may be limited to a certain time of year when trees are dormant, or to a spray which does not come into contact with tree bark or foliage. See product label for details

or cut-stump treatment may be the only alternative; in this situation glyphosate or triclopyr can be used. Gorse and broom are fairly resistant to most herbicides but triclopyr is known to control them. Table 17.1 lists the main herbicides used in forestry.

Special situations

Weeding trees in treeshelters

Treeshelters protect crop trees from physical damage by weeds and from browsing by mammals. They also reduce the degree of moisture stress by maintaining high humidity conditions within the shelter immediately around the tree and minimizing the effect of desiccating winds. But treeshelters are not an alternative to weed control and trees in shelters still benefit from a 1.0 m diameter spot free of weeds in areas with moisture deficits.

Ideally, weeds should be killed before the trees are planted and the shelters erected but this is often impractical because of the difficulty of locating the spots or strips of ground which have been pre-treated with herbicide prior to the time of planting. For this reason most herbicides are applied post-planting. If possible, a residual herbicide should be applied which is tolerated by the trees, but which has the ability to kill the weeds already present and control any new weeds soon after germination, before they begin to compete with the trees.

As the trees are protected by treeshelters, contact and translocated herbicides, which would kill the trees if they came into contact with their foliage, can be used around the base of the shelter in the absence of a residual herbicide or to remove weeds tolerant of the residual chemical. The operation of applying herbicides around trees in shelters is made easier because in addition to being protected each tree can be easily located.

Poplar plantations and stoolbeds

Poplars are highly intolerant of weed competition and survival and growth rates can be seriously depressed unless steps are taken to reduce or eliminate weeds around stools or newly-planted trees during the first few seasons. A pre-plant application and applications for two subsequent seasons after planting are usually required because of the vigorous and invasive nature of weeds on fertile poplar sites. Suitable herbicides include glyphosate and paraquat for pre-plant or directed post-plant sprays. Propyzamide can be applied during the winter for the control of grasses while simazine can be used post-planting as a pre-emergent residual herbicide.

Mulches can also be employed to control weeds after planting and in experiments have been consistently found to improve tree performance more than any other weeding treatment. This is probably because mulches conserve moisture and provide a source of nutrients in addition to preventing weed growth. The mulches which have been found to give the best results in terms of tree growth are those made up of locally-cut vegetation which is laid around the tree as soon as possible after planting and then repaired as required to maintain weed-free conditions for three to four seasons. Imported straw, bracken, sawdust, fresh bark or dung have been found to be reasonably good substitutes.

Some opaque durable materials laid flat on the ground and securely anchored around the base of the tree adequately control weeds during the first few seasons. Black polythene, sisal, craft paper and roofing felt can be used.

Willow beds

Willow species beds cut annually for the production of osiers occur only in a few localities and are usually found on alluvial soil rich in organic matter. Formerly, such beds were regularly cultivated in the first few years after planting to keep weeds down. Growth can also be promoted through the use of herbicides. Simazine can be used to control weeds where land is kept bare. Grass can be controlled by propyzamide and grass and broad-leaved weeds can be controlled by directed sprays of glyphosate or paraquat. Aminotriazole + simazine mixture applied before the stools sprout in the spring gives the most effective control of perennials such as thistle species, docks and common nettle. Oxadiazon can be used to control hedge bindweed.

Stem injection

Many herbicides do not pass easily through bark or waxy foliage, especially aqueous formulations, but translocate readily if introduced into an incision in a woody stem. Very low doses of glyphosate or triclopyr applied by this method are adequate to kill small trees (up to 15 cm diameter). This technique has been used for thinning to waste trees in areas where thinning by stem removal would disrupt the canopy and increase the risk of premature windthrow.

Stump treatment for re-spacing, regeneration and in cleaning operations

Where the stocking of crop trees is excessive, e.g. after natural regeneration, glyphosate or triclopyr usually in combination with a dye, can be placed on the freshly-cut stump, either by paintbrush, knapsack sprayer or special brushsaw attachment in order to suppress any remaining live branches in conifers or prevent coppicing in broad-leaved species.

Application methods and equipment

The choice of application method and equipment is influenced by:

Size of area to be treated
Terrain (slope and roughness)
Type and vigour of weed growth
Presence or absence of susceptible wanted tree species

Degree to which environmental and aesthetic factors
locally influence practice

Complete application is appropriate where partial control of the vegetation
by hand or spot application will not give the desired results, for instance
bracken, bramble, heather or particularly tall vigorous vegetation on
fertile sites normally requires complete application to prevent rapid re-
invasion. Where small container stock is used without intensive cultivation,
complete application may be appropriate. Complete spraying should be
carried out as a pre-plant operation whenever possible as this has the
advantage of greater flexibility in timing without the constraints of sensitive
trees curtailing treatment or access over the site.

In a few circumstances, aerial application can be the most cost-effective
method but this method is currently only approved for spraying bracken
with asulam.

For either pre- or post-planting application in easy terrain, the tractor-
mounted hydraulic or low-speed rotary atomizer boom sprayer will give
greater output per man-day and improved working conditions for the
operator compared with hand-held application. Where the terrain is unsuitable for tractor access, the hand-held high-speed rotary atomizer,
which can treat a 5 m swathe per pass when spraying by the incremental
drift method, offers the cheapest method for overall weed control. The
hand-held low-speed rotary atomizer and the knapsack sprayer are suitable
for small areas and difficult terrain and will treat a 1–2 m swathe per pass
over low vegetation, but this is a more labour-intensive method. Fitting
very low-volume nozzles to the knapsack sprayer will increase the area
treated per tank fill.

Where a complete application is not considered appropriate, a band or
spot application at least 1 m wide should be treated to control the vegetation. The choice of applicators will depend upon whether band or spot
treatment is required and the sensitivity of the crop tree to the herbicide.
Spot treatment has the advantage over band treatment of reducing the
chemical cost and lessening the impact on the environment.

For band application, where the crop trees will tolerate an overall spray
and terrain permits, the tractor-mounted low-speed rotary atomizer, in
band mode, should be the first choice. On other sites, the hand-held low-
speed rotary atomizer or the knapsack sprayer are the alternatives. In
areas where the crop trees are liable to damage and the vegetation is
low enough, the knapsack with the nozzle guarded is the only suitable
applicator but a pass either side of the row will be required. For spot

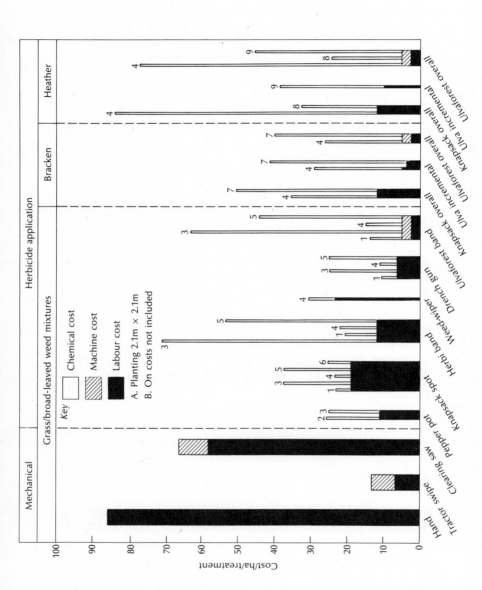

Fig. 17.1 Relative cost of weeding a hectare of lowland plantation by different weeding methods (1986–1987 costs). Herbicides are numbered as follows: 1 = atrazine, 2 = atrazine with dalapon, 3 = propyzamide, 4 = glyphosate, 5 = hexazinone, 6 = paraquat, 7 = asulam, 8 = 2,4-D ester, and 9 = silvapron D. Illustration courtesy of the Forestry Commission.

application where trees are tolerant to herbicide, the knapsack fitted with a cone nozzle or the spot gun or 'herbicator' are suitable applicators. The advantage of the spot gun is the accuracy of the dose and spot diameter together with ergonomic benefits to the operator. When treating species which are very sensitive to herbicides or where guards do not give adequate protection, the rope-wick applicator will give good vegetation control, provided that the flow to the wick is maintained by high standards of cleanliness and crops are treated before weed growth is tall.

Some residual herbicides are available as granules and these are applied by a gravity-feed band applicator or the 'pepperpot' for spot application. Granules can be applied in a moderate wind but tree foliage should be dry at the time of application to avoid the granules sticking to leaves; however, rainfall after application assists in taking the herbicide down to the roots.

Figure 17.1 illustrates the direct cost of labour, machinery and herbicides for a range of vegetation types, applicator and suitable herbicides. Note the high cost of hand and mechanical weedings (mowing) which are unable to relieve the crop trees from root competition for moisture and nutrients.

For the safe and efficient use of herbicides in the forest follow the guidance given in the Code of Practice for the Use of Pesticides in Forestry.

Further reading

Williamson, D.R. & Lane, P.B. (1989) *The Use of Herbicides in the Forest.* Forestry Commission Field Book No. 8.
Williamson, D.R. & Mason, W.L. *Forest Nursery Herbicides.* Forestry Commission Occasional Paper No. 22.

18 / The management of aquatic weeds

P.R.F. BARRETT, K.J. MURPHY* AND P.M. WADE†

*AFRC IACR Aquatic Weeds Research Unit, Sonning Farm. Chervil Lane, Sonning-on-Thames, Reading RG4 0TH; *Department of Botany, The University, Glasgow G12 8QQ; and †Department of Geography, University of Technology, Loughborough, Leics. LE11 3TU*

Introduction

In many UK rivers and drainage channels, it has been necessary for centuries to cut weeds each year. This has been done chiefly because plants obstruct the flow of water causing flooding, impeding drainage, and, through increased silting, channel deterioration. Other locally-important reasons for controlling weeds include navigation, recreation (fishing, boating, swimming), public health and industrial needs. In recent years, agriculture has become more intensive with the resulting need to improve land drainage. There has also been a large increase in the use of water for irrigation, recreation and for potable supply. Urban and road developments have increased the rate of surface run-off which increases the risk of flash floods particularly if drainage channels are clogged with weed. At the same time, the level of plant nutrients in the water has increased through run-off of agricultural fertilizer and through discharges from sewage-treatment works and industry. This increased nutrient loading stimulates more plant growth in waters which are being used more intensively. Weed species are also changing and algae are now recognized as one of the most troublesome weed types. The reason for the increase in algal problems may be partly due to the removal of other weeds but in-

creased nutrient loading is generally considered to be a major factor in
this change. Thus the need to control weed growth has increased while
both the quantity and range of weed species have also increased.

The responsibility for removing excessive weed growth may fall on the
riparian owners, specialized bodies such as Internal Drainage Boards and
British Waterways Board, or on the local water authority. This responsibility depends on the size, use and history of a particular water body
and on the legislation within the different countries of the UK. Riparian
owners and Internal Drainage Boards are usually responsible for maintaining the smaller watercourses and man-made channels but main rivers are
more often maintained by the local water authority. However, riparian
owners may carry out local weed control operations in main rivers if they
require a higher standard of maintenance (for example, for fishing) than is
considered necessary by the water authority. The water authorities also
have responsibilities for fisheries and for maintaining water quality and
they have legal powers to restrict the use of certain weed control techniques both in main rivers and in waters which are not directly under their
control if those waters contain fish or if they discharge ultimately into a
river system.

Techniques for the management of aquatic plants can be classified
generally as mechanical, chemical, biological or environmental. As waterbodies are commonly used for more than one purpose, the management
techniques must be appropriate both to the type of weed problem and to
the uses and functions of the water body.

Types of aquatic weed

A very wide range of aquatic plants can cause problems. For the purposes
of control, it is necessary to differentiate between the various types. One
way of classifying water plants is in terms of where they grow in the water.

Emergent plants

Emergent plants have stems and leaves which protrude above the water
surface. Some of the most important are common reed, common burreed,
bulrush, reed sweet-grass, and common club-rush. All of these have long
narrow leaves and grow to 1–3 m above the water surface. Other common
weeds include the shorter-growing, broad-leaved weeds such as water
plantain, arrowhead and water dock.

It is convenient to include in this group a number of plants that generally grow on banks at or above the water-line. Common weeds in this class include purple loosestrife, great willowherb and reed canarygrass.

Floating plants

Many water plants have leaves that float on the water surface either singly or in rosettes. These plants may be free-floating, as in the case of common duckweed and frog-bit, or they may be anchored in the sediment, like yellow water lily and common water starwort.

Submerged plants

These can be divided into the following:

Submerged plants with leaves: These are often rooted in the sediment, as in the case of spiked water milfoil, Canadian pondweed and fennel pond-weed. Others, such as ivy-leaved duckweed and hornwort, are free-floating below the surface.

Algae: These are simple plants, often growing as filaments, which may form scum on the water surface, slime on rocks and stones, or the char-acteristic entangled mats known as 'blanket weed' or 'cott'. Algae grow and multiply rapidly in favourable conditions and are frequently most troublesome in sluggish or static water.

It should be noted that some aquatic plants can, under certain conditions, produce growth forms which fit into more than one group. For example, arrowhead can produce submerged and floating leaves but is commonly seen as an emergent plant. Algae always start growth in a submerged form but may later form floating mats.

The biology and ecology of aquatic weeds

Seasonal growth and dispersal

Most of the troublesome water weeds are perennials which vary in the way they survive the winter. Although aerial parts are usually destroyed by the first frost, the roots and rhizomes can remain dormant throughout the winter and re-grow when conditions are again favourable. Many of the sub-merged plants, e.g. curled pondweed, produce specialized over-wintering structures which remain in the sediment until the spring. Others, like common water starwort, sometimes over-winter as the whole plant. Filamentous algae can re-grow from spores or from vegetative filaments that have over-wintered on the sediment.

In the spring, water weeds usually lag behind terrestrial plants in starting new growth because water takes longer than soil to reach a temperature that favours plant growth. Once growth has started, it proceeds very rapidly and, within a few weeks, plants such as riverwater crowfoot, for example, can produce stems 6 m long.

The majority of water plants depend for their dispersal on transport of vegetative fragments in the water (and possibly by birds and animals). In comparison with terrestrial weeds, dispersal by seeds is less important.

Factors affecting the distribution of water weeds

The main environmental factors that determine the distribution of water weeds are as follows:

Depth of water

Emergent and rooted floating plants are restricted to shallow water. Light intensity diminishes with depth, so restricting the depth range of submerged plants.

Rate of flow

Some plants are adapted to swiftly flowing streams, e.g. riverwater crowfoot. Conversely, others such as the shallow-rooted Canadian pondweed are not, and are only found where the current is slow or the water static.

Type of sediment

The sediment is a major factor determining the distribution of water weeds. Silts and muds, rich in nutrients, provide good growing conditions for many of the more troublesome species.

Light

Light plays an important role in determining the composition of a plant community and the success of individual species. Emergent species often suppress the growth of submerged weeds and, when the taller plants are eliminated, the submerged species may quickly spread into the cleared area.

Chemical status of the water

The availability of nutrients affects the distribution of algae, floating and submerged plants that absorb all, or some, of their requirements from the water. Other chemical factors which influence the distribution of aquatic plants include pH, hardness of the water and various types of pollution.

Characteristics of problem-causing aquatic plants

While many species of aquatic plants have the potential to cause problems, certain species have particular characteristics which make them more prone to become troublesome. These characteristics include some or all of the following:

Growth rate

Species which have an extremely fast rate of growth and reproduction, e.g. common duckweed.

Physical structure of the plant

Species which produce dense masses of vegetation, e.g. Canadian pond-weed which impedes the flow of water. Others have rigid stems which may impede boat traffic, interfere with access to the water and trap detritus causing the formation of temporary dams, e.g. common reed, which may cause flooding.

Ease of dispersal within the waterbody

Species which can regenerate new plants from fragments broken off a parent plant, e.g. Canadian waterweed.

Mobility

Species which are free-floating, or are easily detached from the sediment, and may move downstream to block culverts or abstraction points, e.g. duckweeds and filamentous algae.

Toxicity and taint

Some plants are toxic to animals, e.g. water dropwort, and others taint
potable waters, e.g. water horsetail. Some blue-green algae produce toxins
which can kill fish and taint the water.

Factors determining the choice of control measures

There are a number of points to be considered when deciding on the best
management strategy to adopt.

Type of control

There is a choice between total control or some form of selective control.
This depends upon the particular management objective. Total control
may be required where land drainage is paramount and the need is for the
permanent or long-term eradication of all weed growth. This form of control is most likely where there is a very high risk of flooding. Selective
control includes a range of options:

1 Control of selected weed species, i.e. the removal of one or more weed
species leaving those which are not troublesome.
2 Control of all species within localized areas.
3 Temporary control when the weed density reaches a nuisance threshold. In practice, this usually means cutting the weed at intervals during
the season.

Uses of the waterbody

Certain weed management approaches may not be compatible with one or
more uses of the water. Flow, water quality and, especially in the case of
herbicides, the rate of dilution below the point of treatment, need to be
considered. Local practical considerations, such as accessibility, the availability of labour and the cost of each operation are other factors to be
taken into account.

The risk of adverse side-effects on the other uses of the waterbody, not
only at the point of treatment but also downstream, must always be given
prior attention. The main points to bear in mind are as follows:

The effect on human safety and domestic water supply: This includes, for
example, toxicity of herbicides to operators and water users, increases in
turbidity owing to mud being stirred up, and also unpleasant odours and
taints resulting from rotting vegetation or herbicide residues in the water.

The safety of farm animals: This could involve the toxicity of herbicides or the possibility of increasing the palatability of poisonous plants by cutting and wilting or through the use of herbicides.

The possible hazard to adjacent crops: This may occur through irrigating with water-containing herbicide residues, spray drift at the time of application and the spread of weeds, such as common reed, by the distribution of rhizome fragments over arable land in the spoil after dredging.

Industrial plant: There is the possibility of damaging or adversely affecting the functioning of industrial plant and so on, for example by blocking water intakes and sluices with cut weed.

Environmental consequences of alternative management options

In general, the more effective the weed clearance, the greater will be the risk of an adverse environmental impact. Particular attention should be given to the following:
1 The danger to fish and other aquatic animals from the toxicity of herbicides and, more importantly, the risk of de-oxygenation of the water by rotting plant material left after cutting or using herbicides.
2 The risk of unintentionally disturbing wildlife or disrupting wildlife habitats, for example by cutting at nesting time.
3 The removal of rare species.
4 The danger of bank erosion following the removal of plant cover.

Management options

Some of the factors which can influence the suitability of a technique and its performance are discussed in the following.

Mechanical control

Cutting, and removing the cut material by raking onto the bank, is a traditional and still commonly-used method of dealing with most forms of water weed. This may be done by hand but, more frequently nowadays, flail mowers, weed buckets or weed boats are used (Fig. 18.1) The weed-cutting bucket cuts and removes the weed in one operation. Most other machines only cut the weed and a second operation is needed to remove and transport it to a suitable dumping site. The extraction of the cut material is necessary to avoid the risk of de-oxygenation as the weed decomposes

Fig. 18.1 A weed cutting boat. The paddle wheels give increased manoeuvrability over conventional propellor driven craft and are less likely to become entangled with the cut weed. The U-shaped cutter has reciprocating knife blades powered by hydraulic motors which also control the height and angle of the blades.

and because rafts of cut weed can drift downstream and block pumps and sluices. This is often a slower, and more expensive operation than the initial weed cutting.

Rapid re-growth of the cut weeds makes frequent cutting necessary throughout the summer to maintain clear water. In practice, the cost of the operation often limits the number of cuts and alternative methods which produce long-lasting effects are being sought.

The mechanization of cutting and raking is possible in some situations but access problems can arise. Many watercourses are flanked by crops or trees, which restrict access along the banks, or are too small and shallow to support boats. Nevertheless, mechanical cutting has an important part to play and a range of suitable plant, including wheeled machines which straddle ditches and boat-mounted submersible cutters, has been developed.

Filamentous algae ('blanket weed' or 'cott') cause particular problems because cutting the tangled fibrous mat of filaments is almost impossible. They are usually removed by rake or dragline but the operation is slow and inefficient as many filaments remain from which rapid re-growth occurs.

Although cutting is not entirely satisfactory as a means of aquatic weed control, it is an established practice to which the ecosystem eventually adapts. Thus, where cutting is used as a regular form of management, the communities of plants and animals are relatively stable and few long-term changes are likely to occur after cutting.

Dredging is seldom employed solely for weed-control purposes but it can have useful effects, particularly on emergent and some floating-leaved weeds which have rhizomes buried in the sediment. These are removed with the sediment and re-colonization is usually fairly slow.

Biological control

There are several forms of biological control used in the British Isles:

Domestic livestock

These can be used to graze banks and so control some emergent vegetation. However, they can cause damage to banks and fencing is required to prevent them straying into adjacent crops. Cattle have also been used in shallow streams where grazing and trampling helps to control submerged weed.

Waterfowl

Waterfowl graze on some species of floating and submerged weed. When present in large numbers, they can have a significant effect on the weed and domestic flocks can be used to manage weed in ponds and lakes. However, it is rare for wild birds to collect in sufficient numbers during the summer months to have much effect on weed growth.

Chinese grass carp

An herbivorous fish, the Chinese grass carp (*Ctenopharyngodon idella*), is now available for the control of species of floating and submerged weed, including some filamentous algae. The prospective user should first approach the local water authority for permission to introduce the fish and, in many cases he will also require a licence under the Wildlife and Countryside Act 1981. Applications for these licences can be made to the following addresses:

England:	Ministry of Agriculture, Fisheries and Food, Great Westminster House, Horseferry Road, London, SW1P 2AE.
Wales:	Welsh Office Agriculture Department, 2nd Floor Crown Offices, Cathays Park, Cardiff, CF1 3NG.
Scotland:	Department of Agriculture and Fisheries for Scotland, Fisheries Division, Chesser House, Gorgie Road, Edinburgh, EH11 3AW.
Northern Ireland:	Department of Agriculture for Northern Ireland, Fisheries Division, Hut 5, Castle Grounds, Stormont, Belfast BT4 3TA.

The grass carp are used only in enclosed waters and are stocked at up to 200 kg fish per hectare of water surface. Small fish are subject to predation by pike (*Esox lucius*). Waters with a high population of these predatory fish should only be stocked with larger grass carp. Grass carp can grow rapidly so that their food demand steadily increases. Hence, culling or re-stocking is necessary periodically to maintain a satisfactory stocking level. There is no evidence that these fish will breed naturally in most temperate countries but they may survive for 15 years or more in favourable conditions.

Chemical control

Herbicides can be used effectively to control a wide range of aquatic plants. They may be applied to emergent or surface-floating weeds by means of a foliage spray, in much the same way as that recommended for land plants. Submerged weeds and algae are treated by adding the chemical to the water, a treatment which may also control some floating and emergent species. The chemicals used to control weeds in or near water are listed in Table 18.1.

As with all other methods of aquatic weed control, thought must be given to the effect the herbicide treatment will have on other functions of the waterbody. This not only involves direct toxicity of the chemical to man, his animals and to wildlife but also the indirect effects caused by the death of the weeds. It is usually desirable to retain some weed growth on the banks and in the water in order to maintain the ecosystem in a stable condition and to provide a habitat for fish and other wildlife. When non-selective herbicides are used, they can remove all weed from the entire waterbody. However, some of these herbicides are now formulated so that they can be used for localized control, to remove patches of weed or

Table 18.1 Chemicals used in or near water

Chemical	Irrigation interval	Formulation*	Target plants
Asulam	Nil	L	Docks and bracken on banks
2,4-D amine	3 weeks	L	Emergent broad-leaved weeds
Dalapon	5 weeks	WSP	Reeds and similar emergent weeds
Dichlobenil	2 weeks	G	Some floating and submerged weeds
Diquat	10 days	L	Some floating and submerged weeds and algae
Diquat alginate	10 days	VL	As diquat, but effective in flowing water
Fosamine-ammonium	Nil	L	Deciduous trees and shrubs on banks
Glyphosate	Nil	L	Water lilies, reeds and emergent weeds
Maleic hydrazide	3 weeks	L	Suppression of grasses on banks
Terbutryn	7 days	G	Some floating and submerged weeds and algae

* L = Liquid, applied as diluted spray. WSP = Water soluble powder, dissolved and applied as spray. VL = Viscous liquid, applied undiluted through special applicator. G = Granule, applied by hand or granule applicator

particularly troublesome plants. This technique also reduces the risk of de-oxygenation by decaying weed; the disruption of habitat; and the cost of weed control. It may also be more acceptable in environmental terms.

Spraying the foliage of aquatic weeds

Recommendations for the correct dose are made, as in the case of normal agricultural applications, in the form of weight of active ingredient of herbicide per unit area. The chemical is most frequently applied by hand-operated sprayers, although boat and tractor-mounted sprayers are sometimes used.

The use of high-volume and low-pressure application equipment delivering a coarse spray with a minimum of small droplets is advisable to reduce the risk of spray drift as the target weeds are often in ditches which are narrow and frequently have closely-adjacent susceptible vegetation including crops.

Localized control and, to some extent, selective control can be achieved by choice of appropriate herbicides and by directing the spray only onto those plants which need to be removed. Water velocity and quality do not affect the treatment as these chemicals are taken up through the exposed foliage above the water surface. However, to avoid any chance of a build-up of herbicide in moving water, it is advisable to proceed upstream while applying the herbicide.

Applying herbicides to water

Recommendations usually refer to the theoretical concentration of active ingredient that would be achieved when the chemical has been evenly dispersed throughout the waterbody but before any adsorption or degradation has occurred. This is usually expressed as the number of parts of chemical per million parts of water (parts per million; ppm). This is the same as milligrams per litre ($mg\,l^{-1}$) or grams per cubic metre ($g\,m^{-3}$). Some special formulations may be applied on a surface-area basis because they sink onto the weeds or the mud before releasing the herbicide and the depth of water is of little importance. In these instances, recommended doses are given in weight per unit area (e.g. $g\,m^{-2}$).

The methods of application depend on the formulation which, in turn, depends on the properties of the herbicide. Those that disperse rapidly in water may be applied in concentrated form at widely-spaced intervals (provided that there is no toxicity risk to the operator) while others must be diluted and sprayed evenly over the water surface to ensure satisfactory distribution. Several granular formulations are available and these should be spread as evenly as possible over the water surface. This operation can be done by hand using a suitable container from which the granules can be ejected but a number of hand-operated and motorized granule spreaders are available and these give a more even distribution.

The length of time that the plants are exposed to the herbicide affects the efficiency of the treatment. In flowing water, special viscous formulations which stick to the plants and give controlled release of the herbicide are needed to achieve an adequate contact time.

The effectiveness of a herbicide can be influenced by water quality. Hardness, pH and dissolved salts can affect performance as can mud,

Fig. 18.2 Localized control with dichlobenil. A cleared strip, 4 m wide and over 100 m long, through a dense bed of mare's tail produced by an application of a slow-release formulation of dichlobenil.

clay and peat in the substrate or suspended in the water. In some situations, recommended doses may be increased under adverse conditions to achieve the desired results.

Time of treatment

Emergent and floating perennial weeds are most susceptible when the herbicide is sprayed in mid-to-late summer, when the leaves are fully emerged. Annual weeds can be sprayed before flowering to prevent production of seeds.

Treatment of submerged weeds and algae is normally recommended in spring and early summer when the weeds are young and actively growing. However, this is the period when many fish and other aquatic fauna are breeding on the weed. Later in the season, when large masses of weed are present in the water, and particularly when the water is warm, there is a danger of causing de-oxygenation by decomposition of the dead weed. Where there is deemed to be an unacceptable risk to aquatic fauna, only localized control techniques should be used (Fig. 18.2).

Safeguarding other interests

Under the Control of Pesticides Regulations 1986, specific approval is given to herbicides for use in aquatic situations and only those herbicides with that approval can be used in or near water. This specific approval is based on toxicological data on the herbicide and on an assessment of the toxicity risk to humans (operators and consumers), farm animals, fish and wildlife.

There are other risks associated with the use of herbicides which should be considered before any application is made. There is a possibility of damage to crops if treated water is used for irrigation. The maximum length of time that the herbicide is likely to remain in water at a concentration phytotoxic to terrestrial crops is printed on the product label and is known as the 'irrigation interval' (Table 18.1). Information must always be sought about any intended irrigation at, or downstream of, the treatment site during the irrigation interval so that satisfactory safety measures can be taken.

A further hazard that must be considered is the possible de-oxygenation of the water caused by the decomposition of large quantities of plant material. This risk can be minimized by applying the herbicide either in the early part of the season when the growth has not fully developed, or to a limited area of the water at any one time. The latter technique is not suitable for all herbicides and can only be used with specially formulated herbicides which carry an appropriate recommendation on the product label.

An indirect hazard to fish arises through possible effects on fauna which provide fish food. These effects may arise either by toxicity to the invertebrate animals themselves, or through the loss of habitat. The loss of habitat can reduce the food available to the invertebrates and increase their exposure to grazing pressure by the fish. However, these effects may be temporary and the available scientific evidence shows no lasting adverse effect following the correct use of aquatic herbicides. The Ministry of Agriculture, Fisheries and Food (MAFF) has issued *Guidelines for the Use of Herbicides on Weeds in or Near Watercourses and Lakes* (MAFF Booklet 2078). This booklet draws attention to the main dangers and the ways of avoiding them.

Environmental control

In some situations it is possible to reduce weed growth by altering one or more factors in the local environment. Factors that offer most scope for manipulation are light intensity, the fertility of the mud and water and the level of the water. Burning and other miscellaneous techniques can be employed in certain situations.

Reduction of light available for plant growth

Attempts to control aquatic plant growth by reducing light intensity have been successful in some instances. There are currently two main approaches.

One approach is to limit the quantity of light that reaches the water surface. Most usually this is achieved by planting trees on the bank. The trees shade the water and are particularly useful on narrow streams and ditches. Care should be taken when planting to ensure that the trees do not impede water flow or hinder maintenance of the channel.

The other method is to reduce light penetration into the water. This can be achieved by the addition of light-absorbing dyes to the water or by deliberately stirring up sediments, for example by stocking the water with bottom-feeding fish or increasing power-boat traffic. Similar effects have been achieved by adding nutrient to the water to induce unicellular algal blooms which shade out the rooted plants. However, this deliberate nutrient enrichment may lead to problems later when the nutrient is recycled into higher plants and filamentous algae. This type of control is more effective in static waters such as canals and ponds where the suspended materials or dye remains *in situ* for long periods. Other forms of shading include plastic or fibre sheeting which can be suspended above the water, floated on the water surface or submerged on the bottom. Such techniques are not used widely but could be developed to provide a useful management technique in particular situations.

Control of nutrient availability

Reduction of the quantity and quality of nutrients entering a waterbody can limit weed growth. A number of possibilities exist, including the following: (1) diversion of farm, sewage or industrial effluents away from the waterbody through bypass channels; (2) reduction of the nutrient concentration in discharges from sewage-treatment works by tertiary treatment; and (3) either by establishing and enhancing food chains in which nutrients are diverted into plants or animals which can be harvested from the system, or by adding chemicals which precipitate essential nutrients in an insoluble form.

Particular attention should be paid to the sediments which can accumulate nutrients and release them slowly over a long time. These sediments can be removed by dredging and, in some instances, by making the channel self-scouring. The nutrients, particularly phosphates, are released more readily under anaerobic conditions and their release can be reduced by inducing a current of oxygenated water over the surface of the sediment. In static waters, this can be achieved by releasing a stream of air bubbles near the bottom which lifts and oxygenates anaerobic water from the lower levels and draws in a current of oxygenated water from the surface.

Alterations in water level

Exposure of part, or all, of the bed of a waterbody by lowering the water level has been used successfully to manage aquatic vegetation. Control is achieved either by dehydration of the vegetation or by exposure to low temperatures. The process is sometimes termed 'drawdown' or 'de-watering'. It can also alter the character of the sediment which may reduce weed growth. However, in deeper water bodies, drawdown can allow weeds to establish in depths below their normal limit. Once established, they can grow up towards the surface, as the level is again raised, to remain in the higher light intensity. In these situations, drawdown can spread a weed problem into areas which would normally remain weed-free.

Burning

Emergent weeds, particularly the stiffer stemmed reeds, do not always collapse in the autumn and can form large masses of dead, standing material. In dry ditches and on banks, this material can sometimes be burned. This can be a useful way of reducing the bulk of plant material which might otherwise collapse during the winter and block the channel. Burning can also be used to destroy cut material after drying. This is a useful method of disposing of poisonous plants, some of which remain toxic after death but become palatable to livestock.

Miscellaneous techniques

Many other techniques for controlling aquatic vegetation have been tried, which do not fit closely into any of the previous categories. These include the following: (1) the use of laser radiation to control emergent or floating weeds and ultrasound vibrations which can disrupt cells of submerged plants; (2) floating oil films which cause floating weeds to sink; and (3)

Table 18.2 The relative efficacy of management options

| | Weed type | | | | | |
| Management option | Emergent | | Floating-leaved | | Submerged | |
	Narrow-leaved	Broad-leaved	Rooted	Free-floating	Leaved	Algae
Mechanical:						
Hand cutting	+	+	+	−	+	−
Flail mower	+	+	−	−	−	−
Weed bucket	+	+	+	+	+	+
Weed boat	+	+	+	+	+	−
Dredger	*	*	*	−	*	+
Biological:						
Livestock	+	+	−	+	+	−
Waterfowl	−	−	−	*	+	+
Grass carp	−	−	+	*	*	*
Chemical:						
Sprayed directly onto foliage	*	*	*	+	−	−
Applied to water	−	−	*	*	*	*
Environmental:						
Shade	+	+	+	+	+	+
Nutrient removal or precipitation	−	−	−	+	+	+

* Effective control lasting at least one season
+ Moderate benefit only, or control lasting less than one season
− No useful effect
 The benefits assigned to each technique are for general guidance only and the results obtained may be modified by local conditions.

increasing wave action which submerges floating weeds and increase turbidity thus suppressing submerged vegetation. There are also several biological control agents, including crayfish, other invertebrates and wildfowl, which have been effective against some weeds. Most of these have been tested in tropical or sub-tropical conditions and, at present, the grass carp is the most effective biological-control agent in the more temperate regions.

Table 18.2 summarizes the range of management techniques and gives rough indications of their uses although specific characteristics may be modified by local conditions.

Legislation

Several Acts of Parliament are concerned with the management of water, the use of herbicides and the movement and introduction of fish. Some of those relating to chemicals are outlined in the *Guidelines for the Use of Herbicides on Weeds in or Near Watercourses and Lakes*. The use of herbicides for aquatic weed control is also covered under the Control of Pesticides Regulations 1986 which govern the sale, storage and use of pesticides, and are enforced under the Food and Environment Protection Act 1985 (Chapter 21). Information about others, such as the Wildlife and Countryside Act 1981, can be obtained from local water authorities, many of whom have produced their own guidelines covering weed control within their areas, and from HMSO (Publications) London.

The Nature Conservancy Council's booklet *Wildlife, the Law and You* is a useful summary of the legislation dealing with rare plants and animals and Sites of Special Scientific Interest. Valuable information and advice on the conservation of aquatic wildlife is provided in *Rivers and Wildlife Handbook* available from the Royal Society for the Protection of Birds.

Further reading

Belcher, H. & Swale, E. (1976) *A Beginner's Guide to Freshwater Algae.* Culture Centre for Algae and Protozoa, Institute of Terrestrial Ecology.

Engelhardt, W. (1973) *Pond life* (Young Specialists Series). Burke, London.

Haslam, S.M., Sinker, C.A. & Wolseley, P.A. (1975) *British Water Plants.* Field Studies Council, Attingham, Shrewsbury.

MAFF (1985) *Guidelines for the Use of Herbicides on Weeds in or Near Watercourses and Lakes.* MAFF Booklet No. 2078. MAFF (Publications), Lion House, Willowburn Estate, Alnwick, Northumberland NE66 2PF.

Nature Conservancy Council (1982) *Wildlife, the Law and You.* Nature Conservancy Council, Attingham Park, Shrewsbury, S74 4TW.

Royal Society for the Protection of Birds (1984). *Rivers and Wildlife Handbook.* Royal Society for the Protection of Birds, Sandy, Bedfordshire.

Spencer-Jones, D. & Wade, M. (1986) *Aquatic Plants; A Guide to Recognition.* ICI Professional Products, Farnham, Surrey.

19 / Industrial weed control

G.G. FISHER, R. GARNETT* AND M. DE'ATH†

Land Capability Consultants Ltd, Times House, Willingham, Cambridge CB4 5LH;
*Monsanto plc, Thames Tower, Burleys Way, Leicester; and †Chipman Ltd, Horsham,
West Sussex RH12 2NR

Introduction

The growth of weeds on pavements, storage areas, railways, petroleum refineries and similar areas presents serious problems to local authorities, government departments, utilities, industry and to other land owners. Unwanted vegetation in such areas can, either directly or indirectly, create serious hazards and cause damage and inefficiency. Although the benefits of weed control in this situation are less easy to assess than those of increased yield or ease of harvesting in crops, they are nonetheless tangible and important. Vegetation can constitute a dangerous fire hazard, it can obscure inspection, impede drainage, encourage metal corrosion and wood decay, or lead to damage of mechanical and electrical equipment vital to distribution, transmission and communication in industry, public utilities and railways. Weed growth may damage metalled and paved surfaces, reduce visibility and provide a source of injurious weeds in neighbouring crops. Apart from removing such hazards to safety and efficiency, weeds may also be controlled to improve appearance in numerous situations.

There are also industrial situations where vegetation can have a useful function in preventing erosion by providing a cover for bare soil, and improving appearance. These include embankments and bund walls liable to erosion, and perimeter areas, not needed for buildings, storage or traffic, and without any particular ornamental value. Rough grassland is the usual choice of cover, and its management by mechanical and chemical means is referred to in Chapter 16.

Environmental factors and characteristic vegetation

Vegetation soon invades bare ground and follows a succession, which in the UK usually develops from mosses and lichens to broad-leaved herbaceous species and grasses, which in turn are replaced by shrubs and eventually by trees. Many variations occur but characteristically on newly-disturbed ground the initial colonization by weeds like groundsel and annual meadow-grass is soon followed by species such as creeping thistle, rosebay willowherb, common ragwort and broad-leaved dock. The later stages of succession can be seen on railway and road embankments which in the absence of cutting or grazing soon become overgrown with shrubs and trees. Weed control or management of the vegetation may be attempted at any stage in a succession but commonly the first attempt at control is at the stage dominated by perennial broad-leaved weeds. Alternatively, complete weed control may be required on land once grazed or cut infrequently and now dominated by perennial grasses such as false oatgrass, cocksfoot and bent. When the initial vegetation cover has been killed, further treatment may be used, at a very early stage in the succession, to maintain the area so that it is not colonized by annual weeds.

The particular plants which develop on a piece of ground depend on the seeds or propagules present, on the species in adjacent areas and on the interaction of climate with factors such as soil structure, pH and nutrient status, which in turn are often very considerably modified by man. Reconstruction of the landscape to create railways, roads, airports and similar features may leave structureless and infertile soils at the surface with the subsequent development of a specialized flora adapted to these abnormal conditions. It is not unusual to have a wide variety of soils and surfaces within a single site, leading to similar diversity in the weed species present.

Infertile soils are frequently colonized by clovers, whilst rat-tail plantain and knotgrass are common on compacted soils. Soils modified by debris and rubble give a variable growing medium in which characteristic species such as coltsfoot will thrive. Deliberate modification of the surface growing media occurs on stone-ballasted railway tracks and gravel-covered areas creating a micro-climate in which ephemeral species including common mouse-ear chickweed, sticky groundsel and annual meadow-grass can flourish. Deep-rooted weeds such as common horsetail also become established in these situations and are difficult to control even with herbicides. Between paving stones, species like field bindweed can support the above-ground parts by root systems which proliferate under the protection of the stone surface, increasing the difficulty of control. Rootstocks of weeds like creeping thistle and dandelion buried beneath asphalt are capable of regenerating and bursting through the paved surface.

Regular control with herbicides giving less than total eradication of both above- and below-ground parts will soon select a residual weed flora consisting of a limited number of hard-to-kill species. Selection may be owing to tolerance of particular species to certain herbicides (e.g. plantains towards substituted-urea herbicides and hogweed towards certain triazine herbicides) or owing to the underground parts having exceptional depth (as in field horsetail), or vigour (as in broad-leaved dock). The continuous use of certain herbicides has led to the development of genetic resistance in certain species, such as that shown by groundsel towards some triazine herbicides, and this trend may be expected to continue (see Chapter 8). Soils may accumulate in urban streets, drainage channels and wall footings and be colonized by weeds such as dandelion, wall barley and plantains. On roads where weed control is required under guard rails or around bridges subjected to continuous buffeting from the traffic slipstream, there will be a preponderance of xerophytic species. Partial contamination from salt used against ice and snow in winter has a further selective effect on roadside vegetation and broad-leaved weeds are confined to a limited number of species of which dandelion, groundsel and mugwort are characteristic, whilst desirable grasses such as red fescue may be encouraged.

Methods of weed control

There is now a wide range of options available for removing, controlling or managing vegetation, using mechanical methods and herbicides. Mechanical methods are commonly employed for total land clearance, such as the preparation of forest fire-breaks or new tracks. Trees and scrub may be cut down, and the stumps removed using a bulldozer or similar heavy equipment, while flail mowers are used to cut back hedges and scrub growing near roads and railways. Land may also be cleared by controlled burning. These methods are labour intensive and allow regeneration of vegetation, which will subsequently need further control, often when access has become restricted and soil cultivation is no longer possible. The main advantage of herbicides over mechanical methods is the ability to prevent new weed growth for extended periods. It is often less costly to use herbicides than hand labour to control weeds, and they can be used where it is difficult to manoeuvre cutting machines, for example around street furniture, under pipelines and cables, or on uneven surfaces.

Types of herbicides

Persistent residual compounds

The persistent root-absorbed residual herbicides, such as atrazine, sima-zine, bromacil and diuron, are the basis of industrial weed control. Herbi-cidally-active residues of these materials may remain in the soil or treated surface for many months, and some such as tebuthiuron, have long residual activity at relatively low doses.

Uptake by plants is primarily through their roots via soil water so their activity is dependent on rainfall. In a well-drained substrate, physical loss by leaching may occur, particularly under high rainfall conditions, and lateral movement may be a problem where the herbicide is applied near susceptible desired species. The availability of the herbicide may also be reduced owing to adsorption on clay, organic matter or cinders, where higher doses will be required than on light, non-organic soils or ground surfaces. High temperatures will increase the rate of breakdown and volatilization.

Some residual herbicides used for total weed control are less dependent on soil moisture. They may be absorbed in the vapour phase by seeds or growing points, as with dichlobenil, or absorbed through weed foliage as well as through the soil as with certain triazines, sodium chlorate and picloram. The persistence of these compounds varies considerably: dichlo-benil and sodium chlorate have moderate persistence in the soil while picloram is more resistant to microbial breakdown, so that residues may remain for long periods.

Foliar-acting compounds

Herbicides absorbed through the leaves are used to kill existing vegetation leaving little or no residual activity in the substrate. Some are largely contact in action, but most are translocated within the plant and many commercial products combine foliar acting materials with residual herbicides.

Of the translocated materials, amitrole, the isopropylamine salt of glyphosate and imazapyr have a broad spectrum of activity and are effec-tive against both annual and perennial broad-leaved weeds and grasses. Glyphosate has no activity in most soils, where it is adsorbed rapidly and degraded by microbial action. Amitrole normally remains active in the soil for 1–2 months and imazapyr for up to a year. Other translocated materials are more specific in their activity: 2,4-D, dicamba, MCPA, picloram and triclopyr selectively kill dictoyledenous species, while dalapon will selectively kill grass weeds. All these translocated chemicals will give the best results if they are applied in good growing conditions.

Woody weeds may be controlled by certain translocated herbicides, including imazapyr, fosamine-ammonium, glyphosate, picloram, triclopyr and ammonium sulphamate. Of these, fosamine-ammonium is particularly slow acting and is normally applied in late summer, before leaf fall, so that symptoms are not observed until the buds fail to develop in the following spring. Imazapyr and glyphosate will control grasses as well as woody weeds, but the other herbicides will kill broad-leaved scrub while leaving a grass ground cover. Selective applications can also be made by stem treatments. Glyphosate and triclopyr, for example, can be injected into a cut made in a growing stem or tree trunk.

The contact herbicides, paraquat and diquat, give a rapid kill of foliage, but do not kill established perennial weeds and some annual weeds may re-grow. Both these herbicides are inactivated on contact with the soil and will not prevent re-colonization from seed in the ground or subsequent invasion by plants from outside the area.

Use of herbicides

When using herbicides in industrial situations, the initial requirement will generally be to kill established vegetation or to manage selectively existing vegetation. If bare ground is desired, the subsequent objective is to maintain the ground free of growth usually for an indefinite period. In the UK this will normally require an initial high-cost treatment followed by annual treatments at lower cost. Care should be taken when planning such a programme that a rotation of herbicides is used to avoid the spread of tolerant and resistant weeds.

Initial treatments

Weeds on uncropped land are often more vigorous and diverse than on arable land where cultivation and crop competition limit weed growth and the variety of species. Treatments used for total weed control should, therefore, be capable of controlling a wide range of grass and broad-leaved weeds and should be persistent in the soil to avoid re-infestation. The residual herbicides are the basic tools for total weed control, but foliar-acting compounds can be mixed with them to give faster kill of existing vegetation and to control species tolerant of the residual herbicide. Since labour costs on industrial sites are often higher than the cost of herbicides, it is not generally economic to make more than one application during the year, although spot treatments may be carried out to kill patches of resistant weeds. Initial treatments are typically in dense vegetation and should aim to kill most or all perennials, since temporary suppression or partial control of some species will reduce the success of maintenance treatments.

Important factors to consider when planning a weed-control programme include the type or combination of herbicides to be applied, the dose and the timing of application. Application techniques (see below) can sometimes be adapted to give higher doses on patches of hard-to-kill perennials. A mixture of herbicides may be more economical, giving control of a wider range of species and avoiding the establishment of a distinctive flora which often follows the use of a single active ingredient. The timing of treatments is important in relation to soil moisture and the growth stage of the plants. Residual herbicides which are dependent on soil moisture for root uptake are most effective when applied in the early spring, whilst those which are absorbed through foliage or growing points tend to give the best results under moist growing conditions. Contact herbicides require a good leaf surface area, whilst translocated materials also require active growth, and most are applied in late spring or summer. Mixtures of residual and foliar-acting herbicides are also usually applied in the spring or summer.

Dead vegetation may greatly increase the fire risk, and in situations where this is important it is better to remove any vegetation before treating the ground with a residual herbicide.

Maintenance treatments

Maintenance treatments are applied to prevent re-infestation in subsequent years and are typically in light vegetation. If the initial treatment has been completely successful, low doses of residual herbicide applied in spring will usually prevent seedling establishment during the season. A foliar-acting herbicide may be added to treat plants which have survived. If initial treatments were inadequate, however, a separate application of a translocated herbicide may be necessary later in the season to control patches of deep rooted weeds. A typical programme for weed control is shown in Table 19.1.

Table 19.1 Typical programme for weed control

	Year 1	Year 2	
		Spring	Summer
Aim:	Clear existing vegetation	Maintain bare ground	Control surviving perennials
Method:	1 Kill foliage 2 Kill roots particularly of perennials 3 Prevent re-infestation	1 Add to residual in the soil 2 Prevent build-up of tolerant weeds 3 Use foliar if necessary	1 Foliar herbicide
Type of treatment:	Overall	Overall	Spot
Type of herbicide:	High dose of residual plus foliar	Lower dose of a different residual (plus foliar)	High dose of translocated herbicide

Temporary treatments

Where ground is required later for growing crops, ornamentals or grass, the contact or translocated herbicides paraquat, glyphosate, dalapon, amitrole and 2,4-D or related compounds may be used to control weed growth. These compounds are all readily broken down or inactivated and do not have long-lasting residual effects in the soil.

Special problems

Railway tracks

Clean stone ballast does not support heavy weed growth and allows easy penetration of herbicides. Consequently, low doses of root-absorbed residual herbicides applied in spring will usually result in satisfactory weed control for a season. When stone becomes mixed with soil and rubbish as occurs on sidings and in cesses, weeds are more prevalent and difficult to control, so that higher doses are required. Standard mixtures include imazapyr + atrazine or amitrole + atrazine + diuron.

Access to railway track for sprayers mounted on road vehicles is usually impractical. For running lines, complete trains are employed consisting of spraying unit, living coach and water tankers. They are capable of applying different doses across the tracks and in the cess with facilities for rapid variation in dose and chemical with changes in weed flora.

Because of restricted access, spraying in yards and sidings is often done with small mobile units or knapsack sprayers. Since water supply is often a limiting factor granular herbicides are also used.

Management of vegetation on embankments is dealt with in Chapter 16.

Pre-surfacing

Where asphalt is put down on a relatively shallow foundation, subsequent weed growth, particularly from deep-rooted perennials, may break up the surface and herbicides, such as atrazine, simazine and dichlobenil are commonly applied to the foundation before asphalting.

Woody species

Woody weeds including brambles are generally resistant to residual herbicides and treatment with herbicides such as picloram, glyphosate, ammonium sulphamate, fosamine-ammonium or triclopyr is necessary. Reference can be made to Chapter 17 for fuller information on the control of woody species.

Footpaths and pavements

Slab pavements, asphalt paths and kerb sides can support a variety of weed species. Treatments include a translocated herbicide such as amitrole, glyphosate or imazapyr plus residual herbicides such as atrazine, diuron or simazine. Allowances should be made by way of dosage rate and application techniques for the difficulty in getting residual herbicides into contact with the roots of weeds in these situations.

Growth inhibitors

Retardation of growth may be preferable to total weed control where some vegetation is needed to prevent erosion of banks or for the sake of appearance. Grass growth regulators like maleic hydrazide, mefluidide or paclobutrazol are used for this purpose. More information on growth retarding, especially as an alternative to grass cutting, is given in Chapter 16.

Application

Total herbicides are available in a range of formulations, each of which has a particular cost or use benefit. Spray formulations available include wettable and soluble powders, suspension concentrates, aqueous concentrates and emulsions. Herbicides may also be formulated for dry application as granules, pellets and powders, in which case they depend mainly on root uptake. The subject is considered generally in Chapter 4.

Sprays

Non-selective herbicides are usually applied, in solution or suspension in water, through power-operated sprayers or knapsack sprayers at medium to high volume. On very small areas a watering can with fine rose may be suitable using a minimum water volume of 0.5 litre m^{-2}. When wettable powders are used, the spraying machine should be equipped with an efficient means of agitation to prevent the chemical from settling out in the spray tank. To reduce spray drift most machines utilize low-pressure nozzles or floodjets which provide a wide angle swath at low pressure. Non-selective herbicides can be applied by controlled droplet application (CDA) equipment delivering $5-20$ litre ha^{-1}. A number of formulations have been specifically developed for this type of application where they are applied undiluted. Additionally, specially-designed equipment is available for particular uses, from footpath sprayers, road-sweeper attachments and weed-wipers to spray trains, incorporating sophisticated machinery for accurate differential spraying at high speeds. Spray-thickening adjuvants, marker dyes and shields may also be used to advantage in reducing wind drift and volatility risks and confining treatment to target plants. In treating large areas there may be patches of deep-rooted perennials which require two or three times the dose applied to the rest of the site and this can be achieved by reducing the speed of travel of the spray over the ground, thus increasing the deposit of chemical. See also Chapter 5.

Dry applications

When using dry formulations, the amount of active ingredient applied per unit area is generally the same, or slightly higher than that applied in spray form. The equipment used for the application of granular herbicides may be relatively simple, but does vary widely in design and operation. Granules can be applied as localized treatments by hand-held 'pepper-pot' type applicators or as overall treatments by chest-mounted spreaders or vehicle-mounted applicators. Dry application is particularly useful where access for spraying equipment is difficult, where water supplies are lacking, on small dispersed sites or where spot treatment only is required.

Precautions

In some situations such as electricity transformer stations, prevention of corrosive effects on metals is important. The possible effects of application on non-target areas and subsequent movement of herbicides by leaching, wash-off, spray drift or volatility should always be considered when treating areas in close proximity to desirable vegetation. On sloping ground, drainage and surface water can carry residual herbicides on, or through, the soil for some distance. The roots of desirable trees and shrubs may be present under areas to be treated or may subsequently grow into treated areas resulting in the absorption of toxic quantities of herbicide. Residual herbicides may be used at relatively high doses for non-selective weed control and, as the name implies, residues can remain in the soil for a number of years. The doses of these chemicals recommended for initial treatments should not be used on land which may be required for growing crops, grass or ornamental plants within 3–4 years. Maintenance doses are likely to affect the germination and growth of desirable plants for up to 2 years.

Further information on the safeguards to the user, public and wildlife in relation to the use of herbicides is given in Chapter 21.

20 / Ecological consequences of modern weed control systems

A.J. WILLIS

Department of Animal and Plant Sciences, The University, Sheffield S10 2TN

Introduction

Many examples are now well known in which man's activities have led to substantial ecological change by disturbing what is often a fine but fairly stable balance in populations of plants and animals. The introduction of prickly pear in Australia, of water hyacinth in Egypt and of the rabbit in the UK from south-west Europe by the Normans, are examples showing the major consequences of disturbing the ecological equilibrium. A less obvious example was that the use of organochlorine pesticides as seed dressings led to the thinning of eggshells and so to a substantial reduction of birds of prey, which in its turn has had other effects.

The use of herbicides, especially in semi-natural vegetation, is clearly likely to bring about changes in both the short and the long term. Some species of plants may have, or develop, tolerance to herbicides and spread at the expense of the susceptible species which become rare or even extinct, with considerable effects on the animals which normally feed on them. Besides these major effects, there are many other interactions with organisms of all kinds, including soil micro-organisms. In using herbicides, therefore, it is desirable to consider the consequences in a broad ecological context.

Weeds in the ecosystem

The ecosystem, made up of the integrated community of all the organisms present and their controlling environment, is, when developed over a considerable period, usually in approximate balance. The interactions of the physical, chemical and biological components lead to the establishment of a dynamic inter-relationship, and a degree of stability.

The interplay between the components of the ecosystem is very complex, because many of the organisms are inter-dependent. Food chains are an obvious link between plants and animals. Also the type and productivity of the vegetation have a marked effect on the animals present. Many butterflies, for example, feed on one or only a few species of plants which means that the existence of the butterflies is entirely dependent on the particular plants. In turn, insects may have a substantial effect on plants, such as the cinnabar moth on ragwort and insect defoliators on oak.

Weeds are especially abundant in habitats disturbed by man, where bare soil can be exploited. With modern control practices, weeds are now much less frequent than formerly on arable land but are still plentiful in many artificial habitats. One of the main characteristics of weeds is their prolific production of seeds and very efficient mechanisms of dispersal, in some species over considerable distances (see Chapter 1). Much resource is partitioned into seed production. Many species of disturbed habitats flower and produce seed over a long period and seed is usually dispersed as soon as it is ripe. On the other hand, many weed species of arable land have fairly short flowering and seeding periods and their seeds often ripen and disperse with those of the crop. A further feature of many weeds is that the seeds may be very long-lived in the soil seedbank. They often show dormancy and so are capable of giving new crops of plants over a number of years, biologically advantageous should seeding fail in some years. In grassland known to have been arable at some time in the past, ploughing may lead to the growth of a substantial number of weeds from seeds long buried; seeds of even the most rapidly declining species are still present after 20 years and there is evidence of seeds surviving burial for 100 years or even longer.

These annual weed species are known as 'r-species' (or r-strategists; for r- and K-selection, see MacArthur and Wilson 1967), continually colonizing ephemeral habitats, a number being themselves ephemerals, and some, such as annual meadow-grass, having several generations per year. These r-strategists are essentially opportunistic, typified by small size, rapid development, high fecundity and a high rate of dispersal. In contrast are the K-strategists, large and long-lived, of low fecundity and a low rate of dispersal. Such species maintain fairly constant populations in contrast to the very variable populations of annual weeds. Perennial weeds often share some characters of both K- and r-strategists, for while they may be large and long-lived, growing in stable habitats, they may be of rapid growth and of high fecundity. They may spread largely by vegetative propagation by underground structures as in creeping thistle and couch-grass. Some species, such as hoary cress, spread extensively by seed as well as vegetatively.

Effect of herbicides on community structure

Selective herbicides, developed to eliminate particular species but to have no or minimal direct effect on the crop or desired species, clearly alter the dynamic equilibrium which normally exists in vegetation. This balance depends on a complex array of interactions between all the plants present. The reduction or total loss of some of the components changes the conditions for growth of those remaining, making available more space, light, water and mineral nutrients. Non-selective compounds, on the other hand, lead to the creation of completely bare areas, which may remain devoid of vegetation for a long period but in due course, if not further managed, would show a vegetation succession starting with the invasion of 'pioneer' species.

As many weeds grow rapidly and are vigorous competitors, their reduction or elimination usually leads to much enhanced growth of the vegetation remaining after treatment with herbicides. Many studies have shown that the effects of neighbouring plants on one another is considerable, some species being very sensitive to competitive pressures which are reflected in their growth and abundance whereas others, often species of low yield, are less sensitive. These sensitivities are reflected in the structure of the new communities arising after herbicide treatment.

In cereal crops two species which lead to loss of crop yield are black-grass and wild oat (see also Chapters 1 and 11). These two annual grasses may be taken to illustrate some general features of the population dynamics (changes with time in the size and nature of populations) of weeds. Like those of many annual weeds, seeds of both these species show two peaks of germination, one in the spring and one in the autumn. In blackgrass, seed from the previous year's crop may form only about one-third of the large buried seedbank, but give rise to about two-thirds of the following year's seedlings; the number of viable dormant seeds of this species per unit area may be as much as 500 times that of the mature plants. Research has shown that although the potential annual multiplication of this weed was 620 times, the population in wheat fields investigated did not increase significantly. To control species such as blackgrass with large seedbanks by herbicides, treatment needs to be repeated for a number of years before the population of viable seeds is substantially depleted. Study of wild oat in barley showed an increase of nearly threefold per year until there was a change to late spring cultivation (cultivation and sowing being delayed) when the populations declined enormously (to about one-hundredth of the starting density after 5 years). This illustrates well the value of weed control by alterations in cultivation practice.

The more similar the weed is to the crop plant or species constituting the desired vegetation, the more difficult it is to control the weeds, especially by herbicide treatment. This is well shown, for example, by the similarity of both the (spring) wild oat and the winter wild oat to cereal crops. An important consequence of the use of any control procedure, notably the use of herbicides, is that plants surviving the treatment are subject to progressive selection which leads to evolutionary change resulting in greater similarity to the crop plant and to increasing tolerance of the control procedure. Tolerant populations which may evolve (see later in this chapter and Chapter 8) are favoured relative to intolerant ones of the species in sites where herbicides are used, and also may spread into unsprayed areas.

Former agricultural practices have sometimes assisted the occurrence and spread of weed species whose seeds were harvested with those of the crop. A striking example is the corncockle whose seeds are not shed from the capsule until the grain is harvested. However, modern seed-cleaning techniques have eliminated the tuberculate seeds of this plant from the grain of the cereals, with the result that this once common annual is now almost extinct in the UK.

Degrees of tolerance to herbicides

The basis of the development of selective herbicides is that some species are much affected by a particular chemical but others little or not at all. The best known and most important group of selective herbicides chemically resemble plant growth substances and includes 2,4-D and related compounds. These herbicides have a pronounced effect on many broad-leaved (dicotyledonous) plants but little or no effect on narrow-leaved (monocotyledonous) species such as grasses. However, there are numerous exceptions to the general rule that broad-leaved plants are susceptible to auxin herbicides and narrow-leaved plants not, and the basis of the different susceptibility may be more physiological than morphological.

A number of broad-leaved species are not susceptible to some or even almost all of the range of auxin herbicides now available. Among the resistant species are lady's bedstraw, goutweed, pineappleweed, corn marigold, field pansy, common fumitory and speedwells. The degree of tolerance to selective herbicides may depend on the species within a genus; for example bulbous buttercup, with its greater powers of regeneration, is more resistant than meadow buttercup or creeping buttercup. The age of the plant also influences susceptibility, tolerance to herbicides being usually least at young stages. Annual knotgrass, field poppy and charlock are all much more resistant when well grown than at the seedling stage. Creeping thistle and field bindweed show low mortality with herbicides at leaf emergence; susceptibility rises to a maximum when the leaves are fully expanded, falling again at flowering. Because of their stage of development some species, although susceptible, may escape effects of herbicides. In the UK, new shoots of field bindweed do not emerge from its extensive deep underground rhizome system until well into the spring and the plant is little affected by early sprays; such late developers are favoured relative to species which make substantial growth early in the season, the ultimate composition of sprayed vegetation consequently depending on time of treatment.

The degree of susceptibility to auxin herbicides varies with the nature of the compound. For example, hemp-nettle and shepherd's needle are more susceptible to MCPA than 2,4-D, whereas red shank and stinking mayweed are more easily killed by 2,4-D than MCPA. Cleavers is resistant to both 2,4-D and MCPA but can be controlled by mecoprop. Relative susceptibility may vary with the stage of development of the plant; dandelions are more susceptible to MCPA than to 2,4-D in spring, but in autumn 2,4-D is more than twice as toxic to them as MCPA. Because of these varying susceptibilities, the floristic composition of vegetation is changed in different ways by different herbicides, with time of spray also having an effect. Loss of susceptible species allows resistant ones to spread, the ultimate composition of the vegetation approaching a new equilibrium dependent on the original nature of the vegetation and the time and type of spray. An important ecological consequence is that resistant species may become more abundant, as found in some places with pineappleweed, chickweed and wild oat.

Herbicide-resistant plants in the UK

The susceptibility of populations of species to herbicides, although unchanging in many species, may alter with the emergence of tolerant forms.
If a small proportion of the population is resistant to a herbicide, selection
of the resistant form can lead to the development of an entire population of
tolerant plants, the susceptible ones being eliminated. This process will be
fastest in plants with short life cycles. Any characteristics which happen
to be associated with the genetic basis of tolerance will be present in the
selected form with perhaps inconspicuous but subsequently important
effects on community structure.

The development of resistance to herbicides has become an increasing
problem in weed control during the last 20 years, after the first clear
demonstration of tolerance, which was to triazine herbicides in groundsel
in the USA in 1968.

While tolerance of groundsel to atrazine has evolved a number of
times, tolerant biotypes now being widespread in the UK, complete
replacement of susceptible forms by triazine-tolerant biotypes has not
taken place, suggesting differences in 'ecological fitness' between biotypes.
In groundsel, susceptible populations show greater growth (total biomass
production), earlier flowering and greater resource allocation to seed
production than resistant populations. A somewhat similar situation has
been found in fat-hen. The growth characteristics of susceptible biotypes
tend to favour them in competition with resistant biotypes, slowing down
complete replacement of susceptible by resistant forms with repeated
spraying.

The evolution of herbicide resistance has been most evident with
respect to triazines (see Chapter 8). Resistant biotypes have evolved, for
example, of annual meadow-grass in orchards and of American willowherb
among bush fruits where sprays have been applied repeatedly over a period
of 5–10 years. Spread of the resistant biotypes obviously depends on the
efficiency of seed dispersal. A range of man's activities, including cultural
practices, can promote the spread of resistant forms which could invade
and change semi-natural vegetation.

Only a few cases are known of the development of resistance to herbicides other than triazines. Examples include chickweed resistant to
mecoprop and populations of annual meadow-grass and rye-grass resistant
to paraquat (see Chapter 8). Instances of cross-resistance, some with
unrelated herbicides, are now, however, also known, mainly from outside
the UK. It seems only a matter of time before resistant forms of yet other

species will evolve in the UK, with serious management and also ecological effects should the resistant forms be more vigorous than the susceptible ones.

Threatened species of the UK flora

It is now well known that many species in the UK which were once widespread have declined to a very substantial extent since botanical recording began, and that losses this century have been considerable. This situation has led to growing concern both for the survival of the species themselves and also for the ecosystems of which they are a part. The Nature Conservancy Council and other bodies involved with conservation have been working towards a halt of this decline and especially the loss of rare species. Important conservation measures include the designation of National Nature Reserves and of Sites of Special Scientific Interest, which provide important refuges for endangered species. Several factors have contributed to the reduction in abundance of many UK plants; of high significance are losses of wetland species resulting from drainage operations but of even greater importance is the decline of weeds of arable land and plants of other open habitats, partly from weed control practices, including the use of herbicides.

Typical of arable land, some 23 species of the UK flora have been classified as nationally rare or threatened. If man-made open habitats are considered, the number is more than doubled. Of species believed extinct, the loss of interrupted brome is probably attributable to better seed cleaning. The same may apply to thorow-wax, a once widespread annual corn-field weed in the more southerly and eastern parts of England; although formerly known from 53 of the 112 botanical vice-counties of Britain, it is now feared extinct. Lamb's succory, a small annual of arable land, has declined to the point of extinction from its occurrence in about a dozen localities in the 1950s. This species, like a number of other corn-field plants, seems to depend on ploughing for its existence, its loss being attributed to the reversion of arable to pasture. There seems no doubt that changes in agricultural practice have led to large reductions this century of some formerly common species and also resulted in some uncommon (vulnerable) species now being classed as endangered. Species in this category are in danger of extinction if the causal factors continue to operate. Included in this group are small alison, false cleavers, downy hemp-nettle, Jersey cudweed, and fingered speed-well. Also threatened and adversely affected by herbicides are corn cleavers (*Galium tricornutum*) and spring speedwell. Vulnerable species, whose decline is probably due to agricultural practices on arable land, include the red-tipped cudweed, the broad-fruited cornsalad, and the hairy-fruited cornsalad.

It seems that species formerly abundant, and strongly controlled in sprayed areas (such as charlock) will, however, long remain as elements of the UK flora, persisting in unsprayed areas such as 'conservation head-lands', and in areas where there is no longer a need to spray because of the diminished number of weeds present.

Changing distributions in the UK flora

Intensive study of the flora of the vicinity of Sheffield indicates that the abundance of nearly three-quarters of the species is changing, and of these about two-thirds are decreasing and one-third is increasing. More generally it seems that the majority of native UK species is decreasing. These changes appear to be closely related to changes in land use and in agricultural practices. In the Somerset flora, for example, about one-third of the losses are those of weeds of corn-fields and other cultivated land. A further finding in the Sheffield survey is that proportional to the number of species associated with different habitats, the losses and local extinctions are greatest in those species associated with arable land; this habitat may be regarded as unique in view of the regular use of herbicides and annual cultivation procedures and also the lack of a semi-natural component of the flora. However, there is as yet insufficient evidence to assign the changes in abundance of a number of the species unequivocably to particular factors.

Application of herbicides has clearly had a marked effect on arable weeds, roadside vegetation and species typical of open sites. Undoubtedly the abundance of plants such as cornflower and corn crowfoot has much declined on arable land; also cow parsley and hogweed have decreased on some roadside verges and daisy and plantains in lawns. On verges and lawns, species diversity has been reduced where selective herbicides have been used, grasses assuming even greater dominance. However, where these verges are wide enough, an unsprayed area adjoining the hedge or fence can maintain diversity and is of nature conservation value.

Changes in agricultural practices (including mode of cultivation, time of sowing, use of fertilizers, seed cleaning) and in land use are unquestionably leading to differences in the relative abundance of some species in different habitats. The typical habitat of the field poppy, for example, is now waste places rather than arable land; its abundant seed production favours its spread. The curled dock, now less frequent on arable land, might be expected to show a trend to a more distinctly perennial habit. Fat-hen and charlock, both much reduced as arable weeds, are now more frequent in broad-leaved than in cereal crops. Thyme-leaved sandwort and small toadflax, both controlled on arable land to an appreciable extent by herbicides, may on the other hand, like cleavers, be increasing elsewhere. Also there is evidence that Venus's looking-glass is favoured by herbicides by reducing competition of other weed species. In addition, minimum cultivation and direct drilling appear to be favouring some species, e.g. parsley piert, common mouse-ear, and barren brome, which are showing increases in some areas. Species such as thale cress may be favoured by the control of perennials by herbicides on railway ballast and in tree nurseries (where it may be brought in from the continent with container plants). Aliens already well established in the UK, such as American willowherb, Oxford ragwort, policeman's helmet and Japanese knotweed, appear to be still increasing, as well as more recent invaders such as slender speedwell (see also Chapter 1 for other examples). In general, species of increasing abundance are found to be competitive or predominantly ruderal, whereas the declining species are slow-growing, long-lived plants of only moderate fecundity (K-strategists), being vulnerable to the increased disturbance and competition of the more productive vegetation stemming from modern land use.

Reductions in the abundance of many aquatic plants have been widely observed. Drainage operations have led to striking reductions and changes in wetland habitats, with accompanying losses of water plants. Another factor is the eutrophication of waterbodies which has adversely affected many hydrophytes. Herbicide treatment of the edges of watercourses is a further factor. Several species of pondweed (*Potamogeton* spp.) have declined, and even the vigorous common reed has decreased; increases are rare, but are shown in places by bulrush and toad rush.

Effects of cultivation practices
on weed control and soil properties

Cultivation systems have a major effect on the population density of weeds and influence the impact of herbicides. With minimum tillage and direct drilling, unlike ploughing, seeds from the buried seedbank are not brought to the surface where they may germinate. The seeds of some species, well represented in the seedbank, persist for many years, whereas the seedbank of other species is more scanty, seeds being short lived. Charlock, for

example, has a long-lived seedbank, and it is estimated that one reproductive phase in 11 years is sufficient to maintain this bank. Barren brome,
on the other hand, has no persistent seedbank. These differences in seed
ecology are reflected in the effect of herbicides under contrasted cultivation procedures leading to important ecological consequences.

Ploughing not only buries seeds shed on the surface but also incor-
porates decaying organic matter and burnt straw residues into deeper
layers. The latter effect results in greater impact of herbicides, because of
their reduced adsorption in the surface soil, than under direct drilling
(especially where there are adsorptive burnt straw remains at the surface).
In winter cereals, now often grown with minimum cultivation, annual
grasses, such as creeping bent, rough meadow-grass and blackgrass, with
small seeds which accumulate near the surface, are encouraged, their
periodicity of germination coinciding with the time of crop establishment.
In contrast, annual dicotyledonous plants, many of which are spring germinating, decrease, succumbing more extensively to herbicide treatment.
Direct drilling may encourage biennials and perennials, diminishing annuals, although there are exceptions; amongst those benefiting are wild
chamomile and pineappleweed. On the other hand, annual knotgrass and
common fumitory are favoured by deep cultivation, and wild oat by tine
cultivation. With ploughing, there is greater control of blackgrass by
herbicides, seed numbers declining rapidly in the soil and many fresh
seeds being buried. Each type of cultivation tends therefore to lead to its
characteristic spectrum of weeds and ecosystem. Addition of fertilizers
also has effects; liming is probably the main factor in the decline of corn
marigold and of corn spurrey.

Effects of herbicides on the soil and soil micro-organisms

A variable proportion of herbicides reaches the soil directly. With foliar-
absorbed herbicides, much of the spray material is likely to be absorbed by
the shoot systems and the herbicides may be partly broken down in the
decomposing litter before reaching the soil. Nevertheless some herbicide
usually reaches the soil quickly by direct run-off and this, dependent on the
nature of the material and the soil type, may penetrate to a substantial
depth, or, like the bipyridyl herbicides, be rapidly adsorbed and remain in
the top centimetre. The phytotoxicity of herbicides in the soil depends on
the proportion in the soil solution as compared to that adsorbed on soil
colloids, the properties varying with water content and climatic factors,
spatial heterogeneity in the soil also being very important (see Chapter 9).

Many soil organisms are affected directly or indirectly by herbicides and changes in their populations may result. As soil micro-organisms are largely responsible for soil fertility through their activities in, for example, decomposition of plant and animal remains, mineralization, nutrient cycling and in nitrogen fixation, changes in the balance of their populations could have important ecological effects.

Under natural 'stress' conditions, such as fluctuations in temperature, water content, pH and physical disturbance, it is known that the populations and activities of soil micro-organisms can be substantially influenced, depressions of 90% occurring not uncommonly. At temperatures of about 15°C it has been found that on average the doubling time for soil micro-organisms is approximately 10 days, giving a recovery time of about 30 days for a depression to about 12.5% of the original population size. With depressions of >90%, recovery is more seriously delayed, and depressions of about 90% take some 60 days for recovery. The effects of herbicides on soil micro-organisms can be appropriately evaluated by comparison with the effects of naturally-occurring changes. On this basis it is considered that when the recovery period of the micro-flora from the addition of herbicides is not more than 30 days, the long-term ecological consequences are negligible; when not more than 60 days they are tolerable, but when the balance is not restored after 60 days the consequences may be critical, with side effects not reversed.

Usually, most seriously affected by herbicides are populations of *Rhizobium* (which fix nitrogen in the root nodules of leguminous plants), nitrifiers, actinomycetes and organisms involved with the degradation of organic matter. Less sensitive are bacteria, fungi (including mycorrhizal species), ammonifiers and aerobic nitrogen fixers. At field rates with many of the herbicides currently employed (but not with some no longer used) recovery of the micro-flora after herbicide application normally occurs within a few weeks, but at high rates of application and highly persistent materials there may be serious impairment of microbiological activity. By eliminating actinomycetes, the bacteria and fungi may proliferate (with reduced competition); also important changes in pathogens can take place. Occasionally herbicides may control disease organisms, as for example dinoseb controlling potato wart disease. In contrast, the action of glyphosate on couch-grass is known to lead to the release of toxins from the killed rhizomes which predispose cereal seedlings to attack by the fungal pathogen *Fusarium culmorum*. These examples illustrate the chain of events that can arise when the ecological equilibrium is disturbed.

Repeated applications of a herbicide may lead to changes in soil micro-flora that increase their ability to degrade it (a phenomenon also discussed in Chapter 9). This so-called 'adaptation' or 'enrichment' may, at least in part, be associated with increased numbers of micro-organisms, especially bacteria, and also with enzyme induction. The phenomenon was first observed in the laboratory and later in the field with the phenoxyalkanoic acids. For MCPA, for example, repeated applications have been reported to increase degradation rate some threefold. As these compounds do not act primarily through the soil the effect is of little practical concern. However recently, several carbamate, thiocarbamate and amide pesticides have been shown to induce enriched micro-flora in the field. These compounds are soil-acting so the consequences can be serious. So far only small effects have been noted in the UK but in parts of the USA some of these compounds are now of only limited effectiveness.

The phenylalkanoic herbicides have only small long-term effects on soils. With 2,4-D, populations of *Rhizobium*, of cellulose degraders and of actinomycetes are reduced, but there is little or no effect on nitrification, ammonification, denitrifiers and overall populations of bacteria and fungi. Atrazine, for example, has greater effects, especially on populations of algae, actinomycetes and of cellulose degraders; however, ammonifiers, denitrifiers and total nitrogen fixers are little changed. Paraquat, which is rapidly adsorbed into the clay lattice, can be decomposed by several micro-organisms and affects soil fungi differentially. Fungi such as *Mucor* and *Zygorhynchus* are more susceptible to bipyridinium compounds than *Aspergillus* and *Penicillium*, and some pathogens, e.g. *Fusarium culmorum*, are tolerant; consequently a different balance of micro-organisms may arise.

The persistence of herbicide activity in the soil depends upon intrinsic phytotoxicity and speed of dissipation. Particular chemicals may be persistent because of features in their structure, such as the presence of chlorine, quite small chemical changes sometimes having considerable effects on biodegradability. In some instances one micro-organism may perform one step in the breakdown and a series of other organisms carry out successive stages. Herbicides of long persistence and having a major effect on micro-organisms are most environmentally damaging because of the disruption of the ecological equilibrium. Although each herbicide may lead to particular changes in the soil micro-flora, recovery of populations, at usual field rates of herbicides, is normally quite rapid. Indeed the effects of herbicides are not as drastic as those of even the most favoured soil sterilants, which may be used routinely in horticultural production, such as methyl bromide. This sterilant results in both chemical and physical changes, with very severe reduction of many micro-organisms including *Rhizobium*, nitrifiers and actinomycetes. On the other hand, ammonification is less affected than nitrification and nutrients (such as nitrogen and phosphorus) are made available from the killed organisms.

Agricultural practices are found to affect the soil micro-flora substantially. Ploughing causes large changes in the spatial structure of micro-habitats and leads to a considerable reduction of microbial biomass as compared with no tillage. Minimum cultivation affects actinomycetes and nitrifiers less than conventional tillage; nitrification is also adversely affected by soil compaction. In many respects, therefore, common agricultural practices such as ploughing and liming, which much affect nutrient availability, and also natural perturbations such as droughting and freezing, have a more pronounced effect on microbial activity than many herbicides at field rates. It is also of note that ploughing leads to a considerable reduction in diversity and numbers of soil fauna such as earthworms and also results in a more even distribution of species vertically than in uncultivated soil.

Effects of weed control on ecosystems and interactions between organisms

The influence of herbicides in the UK is greatest on ruderals of pathsides and waste areas, and on plants of arable land, tree nurseries and on roadside verges. More rarely are semi-natural types of vegetation much affected, although woodland floras may be influenced by herbicides used in tree establishment.

On open areas and on arable land the weed flora has changed where there has been repeated use of herbicides, susceptible species decreasing and resistant species increasing, taking advantage of those declining. Examples have already been given of these altering distributions. Use of herbicides to control broad-leaved species has favoured the proliferation of several grasses as weeds in some crops in the past. Changing husbandry too has clearly affected the weed spectrum, the timing of cultivation being especially important relative to the time of germination of weed seeds; forms developing a different phenology (e.g. late germination) may be selected, evolutionary changes occur and the relative abundance of a species in different habitats may change. The dominance of the crop species has a powerful effect on the composition of the weed flora as also do inter-cropping systems. Altering populations of weed species in arable

fields means a different seed rain in their environs, with possible consequent changes in community structure.

On roadside verges, with the use of selective herbicides to control
noxious weeds and tall herbs, and of growth regulators to control grass
growth, the botanical composition of the sward changes substantially. For
example, long-term studies at Bibury in Gloucestershire show diversity is
reduced by the loss of dicotyledonous plants and grass populations increase
and change. With the regulator maleic hydrazide, the tall tufted grasses,
cocksfoot and false oat grass, decline whereas the short, fine-leaved
rhizomatous grasses, red fescue and smooth meadow-grass, become dominant. With cessation of spray treatment, the short, thick, species-poor
fescue–meadow-grass swards may revert progressively to more diverse
vegetation, but some species take years to re-establish, the availability of
propagules being a major influence. Used on its own, maleic hydrazide can
enhance species diversity for a number of years; small dicotyledonous
plants may be favoured initially by colonizable pockets created by the
reduction of susceptible species. Larger plants, such as ribwort plantain
and crosswort, may be favoured for a considerable period, but ultimately
red fescue dominates strongly, with reduced diversity. Other growth regulators, by their differential effect on species, lead to somewhat different
floristic compositions; mefluidide, which increases diversity, favours common sorrel whereas paclobutrazol leads to an increase in coarse grasses.
Management of verges by frequent cutting favours grasses which tiller
freely, and low growing and rosette species, reducing others. For conservation of rare species, management of the habitat for the particular
requirements of these species is desirable.

The importance of pesticides in influencing interactions between organisms is well illustrated by the decline of birds of prey, such as the
sparrowhawk, known to have resulted from eating birds which fed on
cereals treated with DDT. The effects of herbicides on food chains and the
consequential changes in populations of organisms of different types are
often less obvious and rather little studied. Clearly, however, insects
restricted to a small number of food plants will be lost with the disappearance of the food source.

The project at the Boxworth Experimental Husbandry Farm, near
Cambridge, designed partly to provide information on the long-term
effects of pesticides on animals and plants associated with cereal production, is elucidating some of the interactions which may occur between
organisms. Areas with full treatment of herbicides, fungicides and insecticides showed only small changes in populations of birds and mammals.

However, wood mice tended to be reduced and also starling territories. On the other hand, a large reduction of some invertebrates was found, notably of predatory, poorly dispersive species and those over-wintering on, or near, the soil surface away from field boundaries; amongst arthropods decreased were Carabid beetles (e.g. *Bembidion obtusum*), Collembola and money spiders. Herbivores were reduced, e.g. the Lucerne flea (*Sminthurus viridis*) but cereal aphids increased with the loss of predators. The field-edge flora varied from year to year, but the changes at community level appear to be more related to weather conditions and cultivation practices than to herbicide treatment in the adjacent fields.

Much interest focusses on the wildlife of the hedgerows and hedge bottoms, as the field boundary is of great importance in wildlife conservation. Some two-thirds of the plant species of the field boundary may not occur in the cropped area, and many that are present in this area are within a few metres of the undisturbed field edge. This ecotone is of special interest, its management having far-reaching ecological effects. Studies by the Game Conservancy have attributed the 80% decline in the grey partridge since 1952 to chick mortality associated with the lack of a suitable diet of insects. These insects, on which partridge chicks are almost entirely dependent for the first few weeks of their life, have become much less frequent because their food plants, many of them weeds of cereal crops, have decreased substantially over the years with herbicide treatment. Important host plants for chick-food insects include cleavers, hemp-nettle, fat-hen, annual knotgrass, chickweed and black bindweed. This situation has led to the concept of 'conservation headlands', specially managed areas at the edges of crops extending into the hedge bottoms. An unsprayed zone of 6-m wide at the edge of the crop permits the growth of a number of food plants for the phytophagous insects important as chick food (including beetles, larvae of Lepidoptera and Heteroptera). This regime protects an entire food chain and has led to highly significant increases in partridges where 'conservation headland' practice has been applied in the last few years. A policy of using very selective sprays to try to control particular species (rather than a broad-spectrum herbicide) is sometimes followed on the field boundary. In this way, agriculturally-damaging weeds, such as cleavers, wild oat and couch-grass are aimed to be controlled without affecting the ecologically more valuable species. In this way also, food plants for game birds, butterflies and small mammals are being encouraged, with success.

A further advantage of the unsprayed zone is that this forms an over-wintering refuge for polyphagous predators such as beetles, important in controlling cereal aphid populations. Conservation headlands can also be managed to offset the accelerating decline of rare weed species, especially as preliminary observations have shown that 17 of some 25 species of the recent survey of the Botanical Society of the British Isles of weeds which are in serious decline occur in headlands.

The effects on the soil fauna of many herbicides used at field rates are usually fairly small and transient. Indirect effects caused by the destruction of plant cover, leading to extremes of temperature and drought at the soil surface, and the presence of killed plant material, are generally more important than any direct effects of the herbicides on soil invertebrates (destruction of vegetation by cultivation procedures may give comparable decreases in soil fauna). The abundance of soil micro-arthropods is little affected by 2,4-D and MCPA; however, simazine is known to lead to an initial decrease in predatory mites and in springtails, and has adverse effects on earthworms, enchytraeid worms and insect larvae. Reduction of diversity may subsequently be followed by over-compensation and ultimately adjustment to the level of the control. Monuron can also lead to a decrease in mites, springtails, millipedes and wireworms. Although paraquat and dalapon may result in depletion of populations of soil micro-arthropods (including Collembola and Acari), this is only temporary. On arable land the fairly low diversity of species may be further reduced by herbicides, frequently with large increases in one or several organisms. This is often a saprophagous species which may, with reduced diversity, change feeding habit and become phytophagous, e.g. the collembolan *Onychiurus armatus* causing increasing damage on sugar-beet.

Through interactions of herbicides, pathogens and plants, the effects of herbicides on diseases may be important. Use of herbicides may result, for example, in a change in the biological equilibrium between strains of pathogens. Some herbicides, such as 2,4-D, are known to have anti-fungal activity. Increased root disease has been reported in sugar-beet as a result of herbicides. Also, in Australia, use of chlorsulfuron may reduce the yield of barley and wheat from root rot caused by the fungus *Rhizoctonia solani*, probably the susceptibility of the host plants being increased rather than the virulence of the pathogen. Higher take-all and eyespot disease have been reported in winter wheat from the use of a hydroxybenzonitrile herbicide product, 'Oxitril'; in laboratory tests, triazines have led to reduced mildew (*Erysiphe graminis*) on wheat initially, but subsequently greater disease when the cereal recovered from the herbicide. Another consideration is that some weeds may act as hosts to diseases of cereals and the weeds build-up disease level in the crop (see Chapter 11).

In aquatic communities, treatment with herbicides may lead to a number of changes. These changes may be illustrated by studies involving atrazine, although this cannot be used legally in the UK to control aquatic weeds or those of the banks of watercourses. The amount of dissolved oxygen may decrease and also nitrate-nitrogen, but the levels of these may be soon restored. Productivity can be much reduced, and the dominance by chlorophytes shifted to that by diatoms, large green algae being most affected. Apart from species richness, however, recovery can occur within some 8 weeks. In ponds, there can be large effects on insects, diversity declining. Chironomids (midges) are sensitive and non-predatory insects much reduced, whereas predatory species are little affected. Herbivorous insects may show more rapid emergence. The effects on the insects appear to be indirect, through the reduction of food of non-predators and loss of their macrophyte habitat.

In general, herbicides do not persist long in water, so any hazard tends to be immediate rather than long term. Fish populations may be adversely affected, especially by 2,4-D and MCPA; simazine and paraquat are moderately toxic to fish.

Further prospects for the development of weed control systems and their likely impact on UK vegetation

Besides broad-spectrum and total herbicides, a wide range of chemicals is now available which are herbicidal to particular species or groups of species. The overall effect of continued broad-spectrum herbicide application is that some species are much reduced and others almost eliminated, whereas resistant species remain and new resistant forms evolve. Susceptible species with short-lived seeds, with small seedbanks, are reduced most quickly, the more resistant species persisting, especially if there is a large seedbank. However, by use of other methods of control and in view of the relatively long regeneration cycle of some weeds, the development of resistance may not prove to be an insuperable problem in management. More difficulties arise when the weeds are taxonomically and morphologically similar to the crop species. Use of herbicides affecting only a few or even only a single species holds promise for management for nature conservation, as already found in the 'conservation headland' regime. Results from the study of headlands further indicate the potential of 'set-aside' in conservation, and these findings also show the need to consider the indirect effects of herbicide treatment in respect of the whole ecosystem.

In agricultural practices, the advantages of minimum cultivation and direct drilling are becoming increasingly evident, these leading to a changed balance in weed species. More information is required of the effects of an integrated programme of control involving herbicides, fungicides and other pesticides in view of the variety of interactions between different types of organisms. The establishment of grass swards, for example, is known to be much improved by control of fritfly, leatherjackets, slugs and wireworms. Organic farming, with the increasing use of comfrey (*Symphytum*), is tending to give higher levels of carbon in the soil and increased capacity for water retention and of adsorption. Cellulose xanthate, as a soil conditioner, offers some promise in stabilizing the soil, preventing surface 'capping', so giving better seedling emergence and not seriously impairing the efficiency of soil-acting herbicides.

Biological methods of control seem likely to become more prominent in future (see also Chapters 1 and 2). Among the possibilities are the development of mycoherbicides, used either in the presence or in the absence of conventional herbicides. The fungus *Cochliobolus lunatus*, for example, has been shown to control cockspur when combined with sub-lethal doses of atrazine. Fungal pathogens have already been used successfully in the tropics to control a number of noxious weeds. Another possibility is the use of animal feeders. The cactus moth has successfully controlled prickly pear in Australia; perforate St John's wort, a serious weed in California, has been well controlled by leaf-eating beetles imported from Europe. Great safeguards are needed in the introduction of organisms for biological control; the post-liberation ecological consequences need full assessment, as disturbance of the ecological balance could be considerable and even irreversible. Nevertheless, many countries are now benefiting from biological control procedures, directed much more towards alien rather than native species. Recent research has shown that it appears feasible to control bracken by means of animal feeders, two species of moth from South Africa holding out greatest promise. For stopping the spread of this fern and ultimately reducing it, a combination of methods, mechanical, chemical and biological, seems desirable.

Yet other possible developments include the use of chemical safeners selectively antagonizing the effect of herbicides on the crop but not the weeds, and plant breeding for crop tolerance to herbicides. Integrated regimes may well prove to be the way forward in controlling particular troublesome species, hopefully with minimal damage to ecosystems outside the target areas.

Further reading

Audus, L.J. (ed.) (1976) *Herbicides: Physiology, Biochemistry, Ecology.* Academic Press, London.

Boatman, N.D. & Wilson, P.J. (1988) Field edge management for game and wildlife conservation. *Aspects of Applied Biology*, **16**, 53–61.

Chancellor, R.J. (1979) The long-term effects of herbicides on weed populations. *Annals of Applied Biology*, **91**, 141–144.

Domsch, K.H., Jagnow, G. & Anderson, T.-H. (1983) An ecological concept for the assessment of side-effects of agrochemicals on soil microorganisms. *Residue Reviews*, **86**, 65–105.

Fryer, J.D. & Chancellor, R.J. (1970) Herbicides and our changing weeds. In Perring, F.H. (ed.) *The Flora of a Changing Britain.* Botanical Society of the British Isles Conference Reports, No. **11**, 105–117.

Grime, J.P., Hodgson, J.G. & Hunt, R. (1988) *Comparative Plant Ecology: a Functional Approach to Common British Species.* Unwin Hyman, London.

Hance, R.J. (1987) Herbicide behaviour in the soil, with particular reference to the potential for ground water contamination. In Hutson, D.H. & Roberts, T.R. (eds) *Herbicides.* John Wiley, Chichester.

Harper, J.L. (1977) *Population Biology of Plants.* Academic Press, London.

MacArthur, R.H. & Wilson, E.O. (1967) *The Theory of Island Biogeography.* Princeton University Press, New Jersey.

Marshall, E.J. P. (1985) Field and field edge floras under different herbicide regimes at the Boxworth E.H.F. — initial studies. *Proceedings of the 1985 British Crop Protection Conference — Weeds*, **3**, 999–1006.

Perring, F.H. & Farrell, L. (1977) *British Red Data Books: 1. Vascular Plants.* The Society for the Promotion of Nature Conservation, Lincoln.

Radosevich, S.R. & Holt, J.S. (1984) *Weed Ecology: Implications for Management.* John Wiley, New York.

Warwick, S.I. (1980) Differential growth between and within triazine-resistant and triazine-susceptible biotypes of *Senecio vulgaris* L. *Weed Research*, **20**, 299–303.

Way, J.M. & Greig-Smith, P.W. (eds) (1987) *Field Margins.* Monograph No. 35 BCPC Publications, Bracknell.

Willis, A.J. (1988) The effects of growth retardant and selective herbicide on roadside verges at Bibury, Gloucestershire, over a thirty-year period. *Aspects of Applied Biology*, **16**, 19–26.

21 / Legislation and other safeguards relating to the use of herbicides in the UK

R.J. HANCE

Consultant, Brook Hill, Woodstock, Oxford OX7 1XH

Paracelsus is credited with being the first to suggest that all chemicals are toxic to life if the dose is large enough and that they may show stimulatory effects at lower doses. Herbicides are acutely toxic to susceptible plants at levels that are probably of the order of a few nanograms per plant and susceptible plants are not always weeds. The possibility also exists that they may be acutely or chronically toxic to other organisms, man included. In addition, the possible stimulation of organisms by sub-toxic doses must also be considered. Therefore controls have been introduced on the use of herbicides (and other pesticides) to minimize the possible hazards to non-target organisms of all sorts from micro-organisms to mammals.

The control system in the United Kingdom has changed fundamentally since the previous edition of the *Weed Control Handbook*. The voluntary arrangements of the old Pesticides Safety Precaution Scheme (PSPS) and the Agricultural Chemicals Approvals Scheme (ACAS) have been replaced by statutory controls under the provisions of Part III of the Food and Environment Protection Act 1985 and the Control of Pesticides Regulations 1986. Thus there is now specific legislation concerning pesticides in addition to that dealing with health and safety at work, poisons, and

521

pollution which is wider in its scope and continues to have implications for pesticides.

At the time of writing the process of implementation of the new legislation is not complete so, although every effort has been made to confirm the accuracy of the contents of this chapter, some of the information will inevitably soon be out of date. Therefore what follows should be regarded as indicative, not definitive.

The Food and Environment Protection Act 1985

Part III of the Food and Environment Protection Act 1985 (FEPA) deals with pesticides. Its aims are:

1   The continuous development of means:

to protect the health of human beings, creatures and plants;
to safeguard the environment;
to secure safe, efficient and humane methods of controlling pests.

2   To make information about pesticides available to the public.

The Act gives the government the power to control the import, sale, supply, storage, use and advertisement of pesticides, to approve pesticides, to set maximum residue levels (MRLs) in food crops and feedstuffs, to make available to the public information supplied in connexion with the approval of pesticides and to establish an advisory committee on pesticides. There are also powers to call for data on products, to issue *Codes of Practice*, charge fees, recover certain expenses and authorize enforcement officers.

The Control of Pesticides Regulations 1986

Using the powers provided by the Act these regulations impose statutory requirements on all involved in pesticide use from the supplier to the user. There are two instruments of control, *Consents* and *Approvals*. *Consents* put general conditions on the sale, supply, storage, use and advertisement of pesticides. *Approvals* set specific conditions for individual pesticides.

The term 'pesticide' is defined very broadly as 'any substance, preparation or organism prepared or used for destroying any pest'. It includes those used for: (1) protecting wood and other plant products; (2) regulating the growth of plants; (3) giving protection against harmful creatures; (4) rendering such creatures harmless; (5) controlling organisms with harmful or unwanted effects on water systems, buildings or manufactured products; and (6) protecting animals against ectoparasites. The definition excludes substances being tested only within the research premises of their developer, pesticides intended solely for export and pesticides used in decorative paints, textiles, paper and other industrial processes. Certain products controlled under other legislation, such as the Medicines Act 1968, are also excluded. Also 'organism used for destroying any pest' refers only to bacteria, protozoa, fungi and viruses so that insects are not included.

The approval process

All pesticides must be submitted for approval before they can be put on the market and the sale, supply, storage, use and advertisement of non-approved products is prohibited. An approval is specific for the active ingredient, the formulation, the use and conditions of use. It also attaches requirements concerning the container in which the product is sold and its label. To gain full approval a product must be shown to be safe and efficacious (and if relevant, humane) for its intended use.

Safety

Basically, the requirements for safety data are the same as those formerly needed under the PSPS. The applicant provides information to show that the product is safe to operators, consumers, domestic animals and the environment.

The list of toxicological, metabolic and residue data is formidable (see Chapter 10) but it is important to remember that much of the information is generated in animal studies with a limited number of species, so extrapolation to man, and indeed wildlife, involves a substantial element of subjective judgement. It does not take a great deal of imagination to deduce from the Paracelsus concept mentioned at the beginning of this chapter that any given concentration of any substance is likely to affect some organism or other in the ecosystem. Even if the effect is beneficial to the organism concerned, the consequent environmental balance may be undesirable in some way. Thus the idea of a totally risk-free pesticide is untenable; it is the balance of the risks of unwanted-to-desired effects that is important. For any particular product this balance is not the same everywhere as climatic, agronomic and cultural values are involved. For this reason it does not follow that a material not approved in one country should necessarily be prohibited in another. The converse is also true.

Efficacy

Efficacy is considered in terms of the ability of a pesticide to fulfill the claims made on the product label. To deter the use of unreasonably low label claims for performance, products are required to give a clearly-defined benefit — admittedly difficult for wood preservation where the benefits may take many years to appear. Because efficacy is tied to label claims it follows that it is related to the individual formulation and its method, rate and time of application, not the inherent properties of an active ingredient. Minimum absolute levels of control are not set because there are circumstances when a product of relatively low activity may have desirable features. For example, such a product may have smaller effects on non-target organisms than its more efficacious competitors or it may have a different mode of action and so be an important aid to reducing the risk of resistant strains appearing in the target pest. Relative cost is also a consideration as well as such factors as ease of use, risk of damage to the crop and the interval between application and harvest.

Humaneness

This criterion applies only to pesticides for use against vertebrates so is of no concern to herbicides. Information on the detail of the evidence that must be submitted when seeking approval in the UK is given in *Data Requirements for Approval Under the Control of Pesticides Regulations 1986*, available from The Ministry of Agriculture, Fisheries and Food (MAFF).

Evaluation of data

The initial evaluation is made by the Registration Department of the Ministry of Agriculture, Fisheries and Food (MAFF) if the application concerns agriculture, horticulture, forestry, home use, food storage or water. Applications concerning wood preservatives, masonry biocides, anti-fouling paints and public hygiene products are dealt with by the Health & Safety Executive (HSE). The object of this stage is to decide whether the application should be handled by the Committee procedure or the Secretariat procedure. This decision is made in accordance with guidelines laid down by the Advisory Committee on Pesticides (ACP) a body established to advise the Government on pest control and matters relating to FEPA.

The Committee procedure is usual for products containing a new active ingredient, for applications involving a novel means of application (even if the active ingredient is already approved), for major changes of use, or if a new potential hazard arises (such as a change from a granule to a liquid formulation). All information provided by the applicant, as well as any other relevant information that is available, is considered by the members of the Scientific Sub-Committee (SSC) of the ACP. The applicant may give a presentation at the meeting where the data are discussed. The SSC may then advise the Government directly or, more usually, pass the application together with its assessment to the ACP which then decides on the advice that should go to the Government. The ACP is assisted not only by the SSC but also, when necessary, by four specialist groups: The Medical and Toxicological Panel, The Environmental Panel, The Labelling & Container Design Panel and The Pesticides Application Technology Panel. Inputs are available from other groups such as The Pesticides Usage Survey Steering Group and The Working Party on Pesticide Residues as well as Government specialist committees not specifically concerned with pesticides.

The Secretariat procedure bypasses the Committees though members may be consulted individually. Typically, this procedure is applied to the extension of an approval to a similar crop group and for the rapid approval of an imported product identical to one already approved for use in the UK. The ACP is informed of decisions taken by this procedure.

It is important to note that these procedures result only in advice and recommendations to the Government; decisions on applications and approvals are the responsibility of the relevant Ministers.

Stages of approval

With the introduction of FEPA, the four-phase clearance procedure that operated previously has been replaced by one of three phases: the experimental permit, provisional approval and full approval. Approvals normally specify an individual product and its uses.

Experimental permits: Experimental permits are granted automatically if: the testing is carried out on the premises of the developer (or his agent); only his own (or his agent's) employees are involved; nothing treated with the test materials is consumed by humans or animals; all reasonable precautions are taken to protect human health and the environment. Currently, under these circumstances, trials need not be notified to the Registration Departments, although a simple notification system is likely to be introduced.

If these conditions are not met, for example if work is planned on a grower's land, the Registration Department must be notified and an experimental permit sought. Restrictions may be imposed, for instance on the way the material is handled, disposal of treated crops and the land area used. Specified data may be required at the end of the trial. Experimental permits usually last for one year (or one growing season) and the sale and advertisement of the product in question is not permitted.

Experimental permits must be sought for any work involving the release of genetically-manipulated organisms and must first be approved by the Advisory Committee on Genetic Manipulation of the HSE.

Provisional approval: Provisional approval is given when most of the necessary data have been obtained so that the product may be used safely. It usually authorizes sale, advertisement, storage, supply and use for a stipulated period of up to three years. There may be a restriction on the quantity that may be sold and conditions could be set for further data requirements. Provisional approval may be extended, subject to the submission of interim data, if the applicant is not in a position to apply for full approval at the end of the stipulated period. Details of the label and, for a new active ingredient, the text of the data evaluation, have to be made available to the public.

Full approval: Full approval is granted when all the data requirements have been satisfied. Any conditions necessary to meet the objectives of the legislation may be imposed by Government at any time.

Review of approvals

All approvals are, in principle, kept under continuous review and a full scientific review may be held at any time if new evidence justifies one. Otherwise it is intended that all products be subjected to periodic full reviews on a timetable drawn up in consultation with the ACP. Review details are announced in the *Pesticides Register* and anyone with an interest in the products under review is invited to submit data. Approval holders are contacted directly.

Off-label approvals

There are many uses of herbicides, particularly in horticulture, which would produce sales too small to cover the costs of doing the research and development needed for approval and label recommendations. So as not to deprive growers of a substantial number of valuable pesticides, arrangements have been made for the assessment of unregistered uses which growers wish to use, without the need for either the growers or manufacturers to submit full data packages.

Anyone who wants to use an approved product for a minor use that does not have an appropriate approved label recommendation should submit an application for an off-label recommendation to the Pesticides Registration and Surveillance Dept (PRSD) at the MAFF Harpenden Laboratory, Hatching Green, Harpenden, Herts AL5 2BD. The application must include:

Details of the product, its a.i., manufacturer/distributor, formulation and whether it is regulated under the Poisonous Substances in Agriculture Regulations (as on product label)
Purpose of proposed use
Reason for thinking the product will be effective for the purpose
Details of the crop (type, situation, height in relation to operator)
Proposed method of application and, if a soil application, details of any incorporation procedure
Application rate and spray volume (if a liquid)
Frequency and timing of application(s)
Details of any spray additives or other pesticides to be added to the product
Minimum interval between application and exposure of the public
Any other relevant information

In order to discourage frivolous applications a fee (£50 in 1988) is charged for each application. At the end of February 1988 a total of 179 off-label approvals were announced.

Approval of identical products imported from the EC

There are arrangements for the rapid approval of an imported product that is identical with one already approved under FEPA. They apply only to imported formulations as packaged and sold in the EC and do not include imports of unformulated a.i.'s. To be rated 'identical':

The a.i. in the imported product must be produced by the same company (or an associate) as the product already approved in the UK
The formulation must also be produced by the same company (or an associate) and any differences from the already approved formulation

must be judged by the authorities to have no effect on non-target organisms

Approval of commodity chemicals

Approvals are normally granted for specified uses of specific formulated products. There are, however, a number of substances (currently eight), with both pesticide and non-pesticide uses and which are often sold unformulated, that may be used as pesticides provided the conditions of approval and any other controls under FEPA are observed. The sale of these substances as pesticides is only legal if specific approval had been given for the use of substance as a pesticide under an approval label. None of these currently has a major herbicidal use. The substances are: alpha-chloralose, carbon tetrachloride, ethylene dichloride, ethylene oxide, formaldehyde, methyl bromide, strychnine and sulphuric acid.

Tank-mixes and adjuvants

The consent on use states that:

No person shall combine or mix for use two or more pesticides except in accordance with the conditions of approval given originally in relation to those pesticides, or as varied subsequently by lists of authorised tank-mixes published by the ministers

No person shall use a pesticide in conjunction with an adjuvant except in accordance with the conditions of approval given originally in relation to that pesticide or as varied subsequently by lists of authorised adjuvants published by ministers

An adjuvant is defined as:

A substance other than water, without significant pesticidal properties, which enhances, or is intended to enhance, the effectiveness of a pesticide when it is added to that pesticide

Full details of how these arrangements will operate were not available at the time of writing.

The financing of the approval process

FEPA gives the Government the power to recover the costs of running the approvals system and of some of the work of monitoring and information collection undertaken by Government. Fees are charged in two stages: (1) an initial charge on all applications for approval, which varies with the type of application; and (2) an annual levy on each approval holder of a percentage of his annual turnover of approved products, calculated to make up the difference between receipts from initial charges and the total costs the Government wishes to recover.

List of approved pesticides

The official list of provisionally and fully approved pesticides, *Pesticides*, sub-titled *Pesticides Approved Under the Control of Pesticides Regulations 1986* is available from Her Majesty's Stationery Office (HMSO) Publications. *The UK Pesticides Guide* published annually jointly by BCPC and CAB International gives details of current pesticide products and their uses together with information about application methods, precautions, efficacy and crop safety.

Controls over supply, distribution and use

Control of supply and distribution and use is one of the two major elements of the FEPA legislation, the other being monitoring activities. The basic premise is that everyone concerned with pesticides has an obligation to take all reasonable precautions not to put at risk non-target organisms whether they be man, micro-organisms or indeed any part of the environment.

Labelling

When seeking an approval, the supplier must submit a draft label setting out all the information needed for the product to be used safely and effectively. The label must be agreed with the authorities as part of the approval process and a product can only be supplied under an approved label.

Containers

The container in which a pesticide is supplied must be approved. Factors that are considered include size (and hence weight), effectiveness of pouring arrangements and the ease of opening when wearing the specified protective gloves. It is a criminal offence to store, transport, sell, supply or use a pesticide in a container which does not bear an intact MAFF-approved label. This includes the dangerous practice of decanting pesticides into soft drinks bottles and similar vessels in order to give them to neighbours, employees and so on.

Advertisements

The advertisement of unapproved pesticides or unapproved uses of approved pesticides is prohibited. Advertisements of approved or provisionally approved products must state:

the active ingredient;
any special risk to those who handle the product and, if relevant, the need to wear protective clothing;
any special environmental risk, such as 'dangerous to fish'.

Standards applying to pesticide distribution and supply

Manufacturers, suppliers, merchants and distributors must comply with government conditions with regard to:

The construction and design of buildings
The disposal, drainage and ventilation
Fire, safety and security precautions
Washing facilities
Stock control
Transport

Training of distribution personnel

Anyone responsible for the storage and sale of commercial pesticides (which excludes those for domestic garden use) must have obtained a certificate from a suitable training course which states they have met the required standard.

The British Agrochemical Standards Inspection Scheme (BASIS): BASIS is an independent scheme for the registration of distributors of crop protection products and those who give advice on their use. Its purpose is to ensure that the industry complies with the regulations made under FEPA and the scheme covers two areas:

1 Standards of safety in the storage, distribution and application of crop protection products.

2 Standards of training of everyone selling, applying by contract or giving advice on the use of pesticide products both in relation to crop protection and personal and environmental safety.

The scheme maintains a register of all companies and the staff they employ who are engaged in the storage and sale of crop protection products and of those who give advice on the use of such products. The scheme employs assessors who conduct annual inspections of agro-chemical stores. The relevant Government Department is notified of stores which fail to meet the standards required by FEPA. The scheme also conducts training courses and examinations to FEPA standards.

BASIS is administered by a Board of Management which includes nominees of trade associations and members elected by registered distributors.

Obligations of users of pesticides

Users of pesticides must:

Use a pesticide only for its approved use as stated on the label or in
  published lists of approved uses
Observe any stated maximum application rate and minimum dilution rate
Observe all other conditions of approval
Use only notified adjuvants

General guidance is provided in a statutory code of practice (see below).

Users also have an obligation to ensure that they and their employees are adequately trained, for instance by having completed a course of the Agricultural Training Board. Certificates of competence are required for those applying pesticides on another person's land, or were born after 1964 unless working under the direct and personal supervision of someone holding such a certificate.

Records of pesticide supply and use

Holders of approvals, full and provisional, must notify the Government of the quantity of active ingredients sold in the UK each year. Users are not obliged to keep records (except for aerial applications and for substances covered by the Poisonous Substances in Agriculture Regulation 1984). However, adequate record keeping is recommended in the *Code of*

Practice (see below) and if failure to keep records affects safety it may form part of evidence in a prosecution.

The regulation of aerial spraying

Aerial spraying is controlled by the Civil Aviation Authority (CAA) under the Air Navigation Order of 1985 and only operators holding a current Aerial Application Certificate can do such work. However, aerial spraying operations relevant to FEPA are included in its regulations but the requirements are set out in the Aerial Application Certificate.

Only products with a specific approval for aerial application may be used. Other conditions include the notification in writing at least one day beforehand of the local Chief Environmental Health Officer, the occupants of each building and the owner of any livestock or crops within 75 feet of the boundary of the land to be sprayed, as well as any hospital, school or other institution lying within 500 feet of the flight-path. In addition, the local beekeepers' spraying warning scheme must be consulted at least 2 days in advance.

If the area is within three-quarters of a nautical mile of a nature reserve or a Site of Special Scientific Interest, the Nature Conservancy Council must be consulted at least 3 days before spraying as must the water authorities if the land is adjacent to water. Spraying aquatic weeds on the banks of watercourses or lakes can only be done with the agreement of the water authority.

Many detailed requirements for the actual operation are given in *CAP 414, The Aerial Application Certificate* obtainable from CAA Printing and Publication Services. The rules applying in Northern Ireland are slightly different.

The CAA can suspend or revoke a certificate or take legal action if the conditions are breached and in some circumstances the Agricultural Inspectorate of HSE can also take action. Complaints about aerial spraying operations are dealt with by HSE.

Codes of Practice

Section 17 of FEPA allows the Government to issue *Codes of Practice* giving practical guidance relating to its provisions. Users are required to 'take all reasonable precautions to protect the health of human beings, creatures and plants; to safeguard the environment and in particular to avoid the pollution of water'. Appropriate advice is given in the *Code of*

Practice on the Agricultural and Horticultural Use of Pesticides which was first issued in draft for the 1988 season. It contains sections on:

User training and certification
Planning and preparation
Working with pesticides
Aerial application
Reduced volume spray application from ground-based machinery
Disposal of waste pesticides and containers
Record keeping
Health aspects

There is an equivalent *Code of Practice on the Storage, Supply and Sale of Agricultural and Horticultural Pesticides by Distributors and Contractors.* Its sections are:

Store design and construction
Stock management and stores
Health and hygiene
Transport standards
Training of storekeepers and salesmen
Temporary stores and selling points
Record keeping

Once the pesticides *Codes* have formally been made statutory (and as with the *Highway Code*) 'failure on the part of any person to follow the guidance given in the codes will not of itself render that person liable to proceedings of any kind; but such failure will be admissible in evidence in any criminal proceedings'.

There is also a draft *Code of Practice* for the *Use of Approved Pesticides in Amenity Areas* prepared jointly by the National Association of Agricultural Contractors and the National Turfgrass Council.

Pesticide residues

FEPA contains powers to control the maximum residue levels (MRLs) for pesticides in food, crops and feedingstuffs. The initial stage is contained in the Pesticides (Maximum Residue Levels in Food) Regulations 1988. They implement the Council of European Communities (CEC) Directives 86/362 (for residues in cereals) and 86/363 (for residues in foods of animal origin) and 76/895 (for residues in fruit and vegetables). Over 60 active ingredients are included of which five are herbicides.

There are also MRLs defined by the Codex Alimentarius Commission, set up jointly by the Food and Agriculture Organization of the United Nations and the World Health Organization. These MRLs apply to pesticide residues in commodities moving in international trade. A Codex MRL reflects the maximum residue which should not be exceeded if good agricultural practice (GAP) is followed. It is not a measure of the residue that will be found in every sample of the commodity in question and it does not imply that GAP will always result in a residue.

Where there is no mandatory MRL an estimated MRL based on the UK evaluation of the data in the approval application is used. For imports of commodities not produced in the UK, in which case there would be no UK pesticide approval, the Codex or non-mandatory CEC MRL is used. If no international MRL exists, the limit is set at, or near, the limit of analytical determination (an 'administrative zero').

It is an offence under FEPA to introduce into trading channels a crop, food or feedingstuff containing a pesticide residue which exceeds an MRL. Because MRLs are fixed at levels that would not normally be exceeded if the conditions of approval are observed, there should be no risk of conviction if label instructions are followed. It is a defence to prove that all reasonable precautions were exercised to prevent the commission of an offence. However, those convicted on grounds of excess residues following improper use of a pesticide might also face prosecution for infringing the regulations on pesticide use.

Produce containing a residue exceeding an MRL can be seized and destroyed.

Enforcement of FEPA

Her Majesty's Agricultural Inspectorate of HSE is responsible for enforcing those regulations which concern the supply, storage and use of pesticides on agricultural and horticultural land, premises and vehicles, and forestry land and premises.

Pesticide poisoning of wildlife incidents are investigated by the Wildlife Incident Investigation Scheme operated by the Agricultural Departments (MAFF, the Department of Agriculture for Scotland and the Department of Agriculture for Northern Ireland). The Agricultural Departments are also responsible for the enforcement of the pesticides residues legislation where the Working Party on Pesticide Residues has an important monitoring role.

Enforcement in situations outside agriculture, horticulture and forestry may be undertaken as appropriate by the Factory Inspectorate of HSE, local authority environmental health officers, trading standards officers, the Railway Inspectorate and water authorities.

Disclosure of information

One of the objects of FEPA is to make information about pesticides available to the public. However, this must be done taking into account the interests of those who supply the information (which is often commercially valuable).

Notices of new approvals, amendments, revocations, suspensions and announcements of reviews are published in the *Pesticides Register*.

A full evaluation, based on that prepared for the SSC, is made available on application when a product receives provisional approval. Applicants must undertake not to use the information commercially. This evaluation contains information about the following: (1) chemical properties of the ai; (2) the formulation; (3) its use and application; (4) toxicity; (5) carcinogenicity; (6) mutagenicity; (7) teratogenicity; (8) metabolism; (9) operational exposure characteristics; (10) behaviour and fate in plants, crops soil and water; (11) environmental effects; and (12) efficacy and crop safety. The commercial interest is protected firstly because the notifier has the opportunity to ensure that valuable process secrets are not included, secondly because notifiers of competing or duplicate products must provide their own data or produce evidence that they have the owner's authority to use data already filed, and thirdly because of the undertaking made by those obtaining copies.

Raw data are not made publicly available as a matter of course. Exceptionally they may be made available to an individual who requests to follow a line of investigation in the context of challenging an ACP evaluation that the SSC agrees is scientifically justified. In this case a legally enforceable undertaking of confidentiality must be given by the researcher.

Additional ways of making information available are the reports of the ACP, the Pesticide Usage Survey, the Working Party on Pesticide Residues, the Wildlife Incident Investigation Scheme and the HSE Incident Reports.

**The Health and Safety at Work (etc.) Act 1974
and the Poisonous Substances in Agriculture Regulations 1984**

The Health and Safety at Work (etc.) Act covers the health and safety of
people at work and other people who may be affected by work activities. It
lays obligations for safe working on employers, employees and the self-
employed and also on manufacturers and suppliers of materials and
machines for use at work. The Act enables the government to make
detailed regulations to deal with specific hazards and, like FEPA, to issue
codes of practice to improve standards of protection for workers and the
general public.

The use of the more toxic pesticides in the UK is controlled by the
Poisonous Substances in Agriculture Regulations 1984 made under this
Act. They list the chemicals concerned and require various precautions to
be taken, such as the wearing of specified items of protective clothing, when
carrying out certain operations. The regulations also make employers responsible for ensuring that the appropriate precautions are taken, that any
necessary protective clothing is supplied and that records of usage are kept.
Copies of the regulations are available from HMSO and it is recommended
that anyone involved with the application of regulated products should
also obtain from HMSO a copy of *Poisonous Chemicals on the Farm
HS(G)2* which explains fully the requirements of the regulations and how
they can be met. Users in Northern Ireland, the Channel Islands and the
Isle of Man should consult the regulations applying to their areas.

The Poisons Act 1972

A pesticide recommended for scheduling under the Poisonous Substances
in Agriculture Regulations is considered for inclusion in the regulations
made under the Poisons Act 1972. Paraquat is a notable herbicide included
in these regulations. With some exceptions, chemicals in the Poisons List
Order can be sold only by a retail pharmacist or 'a listed seller of poisons'
(such as an ironmonger or corn merchant) who in registered with the local
authority for this purpose. The *Poisons Rules* specify the labelling, pack-
aging, transport, storage and conditions of purchase of listed poisons. The
current *Poisons List* and *Poisons Rules* are available from HMSO.

Rivers (Prevention of Pollution) Acts 1951–1961

These Acts, the corresponding Scottish Acts of 1951 and 1965 and the
Northern Ireland Water Act 1972, which are administered by the Regional

Water Authorities (River Purification Boards in Scotland) make it an offence to pollute streams. 'Streams' include any watercourse or inland water discharging into a stream. The Acts also give water authorities the power to make byelaws to prohibit or regulate the putting into a stream of objectionable matter whether polluting or not. The Acts apply to certain estuaries and tidal waters so advice should be sought where there may be a risk or intention to discharge in such waters.

The Control of Pollution Act 1974

Part 1 of the Act deals with 'Waste on Land' and makes it an offence, punishable by heavy penalties, to deposit on land any poisonous, noxious or polluting waste (including farm waste) that might be dangerous to persons or animals or could pollute a water supply. Farm waste (including pesticides and used containers) is excluded from the other controls of Part 1 but there is provision to bring specified kinds of farm waste under control if necessary.

Part 2 of the Act deals with 'Pollution of Water' and makes it an offence to cause any poisonous, noxious or pollution matter to enter a stream, controlled water (including lakes and ponds) or specified groundwater. It also includes solid waste and any matter which, either directly or in combination with other matter, impedes water flow so as to lead (or be likely to lead) to aggravation of pollution owing to other causes. However, any pollution arising from an act in accordance with the code of good agricultural practice approved by MAFF for the purposes of this Act is not an offence (unless it has been specified in a notice to desist under the Act). Information concerning the use of aquatic herbicides is included in Chapter 18.

Safe disposal of containers

If the local authority will not collect empty containers, which should have been rinsed out and punctured, they must be buried or burnt under strictly controlled conditions or a waste disposal contractor should be employed. If disposal must be done on the farm the *Code of Practice (Disposal of Waste Pesticide Containers* or the *Code of Practice for the Agricultural and Horticultural use of Pesticides*, which can be obtained from divisional Offices of MAFF or from offices of HSE should be consulted. It is not possible to give guidelines here because at the time of writing some of the details of recommended procedures are being reconsidered.

Civil liabilities of users of herbicides

The law requires that people who deal with intrinsically dangerous things, such as chemicals, take very great care, so that it is often not necessary for a person injured by them to prove negligence.

Employer and employee

Apart from the requirements of the Health and Safety at Work (etc.) Act and the Poisonous Substances in Agriculture Regulations, an employer has a duty to ensure that a safe system of work is adopted. If he fails in this duty he is liable for damages if an employee is injured as a consequence. A worker injured through the failure of his employer to comply with the regulations, for example by not providing specified protective clothing, can bring an action for breach of statutory duty and does not need to prove negligence. Sometimes an employer may have a defence that the worker himself was guilty of contributory negligence. However, this defence is likely to fail if the alleged contributory negligence was the failure to take steps which, under the regulations, it was the employer's duty to ensure were taken.

Duty to neighbours

An occupier spraying chemicals which damage his neighbour's crops or livestock is, in general, absolutely liable; the question of negligence does not arise. Defences that damage occurred through the action of a third party or was owing to an Act of God are rarely successful. If the spraying is done by a contractor the occupier is still liable but could recover indemnity from the contractor if he could prove negligence, assuming indemnity was not excluded from the contract. However, the contractor cannot exclude or restrict his liability for death or personal injury caused by his negligence.

An occupier must fence in his own cattle but has no obligation to fence out his neighbour's. Therefore there is usually no redress for injury to cattle trespassing on sprayed land but the position might be different if a fence was known to be weak and the chemical used was known to be poisonous to cattle (or to make poisonous weeds more palatable) and the occupier failed to warn his neighbour.

Duty to the public

Both occupier and contractor have a duty to the public (persons on a public highway or recognized footpath) to ensure they are not injured by chemicals. The responsibility also includes trade or business (and probably social) visitors. They must also act with 'common humanity' which means that if they know of the likely presence of a trespasser they must take steps to enable him to avoid the danger. Anyone doing something that is an irresistable lure to children (for example spraying by helicopter) has the duty to ensure that effective measures are taken to exclude children. If this is not done then occupier or contractor can be held liable for any injury that occurs.

When a farm changes hands it is up to the new owner or tenant to enquire about the use of persistent herbicides. The outgoing occupier has no obligation to warn his successor unless he is specifically asked for information.

Contractor's duty to occupier

Normally the contractor has the duty to exercise proper care when spraying and is liable if the occupier's crops or livestock are damaged through negligence. Some liability may be limited by the terms of the contract (but not death or personal injury) but this normally requires the advice of a solicitor.

Insurance

The brief summary of civil liabilities illustrates the heavy responsibilities that rest on occupiers and contractors to ensure that every possible care is taken. For this reason it is usual to insure against the possibility of civil claims. Cover can generally be obtained by occupiers, contractors, merchants and manufacturers for their legal liabilities resulting from the use of chemicals in agriculture arising from injury or damage to crops, the property of third parties, livestock and water pollution. Such people should regard insurance as an essential part of their business.

Further reading

The UK Pesticide Guide 1989 British Crop Protection Council & CAB International. BCPC Publications, Bear Farm, Binfield, Bracknell, Berks RG12 5QE, & CAB International, Wallingford, Oxon.

MAFF. *Data Requirements for Approval Under the Control of Pesticides Regulations 1986*. MAFF, Pesticides Safety Division, Great Westminster House, Horseferry Road, London SW1P 2AE.

Pesticides: Pesticides Approved Under the Control of Pesticides Regulations 1986. HMSO (Publications), London.

CAP 414 The Aerial Application Certificate. CAA, Printing and Publication Services, Greville House, 37 Gratton Road, Cheltenham, Glos. GL50 2BN.

22 / Sources of information

Consultant, 36 Sunderland Avenue, Oxford OX2 8DX

Primary sources
  Journals publishing original research
  Proceedings of conferences and
    symposia
  Literature on commercial products

Secondary sources
  Books
  Review articles
  Advisory literature
Names of herbicides

Primary sources

Weed science is a multi-disciplinary subject, and information relating to weeds, weed control and herbicides appears in a wide range of primary publications worldwide. A few key ones particularly relevant to the UK and Europe are listed below. A more comprehensive listing appears annually in *Weed Abstracts*. This abstract journal is published monthly by CAB International (Wallingford, Oxon OX10 8DE, UK). It provides comprehensive coverage of the world literature concerned with the science and technology of weeds and weed control. Thus it both keys into current developments and gives a cumulative indexed source of references to past work.

Journals publishing original research

Weed Research: The Journal of the European Weed Research Society, published by Blackwell Scientific Publications (Osney Mead, Oxford OX2 0EL, UK).

Pesticide Science: Published for the Society of Chemical Industry by Elsevier Applied Science Publishers (Crown House, Linton Road, Barking, Essex 1G11 8JV, UK).

Crop Protection: Published by Butterworths Publications Ltd (88 Kingsway, London WC2B 6AB, UK).

Annals of Applied Biology: Published by the Association of Applied Biologists (AAB Office, AFRC Institute of Horticultural Research, Wellesbourne, Warwick, CV35 9EF, UK).

541

Weed Science: Published by the Weed Society of America (Ed: D.L. Klingman, 2028 Forest Hill Drive, Silver Spring, MD 20903, USA).

Proceedings of conferences and symposia

A number of well-established bodies organize weed control conferences or symposia whose proceedings have become reputable media for the publication of original research and also of reviews. Key organisations are listed below. Additionally there are many national and regional weed control conferences held around the world.

British Crop Protection Council: Organizes 'The Brighton Crop Protection Conference — Weeds' held biennially, and a range of specialist symposia (BCPC, 49 Downing Street, Farnham GU9 7PH, UK).

Association of Applied Biologists: Specialist Meetings often organized by its Weeds Group (AAB Office, AFRC Institute of Horticultural Research, Wellesbourne, Warwick CV35 9EF, UK).

Society of Chemical Industry: Specialist symposia generally organized by its Pesticides Group (SCI, 15 Belgrave Square, London SW1X 8PS, UK).

European Weed Research Society: Organizes Symposia at various venues in Europe (EWRS Symposia Proceedings, Postbus 14, NL-6700 AA Wageningen, The Netherlands).

Literature on commercial products

Firms which manufacture and supply herbicides produce a variety of literature, ranging from technical bulletins and data sheets on new chemicals and formulations to product manuals and leaflets on proprietary products which they market. This material is usually obtainable by direct approach to firms.

Secondary sources

Books

The amount of research on all aspects of weed science has grown enormously from a very low level 50 years ago. The accumulated information has been assimilated into a wide variety of textbooks and manuals for the student and specialist. Some are mentioned at the end of individual chapters in this *Handbook*. New ones are covered in *Weed Abstracts*.

Review articles

Authoritative reviews provide critical interpretation of recent work in particular areas of weed science. No one review journal provides comprehensive coverage for the weed specialist. Many relevant reviews appear in *Residue Reviews* (published by Springer-Verlag, New York). Others appear widely scattered in agricultural journals and the literature of other disciplines.

Advisory literature

The Agricultural Development and Advisory Services of the Ministry of Agriculture, Fisheries and Food produce leaflets and other material on weeds and weed control, some unpriced, for which local ADAS offices in England and Wales should be consulted. In Scotland the advisers at the Scottish Agricultural Colleges should be consulted and in Northern Ireland the Department of Agriculture in Belfast.

Names of herbicides

Common names for herbicides are approved by the British Standards Institution (BSI) and the International Standardization Organization (ISO). A list of those in current use is given in each issue of *Weed Abstracts* and annually in *Weed Research*. Fuller details of common names in use around the world and of complete chemical names are given in *The Pesticide Manual* published by BCPC (BCPC Publications Sales, Bear Farm, Binfield, Bracknell RG12 5QE, UK). This compendium also presents a wide range of other facts on world herbicides. Trade names of approved herbicides on sale in the UK are given, together with much summarized information on their usage, in *The UK Pesticide Guide* published in annual editions by BCPC. Other sources of trade names are *Pesticide Index* and *European Directory of Agrochemical Products, Volume 2: Herbicides* both published by the Royal Society of Chemistry (Thomas Graham House, Science Park, Milton Road, Cambridge CB4 4WF, UK).

Appendix of weed names

Note: The views of taxonomists change from time to time so a definitive list of Latin names is scarcely possible. The list below cites those used in Williams, G.H. (ed.) (1982) *The Dictionary of Weeds of Western Europe*, Elsevier. In some cases, where the Latin name has changed recently, for example with mayweeds, synonyms are included after the name listed in the *Dictionary*. Some authors have used common names other than those in the *Dictionary* which includes only one for each plant. In these cases the common name used in the text is listed with a reference to the corresponding *Dictionary* name. Where the plant is not listed in the *Dictionary*, names are taken from *Flora Europaea* or Clapham, A.R., Tutin, T.G. & Moore, D.M. (1987) *Flora of the British Isles*, 3rd edn, Cambridge University Press.

Alder	*Alnus* spp.
Alison, small	*Alyssum alyssoides*
Amaranth	
common (redroot pigweed)	*Amaranthus retroflexus*
green (smooth pigweed)	*A. hybridus*
Arrowhead	*Sagittaria sagittifolia*
Ash	*Fraxinus* spp.
Aspen	*Populus tremula*
Barberry	*Berberis* spp.
Barley	
foxtail	*Hordeum jubatum*
wall	*H. murinum*
Barnyardgrass, *see* Cockspur	
Beech	*Fagus sylvatica*
Bent	
black	*Agrostis gigantea*
browntop (Highland)	*A. castellana*
common	*A. capillaris; A. tenuis*
creeping (fiorin)	*A. stolonifera*
velvet	*Agrostis canina* ssp. *canina*
Bermuda grass	*Cynodon dactylon*
Bilberry	*Vaccinium myrtillus*

545

Bindweed
  black   *Bilderdykia convolvulus; Polygonum convolvulus; Fallopia convolvulus*
  field   *Convolvulus arvensis*
  hedge   *Calystegia sepium*
Birch   *Betula* spp.
Bistort
  amphibious   *Polygonum amphibium*
  common   *P. bistorta*
Bittercress, hairy   *Cardamine hirsuta*
Blackberry   *Rubus* spp.
Blackgrass   *Alopecurus myosuroides*
Blackthorn   *Prunus spinosa*
Blanket-weed   *Vaucheria dichotoma*
Bracken   *Pteridium aquilinum*
Brambles, *see* Blackberry
Brome
  barren (Sterile)   *Bromus sterilis*
  interrupted   *B. interruptus*
  rye   *B. secalinus*
  soft   *B. hordeaceus; B. mollis*
Broom   *Cytisus scoparius*
Bulrush (great reedmace) *Typha latifolia*
Buttercup
  bulbous   *Ranunculus bulbosus*
  corn   *R. arvensis*
  creeping   *R. repens*
  meadow   *R. acris*

Canarygrass
  bristle-spiked   *Phalaris paradoxa*
  reed   *P. arundinacea*
Capeweed   *Arctotheca calendula*
Caraway, corn   *Petroselinum segetum*
Chamomile   *Chamaemelum nobile*
  wild   *Matricaria recutita*
Charlock   *Sinapis arvensis*
Chestnut, sweet   *Castanea sativa*
Chickweed, common   *Stellaria media*
Cinquefoil, creeping   *Potentilla reptans*
Cleavers   *Galium aparine*
  corn   *G. tricornutum*
  false   *G. spurium*
Clover   *Trifolium* spp.
Club-rush, common   *Schoenoplectus lacustris*
Cocklebur   *Xanthium pennsylvanicum*
Cocksfoot   *Dactylis glomerata*
Cockspur   *Echinochloa crus-galli*
Coltsfoot   *Tussilago farfara*
Corncockle   *Agrostemma githago*
Cornflower   *Centaurea cyanus*

Cornsalad
  broad-fruited   *Valerianella rimosa*
  hairy-fruited   *V. eriocarpa*
Cotton-grass   *Eriophorum* spp.

Couch
  common   *Elymus repens (Agropyron repens)*
  onion   *Arrhenatherum elatius* var. *bulbosus*
Crabgrass, *see* Fingergrass
Cress
  creeping yellow   *Rorippa sylvestris*
  hoary   *Cardaria draba*
  thale   *Arabidopsis thaliana*
Crosswort   *Cruciata laevipes*
Crowfoot
  corn, *see* Buttercup, corn
  riverwater   *Ranunculus fluitans*
Cudweed
  Jersey   *Gnaphalium luteoalbum*
  red-tipped   *Filago lutescens*

Daisy   *Bellis perennis*
Dandelion   *Taraxacum officinale*
Dock
  broad-leaved   *Rumex obtusifolius*
  curled   *R. crispus*
  water   *R. hydrolapathum*
Dodder   *Cuscuta* spp.
Dogstail, crested   *Cyanosurus cristatus*
Dropwort, water   *Oenanthe crocata*
Duckweed
  common   *Lemna minor*
  ivy-leaved   *L. trisulca*

Elder, ground (goutweed)   *Aegopodium podagraria*
Elderberry   *Sambucus nigra*

Fat-hen   *Chenopodium album*
Fescue
  Chewings   *Festuca rubra* ssp. *commutata*
  hard   *F. langifolia*
  meadow   *F. pratensis*
  red   *F. rubra*
  sheep's   *F. ovina*
  sheep's, fine-leaved   *F. tenuifolia*
  tall   *F. arundinacea*
Fingergrass (crabgrass)   *Digitaria* spp.
Fireweed (rosebay willowherb)   *Chamerion angustifolium; Epilobium angustifolium; Chamaenerion angustifolium*

Fleabane
  Canadian   *Conyza canadensis*

hairy   *C. bonariensis*
Philadelphia   *C. philadelphicus*
Fool's parsley   *Aethusa cynapium*
Forget-me-not, field   *Myosotis arvensis*
Foxglove   *Digitalis purpurea*
Frog-bit   *Hydrocharis morsus-ranae*
Fumitory, common   *Fumaria officinalis*

Gallant soldier   *Galinsoga parviflora*
Gold of pleasure   *Camelina sativa*
Goosegrass   *Eleusine indica*
Goutweed (ground elder)   *Aegopodium podagraria*
Gorse   *Ulex* spp.
Groundsel   *Senecio vulgaris*
  sticky   *S. viscosus*

Hairgrass, tufted   *Deschampsia caespitosa*
Hawthorn   *Crataegus monogyna*
Hazel   *Corylus avellana*
Heather   *Calluna vulgaris*
Hemp-nettle
  common   *Galeopsis tetrahit*
  downy   *G. segetum*
Hogweed   *Heracleum sphondylium*
Hornbeam   *Carpinus betulus*
Hornwort   *Ceratophyllum demersum*
Horseradish   *Armoracia rusticana*
Horsetail
  field   *Equisetum arvense*
  water   *E. fluviatile*
Hyacinth, water   *Eichhornia crassipes*

Johnsongrass   *Sorghum halepense*

Knotgrass   *Polygonum aviculare*
Knotweed, Japanese   *Reynoutria japonica; Polygonum cuspidatum*
Kochia   *Kochia scorparia*

Lady's bedstraw   *Galium verum*
Lamb's succory   *Arnoseris minima*
Lettuce, prickly   *Lactuca serriola*
Liverwort   *Marchantia* spp.
Loosestrife, purple   *Lythrum salicaria*

Marigold, corn   *Chrysanthemum segetum*
Mat-grass   *Nardus stricta*
Mayweed
  rayless, *see* Pineappleweed
  scented   *Chamomilla recutita; Matricaria recutita*
  scentless   *Matricaria perforata; Tripleurospermum maritimum* spp. *inodorum* (*T. inodorum*)

stinking   *Anthemis cotula*

Meadow-grass
annual   *Poa annua*
rough   *P. trivialis*
smooth   *P. pratensis*

Medick, black   *Medicago lupulina*
Mercury, dog's   *Mercurialis perennis*
Mignonette, wild   *Reseda lutea*
Milfoil
spiked water   *Myriophylum spicatum*
whorled water   *M. verticillatum*
Mint, corn   *Mentha arvensis*
Moor-grass, purple   *Molinia caerulea*
Morning-glory   *Ipomoea* spp.
Moss   *Bryum* spp.
    Hypnum spp.
    Polytrichum spp.
Mouse-ear, common   *Cerastium fontanum; C. holosteoides*
Mugwort   *Artemisia vulgaris*

Nettle
annual (small)   *Urtica urens*
common   *U. dioica*
dead, red   *Lamium purpureum*
hemp   *Galeopsis* spp.
Nightshade, black   *Solanum nigrum*
Nutsedge
purple   *Cyperus rotundus*
yellow   *C. esculentus*

Oak   *Quercus* spp.
Oat
cultivated   *Avena sativa*
wild, spring   *A. fatua*
wild, winter   *A. sterilis* ssp. *ludoviciana*
Oatgrass, false   *Arrhenatherum elatius* ssp. *elatius*
Onion, wild   *Allium vineale*
Orache, common   *Atriplex patula*

Pansy
field   *Viola arvensis*
wild   *V. tricolor*
Parsley
corn *see* Caraway, corn
cow   *Anthriscus sylvestris*
fool's   *Aethusa cynapium*
piert   *Aphanes arvensis*
upright hedge   *Torilis japonica*
Parsnip, wild   *Pastinaca sativa*
Pear, prickly   *Opuntia* spp.
Penny-cress, field   *Thlaspi arvense*

Pepperwort
    hoary, *see* Cress, hoary
Persicaria, pale   *Polygonum lapathifolium* ssp. *pallidum*
Pheasant's eye   *Adonis annua*
Pigweed, *see* Amaranth
Pimpernel, scarlet   *Anagallis arvensis*
Pineappleweed (rayless mayweed)   *Chamomilla suaveolens; Matricaria matricarioides*

Plantain
    common water   *Alisma plantago-aquatica*
    greater (rat-tail)   *Plantago major*
    hoary   *P. media*
    ribwort   *P. lanceolata*
Policeman's helmet   *Impatiens glandulifera*
Pondweed
    Canadian   *Elodea canadensis*
    curled   *Potamogeton crispus*
    fennel   *P. pectinatus*
Poppy, common field   *Papaver rhoeas*

Radish, wild   *Raphanus raphanistrum*
Ragwort
    common   *Senecio jacobaea*
    Oxford   *S. squalidus*
Redshank   *Polygonum persicaria*
Reed
    branched-bur, *see* common-bur
    common   *Phragmites communis* (*P. australis*)
    common-bur   *Sparganium erectum*
    sweet-grass   *Glyceria maxima*
Reed-grass, *see* Canarygrass, reed
Reedmace, great, *see* Bulrush
Rhododendron   *Rhododendron ponticum*
Rush
    hard   *Juncus inflexus*
    toad   *J. bufonius*
Ryegrass
    annual   *Lolium rigidum*
    perennial   *Lolium perenne*

Sandwort, thyme-leaved   *Arenaria serpyllifolia*
Sedges   *Carex* spp.
Shepherd's needle   *Scandix pecten-veneris*
Shepherd's purse   *Capsella bursa-pastoris*
Silverweed   *Potentilla anserina*
Skeleton weed   *Chondrilla juncea*
Soft-grass, creeping   *Holcus mollis*
Sorrel
    common   *Rumex acetosa*
    sheep's   *R. acetosella*
Sow-thistle
    perennial   *Sonchus arvensis*
    prickly   *S. asper*

Speedwell
  common field   *Veronica persica*
  fingered   *V. triphyllos*
  ivy-leaved   *V. hederifolia*
  slender   *V. filiformis*
  spring   *V. verna*
Spurrey, corn   *Spergula arvensis*
St John's worts   *Hypericum* spp.
  perforate   *H. perforatum*
Starwort, common water   *Callitriche stagnalis*
Stinkweed, *see* Penny-cress, field
Sycamore   *Acer pseudoplatanus*

Tare, hairy   *Vicia hirsuta*
Tarweed (fiddleneck)   *Amsinckia intermedia*
Thistles   *Cirsium & Carduus* spp.
  creeping   *Cirsium arvense*
  Russian   *Salsola iberica*
Thorow-wax   *Bupleurum rotundifolium*
Timothy   *Phleum pratense*
Toadflax, small   *Chaenorhinum minus*
Tor-grass   *Brachypodium pinnatum*
Tormentil   *Potentilla erecta*
Trefoil
  common bird's foot   *Lotus corniculatus*
  lesser   *Trifolium dubium*
Tussock, serrated   *Nasella trichotoma*

Velvetleaf   *Abutilon theophrasti*
Venus's looking-glass   *Legousia hybrida*
Vernal-grass, sweet   *Anthoxanthum odoratum*

Water lily, yellow   *Nuphar lutea*
Waterweed, Canadian, *see* Pondweed
Willowherb
  American   *Epilobium adenocaulon; E. ciliatum*
  great   *E. hirsutum*
  rosebay, *see* Fireweed
Witchweeds   *Striga* spp.
Wood-rush, field   *Luzula campestris*

Yarrow   *Achillea millefolium*
Yorkshire fog   *Holcus lanatus*

Index

Note: Only principal places of consideration of index terms are recorded. This applies particularly to herbicide names. The individual small grain cereals, barley, oat and wheat are not itemized separately. Instead entries are amalgamated under the heading 'cereals'.

553

CLOSING ARGUMENT

CLOSING ARGUMENT

ARGUMENT

DEFENDING (AND BEFRIENDING) JOHN GOTTI

and Other Legal Battles I Have Waged

BRUCE CUTLER

WITH LIONEL RENÉ SAPORTA

Crown Publishers / New York

Grateful acknowledgment is made to Black, Inc., for permission to reprint an excerpt from an article by Jimmy Breslin that appeared in *New York Newsday* (January 7, 1990). Reprinted by permission of Black, Inc.

Published by Crown Publishers, New York, New York
Member of the Crown Publishing Group, a division of Random House, Inc.
www.randomhouse.com

CROWN is a trademark and the Crown colophon is a registered trademark of Random House, Inc.

Printed in the United States of America

Design by Leonard W. Henderson

Library of Congress Cataloging-in-Publication Data
Cutler, Bruce.
Closing argument : defending (and befriending) John Gotti, and other legal battles I have waged / by Bruce Cutler, with Lionel René Saporta.—1st. ed.
1. Cutler, Bruce.   2. Lawyers—United States—Biography.
3. Defense (Criminal procedure)—United States—Biography.
I. Saporta, Lionel René.   II. Title.
KF373.C88 A3   2003
340'.092—dc21
20020139
81

ISBN 0-609-60831-2

First Edition

To my mother and father,

Selma and Murray Cutler, and to my son, Michael

Hoka hey! Follow me!
Today is a good day to fight,
today is a good day to die!

CRAZY HORSE,
ROSEBUD, SOUTH DAKOTA, 1876

CLOSING ARGUMENT

PROLOGUE

IN charge.

That's how I'd describe John. It was March 28, 1985, when he walked through the door behind the bench in District Judge Eugene Nickerson's courtroom, looking like a man in charge—of the other men who walked through the door with him, of his environment, of himself. And it's not easy to look that way when you're in federal custody about to be arraigned on racketeering charges.

I was thirty-six years old, and maybe he should have looked to me like the next ten years of my life, or maybe like the end of my life as I knew it. Maybe I should have sensed in him the beginning of something that would carry me along like an angry crowd moving this way or that, like something bigger than I could control. I'd gotten used to control by then, after nearly seven years as a prosecutor and more than three years as a defense attorney working with Barry Slotnick, who was publicly perceived as one of New York's premier criminal attorneys, having represented, among others, Joseph Colombo, Carmine Persico, Meir Kahane, and Bernard Goetz. But it was just a routine arraignment for me, and all I saw was a man who filled the space that contained him, who I somehow knew would fill any space he inhabited.

I remember it all—every detail, every nuance—a gift and a curse. On the morning of March 28, 1985, I was sitting in my office on the twenty-first floor of the Transportation Building in lower Manhattan, reviewing papers relating to an upcoming trial, when Barry called me on the intercom to ask me to his office down the hall. Nearly four years after coming to work with Barry, the firm banner was Slotnick, Cutler. Barry didn't

1

want an ampersand between our names, and it wasn't because he couldn't stand the added separation; he'd just decided that a lone comma implied additional partners' names and gave the impression of a larger firm. And the partnership *was* in name only.

Barry's corner office was roughly twenty by twenty-five feet, with windows looking south, down Broadway toward Trinity Church, and east, overhanging the East River and the Brooklyn Bridge. The view was impressive. Barry sat with his back to the bridge so his clients would *be* impressed. And he sat behind an enormous wooden desk that curved in an arc around his chair. It, too, was impressive—unless you tried to lift it and found it was balsawood and weighed next to nothing. His clients didn't try to lift it. Neither did I, but once I bumped against it hard and nearly upended it and all its little paperweights, statuettes, and fancy pens—I don't recall ever seeing much paper on Barry's desk. The lighting in Barry's office was situated directly over his chair, so that when a client—or a "junior partner" like me—entered the room in the early evening (when he'd routinely schedule appointments with clients), Barry would be bathed in light, like an oracle with the answers his supplicants so desperately needed. In the morning sunshine, although the lights were on, the effect was minimized.

Opposite his desk was a wall Barry had covered with his diplomas, framed newspaper clippings, plaques, photographs of himself with famous people like Colombo and Kahane, and, in the center of it all a huge portrait of himself. Frank King, who has since passed away, always made fun of the wall. He was a former New York City detective and a figure legendary, in law enforcement and on the street, for toughness and cunning. Frank was thin lipped and balding, with a hard Irish face and a sense of humor to match—a physically formidable man, tall and broad shouldered, with fists "like hamhocks," as my father might have put it when he was a "copper" on the beat himself, shortly before my birth. Unfortunately, Frank also became notorious for his alleged involvement in the bribery scandal that swept the Special Enforcement Unit in the mid-1970s; he was persecuted *and* prosecuted by Tom Puccio and the federal Eastern District Organized Crime Strike Force, convicted, and sentenced to five years in prison, of which he served eleven months hard time. Later he became Barry's tape expert and detective-in-residence. For all the bad publicity Frank received, my experience of him was only as a decent and eminently likable man, who was nothing but scrupulously honest and helpful to me. It was he who introduced me to the street world, to the rough-and-tumble people—the "knock-around guys," the gamblers—whom I would later represent.

In fairness to Barry, as concerns his display wall, I should note that Frank would have made fun of any lawyer's wall of this kind. Moreover, I want to say that Barry has a brilliant legal mind and that I remain grateful to him for giving me an opportunity to work with him and make my "bones" as a criminal defense attorney. But it's never been my thing to have diplomas, clippings, and the like hung on my walls. I figure, if a client is sitting in my office describing his legal problem, he probably knows who I am. He's interested in hiring an attorney who won't buckle under government pressure, no matter how fierce—where I went to school or when I was admitted to the bar won't matter to him. On the other hand, a lot of lawyers I respect have these things hanging in their offices, so maybe I'm off base in this regard. But a larger-than-life painting of one's own face—as though the flesh-and-blood version behind the massive curving desk, under the spotlight, wouldn't be enough? To each his own, I guess.

I knocked on Barry's open door. He swiveled in his chair, turning to me from the window, and asked, "Bruce, are you going anywhere this afternoon? Can you handle an arraignment?"

"Sure. What's it about?"

"A few fellows were indicted—Neil and some of his friends. Neil's in the hospital so I won't be going down there, but the rest of them were locked up. Maybe there'll be a fellow for you. Sounds like a gambling case—nothing too big."

Barry would, of course, represent Aniello "Neil" Dellacroce, allegedly the underboss of the Gambino family, but he was looking for a second client in the case. He'd send me down in the hope I'd pick up another client for the firm. That sort of thing never sat right with me—it felt like prowling the scene of a car accident, dropping business cards on the bleeding as they crawl from the wreckage. But I was there to learn from

Barry then, and I operated out of a sense of firm loyalty, so I'd go where he asked me to go.

That afternoon I took a taxi across the bridge from lower Manhattan to the federal courthouse in Brooklyn. I exited the elevator on the fourth floor, where a bevy of lawyers, all wearing the uniform—conservative suits in varying degrees of gray—stood outside Judge Nickerson's courtroom. Among them were Ron Fischetti, Gerry Shargel, Jeff Hoffman, and Mike Coiro—top-shelf trial lawyers all, the crème de la crème of the New York criminal defense bar and no strangers to "organized crime" cases. I exchanged greetings with a few of them and found out that nine defendants had been indicted, seven of whom had been arrested on racketeering and gambling charges at four-thirty in the morning at the Bergin Hunt and Fish Club in Queens: Neil Dellacroce, John Carneglia, Charlie Carneglia (who was labeled a fugitive), Tony Rampino, Lenny DiMaria, Nick Corozzo, Wilfred "Willie Boy" Johnson, Gene Gotti, and John Gotti.

I entered the courtroom. Judge Nickerson had not yet taken the bench, and a few other people were lounging and chatting: a court reporter, a couple of federal marshals, more attorneys, and a few others. I saw Angelo Ruggiero, another "knock-around guy," whom I later learned was one of John Gotti's closest friends, sitting in the first row behind the balustrade that separates the gallery from the lawyers' tables, the jury box, and the bench; I'd known Angelo for a couple of years by then, from other cases, and I remember thinking that he seemed genuinely embarrassed, or even pained, that his friends had been arrested and he hadn't. I also saw Joe Panzer, an attorney well into his seventies by then, and Richard Rehbock, an attorney who was assisting Joe. Joe was a fixture in gambling cases of this sort *and* at the racetrack. He greeted me warmly and asked after my father, whom he'd known from in and around New York's courthouses over the past three or four decades. Joe told me he and Richie were representing Willie Boy Johnson and asked me who my client was. I said something about being there for Barry in case some question came up about Neil's medical condition. Then I turned away to look at the woman standing at the government's table, leaning over to read her notes, I supposed. It was the assistant U.S. attorney in charge of the case, Diane Giacalone.

I'd seen her before in this case or that, but I'd never worked against her. I couldn't have known then how intimate we'd become—this neatly dressed, slightly built young woman with shoulder-length dark hair and I. I would come to know her better, in some ways, than I knew many of my friends, better perhaps than I knew either of my former wives. That's what it's like when you're working on a case with someone, particularly *against* someone. You see it all. Whether you admire, detest, or just disdain your opponents, you'll surely emerge from trial knowing them: from the tiny details—what they eat for breakfast, where they dine, even how they smell as they stand beside you arguing their objections—to the big things—their strengths, their weaknesses, their breaking points. At times, you'll see opponents raise themselves in the struggle to a stature you couldn't have imagined, and at others, you'll see them sink to a place they *themselves* couldn't have imagined. You'll see them hungry, yearning to win, as though nothing mattered more, as though winning could save their lives—desperately holding on by any means possible, at any cost, even as the case is pouring like sand through their fingers (as some cases will, no matter *who* the lawyer). You'll see them breathless with elation, straining with exertion, and destitute with loss.

Most significantly? They'll see you in the same way.

It isn't war. I've never been to war, thank God. But it's as close as one gets to physical combat. Maybe good trial lawyers are just frustrated warriors. Or maybe they're just looking for intimacy in the only place they know how to find it. For it *is* intimate—the courtroom, I mean. It's the sort of intimacy that the kind of person who would spend a life on trial (especially in criminal cases, where the stakes are measured in that commodity as precious as life itself—time) can seldom find elsewhere.

Anyway, I wasn't in *this* case yet, and as I looked at Giacalone, I found it hard to imagine her being the ambitious, hard-bitten prosecutor I'd heard her to be. But there you have it—people *and* things are seldom what they appear to be.

Then Mike Coiro tapped me on the shoulder and said, "I'm glad Barry sent you. I'll stand up for Gene, and you stand up for his brother, John."

"Okay," I replied.

And right about then the fellows began to file in from detention through a door at the left behind the bench. The atmosphere was relaxed, even jocular, a few of them exchanging banter and laughing, and none were manacled. The lawyers had already spoken to Giacalone, and she'd agreed to bail: All the defendants would be released on personal recognizance bond, which meant that they would each sign an unsecured bond for a million dollars and be permitted to leave and return on their next scheduled court appearance.

John was the first one through the door, the others following him in. He was wearing what he used to call an "ensemble," meaning all his clothes were color coordinated: a pair of fine black slacks, a black knit polo shirt or light sweater, and black loafers, topped off with a silver and black satin Oakland Raiders warm-up jacket. Even his hair matched his ensemble—jet black with a touch of gray here and there. He wasn't terribly tall or large—five ten and weighing about 205 pounds—but he filled the courtroom when he entered. By that time in my career I'd already met Carmine "Junior" Persico, the alleged boss of the Colombo family, Neil Dellacroce, and other alleged high-ranking organized crime figures. I'd never heard of John then, but none of them had John's presence—the easy strength with which he automatically commanded the attention of all in the room, codefendants and lawyers alike.

"Come on, I'll introduce you," Mike said, taking my arm and leading me to where John stood leaning against the jury bar. I later learned that Mike was a longtime attorney and close friend of John's. I had a slight reputation among defense attorneys even then, having left the DA's office as a hard-charger and working with a trial lawyer of Slotnick's stature, and Angelo Ruggiero had briefly introduced me to John and some others in a restaurant shortly before that day; but I'm sure John had barely heard of me.

Mike said, "John, this is Bruce Cutler. He's a terrific lawyer. He works with Barry. He'll stand up for you today, if that's okay."

I said hello and stuck out my hand. John nodded and took my hand firmly in his, not squeezing it too hard, as some will do from time to time to show you how tough they are, as though shaking hands were some silly opening contest of strength or will. It was his eyes, not his hand, that sized me up, deciding what I was made of, whether I could really "stand up" or whether I was just another smart mouth with no real endurance, whether I could be trusted. I learned later that this was his way—to try to assess people at first meeting, to determine their character. I came to know that he was remarkably astute in his assessments and that he trusted his instincts implicitly.

"Nice to meet you, counselor. What do you think of the case?" he said, speaking in his resonant baritone, his mouth twisting in a wistful sneer. "I'm told they pinched me for stuff I've already been to jail for."

He was referring to the fact that he was facing a RICO (Racketeering Influenced and Corrupt Organizations) indictment. I have a lot to say about the RICO Act and its potential, and actual, use and abuse by overzealous prosecutors to curtail the civil liberties of anyone they may decide to target in the furtherance of their own careers, but I'll save it for some other time. For now, it will suffice to note that RICO, among other things, permits the government to point to two or more criminal acts (whether these acts are alleged for the first time or form the basis for past convictions with respect to which sentences have already been served), and allege that these so-called "predicate acts" constitute "a pattern of criminal activity" that furthers the interests of a "criminal enterprise," thereby warranting the imposition of inordinately long sentences. Sound vague? Well, it's a vague statute, with lots of room for interpretation and lots of opportunities for the government to destroy those whom it has decided are worthy of destruction. John wasn't yet a "career-maker," a defendant of sufficient stature to propel his prosecutor to a promotion, higher appointment, or elected office. The government was just flexing its muscles that day, trying out a relatively new toy (RICO was enacted in 1970 but by 1985 had still as yet seen only limited action) against a few gamblers, and John had just been caught up in the government's dragnet, charged largely with crimes for which he'd already served time in jail.

I told him I wasn't yet fully familiar with the indictment but that I would be. He seemed embarrassed by the charges on that first day—small stuff not worthy of him—and his codefendants (except for his mentor, the absent Dellacroce) were clearly not his equals, either in his mind or in theirs. I committed the minor faux pas of saying, "So I understand

7

you're Gene's brother," gesturing toward Gene Gotti, whom I'd met previously.

"I guess you can say that," John laughed. He was obviously used to hearing people refer to Gene as *his* brother.

John was the most gifted of the eleven Gotti children. I heard later that, by the time he was eighteen years old, he'd already been on the streets for six years. At first John ran with the Fulton-Rockaway boys in Brownsville–East New York, honing a reputation for ferocity, protective of friends and family. Eventually he caught the attention of local street bosses who vied for the right to take him under their wings. It was a sort of draft system, something akin to those held in professional sports, but on *these* streets, if you don't get drafted, you're nobody. One of the bosses admiringly dubbed him "Crazy Horse," after the great Sioux war chief. Looking back, at the risk of sounding sentimental, it's sad how prophetic the nickname was. John was the epitome of toughness and independence (which would cause him trouble) and a brilliant thinker. And he had what some have referred to as the "gift of rage," so he was groomed to be the "hoodlum's hoodlum" (as he was affectionately called by a friend in a conversation caught in FBI surveillance).

The courtroom deputy announced the entrance of Judge Nickerson, directing all to rise. The lawyers moved to their places at the broad counsels' tables before the bench, the defendants stood in the jury box, a few other defendants and arresting agents stood in the gallery, and Giacalone stood at the government's table. For that fraction of a second as the judge took the bench, all was frozen at the ready—everyone had his or her place on a stage where the surreal drama of John Gotti's prosecution would unfold over the next six years.

PART ONE

The Lawyer

1

His great hand engulfing mine, *hoisting* my little boy's body up above the waves: That's how I remember Murray, my father. He was a big man, six foot three inches, 215 pounds of heart and brawn. He was not only my father but also what many fathers are not—my father figure.

Murray was also my friend, although it's hard to recall him as such when I was growing up, disciplinarian that he was. I remember him telling me that my only *true* friends in life would be my parents. He was right, wasn't he? I mean, there's no limit to the love and protection afforded by your parents—the love of any other must be limited by self-interest, no? Or is this only the ranting of one paranoid lawyer-cop to his paranoid lawyer son? A legacy of vigilance, passed from centuries of pogrom victims in Lithuania, Hungary, and Austria, to my grandparents Irving and Bertha, and Harry and Sadie in the new world, and from their generation to Murray and Selma, who offered it to me. A legacy of loneliness, a fitting foundation for the egocentricity, pervasive distrust, and maniacal single-mindedness required of a successful trial lawyer.

I was born on April 29, 1948, in Borough Park, Brooklyn, the first son of the first son of the first son. Selma, my mother, was fond of recounting (amid the confirming nods and clucks of my grandmothers, Bertha and Sadie), as she'd bathe me or tuck me into bed, how she'd selected me herself from among all the other children at the hospital, how there'd been no more beautiful infant than I, no child so clearly special, "supernatural even." "I'll show you photos—you'll faint," the women in my family would aver with rolling eyes as I grew up. "You were special, number one double plus." So what can I say? I had a destiny to fulfill, and a kid can't safely turn his back on destiny.

At the age of twelve, in the late nineteenth century, Murray's father, my grandpa Irving, arrived in this country from Vilna, Lithuania, with his own father, for whom I am named. Irving was big like his son Murray. He seemed even bigger to me, as I was growing up. He married Bertha, whose family arrived in New York from Hungary at about the same time as the Cutlers; no Tinkerbell herself, standing nearly six feet, she and Irving seemed perfectly matched to my child's eye—two Tolkien-like giants from a kingdom before the dawn of time, come to the new world to establish a *new* kingdom where *I* would reign.

I recall Irving as a tough man, even tougher than my father, walking around Borough Park where he and Bertha resided (or West Fourth Street in Greenwich Village, where he had a little shop manufacturing leather bands for hats and caps), smoking a big cigar in an orange cigar holder that he'd remove from time to time, with a massive bearlike paw, from where it protruded between huge upper and lower sets of teeth. He cared nothing for how he might appear to others—neither neat nor solicitous—and he was no doting grandfather but rather a gruff, protective presence whose brusque benevolence was mitigated by (as Murray and Selma would tell me) Bertha's warmth. In 1950, when I was only two, Bertha died of stomach cancer (which eventually claimed my father as well in 1994). I have sorely regretted not having had an opportunity to get to know her before she passed away. Years later, when I was a teenager, Irving remarried to Gertie, who was a kindly woman, but she was never my grandma.

My mother's parents, Harry and Sadie, who emigrated from a small town in Austria at roughly the same time (and age, twelve years old), lived nearby in Borough Park. Their house became a second home to me, a place I truly loved, where I would often go after school, a place of candy, cookies, attention—and freedom from my father's discipline. It was a place of learning as well, where Sadie taught me to multiply and to tell time. She was by far the funniest, most loving woman I've met to this day, with a great appetite for food and life in general.

I think Harry was the smallest grown-up I'd ever seen as a child. At five foot nine, he was positively Lilliputian compared to my father's Brobdingnagian family. But even his wife, my grandma Sadie, was five foot ten inches. Harry had gained weight when he stopped smoking sometime before I knew him, and I recall him always with cough drop box in hand; it was one of those antique-looking white boxes, depicting the bearded visages of the Smith Brothers on its front—despite the goyish name (and they might have changed it, as did so many immigrants), they looked like pious Jews to me, and I always associated them with Harry's devotion to Judaism. Harry and Sadie were considerably more devout than Murray's parents, Irving and Bertha. Indeed, Harry had lived in the basement of a synagogue when he was growing up; he spoke Yiddish and English equally well and read Hebrew, too. By contrast, I recall my father coming home one Chanukah season, dressed in a Santa Claus costume and carrying a Christmas tree.

Harry was in the lamp business, a traveling salesman with his office in the house. He would often drink a shot of schnapps before dinner. I remember making fun of this habit, probably because my father frowned on it, and my father's worldview was law then. To my shame, I remember laughingly calling Harry a *shika,* the Yiddish term for "drunk," as I'd heard my father refer to a particular drunken fellow we'd often see stumbling through our neighborhood. In any event, I was utterly unfair in my judgment, for despite his single shot before dinner, I'd never seen Harry act—or dress—in any way but fastidiously. He'd never leave the house without a white shirt and tie; in fact, I never saw him wear a sports shirt.

Harry and Sadie had three children. My mother Selma, the middle child, was the most talented and their favorite, I think. Sadie passed away at eighty-three in 1978, shortly after I broke up with my first wife, Gladys; I never told her about the divorce, for fear it would break her heart. Perhaps she divined it anyway, for a heart attack took her shortly after Harry died of heart failure himself. It was no surprise in Harry's case, given his weight and his excitable nature. Selma was high-strung as well.

Murray and Selma met in Borough Park, where they both attended John J. Pershing Junior High School and New Utrecht High School, although Murray was six years ahead of Selma. They eventually met through mutual friends and began dating. At first Harry and Sadie were put off by how German Murray looked, so tall and blond, with blue eyes, and a policeman by then. They were relieved to find out he was Jewish— maybe not so religious, but at least a *Yiddel*. Murray and Selma would argue a lot, I'm not sure what over, but I was told that Selma would then leave apologetic notes on Murray's windshield and they'd make up. In any event, they soon married, had three children (my older sister Phyllis, my younger brother Richard, who is presently a federal prosecutor in California, and me), and *remained* married until death did them part nearly sixty years later. Upon marrying in 1941, Murray and Selma moved into the six-story apartment building at 4600 Ninth Avenue, where I was born, close by the homes of both Harry and Sadie and Irving and Bertha.

My father had by then made the sergeant's list and shortly thereafter would become a detective, assigned for a time to the squad of the Kings County District Attorney's Office, where both of his as-yet-unborn sons would one day work as assistant district attorneys. Murray graduated from neither Brooklyn College nor New York University, both of which he attended at night for a time after New Utrecht High School, but went directly to Brooklyn Law School (also at night) where he obtained his law degree. His day job, while attending college and law school, was working in Irving's hatband factory for eight or nine dollars a week. Rather than entering into the practice of law upon graduation from Brooklyn Law School, Murray chose to take the test for the New York City Police Academy. Today this might seem an odd career path, but it would have been easier to understand during the Depression, when a law clerk's starting salary was about five dollars a week (if such a position were even available for a young Jewish law school graduate), hardly enough to support a young wife and begin a family.

Murray finished in the top 300 out of some 35,000 who took the police academy test that year. His class at the academy, that of August 1940, was noted for being the first truly integrated one in academy history. Many alumni of the class went on to be extremely successful in public service: Al Seidman, later the New York City chief of detectives; Sanford Garelik, the New York City chief of the Transit Authority Police; and Fred Ludwig, who became the chief assistant to District Attorney Thomas Mackall.

I recall my father reuniting with his classmates each June of nearly every year of his life at German Stadium in the Bronx.

At one such reunion, when I was a small boy, I remember an alumnus recounting a tale of my father as a young twenty-five-year-old police officer patrolling an elevated subway station in Brooklyn. A young girl approached him to request his help with four drunken youths who were accosting her. My father's lifelong friend Dave Silverstone, who was in the navy then and in uniform at the time, happened to have been chatting with my father when the girl approached. Murray's efforts to help her resulted in an all-out brawl; eventually someone handed him his nightstick, and with Silverstone's assistance, he was able to subdue the four, who were hauled into court and ordered to pay for my father's and Silverstone's uniforms, which had been shredded in the fray. The story has remained with me all these years as a recollection from a more innocent time in our city's history, when every misunderstanding didn't necessarily result in immutable tragedy.

Murray remained on the force only seven years, after which he began his career as an attorney. I didn't learn until my college years *why* he'd left the police department—or until later still, how important the events surrounding his departure would be in sculpting my view of the world and ultimately my life.

My earliest recollections of Murray are in Borough Park, where he was a giant—physically enormous and strong but also immensely popular. We'd walk around the neighborhood hand in hand, and everyone would greet him, smile, and exchange friendly words. Often we'd leave the neighborhood, sometimes for such exotic locales as Harriman State Park or Bear Mountain in Orange County, New York, or for a trip on the Day Line up the Hudson River to West Point. My father loved public places, places filled to overflowing with masses of different people doing different things—a great canvas covered with limitless details worthy of observation.

By far our favorite trips were to the beach, and our favorite beach was Jacob Riis Park in Queens. It was a beautiful, pristine beach then, and I suppose it still is, although I haven't been back in a long time, perhaps for fear that it wouldn't live up to my memories. Sometimes Selma would come with us, but she wasn't crazy about it—too much sun and sand, and too many people. When she did, she'd mostly stay under the umbrella. My father and I would change in the Riis Park locker room, and we'd wear our locker keys around our wrists. We'd swim and sunbathe until lunch at the beach cafeteria—my father sticking to some cold chicken he'd bring from home and I gorging myself on hot dogs and fries or maybe even a corned beef sandwich and a potato knish with a cream soda—and then we'd go back to the beach until it was time to go home, when we'd use the public showers. A perfect day in a boy's life—the beach, the locker and key, the cafeteria, the showers.

More than anything else, my father instilled in me that justice was the key—to live justly and to fight injustice wherever I might find it. The stuff comic book heros were made of. I was no reader of comics, growing up. I was no reader at all as a boy. There were other keys besides justice. In fact, too many keys for any normal kid to carry.

My mother was a forceful woman, dominating in a subtle way. In other words, my father ruled the Cutler roost because my mother allowed him to think he did. She was also very protective of her children. Murray was tough on the kids, and he thought I particularly was a mama's boy—I recall Selma sheltering me from my father's slings and arrows. She was lavish with her love, while Murray was parsimonious, making me earn every syllable of praise. No misty-eyed sentimental he, not like Selma and me. Murray was a man's man.

And Selma was a woman's woman. She was always well dressed, even around the house, and I can't recall, as a boy, ever seeing her in pajamas or even a bathrobe. She'd sport one of those cocktail dresses, the kind you might see in an old Ava Gardner movie, just to take out the garbage. In fact, to my eye she resembled Ava Gardner or Rita Hayworth—a broad-shouldered, shapely, and beautiful five foot nine inches and 140 pounds. Perhaps she would have been suited by temperament to a more glamorous lifestyle; she certainly loved going out to parties and dancing with Murray, and people would call them a "striking couple." Although she was a caring mother, she was also a highly nervous woman, for whom being a parent was difficult. She didn't have the patience for the job and would easily become infuriated at relatively minor infractions. I recall that she might slap me (without hurting me, of course—she wasn't strong enough to hurt me) and then break down in tears; I'd wind up feeling awful at getting her so upset. She needed what is called today downtime, and lots of it: time quietly alone or playing cards with her girl-friends. I, too, need downtime. Murray embraced life in all its harshness and beauty. Selma *endured* life, as do I.

I was my mother's darling and rarely ran afoul of her. My older sister, Phyllis, suffered the brunt of her temper and neurotic insistence on neatness. She believed my sister to be extremely untidy, and she couldn't tolerate untidiness. Selma would come into Phyllis's bedroom and find an old dish of ice cream left over from the previous day, or ashtrays in drawers, and the house would shake. Phyllis, then a teenager, would be devastated, crying and apologetic. People would say they looked alike, but they were wrong—they looked nothing alike, which may have been another reason my mother was so hard on her. Unjustly hard, for Phyllis was a brilliant student, beautiful in her own way, and a good and kind girl, if not along my mother's lines. She graduated from Madison High School with a 96.7 average and was admitted to Smith College, Vassar College, and the University of Michigan. Perhaps because it was the far-thest from home, she elected to attend the last for two years, earning top grades, then returned to marry Michael Oliphant, a two-time captain of the Columbia baseball team and a wonderful man.

My mother would tell me to mind the little things: appearance, pre-sentation, syntax, diction, manners, making a good impression, and of course schoolwork. She was concerned with what others might think of us, of me. My father would tell me to mind the big things: physical endurance and strength, discipline, mental toughness, and of course schoolwork. He didn't care a fig about what others thought. Both Murray and Selma were concerned with my schoolwork, as well they would be, for I was surely no scholar as a boy and barely a student. Selma focused on the right side of my report card, which in those days recorded my grades and teachers' comments as to my courses of study—arithmetic, reading, art, science, social studies, and so on. She herself had been a brilliant scholar, graduating at sixteen as salutatorian of her class at New

Utrecht High School in Brooklyn, being offered a full scholarship to Cornell University (which she was not permitted by her parents to accept, because "decent young women" were not then permitted to leave their families prior to marriage), and graduating from Brooklyn College at nineteen, having won the chemistry and Latin medals. She expected nothing less from me.

Murray, on the other hand, was fixated on the left side of the report card, which addressed my conduct among my peers and my comportment and behavior generally. Although my report cards were less than impressive on either side, the left was particularly dismal. While my mother bore stoically her disappointment in my academic efforts, my father was typically more demonstrative. At the entrance to John Jay Pershing Junior High School, a plaque read, "What you are to be, you are now becoming." My father had deep faith in this credo and repeated it to me whenever he thought of it, and he thought of it often. He had no sense of humor when it came to my classroom pranks and little tolerance for disciplinary lapses of any sort. Murray was religious—not in the traditional sense but in that he was upright, steadfast, and resolute. He had the forbearance of a Presbyterian minister, never indulging in gambling, alcohol, or cigarettes (although he would permit himself a good cigar from time to time), and the strict self-discipline of a Jesuit priest; while he bore to his grave a firm belief in personal freedom, he felt that it could operate wholesomely only in the context of strict personal discipline.

Although Murray was not the academic that Selma was, he shared with her a drive to be the best, or at least the best that one could be, and for him this could be achieved only with sweat (and blood if need be), with hard work. This drive went hand in hand with a firm belief in the premise that sparing the rod spoiled the child, and report card day usually brought with it a sound smack or two to the back of my head (at the very least).

But I was an overly active and feisty child, and quiet behavior in the confines of a classroom came hard to me. I would find it difficult to concentrate on word problems in mathematics or on reading at all. As I couldn't get attention in class as a scholar, what else could I do to rightfully establish my presence but disrupt? Now, those familiar with my courtroom style might be tempted to draw parallels, but such analogies would be facile and, more important, wrong. My goal in the courtroom throughout my legal career has been the opposite of disruption and chaos: *Control* has been my aim, *my* control (on behalf of my client), as complete as possible given the presence of the robed figure on the bench and opposing counsel's similar aim. And only in the rare instance that controlled chaos may aid my client's cause will I resort to such, for a jury is usually repelled by chaos. The *universe* may be (as I understand scientists to presently believe) careening toward entropy, but in the courtroom, as in most of our daily lives, there is a great yearning for order. As a child, I found this lesson elusive, despite my father's ever-ready and often heavy-handed reminders.

I recall my grade-school antics to have been puerile and largely harmless, consisting of the usual teacher-baiting and pecking-order squabbles with male classmates. One incident stands out in my memory, however, still evoking feelings of shame. A particular altercation with a classmate resulted somehow in the fellow's being seated on the gym floor crying. I'd been around enough to know that his tears would carry with them the sympathy of the teacher and an inference of my guilt. I rushed to the washroom and rather artlessly splashed water on my face to simulate tears of my own. Needless to say, the charade fooled no one, least of all my teacher who once again duly dispatched me to the principal's office, and my father, who later that night—halfheartedly, I thought—punctuated the incident with the traditional swat or two. But there was more this time, and that's why I so vividly remember the events of that day. I read true disappointment in my father's eyes when my mother described my attempted deception. I recognized then the difference to Murray between disciplinary lapses, which constant drilling might remedy, and failures of character, which he feared were inherent in one's soul and spirit and impervious to limitless marches or whacks. Murray went on to make explicit what I'd already understood, and he told me that the most important part of being a man was honesty, telling the truth to myself and to others, for otherwise I wouldn't know who or where I was, and if I didn't know who and where I was then no one else would either.

Without this moral compass I'd be lost. To Murray, being lost was far worse than being second best, and from then on it would be to me as well. This is not to say that I never again had an ethical lapse, as must we all, but I've since always tried to recognize a moral failure as such and to promptly rectify it whenever I could.

I feel as though I've been fortunate in receiving the best traits of both Murray and Selma. From Selma I like to think I inherited my physical attractiveness (at least when I was younger), loyalty, compassion, and a drive to succeed, to be the best of the best. From Murray I inherited a work ethic, a sense of proportion, endurance, and an ability to take the bitter with the sweet. From both I derived a deep disdain for the insouciant, the indolent, and the indifferent (except occasionally in the women I've been attracted to, where insouciance, indolence, and indifference are sometimes sexy—and always feigned). Interestingly, the people to whom I've been drawn have always had something of my father or mother about them. Survivors of the crucible, all.

In an interview with Mike Wallace of CBS's *60 Minutes,* shortly before my father's death, Murray cited as my strengths: discipline at work and loyalty to friends, family, and clients. As my weaknesses, he cited my overeating, smoking cigarettes, and women. I guess I never attained his standards of discipline, and I imagine I never will. Yet in a world as perfidious and venal as ours, these trespasses are banal, hardly the stuff of legend—I'm almost ashamed of my pedestrian weaknesses. If I couldn't rise to the disciplinary levels set by Murray, at least I might have had a fault of epic proportions. Who knows? Perhaps one of those he cited will wind up a tragic flaw. And maybe Murray was only being kind to his unruly son in a national broadcast. Anyway, no one knew me better than Murray. In the same interview, he told Mr. Wallace he could always count on me to do the right thing. I hope that will suffice, Pop.

2

In 1955, the summer after second grade, Murray and Selma moved from Borough Park to Flatbush. My father had begun working as a lawyer some seven or eight years before then and put down an $8,000 fee he'd received for an estate matter toward the purchase price of $23,000 for a detached house at 1493 East Twenty-fourth Street, between Avenues N and O. I was to transfer from P.S. 169 to another school, P.S. 197. The change to the new school and neighborhood was a frightening one. On my first day in Flatbush I was too embarrassed to leave the house, and my father had to force me out to water the lawn. The lawn, so dazzlingly green—a little bit of Ebbets Field, where the Brooklyn Dodgers used to play—was wide and broad beyond its actual one or two hundred square feet, so *ours*. And we had a *driveway* of our own, where Murray would park his brand-new 1955 aquamarine Buick Roadmaster, with a gleaming white top and four oval jet-age air vents on each front fender.

We had stepped up in the world. It was clear even to my seven-year-old eyes that we were striding forth with the rest of America toward better times. And it wasn't just the house and the lawn. The kids were different, too. I remember my father trying to introduce me to the new neighborhood children when they'd come by—on their *bikes*. I didn't have a bike. Neither did my friends in Borough Park. Where would you bike to in Borough Park? Here in Flatbush, I could ride all the way down Bedford Avenue, past Madison High School to Marine Park—bike-riding heaven.

It was more than just the external trappings of privilege. These kids were smarter and more athletic. To me, it was as if I'd been pulled from P.S. 169 and dropped into Harvard or Yale. My reading ability, which was above average for my old school, was less than average in the context of my new school, although I should note that most of my classmates at P.S. 197 were nearly a year older than I. But according to my teachers, my comportment also still "needed improvement," in the euphemistic language of this new frontier; I recall spending many hours in my new principal's office engaged in strenuous behavior modification. In fairness, I should say that, for the most part, my teachers at P.S. 197 were caring and supportive.

Eventually, I began to fit in slightly better. By sixth grade I was still in 6-5, as opposed to 6-1; the second digit reflected the academic level of the class. Although I was far from being in the in-crowd (the boys were still smarter, six to seven "readers" ahead of me, and more athletic), the girls seemed to like me. Yet it was clear to me that, unlike my brilliant older sister and kid brother, I was at best mediocre at everything I undertook then. Flatbush had taught me the fragility of my psyche—and how important a strong self-image could be in determining how successful a person was in any endeavor undertaken. I was too nervous a child, wanting too badly to do well, feeling too far behind the eight ball in my environment.

Feeling as backward as I did in Flatbush, I suppose it was natural for me to swagger more than I should. This led me to one of my more memorable childhood lessons. It stemmed from a fight I had with the cousin of a neighborhood friend who lived up the block. One day, when I was doubtless throwing my weight around, my friend's cousin stood up to me. It was the first challenge I can recall to my tough-guy status. He was a big fat kid, and I recall that I always had trouble with fat kids (even later when I was wrestling competitively and playing football in prep school and college); their sheer weight made them difficult to maneuver. Well, the fat kid and I fought for a while—which is to say we grappled and slapped, an errant punch here and there, with a few headlocks thrown in. And when I was tired, I just wiped my nose and walked off. I knew it wasn't right, that I'd left the matter incomplete, unfinished. All the kids on the block knew it, too, and they jeered and taunted me as I walked into my house.

My mother asked what it was all about, and I told her. I suppose I'd expected her to pat my head, search me for obvious injuries, and tell me not to mind those silly children outside. To my surprise, she was appalled at my behavior. My father, who walked in from work a few moments later, was actually less disturbed by the incident than my mother, but he seemed to understand my mother's concern that I might be too frightened to defend myself (or perceived by the neighborhood children as such), and at her urging severely chastised me. My grandmother, who was visiting at the time, piped in that I must always stand up to bullies; but of course, it was far from clear that *I* was being bullied here. My father, to whom the final word belonged in all such matters, ruled on a slightly different ground. He said to me firmly: "You gotta finish what you start. You started this one, now you go out there and finish it, one way or another, win or lose."

So I went out again, where all the kids, including my friend and his cousin, still awaited the outcome of proceedings behind closed doors in the Cutler household. Like jurors attending some technical ruling from court chambers, like the People expecting vindication, these uncouth guardians of the scales of kid-justice in the neighborhood stood impatiently waiting, with a catcall here and a deriding whistle there. I walked up to where my friend's smirking fat cousin stood regarding my approach with some suspicion, and I punched his bulbous nose with a right jab. He lunged, wrapping his arms around me, and we fell to the ground, rolling and tumbling for what seemed years but was probably no more than ten minutes, including the long stalemate periods with my arm locked around his enormous head, or with his tonnage plastering me to the sidewalk. When we were both too tired to continue, it was over by mutual agreement. There was no clear winner and no clear loser; unless one could count me the loser in that my stature as "able to lick anyone" had been reduced. But the conventions had been observed, and pride had been preserved on all sides; a trial had ensued, and the public's need to believe that justice prevailed was salved. I'd learned, in *reverse* order of importance: first, to curb a tendency to misuse my strength; second, never to walk away from a fight, particularly one I'd started; and third, and most importantly, *never to underestimate or disappoint the public's need for catharsis.*

In the course of my time at P.S. 197, perhaps at around ten years old, I fell severely ill. It started with a mild sore throat when I returned home from school one afternoon. My parents had scheduled a short trip to New Orleans for Mardi Gras and now considered postponing it, but as I was feeling better the next day, they decided not to alter their plans. When they returned a few days later, my sore throat was still there—as it was a few weeks later, until one Sunday morning I could barely get out of bed. My mother, who permitted *no* absences from school (unless rigor mortis was setting in) and thought I was gold-bricking to avoid Hebrew school (at the East Midwood Jewish Center), of which I was admittedly not fond, forced me to go, but I didn't last through the class. I wound up at Maimonides Hospital with a high fever, tremendous pain in all my joints, and a rash from head to toe. There the doctor suggested that, in view of the symptoms, I likely had rheumatic or scarlet fever. My mother, who questioned nearly everything, pored over medical journals until she came up with a diagnosis of *hennox purpura,* a substantially less dangerous condition, in which Dr. Krim and his colleagues ultimately concurred. I spent a month in the hospital before returning to school. Two things from my stay there stand out vividly: first, a vision of my body's reflection in a mirror, frighteningly skeletal by the day of my departure; and second, a little boy named Niles, about my age, in the bed next to mine, who passed away of leukemia during my stay. I was affected deeply by this, my earliest recognition of mortality—that of other living things and my own.

My many absences from Hebrew school, at a critical point in my bar mitzvah studies, together with my chronic disciplinary problems, caused me to leave the East Midwood Jewish Center. This was of little concern to me or to my father. Murray was Jewish, of course, but his lifestyle, habits, diet, and attitudes were not traditionally Jewish; rather they were molded in the paramilitary ethic of the police department. He was openly irreverent, believing in doing the right thing rather than *davaning*. But my mother, who was ever concerned with appearances, insisted that I have a tutor to teach me Judaic history and the *haftorah* in order that I might have a bar mitzvah like all the other Jewish boys we knew.

So I was bar mitzvahed with a lavish affair at the East Midwood Jewish Center. All my friends were there, including my pals from the 8-1 class (although I was still in 8-5 myself) at P.S. 197. It was a terrific party, but it was no religious event in my mind. I experienced it as an indication of my parents' love for me—and as new proof, sorely lacking since I'd moved to Flatbush, that I was indeed still special. My clearest memory of the day? Posing for that photo, the one where I stand, my smile gleaming, between Joan Steinman and Sherry Danzig, as each plants a kiss on my cheek.

I graduated from P.S. 197, and my life changed abruptly. I was admitted to the Polytechnic Preparatory Country Day School, nestled between Fort Hamilton and the Verrazano-Narrows Bridge in the Dyker Heights section of Brooklyn. I honestly believe this to have been the single most important event in my life.

My father was good friends with Professor Richard Maloney, who was then on the boards of trustees of both Poly Prep and Brooklyn Law School, and after a meeting with me arranged by my father when I was in the eighth grade, Professor Maloney agreed to recommend me for admission to Poly. My father then took me to see the school firsthand, and I was dumbstruck. I can still recall my first impression: twenty-five acres of rolling hills and oak and maple trees, two ponds complete with geese and ducks, a beautifully maintained cinder track with lanes marked perfectly in dazzling white lime, tennis courts, two perfectly trimmed baseball fields, a soccer field, and a football field, all as green as the endless grass at Ebbets Field where my father once took me to see the

Dodgers play. These grounds surrounded an imposing three-story red brick colonial building, trimmed in brilliantly white-painted wood, built around a quadrangular courtyard. The highest point of the building was a white-steepled clock tower, overseeing the paradise that sprawled in all directions from its foot.

The older boys looked like what I imagined Harvard or Yale men to be. All the boys I saw, even the ones my age, were wearing sports jackets, mostly navy-blue blazers, with pressed dress slacks and white shirts and ties, and the girls—well, there were no girls, anywhere.

Entering the brick structure through the white-pillared front porch, I stood in a large foyer, wainscoted to waist-high in dark oak paneling and lined with oak and glass cases filled with bronze and silver cups and trophies, with the names of what I presumed were former students and dates that stretched back to long before I was born, before even my father was born, engraved in a gothic print that I associated with times long past. The hallways that proceeded in both directions from the foyer bore the same dark oak paneling and had an indescribable aroma that I can recall to this day; perhaps it was the odor of aged wood and the oils and waxes used in maintenance over the past half-century since the building was erected, but to me it will forever be the smell of young boys shouting, boisterous and gleeful, running down the halls, eager and bursting through their skins into manhood.

And there were so many more marvels: a competition swimming pool; a dirt-floor gymnasium; a gleaming wood-floor basketball court; rows and rows of locker rooms and showers; and small shiny wooden rooms for a game I'd never heard of before called squash. Nothing typifies Poly, or my impression of it then, more than the squash courts. What an alien concept: the extravagance of rows of tiny wooden rooms for an esoteric sport created by and solely for the upper class. If Poly was squash, I was so stickball. It was a mythical place, far from Borough Park, far from Flatbush and P.S. 197 and Madison High School, a place carved from upper-class America itself, as Eve was carved from Adam. Later, when I went off to college, I would learn that there were other schools like Poly, perhaps still *more* venerable prep schools, boarding schools in New

England, with wondrous grounds and fabulous histories. But *then*, as an eighth grader only recently astounded by the wonders of Flatbush, there could be no other place like this on earth.

Not just anyone could go to Poly. First, your parents had to be willing and able to come up with the then-monstrous sum of about $1,000 a year for tuition, lunches, books, and so on. That was my father's part. Second, there was an admissions test to pass. That was my part. I took the test in the cavernous old fifth-form study hall, and I know I couldn't have done all that well, but Poly agreed to accept me on the basis that I would repeat eighth grade (which was called second form in the British style, seventh through twelfth grades being first through sixth form, respectively). This meant that, rather than being the youngest in my class, as in public school, I'd now be in class with boys my own age. I felt like a Rockefeller, or at least like what I thought a Rockefeller might feel: privileged, empowered.

I remember my father telling me, "Bruce, you've got a chance now, a special chance here, a new lease on life. I expect you to make the most of it."

I made the most of it.

In fact, I became maniacal about succeeding—or I should say, about not failing. I was tireless in my efforts in class and on the playing fields. Poly was a new world for me, one where I could reinvent myself, become the boy I'd always wanted to be, the boy I thought my parents always wanted me to be. It was tabula rasa, a phrase I'd learn that first year in Latin: a clean slate on which I could re-create myself.

A 1950s-style bulbous school bus, freshly painted in navy blue and gray (the school colors), with the words POLY PREP emblazoned on the sides, would pick me up for school at its Cropsey Avenue stop at about seven forty-five A.M. I'd return home, exhausted after athletic practice and a shower, at about seven-thirty in the evening. Then I'd have dinner and begin my homework, which for some of my Poly peers was no sweat but for me was another mountain to climb each night. Throughout my Poly career I had chronic and violent migraine headaches, complete with vomiting and temperature loss; in my early years there the migraines would cause me to leave in the middle of the day about once or twice a month. They were in a good cause as far as I was concerned (although they would worry my mother, who suggested that my head might burst wide open from too much study), because Poly had saved my life. I wouldn't let myself sink there, and I didn't. The migraines disappeared before I graduated.

Predictably, my academic life still had its ups and downs, but as the years went by I improved, until I had all A's and B's in tenth grade. Some subjects proceeded better than others. I found first-year Latin (a subject none of my pals would be taking back in Flatbush) to be wholly alien, like nothing I'd ever had to study before. I, like everyone else, started from square one, with no past failures in tow, no material poorly integrated in past years to drag me down. So it was great fun, and I did well under the kind tutelage of teacher David Winder in eighth grade. But for second-year Latin I would have Kenneth Boyd Lucas, the most feared teacher at Poly, and my Latin career was over.

Mr. Lucas had attained iconic status at Poly, largely for his legendary athletic skill and strength as a young man, particularly on the gridiron, at Poly (class of '17) and later at Harvard (class of '21). I'd heard people say that he was a war hero as well. I imagine he had been in the military, for he had a drill sergeant's manner. It was said that, during World War II, he would run the boys at Poly through parade marching exercises. I recently came upon his photograph as a young teacher and coach in the Poly Prep yearbook of 1934: There he was with matinee-idol looks, a glint in his eye, jaw firmly set.

He was a big man, but his broad shoulders were severely hunched over by the time I attended Poly, and he walked with a slow painful gait that would roll him from side to side, owing presumably to his bad legs, which bowed out visibly at each knee. Every precarious step brought a grimace to his lips. Almost every boy in his classroom had a nickname, usually highly unflattering, that Mr. Lucas would bray out if any excuse were proffered for a less-than-acceptable recitation of Caesar's *Gallic Wars*. Beneath even entitlement to an offensive nickname, I was merged with the mass of nameless boys he would only call Sunny Jim. I recall hesitating in the course of one Caesar recitation and hearing his raspy voice bellow, "C'mon, Sunny Jim," while his big bony knuckles, bigger even than

Grandpa Irving's, banged impossibly loudly in some bizarre rhythm on his desk. Only the best athletes at Poly, WASPs all, were spared his abuse; these he would dub his Praetorian Guard. A mediocre recitation from one of them would provoke nothing worse than a withering glance from Mr. Lucas—what a *good* recitation might have earned one of the rest of us. I allowed Mr. Lucas to destroy Latin for me. My determination and strength, which would be second to none a year or two later, had not congealed by ninth grade. A few years later I would have embraced the challenge he presented, but not then. And I was in the majority, for Poly students abandoned Latin in droves during the Lucas years.

As for Mr. Lucas's Praetorian Guard—his Christian athletes—I'd never seen anything like them before Poly. They were a breed apart in my eyes. There was an effortless elegance in the motion of these blue-blood WASP athletes, whether they were carrying a football, swinging a baseball bat, or laying up a basketball, an easy grace and sangfroid in executing this maneuver or that, unequaled in the schoolyards of Flatbush or Borough Park. They had different mannerisms and speech patterns, too, which I might have seen or heard on television but didn't know existed in real life. Their parents spoke and acted without shouting, without tears, without public displays of emotion. Dave Johnson, who was in my class when I began Poly in the eighth grade, epitomized this type, always bearing himself honorably, dependably, stolidly, and civilly—the sort of behavior I might have associated with British films depicting Eton or Harrow. Johnson left Poly before his junior year to attend the Philips Andover Academy and then Williams College, where he played on the baseball team. I missed him.

WASPdom at Poly was the top rung on the ladder. This fact would be driven home three mornings every week, when the entire student body would congregate in the Poly chapel and sing "Rock of Ages" and other Puritan hymns, from a hymnbook that might have dated from the First Thanksgiving in 1621 Plymouth. Even though I'd not been raised in a strict religious household, and even though I knew there were other Jewish boys in the chapel with me, sitting in those pews, singing those Protestant hymns, accompanied by the traditional organ, towering pipes and all, I felt isolated and out of place. I don't mean to overstate these feelings, for they were subtle and not at the forefront of my conscious-ness at the time, but a part of me felt like a heathen who'd been permit-ted a missionary opportunity to "see the light" and should know enough to convert.

So maybe it was inevitable that I would enter a Christianization stage in my development. Throughout my Poly years and later at Hamilton College, until I returned to Brooklyn for law school and then to the dis-trict attorney's office, I tended to keep my background to myself. I wouldn't lie, if directly asked, but I wouldn't volunteer that I was Jewish, whether at school, or at my summer jobs in construction or among the Irish lifeguards at the beach, or while waiting tables at Leonard's in Great Neck. I am happy to say that I grew out of this phase completely, although I still admire certain qualities of my WASP classmates in prep school and college.

It took some time, but I learned how important it was for me to embrace who I was, and that this would always be my strongest suit. Hiding my origins was not only at odds with everything my father had tried to teach me about honesty and simplicity but was also the antithe-sis of the straightforward demeanor I most admired in my WASP class-mates at Poly.

Certainly there were ambivalent influences at Poly, such as Mr. Lucas and the school's Protestant charter, both of which tended to undermine my sense of self. Another minor negative influence was my physiology teacher (and a professional model), Mr. Bob Bell, who would walk through the classroom pointing out the boy who would have bad skin later in life, the boy who would have a weight problem, the boy who would lose his hair (me). I prided myself on my looks and found his pre-dictions deflating. But other than these items, Poly was an overwhelm-ingly positive experience for me. Harlow T. Parker, Poly's legendary baseball coach and *my* football line coach, stands out in this regard. Mr. Parker was also the director of athletics at Poly and the dean of boys. Always comporting himself in a dignified manner, he was stern but ever there with a kind word of encouragement or apt advice, on the playing field or off.

One of the most influential figures in my Poly life was Ralph Dupee, my eighth-grade math and gym teacher. What Mr. Dupee taught me can't be overstated: how to read; how to study more effectively; how to attack mathematical problems without paralytic anxiety. ("Don't focus on the rope around the goat's neck," I still remember him saying, in connection with a particular math problem. "Focus on the length of rope from the tree to the goat.") Mr. Dupee, a tall slender gentleman of military bearing, saw something of value in me and nurtured it, speaking and working with me hour upon hour. He himself had had an extraordinary lacrosse career at Oberlin College, holding the national record to this day for goals scored in a single intercollegiate game—twenty-three. I understand that he was also a wrestler of some note. It was he who suggested that I try out for the wrestling team in ninth grade and urged me to try weight training as a means of maximizing my strength. "All a matter of confidence," he would tell me. Weight training would give me the confidence for wrestling, and wrestling would give me the confidence for any other endeavor. He was right.

4

I recall going with my father to the York barbell company on Ralph Avenue in Brooklyn, where he bought me my first barbell and free weights. After I'd learned to use the weights at Poly for a while, Mr. Dupee had suggested that I get my own set at home. Improvising with milk crates for a bench at first, I was on my way. Weight training was the key to fundamentally changing who I was. My body took to weight training like a flower takes to water, responding almost immediately, magically. The more I trained, the bigger and more defined I became. Not everyone's body would respond so radically to training; I discerned this from all the others who would work out beside me with far less visible results. The positive reinforcement made me train even more rigorously. I felt special, more confident. At the same time as I was developing the discipline to assiduously keep at the training, my strength, stamina, and agility were also growing exponentially. This was a kind of *physical* studying for me—in my father's parlance, I was paying the price, suffering to get the reward, and I was being visibly rewarded. No one would work harder or be more disciplined than I. I would be the best, and if I could be the best at this, I could excel at anything from sports to my studies.

In sports the results were substantial: I wrestled and played football and lacrosse. (In 1965 I was awarded the Yale Cup for the best athlete in the upcoming senior class.) Wrestling, under Coach Dupee, built up my confidence individually, and football, under Coach Jim Karney, was the ultimate, calling on *all* my assets, mental and physical, in a team effort to prevail. Football was a precursor to my courtroom battles, which required all the strength and cunning I could muster in a team effort with my clients, investigators, experts, and cocounsel, to have our team emerge victorious or unscathed, or at least relatively so.

I was not a great wrestler, but I was so strong and determined that I would win virtually all my matches. I would wrestle in the 157-pound weight class, toiling mightily between football (fall) and wrestling (winter) seasons to drop down from my linebacker weight of 185 pounds. Wrestling gave me a certain notoriety on campus, because of my success and because of the individual and solitary nature of the sport itself. I was on my own on those mats, exposed, with no one to blame for a loss and no one to share my triumphs. It was a visceral sport really, so brutal I felt it must be worthwhile, like lifting weights all day long. From the moment the whistle blows there is no place for shelter or respite, and your opponent's strength and resolve are in your face, his head rubbing against yours, mashing your nose, your ears.

Football was different from any other sport. Football was my life then. It represented the crucible that my father was always speaking of in one way or another, the *ultimate* test: the place where the best and the worst of a man would emerge. It was also the pinnacle of Poly Prep athleticism. Dean Parker would say, "We take our football very seriously here at Poly." In my younger Poly years I recall watching in awe the parade of great stars every year at the award ceremony in the chapel, players like Joe Segal and John Jacobs (both three years my senior and all-city–caliber players). Three years' difference in age was enormous then, but now it seems like moments, and I've remained fast friends with Joe and John over the decades. And the games every Saturday afternoon on Poly's gridiron were major autumn events, filled with girls, parents, teachers, alumni in the stands, girls cheerleading on the sidelines (for the public schools against whom we played, of course), and girls at the parties afterward. . . .

Football was chaos. Amid the utter confusion, rising dust, shouts and grunts, air thick with sweat and testosterone, twenty-one other bodies aggressively mobile, some clearly running hard at you, the challenge was to get your appointed tasks done: on defense, find the ball, tackle someone, hit the opposing quarterback; on offense, find a hole and carry the ball through, or block someone, or run to a certain spot on the field and concentrate hard enough to make that catch. Football was every bit as brutal as wrestling with the added dimension of cooperation. You had to coordinate your every move with those of each of your teammates. There was never a game I played in high school or college in which I said, "Boy, I'm having a good time!" I know there are players who say that. Not me. I was too busy playing to the nth degree, making sure the tackles were right, the hits were right. The moment of greatest exhilaration was the closing whistle. In court there is at least the illusion of order, of being in control—the chaos and confusion are transpiring on a subtler level. In football (as I imagine in war) there are no illusions—you are immersed in the havoc. Those who survive, who prevail, are those who are *so* immersed that they are undisturbed by and become one with the turmoil, those who *are* the turmoil. I was elected football captain for my senior year—King of Chaos.

Coach Karney was a big fellow, maybe six foot four inches or more and heavy-set, who'd starred as a tackle on the Middlebury College football team after World War II. Karney's slow ambling stride and laconism gave him a reputation for slow-wittedness among Poly boys, wholly unmerited in my opinion. He was a talented and inspirational football coach who influenced me nearly as much as Ralph Dupee. Karney reminded me of my father. He knew when you were faking and knocked you (figuratively and in a training sense) from pillar to post. You couldn't fool him when it came down to "on your marks, get ready, let's go." When you beat out a fellow in practice, you started. It was simple and direct. I liked that about him.

At around this time I met Phyllis Kronenberg, who I then thought was the love of my life. To my eye she was an American Indian beauty, exotic with gleaming black hair, full lips, and an aquiline nose. I'd first seen her at Camp Ma-Ho-Ge, in White Lake, New York; her father was the resident physician there, and I'd attended in the summers from eight to eighteen years old (with the exception of my seventeenth summer). She was a distant dream then, but now, in my junior year at Poly, wearing my new-found strength and confidence, Phyllis was attainable—even if only for a little while. We dated through my junior year and attended the junior prom together. She and I spent countless evenings at her mother's apartment, watching television (*Gilligan's Island* being a particular favorite), I trying vainly, but with the indefatigable energy unique to horny teenagers, to scale her walls—or at least unhook her bra. At the end of junior year she unceremoniously dropped me, and I never did get up those walls.

I saw her again many years later, in 1992, when I was trying an attempted murder case in Orange County, California. She would have been about forty-five years old, and she was still beautiful. Phyllis had read that I'd be defending Dr. Thomas Gionnis, John Wayne's son-in-law, on his retrial for the attempted murder of his wife, and she'd come down to see me. We chatted for a while outside the courthouse during a break in the proceedings, reminiscing, when she laughed and said, "You know, Bruce, it was right after we broke up that I started getting serious with boys." All I could muster by way of sad reply was "Thanks for sharing that, Phyllis." And then the jury convicted Mr. Gionnis. Sometimes there's no justice. On the other hand, the Gionnis conviction was later reversed on appeal.

After breaking up with me, Phyllis left for California, and I spent a lonely summer taking a driver's education course with Mr. Dupee, working out like a maniac in preparation for football season at Poly. Upon my return to Poly in the fall, mutual public school friends introduced me to a cheerleader there, Gladys Moross.

The day before our date was Poly's first preseason team football practice. It was scheduled for two P.M. For some reason I went to my room, intending to lie down for a moment before heading to Poly. I fell asleep and didn't wake until it was far too late to attend the practice. I was mortified—the captain missing the first practice—feeling as though I'd let my teammates and coaches down. In a frenzy I recall rushing out to get a haircut, a crewcut. I was a highly energetic young man at that time, hyperactive; I would never take naps. And after a summer of intense weight training, I was eager for football to begin. How could I have slept through a practice, our *first* practice? And why the crewcut?

All I can surmise is that the pressure to succeed in football, in taking my SAT examinations (which were coming up soon), in gaining acceptance to a good college, and even in making a good impression on Gladys were all coming to bear, like lines all drawing together to a single point: the first practice. And something inside me must have unconsciously buckled under the strain. I fell asleep. Upon awakening, the full weight of my breach of responsibility settled on me, and the guilt was overwhelming. For their part, given my profuse apologies, my coaches didn't take the missed practice too hard, other than to worry a bit that I might be coming down with a flu, and my teammates surely needed no proof of my dedication. The ridiculous haircut was mostly about self-flagellation, as was my decision to double up on my workouts until the New Utrecht High School game scrimmage the next week.

Incidently, the punishment extended to my first date with Gladys. I had to show up on her doorstep the following day with that ridiculous haircut (although today I'd be thrilled to have even that much hair). Gladys's friend, Susan Rothstein, came to the door, looked me over, and called upstairs to Gladys that it was okay, I was cute enough. Gladys was a beautiful girl, and we got along very well. She came to the New Utrecht scrimmage the following week, and we stayed together all of our senior year.

Also attending the New Utrecht game were my father and my kid brother, Richard. Murray and Selma had graduated from New Utrecht, where Murray had played on the varsity football team, and he was proud of me, but he thought I spent too much time and effort on football. And wrestling was decidedly not his thing—too preppy and goyish. I didn't invite him to many games or matches. I'm not sure why; probably I thought I couldn't be myself when he was around. I couldn't swagger around my father, and I was a swaggerer. Or maybe I could be *only* myself when Murray was around, and I couldn't be *someone else;* Murray knew me too well for me to feel comfortable in the role I'd decided to play. Or maybe I just didn't want the added pressure of measuring up in his eyes. It was probably all of these things.

In other words, although I'd become a big man on campus, I still felt like a child in many respects. My feelings were confirmed by an incident that occurred in my junior year. The class ring manufacturers made their annual appearance at Poly, and the entire junior class found itself in the fifth-form study hall filling out the necessary form. The manufacturer's representative asked us to fill in our "initials and the year." Well, it was 1965 then, and I knew my initials, so I entered those two items. When the rings appeared several months later, everyone else's ring had his initials printed inside and our year of graduation: 1966. Mine, of course and to my shock, had the previous senior year's graduation date: 1965. It was a sad moment for me, but sadder still was my response to the problem. Rather than rectifying the situation by going to the bursar's office and asking for the ring to be changed, or enlisting the aid of a teacher or my dad, I could only stare down at the ring, seeing in it the culmination of my five-year imposture at Poly, an embodiment of the truth, unmasking me for the unworthy I was in Flatbush (and feared I always might be), before arriving at Poly and pulling the wool over everyone's eyes. How could I go to any authority figure with this proof in hand?

The matter was resolved, if imperfectly, when a classmate saw my ring, and exclaimed how lucky I was because all the girls would think I was a year older, suggesting, "Let's trade." I agreed, and I got his ring with the correct year but the wrong initials printed inside. My shame and fears were once more hidden, now on the inside of my Poly ring. I had the initials changed but lost the ring a couple of years later in college, when I gave it to a girlfriend to wear; she returned the ring, but I then misplaced it. My Hamilton graduation ring would later be lost as well, stolen in a burglary from my apartment in New York. In 1978 I reordered the Hamilton ring, and that, too, was stolen. Did I keep purchasing these rings to remind myself of the gap between who I am and who I think I am? Or do I keep shedding them to forget the distinction between the two? I have a tendency to replay the incident of the Poly ring in my mind. I wish I had it now.

Despite the ring incident, I left Poly feeling great about myself. I was the luckiest kid alive. I'd graduated a sports hero from this wonderful school, I'd been accepted at an equally wonderful college, and I was the handsomest fellow I knew—the girls all told me so. And my mother told me to be proud of it, and to thank God I looked like her, and I did—until I grew older, gained weight, lost my hair, and began to look like my father. My father used to say, "Looks are for idiots and morons who want to play ball or go to Hollywood. Looks are for women and children. Men don't care about looks." "Who's better than I am?" he'd say.

At the time, however, like so many young people, looks meant a great deal to me, and I was utterly taken with myself. I had brought my girl-friend Gladys with me to camp that summer before college, and I was too self-involved to appreciate her presence. We broke up upon our return to New York and went our separate ways to college: she to American University in Washington, and I to Hamilton College in Clinton, New York (forty-five miles east of Syracuse and nine miles from Utica).

I'd applied to Hamilton, as well as Union College and Colgate University. All of my choices were schools that would be extensions of my Poly experience. That is, they were small private schools, with long Protestant histories, tucked away in rural nooks and crannies in upstate New York. I understood Hamilton to be the best of my choices academically, so I applied there on an early-decision basis, which means that, if the college accepts the applicant, he promises to attend that college regardless of the outcome of his other applications. These schools were suggested to me by Alden Carter, Poly's assistant headmaster and college guidance counselor. Although my grades at Poly were excellent (B+/A-average), my SAT scores were not. I would freeze when taking those standardized tests. So I was guaranteed admittance to none of my selections.

My father and I went up together to visit these schools and have the interviews they all required. He let me do the driving in his 1959 Cadillac, as I'd just obtained my driver's license. I was proud to be driving and proud of that car. It was a great trip with my father, reminiscent of those days when we were pals together taking our excursions to the beach or our "marches" upstate. I was getting too old to think hanging around with my father was fun, and I was too young to realize how much I'd one day miss the time we'd spent together. We wouldn't have many more excursions of that sort.

I was ultimately accepted at Hamilton; I decided to attend, pleased to have gained admittance to a school of Hamilton's high scholastic reputation. Happy as I was to have gotten in, matriculating was altogether another matter. In my freshman year I felt lost in what seemed to me to be a cold, dark, dank Germano-Gothic environment. One-third of my class had attended boarding schools prior to arriving at Hamilton and were accustomed to living outside the home. I wasn't. And if my Christianization in a Protestant environment had been a somewhat alienating factor at Poly, it was twice so at Hamilton, where there were few Jews, and in my freshman year I couldn't find a central activity in which I excelled to anchor my self-esteem. At bottom, my problem was not Hamilton but simply that I was lonely and homesick, and I missed the structure that had heretofore nurtured and encouraged me: my mother and father, my coaches and the football and wrestling teams, Gladys. My suite mates were great fellows and would become lifelong friends: Greg Hoch (a football player from Colorado), Glenn Craft (a basketball player from upstate New York), and Steve Linette (a basketball player from Lafayette High School in Brooklyn). And I became inseparable from one of the first fellows I met at Hamilton, John Pitaressi, my best friend in college. But until I could find some major endeavor that I could dominate, I would be disconsolate, taking refuge in exercising and obsessive neatness.

There was then no formal freshman football team at Hamilton. In winter, when I joined the wrestling club (by then wrestling no longer held team status at Hamilton), the fraternities began to rush me and I started weight training again, I became slightly more acclimated, but I was still unhappy. Later, in spring, I joined the freshman lacrosse team; I'd played my junior year at Poly, at the urging of Mr. Dupee, who was also Poly's lacrosse team coach, and I found that I actually played pretty well and enjoyed the speed and combativeness of the sport.

And there were no women at Hamilton. Like most top-notch small private eastern colleges of the time, Hamilton had not yet adopted a coed student body. Neither had Poly then, but somehow an all-boys' school made more sense to me—immature postadolescent boys can lose sight of their academic and athletic goals, drowning in their self-conscious posturing for the girls. But immature college boys *need* women around to help them *feel* like the mature men they're supposed to be. I suppose the administration understood this and tried in some measure to address the problem. Hence there were mixers galore with all the women's colleges:

Skidmore, Vassar, Wells, and on and on. But all I met were these Christian girls—and I don't mean Catholic girls but *real* Christian girls, WASPs, who all looked like boys to me, with their crew-neck sweaters, plaid skirts, and penny loafers, with their short blond haircuts and handsome athletic features. I was in a completely different world, and it wasn't for me. I thought I was pining for Gladys, but I was really pining for some sense of self-esteem.

My emotional malaise surfaced nowhere more prominently than in my academics, where after one semester I was failing three courses (geology, English, and philosophy). I dropped philosophy and cleared up my problems in English by the end of the year, but I could do nothing about my four-credit F in geology. So that summer, while waiting on tables at Leonard's in Great Neck, I had to attend Brooklyn College to make up the course I'd failed at Hamilton; I took and passed an economics course that was recognized by Hamilton. But I was so disenchanted with my freshman experience that I tried to transfer to New York University, where I was rejected, no doubt in part due to my freshman academic performance at Hamilton, and also because my application was filed too late to be considered for transfer.

When I returned to college for sophomore year, I was ready to play varsity football and all was once more right with the world; everything was in its place again. I played offensive guard at an unusually light (but extremely muscular) 180 pounds. My grades improved considerably, I was rushed and initiated into the college's premier fraternity, Delta Kappa Epsilon, and in the spring I started on defense on the varsity lacrosse team. I even met a few young women whose company I enjoyed from local junior colleges in the area, such as Casanovia.

That summer I returned to Brooklyn to train like a man possessed, lifting weights twice a day, isometric training once a day, and running at night on the Madison High School track, determined to be the strongest and the fastest I could be. I gained more than thirty pounds of pure muscle, eating five meals a day, with cans of Nutriment liberally interspersed throughout. My mother was frightened by the changes in my physical structure. I suppose I must have seemed a distortion of her little boy. So I had to hide the Nutriment from her. My father objected as well, on his usual grounds: Weight training was narcissistic, and football was a kid's game. But he let me do as I wished. I weighed in at the first football practice of my junior year at 212 pounds. I had a terrific year of football (and was named captain for my senior year, jointly with John Pitaressi and Richard Brown), under Coach Donald M. Jones, the handsome, articulate, former Rutgers All-American who coached Hamilton football for some thirty years. I also had a great year of lacrosse (and was named cocaptain of that team as well), under the nurturing Coach Manfred von Schiller. I had a pretty good year academically to boot.

During spring break of my junior year I saw Gladys again. Mutual friends had told her that I was still heartsick over our breakup. She intimated that she would see me again if I called, so I did so over the Christmas vacation, and we agreed to get together when we returned from college over the spring break. I picked her up at her house wearing a pair of bell-bottom pants I'd borrowed from a Hamilton pal. They'd become fashionable in the fall of 1968 at Hamilton, which meant the rest of the free world had already been wearing them for five years. I took Gladys, together with a couple of other friends, to Doris Chan's on Ocean Parkway for beer and Chinese food. Doris Chan's was a spot frequented by Poly Prep kids and their dates, even when I was still at Poly, largely because the management was too accommodating to get shirty when a fellow or his date ordered an alcoholic beverage without the proper ID. Looking back and recalling how many drinks we would order and how sober we were when we left, I wonder who was fooling whom.

By the end of my junior year at Hamilton, Gladys and I had become engaged. Gladys was the first girlfriend with whom I'd ever been intimate, and here we were getting engaged. It may seem romantic, but it was all wrong. Mostly for Gladys—a beautiful, talented, caring, and supportive woman and wife—and I'm truly sorry for that. I wasn't ready for marriage; I was far too immature and egotistical to appreciate the responsibility it entailed, and my father told me so. Although he thought the world of Gladys, he said I was wrong to have gotten engaged so young. But as Murray would have said, "There are none so blind as those who will not see." Being the stand-up guy he always was, despite his personal opinion, he gave me the $1,200 I needed to purchase an engagement ring, and Gladys and I agreed to marry upon our respective graduations the following year.

That spring of 1970, like all springs I suppose, was an ending to many things in my life and a beginning to others. It was the end of my college life at Hamilton and my boyhood. I wish I could return now to take advantage of the extraordinary opportunity afforded me to learn about anything I wanted, anything in the world, without the incursion of daily adult responsibilities. I certainly didn't take advantage of it then. That spring was also the onset of those responsibilities, which I bore all too lightly: On June 18, 1970, Gladys and I were married at the East Midwood Jewish Center, the same place where I'd been bar mitzvahed nine years before.

Something else was revealed when the snow melted that spring, disclosing what had lain dormant. It was on Easter vacation of my senior year at Hamilton, visiting Grandpa Irving in the adult home where he was then living, that the information began to sprout through, the information that would impel me to be a lawyer.

5

I picture a cold autumn night: November 18, 1946. The war in Europe ended more than a year ago, and the world is safe for democracy once more. It's back to business in a bankrupt New York City. A wind howls across Wall Street, over the East River, whistling through the bridge's sweeping steel suspenders into the cobblestone streets of Brooklyn Heights. Maybe it's a night foreboding an early, harsh winter—the kind of night that drives Tedoro Marquis, a fifty-year-old recent Filipino immigrant on his way home from his job as a cook, into a bar near his apartment at 79 Henry Street, just around the corner from the Poplar Street police station. I'm not even born yet, although I will be about a year later, not too far away, in Borough Park. Yet my professional life is perhaps fixed in that moment when Marquis decides to step in for a few drinks and conversation rather than turn in for the night.

On the same night Patsy Miglio, a former Department of Immigration and Naturalization officer, and Arthur Friscenda, a former Veterans Administration employee, step into the same bar and notice the large bankroll Marquis flashes when paying for his next round. The newspaper accounts will never explain what a short-order cook was doing with $7,400 in walking-around money in those hard times, but perhaps it doesn't matter. Perhaps Miglio and Friscenda just think they're more fairly redistributing ill-gotten gains when they befriend the tipsy Marquis and ask him if they can borrow a hundred bucks—maybe they'll pick up a C-note, and maybe they'll get a better look at that roll in the process. Later they walk Marquis home to 79 Henry Street, where they "find" the handgun they've just planted and then, advising him that they're "police detectives," threaten to arrest him and have him deported. Of course, they say, the whole unfortunate circumstance can be forgotten if Marquis pays them $4,800. Marquis pays. But the cold morning light brings Marquis a healthful measure of resentment, which no doubt is what drives him to the nearby precinct house, where he files a complaint. Two officers are assigned to the case: Detective William Neubauer, a twenty-six-year veteran of the force, soon to retire, and his young partner, Detective Third Grade Murray Cutler.

Based on Marquis's description of the two fake detectives and further investigation, the *real* detectives arrest Miglio and Friscenda a few days after the crime, and they also retrieve the $4,800. Miglio and Friscenda are prosecuted and convicted. A job well done—it should be the end of the story. But it's not.

This was postwar New York. Prior to November 1945, the Kings County (Brooklyn) district attorney had been William O'Dwyer, a charismatic Irishman who'd once worked as a longshoreman and a coal shoveler—a personal favorite of my father. Before his election to the job of Brooklyn's district attorney, he'd also been a policeman, a lawyer, and a magistrate. Although O'Dwyer had a national reputation as a rackets-buster, he was locally tainted by his failure to indict alleged organized crime figure Albert Anastasia, as well as by the mysterious death of Abe "Kid Twist" Reles. Reles was a self-confessed murderer who, having agreed to testify against his Murder Inc. associates, had been sequestered awaiting trial in the Half-Moon Hotel in Coney Island, in the around-the-clock custody of some six of New York's Finest. Nevertheless, when the sun rose on November 12, 1941, he was discovered splattered on the parapet six stories below his ninth-floor window. All the policemen guarding Reles claimed to have been asleep at the time of his demise. Allegedly several sheets tied together were found hanging outside the window from which Reles "escaped," but the distance between the Half-Moon's exterior wall and the spot where his corpse was discovered (a full eight to sixteen feet) was so great as to cause speculation that Reles must have been forcefully projected from his window. It was never determined with any finality whether he had been murdered by his former associates or by his police escorts (whom, it is rumored, he abused mercilessly during his stay at the Half-Moon), or whether he killed himself. Upon Reles's death, it was said by some that "this bird can sing, but he can't fly." Besides the puzzle and the "witticisms," Reles's death also left behind a stench that could not help but engulf the police department and DA O'Dwyer himself. In 1945 a special grand jury publicly held him guilty of gross laxity and poor administration, but O'Dwyer managed to convince the public that the report was an election-year smear, and inheriting a financially stricken city from Fiorello LaGuardia, he was elected the hundredth mayor of New York.

And there was a new district attorney in Brooklyn. In March 1945 Miles F. McDonald, an assistant district attorney under O'Dwyer, was named by President Franklin D. Roosevelt (just fifteen days before FDR's death), U.S. Attorney for the Eastern District of New York (including Brooklyn). McDonald held the post for only a few months until he ran for Kings County district attorney and was elected in November. The ambitious scion of an old established Brooklyn family, he was determined to show the world that he was untouchable and that political expediency would no longer motivate prosecution or the lack thereof in Brooklyn. In short, he was prepared to be a zealot when it came to administrative targets. This is not to say that he was apolitical. McDonald was vocal and candid about his ultimate desire for a judgeship (which he obtained in 1952), and he was savvy enough to know that in a Democratically run borough, you had to play ball with the party machine to attain your ambitions. So, as reported in the newspapers of the time, McDonald freely admitted that the $500,000 in patronage that the Kings County District Attorney's Office controlled belonged to longtime Brooklyn Democratic boss Frank Kelly, McDonald's political mentor. After Kelly, whom McDonald had said he revered only next to his father, successfully backed McDonald for DA, he told him, "I want every job in your office, but I'll never embarrass you." McDonald was quoted as responding proudly, if somewhat lamely, and without so much as a nod to the appearance of impropriety, "You can have every job, but I won't hire incompetents." McDonald noted further, "Kelly has the answer to politics. He is honest. Sure he gets jobs for dunderheads, but not where they could harm the party."

Presumably, the self-avowed crusader McDonald had the blessings of the Brooklyn Democratic bosses when he embarked on a career-long attack on corruption in the police department, which incidently had the effect of destroying O'Dwyer's political aspirations (for the New York State governorship and beyond), and McDonald's crusading predictably did not extend to the hallowed halls of Brooklyn's political machinery. In any event, when the roughly sixty-year-old Mayor O'Dwyer returned from his Christmas 1949 honeymoon with his second wife, the lovely thirty-three-year-old socialite model Sloan Simpson, he found that McDonald had begun a grand jury inquiry into the connections between New York's bookmaking rings and the police department—an inquiry that would occupy the front pages of New York's tabloids for the next year or two.

O'Dwyer was at first highly critical of the investigation, calling it a political witch hunt. When the commander of the Fourth Precinct, Captain John G. Flynn, shot himself after testifying before the grand jury, O'Dwyer led the funeral procession of some six thousand officers (about a third of the force), and although a note left behind by Flynn stated that his suicide had nothing to do with the probe, and although Flynn was never indicted, O'Dwyer and Flynn's wife suggested that McDonald had no less than murdered a good man. But police officers were resigning in droves, and as it became increasingly apparent that New York's police corruption scandal was a conflagration raging out of control, the newspapers reported that Democratic Party boss Ed Flynn asked President Harry Truman to "promote" Mayor O'Dwyer to the ambassadorship to Mexico. Two weeks after O'Dwyer and his glamorous wife hurriedly left town, McDonald's office arrested Brooklyn bookmaker Harry Gross, who confessed that his $20 million-a-year operation had been paying about a million dollars a year to all echelons of the police department. This caused U.S. Senator Estes Kefauver to announce that his Senate Crime Committee would be investigating alliances between professional gamblers in New York and the police department; both McDonald and O'Dwyer would ultimately testify before the committee. By the time the crusading fervor died out at the end of 1951, nearly five hundred New York City police officers had resigned or retired or had been fired.

But we're talking about November 1946, a year after DA McDonald first took office, the dawn of an era in which the *appearance* of graft-busting would be good politics. Who knows what conversations took place between my father and his partner and the young assistant district attorney assigned to the Marquis case in Brooklyn? I certainly don't. But somehow out of those conversations arose a decision by DA McDonald to seek an indictment (and obtain one, for as everyone in the profession knows, a grand jury will do anything the DA wants it to) against my father and his partner for extortion, bribery, and obtaining unlawful fees.

Until my conversation with Grandpa Irving at the adult home in Long Beach, years later while I was at Hamilton, I would hear my father's "trouble" referred to in only a murmur from one relative to another, around the dinner table on some rare occasion. It was too shameful for discussion aloud and certainly not for the ears of kids. I can't even imagine the humiliation to my father and the family (particularly my mother's side) of an indictment—for immigrant Jews? It was inconceivable! It had no place in the upward-mobility credo that my family lived by, no place in the expectation that we would live righteously, lawfully as model citizens. Murray was a policeman, duty bound to uphold the law, and an attorney to boot. One newspaper account reported that he was moved to tears when it was all over and the jury's verdict was read.

But it was on that day in the early spring of 1970, during one of my regular visits with Grandpa Irving, that I finally found out about the indictment. I can't say exactly how it came out or even how the topic arose—at that point in his life one idea didn't necessarily have to follow directly from the one he'd had before, and anyway he had become more voluble than I'd remembered him in earlier days, as if he knew that time was running out and he had a lifetime of things stored up to say. I remember nearly choking up a mouthful of the chicken sandwich I was eating when the words *indictment* and *your father* came from his mouth. I swallowed hard and hit him with a barrage of questions. He clammed up, and I got nothing more out of him. Maybe he knew he'd inadvertently broken some family vow of silence, whether express or implied, and he'd say no more. Or maybe the shrewd old man figured it was time I was goaded into asking my parents the right questions.

Directly upon returning home that evening, I asked my mother and father what he'd been indicted for. She hit the roof. How could Grandpa have mentioned such a thing to you—he truly must be senile—what could he be thinking? In short, I didn't get much in the way of answers from her. My father was equally reticent in responding to my questions, and to be truthful I didn't ask too many, for I was obviously probing wounds that would never fully be healed. Even to date, between me (with a career in criminal prosecution and defense), and my younger brother (a lifelong state and federal prosecutor), we still know remarkably little about the affair.

What we *do* know, largely from newspaper archive accounts of the events, was that the indictment alleged that Murray and his partner had tracked down the fake detectives, Miglio and Friscenda, who had extorted $4,800 from Marquis, and recovered the funds in question but returned only half or $2,400 to Marquis, giving rise to an extortion count and a count of unlawfully taking fees to perform their duties. They were also alleged to have attempted to bribe the assistant district attorney assigned to the Marquis case with $100 to drop the case against Miglio and Friscenda. My father did tell me that they were offered pleas of guilty to a misdemeanor, which they both rejected, deciding instead to vindicate themselves at trial. In April 1947, after a trial in which Murray was represented by Leo Healy and Bill Neubauer was represented by Hyman (later New York State Supreme Court Justice) Barshay, the jury returned a verdict acquitting my father and his partner of all charges.

Frankly, I don't know what to make of the tale. Perhaps it's a case of being too close to my father and too close to the criminal justice system, but none of it makes any sense to me. First of all, the nature of the allegations bespeaks a man other than the one who raised me, who drilled into me an almost military sense of honor and pride in the way a man performs his duty. Second, the decision to obtain the indictment makes no prosecutorial sense in view of the reports of the trial and verdict. The newspapers state that Marquis admitted under cross-examination having *offered* Murray and his partner half the extorted funds if they could recover them. This explains the not-guilty verdict on the extortion count, but wouldn't the prosecution have discerned this bit of information before submitting the facts to the grand jury? And if so, wouldn't this fact have been submitted to the grand jury, obviating that count? Moreover, why did the trial jury return a not-guilty verdict on the "unlawfully taking fees" count, unless there was no evidence submitted at trial that Murray and Neubauer ever received $2,400 from Marquis or the fake detectives? Finally, how could the prosecution have brought the bribery count to trial without submitting the testimony of the assistant district attorney to whom the alleged bribe was offered? And if the assistant district attorney in fact *did* testify, how odd that a jury would reject his testimony and return a verdict of not guilty on this count as well.

Which brings me back to those initial conversations between Murray or his partner and the assistant district attorney whom they were alleged to have bribed. Was it something one of the two police officers might have said that was misunderstood—that DA McDonald was *eager* to misunderstand in his haste to commence, in 1946, the crusade against police corruption that would make his career a few years later? From any perspective, how cavalier a decision to destroy the lives of two men with an indictment based on evidence which couldn't stand the scrutiny of trial.

In any event, I believe the impact of the information I'd received from my grandfather and thereafter about the Marquis indictment was profound, although perhaps I couldn't have said so at the time. Perhaps the effects lay dormant until I reached professional maturity. I know for sure that from that time on I wanted to be an attorney. Yes, because my father was an attorney and I admired him, although he never pressured me to follow in his footsteps. But I think more directly because my father's experience fundamentally jarred my worldview.

Grandpas Irving and Harry and Grandmas Bertha and Sadie, or their antecedents, arrived in this country from Lithuania, Hungary, and Austria several decades before the Nazi genocide in Europe, but all knew families who'd lost relatives, and anti-Semitism was rampant here in the United States as well, certainly throughout the first half of the century, even, and sometimes especially, in New York. As great as this country had been for Jews and for my family specifically, there was a distinct awareness of the "otherness" of being a Jew. I recall dinner table discussions about how the United States had refused entry to boatloads of Jewish refugees who returned to Europe to die at the hands of the Nazis. My father said it had been almost impossible for a young Jewish man to get a job in a law firm before the war, and that this was at least in part why he'd become a policeman with a law degree before he'd become a practicing attorney. And I recall being surrounded by the largely well-to-do, largely Christian student body of Hamilton, feeling somewhat an outsider, despite my success in football, and trying to Christianize myself.

My father's indictment was proof, if any were needed, of the fragility of our lives, of the socioeconomic structure we inhabited, even here (although clearly less so than nearly anywhere else on earth)—proof that the whim of government hung ever as a sword of Damocles above our heads, threatening to end it all: the privileges, the prosperity, the job, the degrees, the standard of living, the home, the freedom, the very air around us. Even the best government, and I was sure we had the best form of government in the world, could be perverted in certain instances by the agendas of individuals in control who would stoop to obtain some personal advancement or gain, whether it be in the form of fame, power, or money. And such encroachments would often occur at the expense of the disenfranchised: the poor, the refugees, and the minorities without loud enough voices.

I don't believe I was mature enough to have understood my own feelings at the time, but I clearly perceived that there were those who were enfranchised from birth and those who stood on the fringe. Perhaps for the first time I sensed that whatever privileges I attained at home, at school, or on the playing field were temporary, ephemeral, and would have to be *re*attained every day by the sweat of my brow.

My parents had always taught me that with discipline and hard work I could do anything, be anything, and they were right—to a point. There are limitations imposed from within. Sure, I could build my strength, self-confidence, and self-discipline through weight training, but there were those around me whose bodies didn't respond so readily to the efforts invested, and even I had occasion to meet someone stronger and more skilled on the wrestling mat. And there were surely other physical and mental limitations imposed by genetics. You couldn't do much about them, but you could go a long way with courage and hard work. I'd already proved that to myself.

And then there are the limits imposed from without.

First, there's luck or fate—a brick falling from a construction site, a sudden illness. Rather than a step ahead of the blindly wielded sword, or a step behind, you're right there. It's timing. You're in the wrong place at the wrong time. There's nothing you can do about it, except perhaps try to live honorably, and although I know it's an old-fashioned concept, I believed then, and I do still, that luck, fate, or God's hand favors those who live uprightly. Maybe it's just that my timing is better when I believe I'm doing the right thing. Anyway, I'm a very old-fashioned person.

Finally, there are the whims of those in power—their sudden exigencies, purposes, and schemes, grand or petty. That's how I perceived what had happened to Murray. McDonald had an itch, and my father was available to help him scratch it. He could have been more accommodating, no doubt, by rolling over and taking a plea. But that wasn't who he was. I don't know what the experience taught *him;* for as I said, he hardly ever spoke of the incident. But when *I* learned of his trouble at the age of twenty-one, I learned that just because the government says it doesn't make it so. And somewhere deep in the recesses of my consciousness I think I learned to seek out the motivations of those in power in assessing the stances taken by the government. The whims of those in power— here was a potential menace I could do something about. I could learn how to defend myself, the way my father defended himself—and defend others, the way my father was defended by Leo Healy in the Marquis case—on the playing field where he'd ultimately been vindicated: the courtroom.

As for Murray, it all worked out right. He resigned from the police force and took a job clerking for Leo Healy. A few years later Murray represented Harry Gross in an early indictment brought against him by DA McDonald. He won an acquittal in that case, too (although as I mentioned earlier, Gross wound up the cornerstone of McDonald's career, providing critical testimony in the police corruption scandal of 1951). Time has a way of erasing all these tribulations (and the trials, too).

Murray probably tried hundreds of cases against the Brooklyn DA's office over the course of a career spanning nearly a half-century, and there was nothing but mutual respect between him and that office. And of course, I spent almost seven years honing my trial skills in the Homicide Bureau of the Brooklyn DA's office, under District Attorney Eugene Gold from 1974 to 1981. My kid brother, Richard, a federal prosecutor in California, also worked there for a few years in the Appeals Bureau. In the DA's office one of my most trusted detectives assigned to the DA's squad was Sergeant Ken McCabe, whose father, Kenneth McCabe Sr., was Miles McDonald's chief assistant district attorney at the time of the Marquis case.

All is forgiven, even forgotten. But I remember.

6

BECOMING a lawyer is nowhere near as easy as *deciding* to become one, particularly when you have a C+ course average in college and when you manage a mediocre LSAT score, as I did. I had little or no chance of being accepted to any respectable law school upon graduation from Hamilton. Once again my father's friend Richard Maloney of Brooklyn Law School came to my rescue. He suggested that if I were to take a year off from school and endeavor to get my LSAT score to a respectable level on a second attempt, he could help me gain admittance to Brooklyn Law School. I took his advice.

Gladys and I embarked on our honeymoon in Spain under the cloud of my failure to get into a law school after graduation. I felt inadequate as a husband and as a person. While I desperately needed Gladys's support and affection in view of the uncertainty in my professional future, ironically the low self-esteem generated by that uncertainty made me less able to act responsibly toward the marriage. I felt constricted by, and resented, the very support I craved and needed.

As soon as we returned from Spain, my father (who resented our trip to Spain because of his abhorrence of its dictator Franco) obtained employment for me with Monroe Office Products, selling business machines. I kept this job through the summer into the early fall and then taught language arts at Stranahan Junior High School. When the school year was over at Stranahan, I began work in construction for Sam Singer's sheet metal factory in Long Island City. From the time I graduated from Hamilton, I'd begun a program of reading the *New York Times* every single day, from the first word to the last. I was determined to improve my reading comprehension, vocabulary, and speed, so as to substantially better my LSAT score. And it worked. When I took the LSAT again, my score was more than a hundred points higher, and true to his word, Professor Maloney helped obtain my admittance to Brooklyn Law School in the fall of 1971.

During that year before entering law school, I began a pattern of infidelity that would ultimately destroy my marriage. Due to my immaturity, and stupidity, I was repeatedly distracted by other women, including a very seductive thirty-three-year-old teacher at Stranahan. In the meantime Gladys and I moved to an apartment on Fleet Street in Rego Park, Queens, from which I drove to work and then law school each day, until the summer of 1972, when we moved to an apartment at 96 Hicks Street in Brooklyn Heights (an increase in rent from $207 to $225 per month), just a few blocks from Brooklyn Law School. While I was in law school, Gladys was employed by the Garolini Shoe Company and eventually got a job teaching at Stranahan as well. Throughout this period, indeed throughout our short-lived marriage (seven years), she was patient, understanding, supportive, and generally took good care of me. She deserved far better than what she received from me in return.

A year off from schoolwork and a dose of the working life—when that work does not inspire you—was a salutary thing. It planted the fear in me that I might spend a lifetime trudging to and from a job for which I had no particular aptitude, a job that left me feeling empty and exhausted at the end of the day. I'm not suggesting there's anything wrong with teaching, or construction work, or sales; they're honorable careers all but are fulfilling only *for those who are called to them.* I wasn't. At that point in my life I believed my calling was the law—and I was right.

So from my first day of law school I set myself to the work, trying to ensure against failure by limiting my schedule to walking to school in the morning, going to my classes, walking home in the evening, and studying at a little Parsons table in our bedroom. At first I didn't even exercise, but later I brought some free weights to the apartment and used a steam pipe in the hallway as a pull-up bar. I treated law school as a kind of military stint, perhaps the military stint that was foreclosed to me. I was single-minded and had absolutely no social life, but on the bright side my extramarital dalliances ceased as well. Thanks to Gladys, who took care of all my everyday needs in life, I had complete freedom to bury myself in my course work, without worrying about paying the rent, making dinner, or cleaning up.

At the end of my first year I received all A's, except for contracts, in which I received a C, and I qualified for participation in the *Law Review,* an honor reserved for the top ten percent of the class. That summer I took a great-paying job as a clerk, docketing civil judgments in Queens County Court. Then it was back to my second-year schoolwork, and again I received straight A's, even in the legendarily tough courses like federal taxation and corporations.

I recall being completely at peace with my father in those days (a rarity after I'd reached adulthood and developed ideas of my own). Murray had always believed I should be in monastically rigorous training for one thing or another at *all* times; my legal studies fit the bill perfectly. One Saturday afternoon in the fall of second year, he even took me to lunch at Peter Luger, a famous Brooklyn steakhouse, to celebrate my law school success. He was proud of me, and after all Brooklyn Law School was the only school from which he'd graduated. It was around this time that I first discovered that my father was a criminal lawyer. My mother would always tell me he "mostly handled accident cases." For reasons I don't entirely understand to this day, just as Murray's 1947 indictment was never mentioned, she never wanted me to know he was involved with the criminal element, even as counsel, and he himself rarely discussed his work at home. Murray did, in fact, handle some tort matters, but everyone at law school (and elsewhere) knew him as a criminal lawyer.

I interned in the Kings County District Attorney's Office (the same office that had indicted my father a quarter-century before) in the summer of my second year. This was my first taste of the excitement of criminal practice, with the police squad rooms, the detective rooms, the Homicide Bureau, and the energetic young assistant district attorneys like Tom Martin who were riding homicide cases. I was hooked. I returned to school to finish my third year of study with all A's once more, firm in my desire to be an assistant district attorney trying homicide cases. I could tell from the clinical programs I'd taken in school and

from what I'd seen in the DA's office that I wanted to be on my feet trying cases like Leo Healy, and there was no better boot camp for trial
lawyers than a district attorney's office or legal aid, where recruits were
thrown into court on trial day in and day out. To stand up to government, I would first have to understand how it worked—and to understand, I would have to be a part of it.

To my amazement, all my hard work at law school was translated into
good grades. For the first time in my scholastic career, the concepts were
all coming easily to me. The law was so clean, practical, and organized—
not theoretical and esoteric, like most courses I'd taken in college.
Unlike English, philosophy, science, or nearly every other subject I'd had
to study at Poly or Hamilton, the law was all making sense to me, as if the
paths of legal reasoning had been hardwired in my brain at birth. Maybe
it was all just the work I was putting in, but I don't think so. To me then
as now, the way my mind embraced legal study (just as my body had
embraced weight training) was a sign that I was meant to be a lawyer—as
though my father's 1947 indictment, and Grandpa Irving's timely comment to me about it, had occurred for the sole purpose of putting me on
my path. My parents and Gladys attended my graduation from Brooklyn
Law School cum laude in June 1974. I'd graduated thirty-fourth out of
some four hundred students in my class. I'm not generally considered a
"law" man but rather a "trial" man, with good forensic skills in the courtroom, theatrical appeal to jurors, and a tenacious cross-examination. I
laugh when I hear that, and it never fails to work in my favor when I'm
underestimated on the law by an opponent.

I applied for an assistant district attorney's position in Brooklyn,
Queens, Bronx, and New York Counties. Frank Hogan, the district attorney in Manhattan, probably the most prestigious office in New York City,
had been a friend of Murray's, but he'd just passed away. I didn't think
I'd be offered a position there without his help. District Attorney Eugene
Gold's office in Brooklyn had grown enormously in stature since he'd
taken over, and I'd already become friendly with several ADAs, including
Robert Keating (then the chief of the Supreme Court Bureau, later first
assistant district attorney, and still later chief administrative judge of New
York City under Mayor Edward Koch), at the Shelton Gym around the

corner from the district attorney's office, where I'd begun to work out in 1973 and where Gladys was teaching an aerobics class. So when the Brooklyn office called to offer me a job in June 1974, I accepted without waiting to hear from Queens or the Bronx, or from Legal Aid, where I'd also applied for a job. It wasn't a difficult decision for me, for convenience's sake—I lived a five-minute walk from the office and the Kings County courthouses—and because I'd interned in the Kings County District Attorney's Office my previous summer and made a lot of friends. The office had a fraternitylike feel to me, an extension of my Poly and Hamilton days, as it were.

I spent the rest of June and July preparing for the New York State bar examination in the grueling Marino Bar Review course. I took the exam at New York University at the end of July. In August 1974 I began working in the Kings County District Attorney's Office, at a starting salary of $11,500 a year, which would automatically go up to $13,000 upon passing the bar exam and being admitted to the bar. If I passed the bar exam on my first try, I would be admitted to the bar in February 1975.

In the meantime I was a criminal law investigator in Eugene Gold's office. "Criminal law investigator" was the job title until a young law school graduate was admitted to the bar; then the job title became officially "assistant district attorney." Gold's mental and legal difficulties late in life have obscured the fact that he took over a small-time patronage-shaped institution and turned it into one of the greatest prosecutorial offices in the country, drawing talent from the finest law schools. When I arrived in the office, I was one of some 350 lawyers working there, making it the second largest local district attorney's office in the country, after Los Angeles County.

I began working in the Investigations Section of the Homicide Bureau. The ordinary initial assignment for an ADA was the Criminal Court Bureau, where the lowest-level arrests are first processed. I assume I was given a taste of the Investigations Section because I'd previously interned there. For the most part in the Investigations Section I dealt only with the ME (medical examiner) cases, or cases of deaths that were unexplained but not really homicides—the suicides, the accidents, the "causes unknown." The more experienced I became as an ADA in the

Homicide Bureau, the less I believed in causes and the more I believed in wrong places, wrong times, and wrong states of mind—a confluence of serendipitously insidious circumstances, with interchangeable characters filling the roles of perpetrator, victim, witness. But I was new then, and I loved visiting the precinct houses and interviewing the witnesses, even if only in ME cases.

After only a few months, when I received the good news that I had passed the bar exam, I also received the bad news that I was reassigned to the Criminal Court Bureau to learn how a criminal prosecution is developed at the intake stage. I hated it there. To me the Criminal Court Bureau was a massive amorphous bureaucracy devoid of esprit de corps, where complex procedures, a multilayered system of supervision, and tons of paperwork bogged one down. Many of the judges were brusque and tired by the vast numbers of matters they had to address in a day. An overall sense of futility reigned, and I was a cog in the relentless gears of justice, grinding from morning until night. There were no cases to be tried, no opportunities for me to excel, to break out—no opportunities for me to be Bruce.

A young ADA in the Criminal Court Bureau was supposed to assess the value of a case at intake, then dispose of it with a plea if at all possible; the survival of the system, with its limited financial resources and courtrooms and ADAs, required the pretrial disposal of more than fifty percent of the cases that arrived in the office. But I had little idea how to evaluate a case: What crime is worthy and what crime isn't? It was particularly uncomfortable to assess a case when you knew that layers of supervisors were waiting with bated breath to second-guess your decision and cover their own behinds. For example, many cases had a little notation on the flap indicating that the Rackets Bureau, or some other elite bureau, had a particular interest in the matter (it seemed often for as mundane a reason as that the defendant had an Italian surname). And there were the reports: the intake reports, the felony dismissal reports, the reports on reports.

In retrospect, the most important thing I did in the Criminal Court Bureau was argue for the setting of bail. But at the time I had no comprehension of the importance of bail—it was no more than an assertion of the power of the district attorney (whose mantle I wore), an abstract exercise, haggling over whether the docket number on the file in question would be accorded some attainable bail and allowed to go home or whether those faceless digits would languish in jail for the many months before trial. Invariably, a young ADA would demand some ridiculous bail, knowing that to do less would likely draw the criticism of a supervisor (and in some instances a judge), seeking deniability in the event the defendant should skip bail or, worse yet, commit *another* newsworthy crime while out on bail. Often the judge would see the big picture and set bail in an amount less unreasonable than that demanded by the ADA but unattainable nonetheless. Only when I became a defense attorney did I understand the importance of this stage of the proceedings, when I felt responsibility for clients and their families who might not see each other for a year while awaiting trial on charges that might be wholly unfounded. The brief moments of barter-by-rote among the prosecutor, the defense counsel, and the court often determined whether the client would be able to effectively aid counsel in his own defense, whether he would retain any vestige of family unity, and whether his family would be able to survive financially. Although I should have known better, given Murray's 1947 indictment, it was hard to live and breathe the air of authority and not develop a smug sense of self-righteousness, the vague faith that any importunity foisted on those who fell under the wheels of justice (as meted out by fatuous, if well-meaning, young ADAs) was justifiable in the name of God and the greater good. Perhaps it was *because* of Murray's indictment that I found the *souk*-like trading and bargain-basement justice of the Criminal Court Bureau so intolerable.

So I was enormously relieved when, after about six months, I was sent back to the Homicide Bureau. This time I was given a special "riding assistant's" position—a post given to only two lawyers from each entering class of ADAs whom the Homicide Bureau wished to prime for future membership in the homicide fraternity. The riding assistant would do everything the other homicide ADAs would do *except* try cases. The station house, the detectives with their gruesome jocularity keeping them sane, the crime scene, the body on the floor, the gore on the walls, the stunned witness, the gathering of evidence and clues—I was neck deep in the action of city life. Here was the flesh and blood of the street, the teamwork, the esprit de corps lacking in the Criminal Court Bureau, where people's lives were pushed here and there like the sheaves of paper that arrived with them and that would accompany them through the system. And the Homicide Bureau *did* feel to me like my football teams—the rough-and-tumble camaraderie, the locker room at the station house. There was even an organized week-end homicide detective investigators' touch football game, where Chief Assistant Bob Keating, Supreme Court Bureau Chief Mike Belson, a couple of other homicide ADAs, and I would mix it up with the homicide detectives, no holds barred, full blocking, without regard to rank—just the boys letting off steam.

I liked the detectives. They reminded me of Murray, who after all *had* been a policeman. I definitely liked some of them better than others, but they were all decent fellows. They were sometimes crude, and I can't say I always approved of their humor, which often consisted of insensitive comments about alleged perpetrators or even victims, but I wrote this off as largely a defense against the everyday brutality of the job. Despite what I read all too frequently in the newspapers nowadays, none of the detectives I met were the type to bully someone weaker or to physically abuse a suspect. Maybe I was just lucky. There was definitely a team sensibility: There were those on the team, and there was the rest of the world. And I was on the team. Not many ADAs were, certainly not the ones who, with a couple of years or less in the DA's office, thought they knew enough about the law and the street to pull rank on the detectives they rode with. Even after I had fifty homicide trials under my belt, I knew the street was their turf and the courtroom was mine, and when on their turf I would stick to my role: preparing the case for trial.

My stint as a riding assistant was short-lived, however, and after a few months I was transferred to the Supreme Court Bureau, headed by Mike Belson. The vast majority of felony trials were handled here. This was the true training ground for the ADAs, and I was determined to make the most of it.

Judge William "Billy" Ballard presided over the first trial part (courtroom) to which I was assigned. Judge Ballard had been a fireman in

Brooklyn when my father was a policeman on the beat. He would reminisce about the days when cops on the graveyard shift (midnight to eight) would stop in at the firehouse for a catnap (certainly not *my* father). On one of my early appearances before Judge Ballard, he beckoned me to approach and said, "Don't be a lazy bum like most of these young lawyers," waving a hand vaguely around the courtroom. "Some cases will fold, some you're gonna want to dismiss. Try all the others, every one you get your hands on. What's the worst that can happen? The jury won't agree with you? You'll learn from every loss, every situation. Get in there and mix it up!" And I did. I became a trial machine, trying everything that should be tried. In the Supreme Court Bureau we worked in teams of two ADAs per trial part. I was teamed up with a lawyer who didn't particularly like to try cases; he was tailor-made for me, handling the paperwork (which I detested), while I tried the cases.

At one point I tried twelve cases, one right after the other, without any days (other than weekends) in between. And every case wound up going my way. No other ADA tried cases that way. Most were afraid—afraid of the exertion, afraid of losing, afraid of not being adequately prepared, afraid of the ridicule or the disapproval of supervisors if they lost the cases. At bottom, they didn't want to climb into the ring and fight, because fighting always entails risk. And risk avoidance is as natural to the ordinary lawyer as spots to leopards.

But I didn't want to be the ordinary lawyer. And I couldn't countenance fear. I certainly don't hold myself out as a paragon of bravery—there are many things I fear. *I simply can't allow myself to act on those fears.* This is me, this is my raison d'être. If others were afraid to try cases, so much the better—I'd build trial muscles by trying case after case.

My first trial involved a defendant named Raines—I forget his first name. It was a simple burglary. In the course of my summation, I tried to re-create the crime for the jurors, and in doing so I threw the burglars' tools (exhibits in the case) on the floor against the jury box rail, just as Raines was alleged to have done on the rooftop of the building he'd burglarized. The impact on the jury was immediate and visceral: They were *there* on the rooftop with Raines, and I was *there* in the jury box with them—where I needed to be. I understood then that I could do this thing and do it well. It was more than just a facility with legal concepts; my success in law school revealed that to me. On my feet in the courtroom, before a jury, I could become transported. I could enter a zone of consciousness where nothing existed but the facts of the case on trial— more than the facts, the events as I imagined them (based on the evidence) to have transpired. And I could convey this excitement, this fervor to the jurors; for if an attorney brings no energy into the well between counsel's table and the jury box, he almost certainly will lose. And I didn't have to do a thousand push-ups, or lift weights for years, or run a thousand wind sprints, or study a thousand hours. All I had to do was know the facts of my case, get up on my feet, and open my mouth. Of course, I had to know the jurors, too, but somehow I knew these people. I knew what they were thinking, what they wanted to hear me say. It was effortless.

Or at least it seemed so then. I've learned since that the effortlessness was a gift, and some days it just wouldn't be there. But for the most part over the course of my career, if I care about a case, I can convey this to the jury. As a result, I've tried to take on only cases that I care about, cases in which I can muster the necessary indignation and empathy.

This ease in relating to the facts of my cases and to the jurors did not mean I had no forensic skills to acquire. In the Supreme Court Bureau I had to learn it all. First, there was direct examination—putting the facts before the jury through the testimony of your own witnesses. The rules of procedure do not permit an attorney to "lead" his or her own witness; that is, your question can't indicate to the witness what answer you'd like to hear. This is only fair; the idea is that the witness, not the attorney, should tell the story. Of course, with adequate preparation, the witness will still be telling the story the attorney wants to hear, most of the time. I've seldom tried a case where I haven't been surprised by something that pops from a witness's mouth, no matter how much I've prepared him. In any event, eliciting the facts to support your case from your witnesses in an interesting and fluid form is an art. I learned the technique from watching seasoned ADAs like Barry Schreiber and (now New York State Supreme Court Justice) Neil Firetog.

Cross-examination—questioning the other side's witnesses—came naturally to me. On cross an attorney is permitted to lead a witness; in fact, on cross one should almost never ask an open-ended question that might allow the usually hostile witness to deviate from the path you want him to travel. I would become immersed in the process of moving the witness, question by question, to the desired destination, as doggedly as the wrestler I'd once been.

My performance in the Supreme Court Bureau caught the attention of my supervisors, and after a short while I found myself being sought by the two most prestigious bureaus in the DA's office: Rackets and Homicide. It was like being rushed by the fraternities in college again. The choice was easy for me. Working in the Rackets Bureau involved painstakingly investigating organized crime cases: It would characteristically take years to put together the evidence, usually from wiretaps and informants, and a case would very seldom reach trial, either because it was flimsy and would be dismissed after complex motion practice; or because (if the tapes were good) a plea would dispose of it; or the experienced defense lawyers would mire it in procedural tar babies (if the defendant was out on bail) for so long that the ADAs on the case would leave the office before it came to trial. By contrast, being in Homicide was being in the trenches. There were seldom delays in trying the cases, as the defendants were usually incarcerated pending trial, and the cases themselves were usually uncomplicated. Rackets was intrigue and subterfuge; Homicide was combat and therefore more my style. Most of my success as a defense attorney has come by eschewing the subterfuge and intrigue used (or abused) to prosecute my clients. Sure, I felt at home in Homicide, from having interned there as a law student and from having worked there as a criminal law investigator and a riding assistant. But even without that, it was natural for me to choose the forthright style and the active trial schedule of the Homicide Bureau over the more esoteric and paper-intensive Rackets Bureau.

7

WHILE in the Homicide Bureau, I took over fifty murders to trial and verdict. There were very few acquittals. In fairness, my record of convictions was a reflection less of my litigating skills than of the crushing pressure the authorities bring to bear on a defendant in any prosecution, both financially and because of the widespread public perception of the government's moral superiority; the latter is something that naturally taints every jury from the outset. Besides, murder cases are usually simple affairs in which the evidence against the defendant is overwhelming. The convictions run together in my mind, an endless sea of mutilated corpses, speechless hollow-eyed spectators of the worst our species can muster. I recall horrific crime scenes, rendered more horrific still by the impersonal black and white crime scene photos, the kind that the photographer Arthur "Weegee" Fellig made famous. And there were the befuddled witnesses, dazed and only slightly less mute than the victims, wondering how they came to be sitting in the gray metal government-issue chairs, in the fluorescent light of the homicide detectives' questioning. There those witnesses sat, too often interchangeable with the accused, or even the victims. The witness chained to the detective's desk, the accused eating a sandwich in a room a few feet away—at some point in the not-too-distant past or future, perhaps their places would be reversed, and later maybe either one of them might be lying beside the victim. It's tragic that the convictions have largely become one in my memory: one defendant, one witness, one victim.

But the acquittals are carved into my brain forever. Is memory no more than a catalog of our failures, a handy briefcase of regret?

The general rule that emerged from my experiences in Homicide was that the difficult prosecutions were those in which the victim or the prosecution witnesses were unsympathetic or unsavory. This is unfair and illogical, and I would struggle mightily against the tendency in jurors to assess a murder with respect to how likeable the victim or the witness, but there's only so much an attorney can do to overcome human nature. For example, in the case of Charles Vitello, accused of killing a prostitute in a car parked near Coney Island, a fine job by Vitello's attorney Franklin Gould won him an acquittal. Vitello was white and of Italian descent: two statistical points that, in Brooklyn at least, greatly (and unfortunately) enhanced a murder defendant's chances of acquittal. And his victim was a known prostitute. Tough case.

There were also those cases where a witness, overly intimidated by the homicide detectives and the system, tried too hard to please the prosecutor with small lies that wouldn't hold up under the defense's cross-examination. I recall once having insufficiently grilled a witness who later fell apart in court and won the case for the defense. It was a lesson that was well worth the loss, for in my later work as a defense counsel, I would be ever-vigilant for the witness who was too willing to please the prosecutor and would win many an acquittal by sensing such a witness and taking him to task on the stand.

I won't say that my losses as a prosecutor weren't disappointing, but they should have been predictable. Nevertheless I would still be mildly surprised by each not-guilty verdict. When I'm on trial, I feel invincible. Certainly I felt so as a prosecutor. In the preparation for trial and in the course of the trial itself, I would necessarily develop a deep belief in the strength of my case and unshakable confidence in my power to convince the jury. I didn't take the acquittals hard or personally. I had just the briefest moment of vulnerability, and then I would move on and prepare for the next trial as quickly as possible.

In June 1977 I first saw Barbara Wolf Patterson, and she took my breath away. I mean that literally—I gawked for a moment and then noticed that I hadn't been breathing. I'd never seen anyone like her before. Even as she first appeared, tall and slender, in a simple sundress and flat open sandals with laces that rose around her calves, she was stunning, with an understated elegance. She was like an Audrey Hepburn or a Capucine, a *Jewish* Hepburn or Capucine, although she played at not being Jewish. (Her grandfather was from Britain, and her surname—her soon-to-be-ex-husband's—was Patterson.) But she played at so many things, lightly, floating. Barbara had so many airs, if she'd been a balloon, she'd have floated to Pluto, but in her distracted, and distracting fashion, I found them all charming. What in another I might have found silly or artificial, I found captivating in her. I'd been exposed to the meticulously well-tended Long Island girls at Camp Ma-Ho-Ge, who were until then my standard for femininity, but I'd never encountered anyone as stylish as Barbara. Even the summer cold she had when I returned from that July Fourth weekend with Gladys at the Forward Watch of Gurney's Inn in Montauk—a simple cold, with its delicate sniffle of her stuffy pink nose, dabbed lightly with her lacy handkerchief—was a graceful accessory for Barbara. In short, I was smitten. I was also married.

Barbara had arrived as a legal intern assigned to the Investigations/Homicide Bureaus, and she was instantly the heartthrob of the office. Rumor had it (correctly) that she was married but separated, and I would never have had the temerity to speak to her, but she had herself assigned as my intern when I returned from that July Fourth weekend, presumably because I was the most active trial ADA in the bureau at the time. I was beginning a case I would lose, prosecuting a man named John Smalls, who was defended by the very able John Corbett. Barbara tried to help me prepare, but all she did was distract me. She would buy my cigarettes for me—True Greens because she decided that they were milder than my usual Newports—then she would smoke them, and I'd be lost watching the smoke wafting from her mouth to the ceiling. A phone call from a woman other than my wife would spark a worldly remark from Barbara. The days of that week and into the next, and the Smalls trial, were one long flirtation, and by the time I was having dinner and drinks with her at Capulet's on Montague Street, I was sinking deeper and deeper into trouble. I wonder how many would-be Romeos wooed their Juliets at (the now long-closed) Capulet's.

I had no business engaging in this sort of thing with Barbara. I knew my feelings for her were bordering on the obsessive and were developing into a real threat to my marriage, and there was nothing I could do to halt the process, yet I was revolted by the subterfuge. Finally on July 12, as Corbett wrapped up the defense case and we were to prepare our summation arguments, I went to dinner at Barbara's apartment on West Seventieth Street in Manhattan. At first I refused to stay the night, but Barbara said it would cause irreparable harm to our relationship if I left. *Irreparable harm*: a lawyer's phrase, meaning harm that cannot be repaired, damage that cannot be undone. The damage I caused Gladys can never be undone. Nor can the damage I caused myself. I probably even irreparably harmed Barbara—but she was a willing, demanding, accomplice. I can only hope John Smalls was an innocent man. I stayed at Barbara's that night.

I prepared my summation on the subway ride downtown to Brooklyn. This was in itself not as dilatory as it might sound. I often worked on murder cases while going to and from the office or court. Murder cases are not overly complicated. That said, the Smalls summation was not one of my best. On July 13 the jury deliberated through the night of the great New York blackout of 1977. What was on *my* mind? Public transportation was suspended, and I couldn't get to see Barbara. I called her that night frantically and heard the anecdotal "other man" in the background. There were candlelight parties in her building because of the blackout, and people were staying over—I understood, didn't I?

On the fourteenth, Bastille Day, Barbara came by in a checkered cab to pick me up for court. The lights were still out there, and the judges were seated in lawn chairs in front of the courthouse on Court Street. The jurors, who'd been working inside by candlelight, returned their verdict: Smalls was acquitted.

I was ashamed of my behavior toward Gladys and unable to sneak around behind her back. I disclosed my affair to her, which was the last straw as far as she was concerned, and she moved out. I recall a meeting with my father at about that time, during which I asked his advice as we strolled down Montague Street to the Promenade. Murray didn't lecture me. Rather, he told me a story. It was about a fellow who meets a beautiful woman, takes her to a fabulous restaurant, stops at a jewelry store and an expensive clothing boutique, and takes her to a sumptuous suite in a top hotel where they stay the night. In the morning after breakfast they pass the same stores, and the woman inquires about the jewelry and clothing the man said he'd buy her the day before. The man replies, "Darling, don't you know that when I'm hard I'm soft, and when I'm soft I'm hard?" What my father was, perhaps indelicately, trying to convey was that I shouldn't allow sexual attraction to undermine my marital vows. He'd always believed I'd married too young, but having done so, sneaking around was not his style. A solid marriage—a social institution—to a good woman was not to be lightly discarded. And he liked Gladys.

Based on his advice, I made a lame attempt to reconcile with Gladys, and she moved back into the apartment before Labor Day. But as Barbara and I were driving back from what was supposed to have been our final weekend together (in Montauk), I told her of my decision, and she became upset and then grew distant. I couldn't bear the thought of not seeing Barbara ever again, and within a few days I'd packed my things and showed up on her doorstep. I said, "You're going to regret this, but here I am." She was tearful and happy then, but I'm sure she did regret it later, although we spent a marvelously romantic time together and were inseparable until 1982. And I was lost on the West Side of Manhattan—where I've been lost for the past twenty-four years. At first Barbara found me a small studio apartment at One Nevada Plaza for $546 per month (more than double what I'd been paying in Brooklyn Heights), a figure I couldn't afford, but Barbara's apartment was too small and I wasn't ready to move in with another woman while I was still married to Gladys. I sold my car, a Peugeot that I adored, because I could no longer afford the loan payments. Barbara and I didn't take an apartment together (at One Lincoln Plaza) until 1980, when my divorce from Gladys was finalized amicably. On February 17, 1981, Barbara and I were married. I honestly believed there'd be divine retribution for my marital failure with Gladys. There wasn't—unless you count Barbara and I as each other's retribution. My relentless trial schedule and my nightly dinner meetings with clients prohibited my paying Barbara the attention she demanded. I wasn't as responsive as I'd been earlier in the relationship, and her once-charming and sexy idiosyncrasies were now distracting. By

Mother's Day 1982, after months of bitter squabbling, Barbara and I finally and permanently broke up.

But looking back to 1977, after I'd separated from Gladys and put the Smalls case behind me, I regained my focus with the Tucci case—a headline-grabbing, brutal double murder in the Italian, working-class Bensonhurst section of Brooklyn. A well-liked elderly couple was found brutally murdered and robbed in their home in a quiet community known for its low crime statistics. It was the kind of crime that the white establishment would want to attribute to a marauding gang of minority kids, but it didn't work out that way. Instead, the eventual suspects were three clean-cut local Italian teenagers: Tamilio, Santanella, and Cappiello. The lead defendant, sixteen-year-old Tamilio, was represented by Al Brackley, one of the great all-time New York criminal lawyers. Al, who sported a great bushy head of snow-white hair even in his youth, was and remains to this day a man of the old school, never taking a day off and doing all his research and preparation himself. I would see him at the courthouse archives obtaining all the relevant felony complaints and other documents that would be his ammunition for cross-examination. He's a loner and a renegade, famous for his prodigious memory, never taking or using any notes in the course of his scathing cross-examinations or summations. His ready temper would flare at the perfect moment for dramatic impact and dissipate as abruptly as it arose. As a prosecutor, the problem with working against Al (other than that he was so effective) was that I was tempted to sit back and enjoy his performance rather than focus on my own case. And he is always very friendly, with ready words of advice for younger practitioners. In fact, after the Tucci trial he took all of us, detectives and lawyers alike, to dinner at the Peter Luger Steak House.

Santanella was represented by Joe Fontana, and Cappiello, who was the captain of the Tilden High School football team, was initially represented by Ron Veneziano and later by another great New York criminal lawyer, Jack Evseroff. As a patron of the flower shop run by Tamilio's father, Mr. Tucci was acquainted with young Tamilio. So when the three boys showed up on the Tucci doorstep, they were readily allowed access. Evidently the boys had agreed only to tie the Tuccis up and then ransack the house for valuables. But as is so common in crime, the original scheme went awry and ended in senseless carnage. At the crime scene Mr. Tucci's glasses lying broken and blood-spattered on the floor, and the bloodstained ropes that were used to bind the elderly couple, spoke volumes to the jury. In sum, it was an old-fashioned murder case in all its horrific details.

Justice John Starkey, a seasoned homicide judge, was assigned to the case, and all three defendants were convicted (Tamilio as the killer, and Cappiello and Santanella under the "felony murder" law that holds accomplices in a felony—here, robbery—equally responsible with the killer for any deaths that arise from the crime) and sentenced to from twenty-five years to life in prison. On appeal, new trials were granted, in part on the basis of a faulty jury charge relative to Cappiello's and Santanella's affirmative defenses to felony murder: They claimed not to have been aware of Tamilio's intent to commit any violence or of the fact that he was carrying a gun. The second trial was also before Justice Starkey, and the defendants waived a jury this time. Santanella and Cappiello were again convicted of the felony murder but received reduced sentences.

The case was important because of the tumult it caused in its community; indeed, Homicide Bureau Chief (later New York Supreme Court Justice) Ronald Aiello assigned the case to me with the admonition that it was a "high pressure" case and implied that had the defendants not been arrested expeditiously, neighborhood toughs might have taken the law into their own hands. The case was also important, as is so often true, because the media chose to treat it that way. But that meant little to me in those days—I was oblivious to the pressure of high-publicity cases, thriving on the broader audience.

8

My career as a trial lawyer in the DA's office effectively ended with the spectacularly controversial 1978 Crown Heights vigilante case. Shortly after midnight on June 16, 1978, in the Hasidic section of Crown Heights, Brooklyn, a sixteen-year-old black youth named Victor Rhodes was savagely beaten by a mob of some thirty to fifty Hasidic Jews. The Rhodes family and the black community claimed that Rhodes was set upon without provocation as he was walking his girlfriend home; the Hasidic community held that Rhodes had assaulted an elderly Hasidic man, knocking the hat from his head. Louis Brennan and Jonathan Hackner, two Hasidic men in their early thirties, who were driving in the vicinity a few minutes after the Rhodes assault and who were identified to police as having participated in the beating, were arrested and charged as acting in concert with others in the beating. Mayor Ed Koch called the Rhodes incident "a vicious assault" that threatened the "fragile thread of inter-group harmony and tolerance" in New York City.

The impact of the incident was exacerbated by the fact that it followed only two days after the death of Arthur Miller, a black businessman and civic leader who was strangled during an altercation with the police, and Miller's death already had the black community in an uproar. As Brooklyn District Attorney Eugene Gold was convening a grand jury to investigate the Miller case, hoping to defuse a potentially explosive situation, Rhodes and the Hasidic community struck the match and lit the fuse on the powder keg.

While Rhodes lay drifting in and out of a coma for two months in Kings County Hospital, the city was roiled by racial and political strife.

Obviously the Hasidic and black communities in Crown Heights, which theretofore had coexisted uneasily, were now pitted against each other: The blacks were calling for the abolition of the Hasidic civilian anticrime neighborhood patrols or the creation of similar black patrols to keep the Hasidim under control; and the Hasidic community was decrying the terrorization of their neighborhood by marauding black criminals. But the New York Jewish community was riven as well, with some Jewish leaders deploring the Hasidic bystanders' horrific overreaction to what was at worst an adolescent's mischievous and disrespectful behavior, and other Jewish leaders suggesting that certain politicians were all too eager to pander to the anti-Semitic sentiment prevalent in the black community. In the black community itself, radical spokesmen like the Reverend Herbert Daughtry (leader of the controversial Black United Front) struggled with moderate black leaders like Congresswoman Shirley Chisholm for control of the situation.

The case was initially handled by me for the Homicide Bureau, because Rhodes was sufficiently injured to create the expectation that he might die. When he survived, the defendants were charged only with assault and attempted murder, and the case was sent down (from the Homicide Bureau) to the Supreme Court Bureau. As I'd presented the case to the grand jury, I retained it for trial. Justice Elliot Golden, a former chief assistant district attorney under Eugene Gold, was to preside.

I can only imagine the pressure brought to bear by the Hasidic community on Eugene Gold—he must have felt some ambivalence about the case, which in turn must have somehow infected the rest of the office. There is no question that the case was not investigated and prepared as fully as a homicide might have been. But in late 1979, some eighteen months after Rhodes was assaulted, the case finally staggered to trial. In the course of an early pretrial proceeding of some sort, I made the mistake of asking rhetorically how many tall Hasidic men with red hair and beards (like the defendants) there could be in Crown Heights. On the next day the gallery was filled with tall red-headed, red-bearded Hasidic men. And they stayed there at every pretrial proceeding and throughout the trial. The defendants waived their rights to be seated next to their attorneys at counsel's table in the courtroom well and sat among all the other tall red-headed, red-bearded Hasidic men until I rested the prosecution's case. Barry Slotnick and his co–defense counsel, Irving Seidman, had moved the court, over my objection, to permit this arrangement in order to avoid facilitating the prosecution witnesses' knee-jerk in-court identification of the defendants as participants in the assault. Justice Golden granted the defense motion, which proved a brilliant strategy for the defense—none of the ten eyewitnesses I called could identify the defendants from among the crowd of Hasidim in the gallery. In the words of one witness, black-garbed, red-bearded Hasidim "all look alike."

Even the police witnesses, including arresting officer Michael Costello, were unable to identify the defendants without the assistance of the mug shots taken at the precinct house after the arrests. The situation worsened when, as the trial was proceeding, Officer Costello was suspended without pay for pumping gas at a Brooklyn service station when he was scheduled to be on regular foot patrol on Empire Boulevard. The suspension provided the defense with another means of attacking the already shaky credibility of the police testimony.

There was no ambiguity in the case for me. I excoriated the disproportionate reaction of a vigilante mob that had beaten a young man for the relatively minor offense of knocking a hat off a man's head. The theme of my opening and closing statements was that membership in a fundamentalist religious group, the Hasidim, of the Lubovitch sect, and wearing solemn black garb, accorded the defendants no special dispensation to perpetrate vicious and violent acts—the acts were cowardly and base regardless of the religious trappings of the perpetrators. It was a media circus worthy of a Tom Wolfe novel, with vocal members of the radical United Black Front in attendance at trial each day "to see that justice is served" and the courtroom gallery filled with Hasidim growling at my every comment. The newspapers fed the anger of the Hasidim who were *not* in the courtroom with a running account of my remarks at trial the day before.

Outside the courthouse, the black and Hasidic communities conducted protests, and the Hasidim became particularly vocal whenever I ventured forth, vilifying me as an anti-Semite and causing District Attorney Eugene Gold to have the American Jewish Congress send a letter to the media in my defense. It likely wasn't popularly known then that I am Jewish. I certainly did nothing to advise them or the media of that fact—my religion was no one's business but my own. In any event, it bore no relation to how I felt as a human being and as a working prosecutor about the crime that had been committed.

In the end, on February 27, 1980, after a two-month trial and two days of deliberations, the jury announced that it had reached a verdict. Louis Brennan and Jonathan Hackner were acquitted of all charges in the Rhodes case. Despite the acquittal, the case had a bright side. Until that trial my public appeal had been to the Irish and Italian working-class juror; thereafter I was also embraced by the black working-class juror.

During the case I welcomed the opportunity to see Barry Slotnick in action; he was described at the time as one of New York's top criminal attorneys. It was an important case for him, as he drew considerable business from the Hasidic community. I knew he'd represented Jewish militant Meir Kahane, as well as Joe Colombo, alleged former boss and founder of the so-called Colombo family. (In fact Barry had helped Colombo found the Italian-American Civil Rights League in the late 1960s and was present when Joe Colombo was shot down in Columbus Circle in 1971.) Barry and I became very friendly in the course of the trial. I respected his acumen as an attorney—his style wasn't mine, and he wasn't necessarily as forensically skilled as some of the best I'd seen, but he was extremely shrewd and knew how to align all the out-of-court aspects of a case and bring them to bear in his client's interest.

As we tried the Crown Heights case, Barry flirted with asking me to join his firm when the trial was over. This can be a ploy of experienced defense counsel when trying a case against a young assistant district attorney who will shortly be seeking employment in the private sector. The fact that both parties know that no such arrangement can ethically be consummated during the trial justifies the defense attorney's reticence about being specific in his offer, and when the trial is over, of course the shadow offer evaporates. In the meantime the young prosecutor is supposed to take it easy, be less aggressive than he might otherwise be, and take his mind off what's important: the trial. I was not surprised when no definitive offer followed the trial. I *was* surprised when he *did* offer me an "of counsel" arrangement (a salaried senior attorney position) approximately one year later. He had a sterling reputation and a good, solid criminal practice. I joined him in the Woolworth Building on Monday, July 13, 1981.

During the period between the Crown Heights trial in late 1979 and my acceptance of Barry's offer, I was a senior supervising trial attorney in the Homicide Bureau and then deputy chief of the Criminal Court, trying no cases for two years, getting rusty and lazy. My friend Bob Keating, Eugene Gold's chief assistant district attorney, had lined up a top job in the inspector general's office, which would have meant more government work but far less trial work. I decided that it was trial work that I needed, that I wanted, and I had to go where I was going to get it, be it on the other side of the law. Barry was offering me a true challenge, and I would have been crazy not to accept it. Not that I made the decision lightly—Barbara, to her credit, urged me along when I had my doubts.

9

I had been in the Woolworth Building for only two weeks when Barry moved his offices across the street to the Transportation Building. In either location I could tell right away that my lifestyle was going to change dramatically. On my first day of work, at about five-thirty in the evening, one of Barry's associates brought in coffee and snacks to tide us over into the late-night work. Yes, the hours were different from those of the DA's office. This was the business of law, and the business of law was all about bringing in cases and making money. And if Barry was good at anything, he was good at the business of law—the best, in fact. I joined the gym at the Woolworth Building, but I was lucky if I could get there once or twice a week. Within two months of joining the firm, in October 1981, I tried my first murder case for the defense.

It was a Flying Dragon case—that is, the defendant, Fun "Bill" Tok, was an alleged member of the Chinese youth gang of that name. He was alleged to have shot a man in a gang fight ten years before and had been a fugitive ever since. Fortunately for me and Fun Bill, the coroner's report said that the victim had died of a knife wound; in addition, I was able to create enough doubt as to the police custody of the gun to severely impact the authorities' credibility. The prosecutor in the New York District Attorney Robert Morgenthau's office, Dennis Wade, was about the best that office had to offer—he was implacable, highly professional, and an excellent trial lawyer. After about three and a half days of deliberations, the jury returned with an acquittal. But I don't recall doing a particularly great job. Certainly I would later try, and lose, far better cases against Nancy Ryan (also known as "Ryan's Express," for the way she would roll over opponents in court), Morgenthau's then chief of the Chinatown Gang Section (his special "Jade Unit"). What I was learning early was that it was nowhere near as easy to win a defense case as it was to win a prosecution case. Sure, you'd win some because the government's proof was simply and clearly deficient, or its witnesses bad, or its victim a scoundrel, but to be in the business of winning, I'd have to do something different at every trial, something dramatic, memorable.

I was learning from Barry that the business of law is about winning because *money* is about winning and the business of law is about money. This was and remains a tough lesson to accept; the concept is counter-intuitive to me. I'm not saying I'm holier than other attorneys. At its most basic, winning was for me about meeting the expectations of those who loved me, about gaining acceptance, and about proving that I could excel, prevail, even gain some fame. Later, especially in private practice, when I gained a deeper appreciation for what was at stake for those whom I represent, those upon whom the full weight of government power is brought (often unfairly) to bear, winning became about saving a client's life—and finally, yes, about justice.

Money has never been the sole motivating force for me, and it never will be. To this day, after all my years of practice, I still rent my home, and I've never replaced the old Peugeot I was forced to sell when I moved to Manhattan. Compare this to the other top-flight criminal law practitioners around. Sadly perhaps, I was never able to learn the business of law. All I know how to do is go into a courtroom and fight as hard as I can for my client—no compromise, no quarter asked or given. The client, no matter how well-heeled or willing to pay high fees, who is inclined to negotiate a reduced charge or sentence recommendation with the government (which can mean providing evidence—sometimes true and sometimes not—against another prosecutorial target), will probably not be seeking my services. This is not what I do best. My client is the defendant who plans to go the distance to vindicate himself. He wants an attorney who will not weary of the fight and who is willing to scorch the earth on which *even his own house* stands to achieve his client's liberty.

This means I have to believe in my client. It has become fashionable for lawyers today, especially in criminal law, to say that they don't ask a client if he "did it." This is knowledge they don't need, they say, because it would only interfere with their ability to represent the client. But I can't leave as much of my blood on the courtroom floor as I do unless I believe in the cause. And if I can't convince myself of my client's innocence of the charges against him, or of the inherent unfairness of the case brought against him, I surely can't convince the jury. I believe that jurors are remarkably canny, and this belief has brought me what success I have enjoyed. I never underestimate their ability to discern the truth in courtrooms that are often filled with the subterfuge of counsel and witnesses, the camouflage of the process, and outright deception practiced by both sides, so I do not condescend to them. They're not the "great unwashed" sitting there in the well—they're me.

After the Chinese gang cases I began to receive matters referred by members of the police department whom I'd met when I was an assistant district attorney. For the most part I viewed the officers I had met while I was a prosecutor as my friends, and they saw me in the same light. This was a special sort of relationship that I found most prosecutors didn't foster. As soon as a homicide reached my desk, I always endeavored to have a detective assigned to it; this option was open to me pursuant to office policy, and I would invariably take advantage of it because the detectives were able to help me better understand the cases. Besides, they and their families appreciated the opportunity to earn overtime pay. I related to the police officers: At bottom I suppose they reminded me of my father, and their families reminded me of my own. Police officers are not FBI agents, recruited nationally, with no local roots. They do their jobs in their own neighborhoods, where they and their families live. Local police officers are for the most part community-minded people who *know* everyone else in the neighborhood, many on a personal level. They know neighbors who get in trouble with the law, and they often will want to help out, offer advice, or recommend a lawyer. I became that lawyer. Bit by bit I built up a following in the police department and in the Patrolmen's Benevolent Association.

Jimmy Shea was a well-known sergeant of detectives in the Tenth Homicide Zone (Coney Island), smart, well spoken, and polite, a really tough old-fashioned Irish cop with a Jimmy Cagney look about him.

Police blue through and through, he lived on the New York policeman-fireman Riviera, Breezy Point. When Jimmy's son, who was also a police officer, got in trouble in front of Nathan's in Coney Island, and I was in my early days of private practice with Barry, Jimmy called me.

Jimmy's son, who was off duty at the time, got into an altercation with a young black man and allegedly made racially offensive comments. The young man went across the street to the Transit Authority Police station and filed a complaint, and the transit police responded with alacrity. Jimmy's son advised the transit police that he was a police officer himself and that the complainant had offered to sell him marijuana, but the transit cops said there was nothing they could do—they had a complainant—and so Jimmy's son was charged with assault. Jimmy called me, and I tried a great case for his son, who was acquitted. I went to the police commissioner and succeeded in saving his job as well.

Then Jimmy Shea sent me a real "wrong man" case. The client was a hard-working, six-foot-five-inch eastern European named Zadik (phonetic), balding, with a massive bushy mustache, who worked as the manager of an Entenmann's bakery in Long Island and lived on the Queens-Brooklyn border. He, together with an unknown black man who was never identified, was accused of a series of liquor store robberies. Our investigator, Les Wolf, tracked down another person who was a dead ringer for Zadik, who was then in custody for armed robbery but was at large when the robberies in question had occurred. In fact, the liquor store robberies had stopped when the look-alike began his incarceration. Obviously Zadik was an unusual-looking man, and it was hard to imagine a misidentification. I suppose that was the Brooklyn DA's problem, for when we brought the look-alike to the prosecutor's attention, he was unimpressed. I felt so strongly about the case that I decided to advise the client to waive a jury. I'd never done that before, and I've rarely done it since. I usually have more faith in a not-yet-jaded juror than in a seasoned judge who's heard it all, to find reasonable doubt in a case. Moreover, in a trial where the judge is the trier of fact, *all* the facts—even those that are ordinarily withheld from jurors as unduly prejudicial, like certain of the defendant's prior misdeeds—are before the court on the assumption that the judge will be better able than jurors to find facts objectively and in accordance with the law, no matter how prejudicial the things he may hear. The latter assumption is not always valid, for a judge is no less human than a juror and thus is also sometimes subject to prejudice. Zadik was tried before Justice Gloria Goldstein, who acquitted. The case reinforced a lesson I should have learned from my father's prosecution in the Marquis matter but that had been diluted by some seven years as an ADA: Just because the government brings charges doesn't mean the defendant is guilty, no matter how patent the proof against him may appear at the outset.

My first Brooklyn murder case for the defense was against Ed Boyar, the best trial assistant in the borough. I had once represented a fireman named Jimmy Thompson, referred to me by my Police Department friends, on bookmaking charges. The case was disposed of with a fine. When Thompson's son Charlie was arrested for murder, he brought him to me. There was a packed courtroom at Charlie's trial every day, filled with Brooklyn assistant district attorneys who wanted to see me try a case against the Homicide Bureau. Charlie Thompson was acquitted. Six months later I attended his wake after he was murdered. Maybe if he'd been convicted and incarcerated, he'd be alive today. I guess these are things lawyers aren't supposed to ask themselves. I can't help but ask.

These cases—Shea, Zadik, Thompson—were big cases in Brooklyn, and they were particularly important for me at the time, too. I was back on my old stomping ground, and if simple assaults, thefts, and murders charged in the state courts in Brooklyn were not the play-offs or league championships, I was still playing in the major leagues. These cases were my validation, the proof that I could succeed in the transition to private practice in a way that was unusual in that era. I attribute that success in equal parts to luck—in teaming up with Barry, in the cases that were referred to me, in the verdicts that came in—and my own efforts, natural talent, and flair. There's no doubt you need all. But with those first cases in Brooklyn I regained my equilibrium. I was back in a routine, working out regularly. I was myself again.

Thereafter I tried two relatively minor so-called organized crime cases. Both were important steps in my career. In the first the client, Alphonse Merola, alleged by the government to be a lieutenant of Anthony

Colombo, Joe Colombo's son, ran an after-hours club in Manhattan and was accused of paying off police officers to look the other way. Merola was charged by the special prosecutor's office (which would handle governmental corruption cases) on the basis of hours of tapes purporting to reflect the payoffs. The Merola case was my first foray into the community of gamblers and street fellows who constitute what the government likes to call organized crime and who followed the case closely. Had I won the case, it might have jump-started my career in that milieu. I didn't. It was a setback for me, but it wasn't life or death for the client, who received a light sentence and later became an informant for the government anyway.

The second matter was the Angelo Ferrugia case, my first federal trial, and I was nervous, not knowing what to expect. It took place in the Eastern District of New York in Brooklyn Heights, and the proceedings were held in a clean, well-tended building with only three or four floors and three or four beautiful courtrooms per floor—a vast change from the relative squalor of the state court system to which I was accustomed. The prosecution, brought by the Organized Crime Strike Force for the Eastern District of New York, alleged that Ferrugia, who had a thriving freight-hauling business and whose *grandfather* (Salvatore) was allegedly a member of the so-called Bonanno crime family, was generating money for organized crime syndicates by fraudulently billing for haulage, storage, and the like at John F. Kennedy International Airport. He was charged with federal tax violations, as well as perjury and obstruction of justice that arose from his testimony before the federally empaneled investigative grand jury (where he was represented by another well-respected New York criminal lawyer, Frank Lopez).

U.S. District Judge Henry Bramwell, a Republican-appointed jurist, presided over the Ferrugia case. On the bright side, he was very fond of my father, but on the other hand, he was not fond of Barry Slotnick. In the end he tried a fair case, dismissing all tax charges on pretrial motions (and severely chastising Edward McDonald, the then chief of the Organized Crime Strike Force, when the latter appeared in court asking to have the tax charges reinstated). At bottom this was one of those manufactured victimless cases, in which the government creatively constructs the case on questionable evidence and perhaps subpoenas a person to testify before the grand jury in the hope that the witness might be trapped in some minor evasion or falsehood that would then form the basis for a contempt, obstruction, or perjury indictment. The true government goal in the whole process is to pressure the witness into providing information against the government's real target in exchange for dropping the contempt, perjury, or obstruction charge. Ferrugia was no criminal, and I believe the government knew that, but it wanted him to provide information it believed he had against one Henry Bono (represented by Gustav Newman, another luminary of the New York criminal bar).

The perjury and obstruction charges against Ferrugia remained for trial, and I tried a good, strong case, along with an excellent Nassau County lawyer, Ken Weinstein (who assisted me with his tax law expertise). I felt like myself, even in the new federal surroundings, and was able to control the courtroom. John Carneglia, a close friend of John Gotti, who had another matter in the same courthouse, was present during my cross-examination of the government's forensic accountant, and I understand that he was sufficiently impressed to later rave to John about my performance. Despite all this, Ferrugia was convicted on the testimony of Douglas Behm, the assistant U.S. attorney who'd submitted Ferrugia's testimony to the grand jury. As we awaited the verdict at a nearby restaurant, Ferrugia had a couple of martinis. After the verdict was pronounced, in his frustration Ferrugia spit at Behm in open court. Perhaps I should have read or anticipated the animosity he held for Behm. Ferrugia was ultimately sentenced to no jail time and probation only. Purely as a result of the spitting incident, however, he was remanded by Judge Bramwell between the date of his conviction and his sentencing, thus needlessly spending some six weeks in jail.

As I still hadn't tried enough cases for the defense by then, certainly not in federal court, my scorched-earth approach and style of strong identification with the client were not yet familiar to the prosecutors. Norman Block, the assistant U.S. attorney who tried the Ferrugia case, later called me a "partisan" for my client. I don't think he intended the remark to be complimentary.

I believe the government felt I condoned Ferrugia's courtroom behavior or even knew of his intentions beforehand. I neither knew of his intent nor condoned it, particularly not with respect to Doug Behm, with whom I'd worked at the DA's office and for whom I have only respect professionally and personally; and I deeply regret any residual impression that I would condone any such show of disrespect for him or the court. Moreover, I don't subscribe to the notion prevalent in the government that an attorney should only go *so far* in the defense of his client, that he should assume the good faith of prosecutors and accept that any governmental tactic is fair play in what the prosecutors may view as the public interest. The government, and individual prosecutors, are *not* always right, and their goals are *not* always in the best interest of the public. *My* job is not to make *their* jobs easy. My job is to keep them honest, when it comes to my client's rights. If not rolling over for the government makes me a partisan in my client's interests, then so be it.

In 1983 Barry Slotnick was retained by a man named Mark Reiter when he was indicted in a major federal narcotics case brought by the Organized Crime Strike Force in the Eastern District of New York. By all accounts Reiter was a tough Jewish guy; Barry had represented him on a previous charge. Indicted with Reiter were Gene Gotti (John's brother), Angelo Ruggiero, John Carneglia, Arnold Squiteri, Edward Lino, Anthony Moscatiello, and Mike Coiro (an attorney who had previously represented his codefendants in other matters). Barry saw to it that I was nominally retained to represent Moscatiello—the law required that he be represented separately. Moscatiello was a legitimate businessman who had been erroneously caught up in the vast sweep of the government's case; the government would voluntarily dismiss his charges years later, and I would eventually step in (after I'd already left Barry's office) to defend Mike Coiro (who would unfortunately be convicted) at the trial. More important, the 1983 federal narcotics case was another step in the dance of ever-shrinking circles that would ultimately bring John Gotti and me together.

John Carneglia, who'd seen me cross-examine the forensic accountant in the Ferrugia case and unknown to me had lauded me to others, was alleged to be a soldier in John Gotti's crew. Carneglia, who was a very tough and formidable person, was also a successful businessman who owned a junkyard. He'd previously been indicted for any number of offenses and been acquitted of those charges, and he was very knowledgeable about courtroom procedure. He'd seen all the best lawyers in action, so I surmise that when he was impressed by my performance, it meant something to John Gotti. But of course, I hadn't yet met John Gotti and had barely heard his name by then. Although he may have heard of my performance in the Ferrugia case from Carneglia, he certainly knew little about me either.

Angelo Ruggiero was John Gotti's *goombara,* a Sicilian word derived from the Italian *compare,* which sometimes means "godfather" but more often is used, certainly on the street, to signify an old friend of unquestioned loyalty: a brother. John and Angelo had known each other since they were no more than twelve years old, although John was from a dirt-poor background and Angelo was not. The way the story came to me, Angelo was walking home one day in the neighborhood (Brownsville–East New York in Brooklyn) with an arm full of dry cleaning (and eating a huge sandwich at the same time—Angelo was legendary for the size and frequency of his meals) when he was confronted by three guys who were going to rough him up. John, who was across the street at the time, witnessed the incident and stepped in, single-handedly "starching" all three. Angelo and John were inseparable thereafter.

When they were about eighteen, John introduced Angelo to Neil Dellacroce, who was called "Uncle Neil" in deference to his age and stature. The government once overheard Angelo call Dellacroce "Uncle Neil" and assumed that he was in fact Angelo's uncle, that Angelo had introduced John to Dellacroce, and that the friendship between Angelo and John had led to John's swift rise in the so-called Gambino family. Angelo was no relation to Dellacroce, however, and made his acquaintance only through John. Like so many of the government's theories with respect to these people with whom it was so obsessed, a full-blown fantasy would arise from a single word overheard in the course of electronic surveillance.

The 1983 narcotics case was built almost entirely on the transcripts of electronic eavesdropping tapes from taps on Angelo Ruggiero's home telephone and bugs planted in his house. Angelo's brother, Salvatore, was a fugitive in a tax prosecution arising from his reputed narcotics-dealing business. Salvatore Ruggiero was himself never accused of being a member of any crime family, because in the government's opinion his alleged substantial narcotics-dealing involvement disqualified him from membership in old-style Cosa Nostra families who were known to eschew narcotics dealing. It was said that they believed drugs would undermine the moral fiber of their own communities and affect discipline among those in their own ranks who might themselves become addicted. In 1982 Salvatore Ruggiero and his wife were killed in an airplane crash. According to the government's theory in the 1983 narcotics case, based on taped conversation in the course of which Angelo discussed seeing that his brother's children were provided for by the liquidation of Salvatore's "assets," Angelo was referring to Salvatore's *narcotic* assets. Angelo's defense and that of his codefendants was that the assets in question were Salvatore's legitimate financial and real property assets. Of course, the government had no actual narcotics to submit in evidence but merely proffered the ambiguous taped conversations. The case was tried twice, with both trials resulting in deadlocked juries; then, after the defendants were severed and tried in separate cases, the government obtained convictions against most of them. Lino was acquitted; and Angelo Ruggiero's case was mooted by his death of lung cancer. Before reaching its conclusion in 1989, the 1983 narcotics case was the longest-running federal criminal case of all time.

My client Moscatiello's case was severed from the government's main case, and, as I said, was ultimately voluntarily dismissed by the government, so I had little if anything to do with the Ruggiero trial. But in the course of reviewing the transcripts of the electronic eavesdropping tapes, I often read John Gotti's name, and I began to develop a feel for the close relationship between him and Angelo. At some point late in the series of Ruggiero trials, I went with Angelo to Taormina, a restaurant on Mulberry Street in Little Italy, where I was supposed to meet an investigator whom we might possibly employ in the case. The codefendants in the 1983 narcotics case—Ruggiero, Carneglia, Reiter, Squiteri, Gene Gotti—were a brawny, sharply dressed, and thoroughly redoubtable group. As I stood at the bar, Angelo took me by the arm, said he wanted me to meet somebody, and led me to a table nearby where several men sat eating dinner. He introduced me as the young attorney working with Barry whom Carneglia had seen try the Ferrugia case. John Gotti, who was seated at the head of the table, shook my hand, laughed, and joked about how Angelo was spending all his time in lawyers' offices. The men around this table were in a different world from the defendants in the 1983 narcotics case, an entirely new level of toughness. They were older looking and more mature, far better dressed in their expensive Italian suits and ties and custom-tailored shirts—a clear leap in every way from Angelo and his codefendants. And despite the jocular tone, the visceral sense of power that emanated from John and those at his table (whom I do not recall, even though I probably had occasion to later see or meet them) was palpable and almost overwhelming to me.

At around this time Angelo brought a new case to Barry, specifically requesting that his "junior homicide partner" handle it. Angelo's future son-in-law's brother, one John Gurino, was the client. Gurino's father was a friend of Angelo and John Gotti and owned a Zabar's-like food and housewares emporium in Howard Beach, Queens, called Ragtime. Ragtime was also a local general store for the Howard Beach community, including some of John Gotti's friends. Of course, John and his friends' true hangout was in Ozone Park, the Bergin Hunt and Fish Club (a reference to Bergen Street, in Brownsville–East New York, where John was raised). The client, John Gurino, was a tough young neighborhood fellow who would run errands—pick up the phone, order lunch—for John Gotti at Bergin. John Gotti was already an icon in Queens and in particular Howard Beach, the neighborhood where he'd settled his family. Angelo Ruggiero also lived in Howard Beach for a time, as did Joey Massino (a reputed boss of what the government would call the Bonnano family and an old friend of John's), "Gentleman Jimmy" Burke (suspected of but never charged with the Lufthansa payroll heist at John F. Kennedy airport and unflatteringly portrayed by Robert De Niro in the Martin Scorsese film *GoodFellas*), and many other reputed underworld figures, as well as policemen and professional athletes. Just as policemen like Jimmy Shea would be concerned about their neighborhood and neighbors, so were John Gotti and his friends concerned about theirs. The Gurinos, my client's parents, were prominent and well-liked denizens of Howard Beach. John Gurino was about thirty-one years old, and I wasn't much older at thirty-five. The case would take my career to a whole new level.

The Gurino case was my first murder trial in Queens. It was against an excellent local prosecutor named Bill Quinn and before Justice John J. Leahy, who had a strange connection to my father. Evidently, as part of an investigation by the special prosecutor's office into allegations of pay-offs to judges, an investigator posing as a client was sent into my father's office wearing a wire, and he offered Murray a certain amount of money in cash. When Murray asked if this was supposed to be his fee, the fellow replied, "No, it's for Judge Leahy." Murray refused the money and said, "First of all, this is not something I do, and secondly, I know Judge Leahy, and he's the most incorruptible judge there is." Eventually, when the investigation was completed, the tape of the conversation between my father and the investigator found its way to Justice Leahy, who was under-standably grateful. Later, after I represented John Gotti in his acquittal in the 1985 federal RICO case in the Eastern District of New York, Justice Leahy wrote me a letter, marked "personal and unofficial," congratulat-ing his "friend Murray's little boy" and urging me to "keep up the good work." I framed the letter and have it to this day.

Gurino, whom law enforcement viewed as a disrespectful kid and a mob wannabe, was said to have had a previous argument with the victim, who was a rough-and-tumble fellow himself. The victim had been outside in front of his house working on his car when someone, alleged to be Gurino, drove by and shot him. In an emotional scene in the ambulance driving the victim to the hospital, the victim moaned to a detective, "J.J. did it," referring to Gurino by his nickname. Against the odds the victim appeared to recover but then suffered a setback in the hospital and passed away.

Justice Leahy was notoriously tough on sentencing; it was his practice to revoke bail on the eve of trial, thereby avoiding the possibility that a defendant whose case was not proceeding as well as he'd hoped would choose to flee rather than await the jury's verdict. This practice is effected by some judges upon a strong motion on notice by the prosecution and a full-blown hearing, or when the prosecutor has rested after putting in its case, or sometimes when the jury is out deliberating. To remand a defendant just before a trial commences, without a prosecutor's prompting, is highly unusual, but Justice Leahy justified his practice on the basis of his frightening reputation for heavy-handed sentences. So Gurino, who'd been out on bail awaiting trial, was remanded as the trial began. I went a little wild in the courtroom, or at least as wild as a lawyer can go without being incarcerated for contempt of court. I demanded a hearing, the judge granted it on the spot, and when it was over, the judge peremptorily ordered Gurino remanded into custody.

The trial was hard fought, the prosecution relying heavily on the victim's "dying declaration" to the police, which would ordinarily qualify as an exception to the well-known hearsay rule. The hearsay rule prohibits the admission in evidence of out-of-court statements. The idea behind this ancient exception to the hearsay rule is that, if you're dying and on your way to meet your maker, you're less likely to be lying. In the Gurino case, the matter was complicated by the fact that the victim survived long enough after the statement was uttered to make a contradictory statement to his girlfriend. Moreover, the bad blood between him and Gurino would have given him a motive to lie and implicate Gurino, even if he believed himself to be on his deathbed. The statement was ultimately admitted in evidence but as an "excited utterance"—yet another exception to the highly porous hearsay rule. Most important, I brought out that, as soon as the police heard the victim's incriminating statement, they hurried to Ragtime where they found Gurino was busy at work, not in a sweat, and where the hood of his car was stone cold. In short, there was room to create reasonable doubt. Gurino was acquitted. For the first time in my career, I felt the exhilaration of walking a client, incarcerated during trial (as opposed to out on bail), to the street and freedom.

As I described earlier, just two months before the Gurino verdict, after the Ferrugia case, I'd met John Gotti briefly at Taormina. Then I conversed with John for the first time in federal court in the Eastern District (Brooklyn) when, on March 28, 1985, I was sent in by Barry to represent someone at the arraignment on the prosecution brought by Assistant

U.S. Attorney Diane Giacalone. Now, on the first Saturday night after the Gurino acquittal in May 1985, at a lovely waterside restaurant in the Bronx, Angelo Ruggiero's daughter married Gurino's younger brother. I was invited, as was virtually every top criminal lawyer in New York, including Barry Slotnick. Also present at the wedding were many men whose names I later found to appear on so-called Cosa Nostra charts hanging on the wall of the federal Organized Crime Strike Force's offices in the Eastern District of New York—including John Gotti. I left early, after the main course at dinner; I heard later that John and the rest of the fellows stayed until daybreak, singing songs through the night until breakfast. But while I was there, I basked in the glorious afterglow of Gurino's acquittal. I was the toast of the wedding, at least as much as one may be at a wedding where you're neither bride nor groom. I felt invincible on that Saturday in May. John spoke to me briefly, congratulating me on the victory. Little did I know at the time how much our lives would be intertwined; how dependent, for better and worse, we would become on each other. It would change everything.

PART TWO

The Client

10

IN late 1984 John Gotti stopped at the Cozy Corner restaurant in Maspeth, just east of Ozone Park. Every now and then he'd meet friends there for a seafood meal. On this occasion he'd driven over with Frank Colletti. They double-parked, and John entered the restaurant while Frank waited for a parking spot to open up. As Frank waited, a van pulled up behind him. The driver leaned on his horn and began to shout and curse Frank, directing him to move the "fuckin' car." The driver of the van, Romual Piecyk, a thirty-five-year-old refrigeration mechanic who stood six feet two inches and weighed over two hundred pounds, had previously been arrested for assault, public drunkenness, and weapons possession. When Frank suggested he take it easy and wait a minute, Piecyk, who thought himself something of a hard case and had already had a couple of drinks that day, bolted from his van and rushed toward Frank. John emerged from the restaurant and starched Piecyk, or administered a few slaps by way of attitude adjustment. Chastened but indignant, Piecyk returned to his van and drove to the nearest police station. It may have been the one and only time he would ever seek help from a man in blue, but in any event on that day Piecyk, now in the company of two police officers, returned to the Cozy Corner looking for the man who'd "assaulted him." He also claimed he'd been relieved of $325 in cash that he'd been carrying in his shirt pocket.

When the police advised John of Piecyk's accusation regarding the cash, in an effort to demonstrate how ridiculous the claim was John said, "Are you guys nuts? Do you know who I am?" And he hauled out a roll of some $6,000 from his pocket. There was a naïve sincerity about John, a pride in the accomplishments that carried him from a youth of abject poverty to a man with a fine suit and pockets full of cash. This may sound as though *I'm* naïve—or worse yet, disingenuous. But you had to see and hear John Gotti yourself, not through the fog of government propaganda or media adulation, to understand the common bond he had with the man on the street. He had a very powerful sense of the present and was incapable of acting any other way than directly. His comment to the police at the Cozy Corner would only create problems for John two years later, when Diane Giacalone would use it to argue at the bail revocation hearing before Judge Nickerson that John had arrogantly admitted he was a big-time gangster and that the large sum of illicit pocket money was further confirmation of such.

"If you gotta take me, take me," John said to the police officers at the Cozy Corner, holding out his wrists for cuffing. And they did. At this time, of course, Paul Castellano, the alleged boss of the Gambino family, had yet to be assassinated, and the government position on John was that he was still only an "acting *capo*" under Neil Dellacroce, the cancer-stricken Gambino family underboss. I'd not yet met John then, and he was to be represented in the Piecyk case by Mike Coiro, his longtime attorney, who appeared at John's arraignment and bail hearing in the matter.

Everything changed for John—and for me—when Neil Dellacroce died of cancer on December 2, 1985, and when Castellano was gunned down two weeks later, as he emerged from Sparks Steak House on East Forty-sixth Street.

Thereafter the government claimed repeatedly, via media leaks and in its much later 1990 federal RICO indictment, that John Gotti shot Castellano. The government-organized crime pundits opined that when Neil Dellacroce, John's mentor, passed away, John had to kill Castellano to avoid being killed himself as a threat to the power structure of the Gambino family. The government's initial theory was that John ordered, orchestrated, and personally participated in the shooting, with the express approval of the "commission" (legendary meetings of representatives of New York's five alleged organized crime families and other criminal organizations across the country). The government's principal informant and witness at the trial of the 1990 RICO charges, Salvatore Gravano, on the other hand, would testify that Castellano and his driver Tommy Bilotti were killed by a number of gunmen waiting for him in front of Sparks, but that John was not among them; rather, Gravano said, John ordered the shooting but was one block away at the time and only later drove past the carnage with Gravano, heading east on Forty-sixth Street, making a right turn on Second and heading south. Which version is true? Which version does the government *believe* is true? Is *either* one true? Does it *matter* to the government? John told me he had nothing to do with Castellano's killing. I believe it—from what I knew of John, if he'd had a problem with Castellano, he would have taken care of it himself.

Perception is reality, though, isn't it? At a certain point perhaps it doesn't matter what *really* happened, and the truth is forever supplanted by legend. Ultimately, this is far more convenient. Unhampered by the banal truth, history can be revamped, even re-*created* at will—a time-honored governmental practice that we tend to associate with "bad" governments but in which even "good" governments indulge regularly.

Over time, our statues are erected, and torn down and resurrected, or replaced by new ones. Truth would only hinder all the demolition and construction. What are little historical revisions beside the erection of important governmental careers and the pumping of newspaper circulation? This is the post-Mafia clean-up era, now that former Mayor Giuliani and the other federal prosecutorial teams have rid the city (and maybe even the country) of the scourge of Cosa Nostra and rendered it once more safe for democracy, for honest industry, and for the court-appointed monitors (drawn from the ranks of former prosecutors) who oversee the Fulton Fish Market, the sanitation industry, the garment district, the construction and concrete industries, the Feast of San Gennaro, and various unions, and who cost more than the "organized crime" they replaced ever did.

The usual assault charge arising from fisticuffs, where there were no serious injuries alleged and the victim was equally to blame for the altercation, would ordinarily have been disposed of upon intake at the DA's office or at worst handled peremptorily as "disorderly conduct" at arraignment. In fact, at the time of John's arrest in the Piecyk affair, the officers (who didn't know who he was as yet) asked him if he wanted to file a cross-complaint against Piecyk, figuring both complaints would be thrown out, but John refused to file a complaint against anyone. And the Queens DA, who even at the time of the incident knew that John was asserted by the government to be an acting *capo* of the Gambino family, wasn't about to throw out *any* charge against him. Piecyk, who of course had no idea who John was when he agreed to testify before the grand jury, was told at that time that John was not a person of major note. So a Queens grand jury returned an indictment on assault and theft charges.

But in the eyes of the world, after the Castellano murder, when the government turned a double homicide into the media event of the year, John Gotti became the *capo di tutti capi,* the boss of bosses in the Gambino family, which in turn was touted as *la famiglia di tutti famiglie*— the family of families. The 1985 Eastern District RICO indictment, on which John was arraigned some nine months earlier, was suddenly the most important federal prosecution in the country. And the publicity-starved Queens DA's office couldn't *believe* its good luck, as its ridiculous Piecyk assault indictment of more than a year before had just as suddenly become the most important case in *that* county.

The young Brahmin assistant district attorney handling the Piecyk case in Queens, one A. Kirke Bartley, had a problem, though. The same news reports that transformed the silly who-hit-whom-first into a public-ity bonanza also scared the life out of Piecyk. The DA's office didn't help matters by insisting that Piecyk be protected around the clock and hid-den in safe houses. As a result, Piecyk was an understandably reluctant witness.

It was around this time, between Castellano's death on December 16, 1985, and the Piecyk trial in March 1986, that Angelo Ruggiero, who was working very closely with Barry then on his 1983 narcotics case, was in the suite and stopped into my office to say hello. I was being inundated with phone calls from the press asking about John Gotti's relationship to ARC Plumbing, with which John was alleged to be employed. I said I didn't know anything about ARC Plumbing or John's connection to it.

Angelo suggested that I meet with John—after all, he said, "you've got a trial comin' up where you're defendin' him—you oughta get to know your client." I agreed, so Angelo called the Ravenite Social Club on Mulberry Street in Little Italy and arranged for me to meet John there later that day. I'd heard it was a favorite spot of Neil Dellacroce, and by the time I'd met John, it was also a place where he and his friends met to socialize, play cards, and hang out.

John hardly ever went to lawyers' offices for meetings, preferring to meet in restaurants, other public places, or at his offices. It was a peculiarity of his that dated back even to the days when John was an unknown, represented by the infamous Roy Cohn in connection with John's 1973 indictment with two others for the murder of James McBratney. (John, who was *not* alleged to be the shooter in the murder, ultimately pleaded guilty to attempted manslaughter, on the advice of Cohn, and was sentenced to four years in prison.) In all the time I've known John, through all the cases in which I have represented him, I recall him appearing only twice at 225 Broadway, where I maintained offices until 1990, and once again at my offices at 41 Madison Avenue. This aversion to lawyers' offices, however, wound up affording the government a golden opportunity to strip him of his attorneys, including me, in John's last trial in the Eastern District on the 1990 federal RICO indictment.

The government would later make much of my appearances at the Ravenite and my dinners with John and suggest—in the hearings to disqualify me from representing John in the 1990 RICO trial—that by agreeing to meet John in places other than my office or, in particular, at the Ravenite, I (and two other of New York's highly regarded criminal counsel, Gerald Shargel and John Pollok) had crossed the line between client and counsel, had gotten too close to my client, and had become house counsel to the alleged Gambino family. If I were a civil lawyer and John were a corporate captain of industry and my most important client (either financially or from the point of view of publicity), no one would suggest that visiting him to work on the matters entrusted to me was inappropriate. Nor would anyone suggest that dining with my client and even socializing with him was improper; indeed, it would be called "client development and entertainment," tax deductible, and encouraged by the firm. Even if John were a corporate client charged with a white-collar crime (like insider trading, antitrust violations, even some sort of fraud), no one would frown on meeting with him at his offices or in restaurants. Microsoft and Merrill Lynch *have* house counsel who work exclusively for the client, whose salary is paid exclusively by the client, and whose offices are on site with rent paid by the client. It's never been a problem for the government. I always had many clients besides John and those alleged to be involved with the Gambino family.

So then is it the nature of the crimes alleged that determines whether an attorney may see a client outside the attorney's office or dine with the client? Should counsel assume his client is guilty as charged, and must a criminal attorney defending a client charged with a blue-collar crime like gambling, or a crime of violence, maintain a certain disdain for the client? Or rather, is it the perceived social status of the client that determines whether an attorney is legally or ethically permitted to confer with him at the client's office or socialize with him?

In the government's view, I wasn't supposed to *like* John Gotti. Maybe a criminal defendant has a right to counsel, and maybe a criminal lawyer has a right to earn a fee, but the lawyer is not obligated to *like* his client, and the client has no constitutional right *to be liked.* And the government believes there should be a limit to the vigor used in defending certain clients, because the government firmly believes it is doing God's work, and the criminal lawyer should acknowledge this as well, as he drags himself through the charade of a defense before collecting his fee and bundling his client off to jail. I have two problems with this view. One: I defend anyone I represent with every ounce of strength in my body. I was driven as a prosecutor as well, but now, as a defense attorney with a client facing loss of liberty and more, I couldn't look in the mirror if I felt there was something I could have done to defend the client and didn't. Two: I did like John Gotti. I found much to admire in the man. This, in the government's eyes, was my unpardonable sin.

The other reason the government called me house counsel and raised the specter of my alleged involvement in what it called a criminal enterprise and specious issues of conflict of interest should be obvious. I'd never lost a case in defense of John Gotti. He had faith in me and that faith was heavily rewarded. What greater advantage to the government than to remove a criminal defendant's choice of counsel, particularly if that counsel has successfully defended him on several prior occasions?

So on that day in early 1986, when Angelo arranged for me to meet with John, I left my office in the early afternoon, took a cab to Mulberry Street, and walked into the Ravenite. It was a work day in the office for me, so I didn't anticipate having to appear in court or see clients. I was casually attired in blue jeans, a short bomber-type jacket, boots, and a skimmer cap. John, dressed in one of his informal ensembles (a pair of gray slacks and a black knit pullover), was seated at a table chatting with Frank DeCicco. They were alone in the place, as far as I could tell. I'd been to the Ravenite years before with Barry, when Barry was representing Neil Dellacroce on some legal matter or other, but it had always been packed with people and alive with activity. I'd never seen the Ravenite when it was empty. John's eyes crinkled up in a smile as he looked me over. Then he turned to Frank and, with a laugh and a nod in my direction, remarked, "He looks like he's dressed to go out and do a piece of work," meaning I looked like I was ready to do an illegal or clandestine street job. I said, "Yeah," somewhat embarrassed, and started explaining that I hadn't anticipated seeing clients today. John laughed again and said, "Take it easy—I'm only kiddin'!"

John nodded toward DeCicco and said, "You know Frankie." I replied that we hadn't been formally introduced. John turned to DeCicco and, gesturing with his hand, said, "This is Bruce Cutler, the counselor." DeCicco, a husky fellow with black-gray wavy hair wearing a sports shirt and a pair of khaki pants, flashed a broad, pleasant smile and shook my hand. At that time the government believed John had appointed DeCicco to the post of Gambino family underboss. Only a month or two later, in April 1986, DeCicco would be blown up by a bomb planted under his car parked outside the Veterans and Friends Social Club in Bensonhurst, Brooklyn. The government's view of the bombing was, depending on who you asked and whom the government was prosecuting at the time, that the bomb was the revenge of relatives of Tommy Bilotti (murdered with Paul Castellano outside Sparks) or the work of Anthony "Gaspipe" Casso (an alleged Lucchese family underboss who later would become a government informant) acting in collusion with alleged Genovese family boss Vincent "Chin" Gigante to kill John Gotti and Frank DeCicco for the unsanctioned murder of Castellano. Of course, both theories are only that: theories. The government also put an elderly man, "Fat Tony" Salerno, in prison for life on the basis that *he* was the Genovese boss. Although "Gaspipe" Casso would later agree to testify in the prosecution of Gigante and would be given a government cooperation agreement, by the time that case came to trial the government would decide against using Casso because his stories didn't dovetail with the tale that Gravano would tell in the government's 1992 trial of John Gotti. Whoever killed DeCicco, and for whatever reason, it was a vicious and cowardly act.

On that afternoon at the Ravenite Frankie graciously nodded when John rose, excused us, and steered me by the elbow toward the door I'd just entered. He said, "Why don't we take a little walk around the block?" I said, "Sure," although I knew he wasn't really asking my opinion. As we walked out, he asked if I was related to Barney Cutler, a gambler from Brownsville–East New York. I told him I wasn't. Once outside we began to walk down Mulberry Street toward Spring Street—past the thin, elderly Italian men in battered fedoras smoking stogies, past the robust aged ladies in gray and black, chattering and gesturing with meaty arms in the doorways and on the sidewalk, and all waving and nodding to John— while John talked to me. He asked me a few questions, and I asked him a few of my own, but mostly John talked for an hour or two. It was my first exposure to one of his operatic soliloquies, in the course of which he would hold forth on his views on life: what was important, what was expected of him, and what he expected of life. The phrase that would currently be applied to our meeting would be a "bonding experience." John would have called it "getting to know each other."

"Some guys, money is their idol—a buck or two, I love you. Not me. My family, my friends, they come first," he told me. "I've devoted myself to a way of life that I believe in. There's a dignity to it, principles to follow, ways to comport yourself, things to believe in. That's what I live for."

There were no other choices for him, he told me. As DiMaggio was born to swing a bat, he was born to be a tough guy on the street and to lead others in that life. If there were no baseball, who would DiMaggio have been? Without admitting any wrongdoing, if there'd been no street life, if he hadn't been a "hoodlum's hoodlum," John asked, who would John Gotti have been?

Perhaps, had he not been so effective in the life chosen for him, or had he *my* advantages growing up, or had his shrewd toughness and tenacity been directed elsewhere, or had he been born into a time when warriors rather than paper-shufflers and politicians were held in society's esteem, John might have achieved true greatness. But John was extraordinarily quick, incisive, and insightful, even on the often complex legal issues with which I was assisting him. Although he tended to leave the lawyering to the lawyers who were getting paid for it, he would occasionally ask a question that showed his grasp of subtle nuances of such terms of art as *predicate acts, enterprise,* or *continuity,* in the context of the federal RICO Act, which has befuddled attorneys and judges since its enactment. His own command and use of language was remarkably concise, dramatic, and forceful, when he wished it to be, and he could be highly articulate (when he chose not to use the four-letter words that are the communicative staple of street life). He would tell me later that his time in jail was spent focusing on exercise and reading—voracious reading.

"Even now," John said as a grown man, a very independent man who'd achieved a certain stature, "my life has been dictated for me. I didn't dictate my life. The government may claim this or that about me, but I want *you* to know what I'm about." And then he spoke to me of the hundreds of people and their wives and children whose livelihood and well-being depended on him. This was a responsibility he carried with him wherever he went, whatever he was doing, he said. It wasn't something you could put away when you left the office. A man had a duty to reach as high as his abilities allowed him, and when he attained those heights, he had a duty to those whom he led. That was why it was important for him to be out on bail pending trial, and why beating the Piecyk case and the pending Eastern District case was even more important—people depended on him.

He felt it particularly important for jurors to understand, though, that he stood on his own, part of no "mob." He was a man who always tried to live in the present, to enjoy everything in the moment, to taste life as if there'd be no tomorrow. He was his own man, answerable to no other, he said. But if you led other men, he said, you couldn't always think of only yourself in the present—you had to think ahead for the good of all those who depended on you.

To whom was he referring, for whom was he responsible? I was his lawyer, and the attorney-client privilege of secrecy attached, but the contours of the privilege are overgrown with fine legal hairs best left unsplit by the attentive lawyer—and any *good* lawyer is an attentive lawyer. Perhaps I should have felt some trepidation as I walked west on Prince Street shoulder to shoulder with John, turning left on Lafayette Street, past the firehouse and the firemen who all exchanged friendly first-name greetings with John. I was surely crossing a threshold into a caballike shadow world of secret numbers, ancient hierarchies and rituals—a world on the edge, a world in the moment. But I felt no trepidation.

Even though the stakes were so much higher, there would be no fear with John, here on Mulberry Street, before trial, or in the courtroom. There was a totally different state of mind: You set your jaw, did your absolute best, and accepted the outcome. As I sauntered through Little Italy alongside him, turning left now onto Spring Street, I felt singularly alive and on the verge of a great adventure. That was how John made me feel, as though I were worthy, as though I could get the job done. I would learn that he made everyone feel that way.

As we strolled, I asked John about ARC Plumbing. He told me that he'd known ARC's owners, the Gurino brothers, Anthony and Cesar, all his life from Ozone Park. The Gurinos respected John enough to provide him a position there that was not very active and certainly not a nine-to-five job; John would merely let them know whenever, in the course of his other endeavors, he came across construction opportunities in which ARC might be able to get involved. It was a mutually beneficial relationship. The ARC position produced a salary, and John's wife had her own funds from an inheritance, so for the most part John had enough legitimate income to support his rather modest lifestyle—clothes were his most ostentatious expenditure, but his home in Howard

Beach was understated to say the least. However, John gambled and lost prodigiously, often leaving his finances in arrears.

The press, in whose ears the government had whispered, loved to coyly inquire about ARC, play the government shill, and take the cheap shot. "So the don's a plumbing supply salesman, right?" They never tired of it. I suppose John's ARC connection seemed a weakness, an angle, a front-page irony. I guess John couldn't complain, and neither could I— the press did him his share of good turns, too. "The play's the thing," even for the dailies. Short term, the comedy; long term, the tragedy. Build him up, pull him down. It's a trite and true formula that still sells and always will: LeRoy's *Little Caesar,* Wellman's *Public Enemy,* Hawks's *Scarface.* The public identifies with the gangster's angry response to the harsh lot into which he was born, and his meteoric, violent rise—a choleric version of the American dream. But our Puritan ethic, and our need to feel superior in our small, safe lives, are satisfied only by his tragic demise. Anyway, John's life was obviously not about ARC Plumbing, and he was no plumbing supply salesman.

Mike Coiro, who'd been John's lawyer for many years, had been representing John in the Piecyk case; but as Mike was a defendant in the 1983 narcotics case with Angelo Ruggiero, I was asked to "assist" him in John's defense. And as the Piecyk trial began, it was clear that I would be the primary spokesman for John. As for the sprawling federal RICO case brought in 1985 in the Eastern District of New York, just because I'd represented John at his arraignment nearly a year before—before the press and the government had anointed John *capo di tutti capi*—did *not* mean John would still have me trying the case for him.

So on that same stroll outside the Ravenite I said, "You know, John, we've got the federal case coming up. I don't want you to feel you're obligated to stay with me as your attorney. I mean, I've only tried two federal cases in my life. I'm only thirty-six years old. There are more seasoned, more experienced men around—"

"No," he interrupted me with an abrupt wave of his thick hand. "If it's all right, I'd like to stick with you, and keep the bearded guy [Barry Slotnick] in the background. Let him help with some ideas. I wanna stay with you."

As we spoke further about my defending him, I realized that he appreciated my youth, exuberance, and inexperience, with little exposure to organized crime cases and without preconceived notions about the Cosa Nostra. He was comfortable with me championing his cause. It's almost impossible for me to express here how strong an impact John's faith had on me at that moment on Mulberry Street, and thereafter: the pride, the self-confidence, the loyalty, and the will to prevail on his behalf.

11

WHEN the Piecyk trial began in March 1986, I delivered the opening statement.

Mike Coiro, as I said, was John's counsel of record in that case, and he could have tried it and won in his sleep. But Mike was somewhat emotionally and physically depleted by the pressures of his own indictment and pending trial, with Angelo, in the 1983 narcotics case. In addition, during the preceding few months, I had become John's spokesman in the media, responding to the government's regularly leaked allegations designed to prejudice John in the pending federal Eastern District RICO case. And John had gotten comfortable with me as we prepared for upcoming pretrial hearings in the federal case. Everyone was of the view that Mike Coiro could use some help at the Piecyk trial, but I knew Mike as an extremely capable attorney who would have needed no help from me, and I didn't feel good about injecting myself into a case he was handling quite well on his own. Nevertheless, it was decided I'd be the point man. John had heard that I'd been dramatic and effective in the Ferrugia and Gurino cases, so he decided that I could start the Piecyk trial off with a bang. I did. With Mike Coiro and John Gotti sitting beside me in the well at the defense table, before the ever-full-to-bursting courtroom gallery, I delivered the opening statement for the defense.

The broad theme of it was that a drunk and belligerent Piecyk had threatened serious physical injury to Frank Colletti, who had a heart condition and was half Piecyk's size, and that John came to Frank's aid in a relatively restrained manner that caused no injury to Piecyk other than to perhaps his pride. I began the process of introducing the jurors—in the Piecyk case, and through the press coverage of the trial, in future cases—to John Gotti as a person rather than as the government-created fictional evil presence portrayed in the media. And a nascent subtext was emerging as well, one that would mature into John's most potent defense in the trials to come: that of the overweening and even dishonest government, willing to go to any lengths to portray imaginary monsters in the media and then "rescue" society from them, all in furtherance of the personal ambitions of individual prosecutors. In the Piecyk case the DA would never have forced such a silly squabble through the grand jury to a felony indictment and then to trial, had it not lusted for the glory of jailing John Gotti. Still, the Piecyk trial bore none of the virulent animosity between defense counsel and the prosecution that would characterize the Eastern District case in federal court a year later. John was pleased with my opening comments. If he had had any doubts about placing his legal fate in my hands before then, after my comments at the outset of the Piecyk case, he never looked back.

As sharp as John was, and as insightful as his suggestions could be, he generally did not interfere in his own defense. I know the contrary has often been said, that he hypermanaged the courtroom tactics employed by his counsel, but it simply wasn't true. It may have seemed that way because from our first moments in court together we were so synchronous. From listening to his views of life, his credo, and his factual renditions of the events at issue in the proceeding, his voice would flow from me. He never had to direct me, because in court he *was* me.

In the Piecyk courtroom John himself changed the ambiance of the proceedings. I noticed that everyone in the courtroom, from the clerks to the attorneys, from the judge to the jurors, started to dress a little better—a newer tie, shinier shoes, a white shirt instead of a pastel, slacks well pressed—and stand a little straighter. John would chuckle and say, "Look, everyone's in their Sunday best when I come here." His presence seemed to affect everything. And he was pleasant to everyone, joking and making friends with the court officers and clerks as well as the judge, Justice Ann B. Dufficey, who was even-handed and cordial throughout the trial.

One day, on the way to the the Piecyk trial from a pretrial conference in federal court in Brooklyn, John picked me up in his car with his friend Bobby Boriello at the wheel. Bobby would be murdered a few years later in his own driveway (when John was in jail awaiting trial on the 1990 federal RICO indictment); the murder case was never solved, and I understand that even the government believes it resulted from a private disagreement between Bobby and his killer. But on this morning before court, we were lost trying to find an entrance to the Brooklyn-Queens Expressway in downtown Brooklyn. John said to Bobby, "Bobby, pull over, will ya. I'll ask the hack." *Hack* is jailhouse jargon for a prison guard, and there was a police officer on a nearby corner. Bobby stopped near where the officer stood, and John got out of the car and asked him for directions. The police officer pointed the way, John thanked him and hopped back in the car, and we drove off. The picture of the so-called boss of bosses approaching a policeman to ask directions struck me as ironic. I don't think the officer recognized John, or if he did, he hid it well. John has a very simple side to him; its emergence has always been dramatic for me.

The resolution of the Piecyk case proved highly anticlimactic. When Assistant District Attorney Bartley called Piecyk to the stand, Piecyk clearly would rather have been anywhere else but in that courtroom. He appeared literally to be in physical pain at being forced by the government to testify. When Bartley asked him if the men who had "punched and smacked" him were in the courtroom, after looking all around, he said he recognized no one. Bartley tried to rehabilitate him by referring to the minutes of his grand jury testimony, but it was no use. The case was over. When the prosecution rested, I moved to dismiss the charges, and the motion was granted. The next day's *New York Post* headline proclaimed I FORGOTTI!

After the trial Piecyk became a major fan of John's, and I understand that he regularly wrote to John in prison.

Around the time of the Piecyk case the National Collegiate Athletic Association basketball tournament was in progress. John, who loved to gamble on sporting events, asked me who I thought would win. I said the University of Kansas. John said he was betting on Duke. Our friendly sports wagering had begun. The stakes were high: push-ups. We would bet push-ups—in sets of a hundred—on each game. When I lost, I'd have to do the push-ups; when John lost, he'd have to do them. Later we graduated to higher stakes still: David Nadler's custom-made shirts. Nadler was an elderly Jewish gentleman who'd emigrated from eastern Europe, where he'd survived a Nazi concentration camp, and had a custom shirt shop under the elevated B line BMT subway tracks on Eighty-sixth Street in Bensonhurst, Brooklyn. John had all his shirts made there, and he'd bet me a couple of shirts on the outcome of a game. Shirt stakes betting became a tradition with us, and I found my shirtmaker for life in David Nadler.

At around this time, while John was out on bail on the pending Eastern District charges, he suggested that I allow his friend Bob Pellegrino to take me to DeLisi Clothing on East Fifty-fifth Street, off Fifth Avenue in Manhattan. That was where he sometimes shopped, and he thought I might like their selection as well. *Veni, vidi, vici.* I agreed, went to DeLisi's, and picked out my first double-breasted suit. Until then I was hardly an Ivy League dresser, as many in the press have intimated. Yes, before that first Eastern District trial, my sense of fashion consisted largely of the off-the-rack offerings of French designers, except for the occasional Kilgour, French, and Stanbury suit on sale at Barney's, but my conversion to higher-quality Italian suits was not exactly revolutionary. It was rather the natural progression of a New York lawyer rising professionally and financially. I was doing better in business, and I was dressing better.

Nevertheless, the more press coverage I fielded for John in that post-Piecyk, pre-Giacalone era, the more I would read and hear that I was beginning to look and sound like my client. I was in pretty solid physical shape then, as was John, both of us strongly built about the arms, chest, and shoulders, both about the same height and weight. Even though I was balding and John had a full head of hair, when we were both attired in double-breasted suits, I suppose comparisons were inevitable—particularly since my uncompromising courtroom stance, often perceived as belligerent, seemed to mirror John's persona. In response to a newscaster's comment, I once said that John was a good-looking fellow, and I was flattered if people thought I resembled him.

The resemblance was noted by more than the media. When my father showed up unexpectedly during jury selection in the Eastern District in April 1986, I introduced him to John. Murray's appearances in court when I was in the well were very rare, in large measure because (as I mentioned previously) he made me uncomfortable—pleasing him was an additional pressure that I eschewed for the most part. But on this occasion I was glad for the opportunity to introduce him to John. John said he was happy to meet Murray and that he'd heard a lot about him. My father responded in kind and wished John luck in the pending case. We three stood there during a break in the proceedings in Judge Nickerson's courtroom gallery, surrounded by the roiling crowd—John's friends and associates, reporters, attorneys, and onlookers of every stripe—exchanging jokes and small talk for a while, until Murray was about to leave for a court appearance of his own. That instant, standing there beside John, Murray towering over both of us, was a moment of happiness. No doubt my sense of contentedness was somewhat guarded, as there was so much work ahead of me and John, so much danger, but I recall the moment as one of remarkably few instances of true happiness. I suppose Murray's approval, always so hard-earned, was of great importance to me, and standing there in the media spotlight, basking in the aura of my larger-than-life client, was a sort of arrival for me, both professionally and as a man—a coming of age. On the other hand, I was pleased to open this personal side of my life to John, who was an influential client, the source of my newfound prominence, and someone with whom I was becoming close. It was a minute or two of grace.

As Murray was shaking John's hand and turning to go, his eyes darted from John to me and back to John again. Then his eyes crinkled up, and he said with a chuckle, "You know, you two look alike." John snorted, I shook my head, and Murray took off.

But the constant media attention to our likeness soon became tiresome and uncomplimentary. The implication was that I was consciously or, perhaps worse, subconsciously mimicking John; I was sometimes cast in the press and on television as a caricature of my client. One reporter even came up with the stupid and false suggestion that I'd taken up weight training to look more like John. These barbs hurt my pride, but I was certainly not about to change my tastes or my mannerisms.

On the other hand, looking back at that period in my life, I have to consider how much I was responsible for the attention that my resemblance to John received. Something *was* happening to me. I was young and new to the dramatic world into which I was suddenly thrust, new to working so closely with as strong and dynamic a personality as John, and new to the massive publicity barrage with which we were dealing every day. And yet it all seemed right. I can't wholly explain it, but I felt mysteriously calm and at home in the eye of the hurricane. I clothed myself in the chaos around me. I stepped completely into the role of hard-nosed advocate for a client who, because of an alternate and rebellious lifestyle, was an easy target for overreaching prosecutors seeking a scapegoat for all society's ills and an easy route to judgeships, mayoralties, referee positions, and so on. I suppose it was something vaguely akin to my weight training in prep school—my investment of energy and effort on John's behalf was reaping enormous, even disproportionate, benefits for me and my client. The thrill was addictive, and as we proceeded from the Piecyk case to the victories in the Eastern District federal case and the New York County case, I threw myself into it more and more, in attire, in loyalty, in body and soul. I embraced the whole milieu: the aura of respect and immediate name and visual recognition accorded me wherever I would work, dine, shop, and even stroll; and, yes, the excitement and danger of John's world—as his counsel, faced with the mounting overt hostility of the government, which began to focus its venom on me personally; and as a man, faced with the inescapable fact that people who'd chosen his life, people I'd meet from day to day, would often die suddenly and violently. It's been said that a man's sexual potency and feeling of physical invincibility are inextricably linked with his sense of his own power, and I felt undeniably powerful in that milieu. Whatever spot in my brain these heady times were touching was dangerously close to where my sense of true virility lies. I can only imagine what life must have been like for John.

The nether side of that world—the violence—I never understood. I didn't generally talk about specific acts of violence with my clients; as a lawyer I'd understandably be insulated from this kind of conversation, except insofar as it involved defending a client in court against allegations that he had committed such acts, justifiably or not. My own exposure to violence was limited to playground scuffles as a kid, the wrestling mat and the gridiron, and short youthful bursts of temper in high school, college, and thereafter. Violence is not the way I solve my problems, and I confess I didn't allow myself to do much thinking about it. As counsel, this is fine. My duty to highlight the deficiencies in the government's evidence of a client's involvement in any violence doesn't require me to dwell on the violence itself or the reasons for it. But as a man?

I would, and still may only, address it as a decision made by each man who chooses a particular lifestyle. The options echo, even if remotely, those of my generation deciding whether to follow the well-worn paths (college, professional school, the proper job offer) to financial security and success, or to drop off the fast track and enjoy each day to the fullest as a surfer or ski bum and thereby accept the attendant future risks. But how can one born to privilege weigh and pass judgment on the options facing ghetto kids (whether black, Italian, Jewish, or Hispanic), without being condescending or trivializing? Growing up in East New York, John and those around him observed that the well-dressed, respected men on the street were those involved in underworld life, while those who worked hard or, like his own father, worked sporadically at best were barely surviving and were at the mercy of the politicians and the rich: the powerful. The texture of life has a different feel to John and those raised as he was—on the one hand, the heady *strength* of acting in the moment and expressing oneself without filter, but on the other hand, the *fragility* that comes as a consequence of those unfiltered acts and expressions. John and those in his world lived for today and took the risk of tomorrow. One of the risks was the violence that attends taking what one wants and defending what one owns from those who would take it. The violence is a perversion no doubt, but one evidently accepted by those who choose the life.

The criminal lawyer's lot is to face the often unspeakable allegations against his client, put them from his mind, and continue fighting. Sometimes attorneys have the luxury of choosing their clients or cases. This is what I do whenever I can. Life is easier that way, avoiding the ambivalence and the internal contradictions between the lawyer's suit and the man inside it. I'm not saying I don't occasionally make errors of judgment about a client or a case and suffer through my mistake all the way to the entry of judgment. And sometimes an attorney deludes himself into believing *he's* chosen the client, but the client or the case chooses him, and destiny takes over. I can honestly say that in representing John Gotti, I never felt ambivalent. Despite John's operatic manner of speaking for effect (caught on tape by eavesdropping bugs) or the "testimony" of admitted reprobate-turncoats paid by the government to say whatever it wished to hear that day, I never believed John guilty *of the things with which he was being charged.*

12

WILFRED "Willie Boy" Johnson, whose mother was Italian and whose father was an American Indian who did high-altitude construction work, had been a close friend of John's since they were teenagers in the 1950s with the Fulton-Rockaway boys in Brownsville–East New York. At five feet ten inches tall and weighing about 250 pounds, Johnson was inseparable from John until 1985, when prosecutor Diane Giacalone revealed him to have been an FBI informant. As the government story goes, Johnson never forgave the failure of Carmine Fatico, a Gambino family *capo* for whom John and Willie Boy were working in the 1960s, to financially see to Willie Boy's family when he was away in prison for three years on a charge of armed robbery to which he'd pleaded guilty at Fatico's urging. Mrs. Johnson was forced to the indignity of seeking welfare to feed herself and her two children. Allegedly, Johnson could never become a full-fledged member of Cosa Nostra because his father was not Italian and membership was reserved for purebred Italians. According to the FBI, which tries to keep abreast of this sort of dissatisfaction in the ranks of those in the street life, it approached Johnson in 1967 and somehow convinced him to become a "top echelon" informant for the government. How, we'll never really know.

Johnson's alleged function was to have been purely informational, providing the government with intelligence as to the workings of Cosa Nostra. He'd always been assured that he'd never be called to testify against any of his friends or associates in any prosecution because his usefulness as a source of information would thereby be ended and his life would be endangered. On the other hand, although Johnson was permitted to continue his involvement in certain criminal activities, such as gambling, he was supposed to refrain from violent crime such as robbery or murder.

The government claims that it did not keep its promise to Willie Boy. As early as 1978 the Queens County district attorney's detective squad had supposedly by chance gotten wind of Johnson's arrangement with the FBI. The Queens squad had targeted certain members of the Bergin entourage, Johnson included, for random surveillance, and in the course of such Johnson was seen entering a vehicle whose license plates were easily traced back to the federal government. Thereafter Johnson was allegedly providing "intelligence" to the Queens DA's office *and* to the FBI. But Johnson's real problems had yet to surface.

As I've noted previously, Mafia hunting is big business in the United States, whether for prosecutors seeking advancement or lucrative refereeships; for politicians seeking higher office; for the many attendant cottage industries like electronic eavesdropping equipment manufacturers, prisons as human warehouses (with nearby hotels for visitors, rental car businesses, and restaurants); for organized crime experts; and, yes, for criminal lawyers. As with so many other businesses, New York City is the pinnacle—or the nadir—of the Mafia-hunting business. New York has by far the greatest concentration of organized crime fighters in the country and probably the world, all standing in line for their bite of the apple.

On the federal side there is, first, located in downtown Manhattan, the office of the U.S. Attorney for the Southern District of New York, from whence sprang former U.S. Attorney Giuliani. The Southern District's geographical jurisdiction includes Manhattan, the Bronx, and Westchester County. Second, in downtown Brooklyn, just across the Brooklyn Bridge over the East River, there's the office of the U.S. attorney for the Eastern District of New York, whose geographical jurisdiction technically includes Brooklyn and Queens, those legendary hotbeds of Cosa Nostra activity. What the Barbary Coast is to piracy, what Yale is to spies and presidents, Brooklyn and Queens are to organized crime, as far as the government is concerned. Third, there was at the time in downtown Brooklyn the Organized Crime Strike Force, one of fourteen independent strike forces across the country, all of which answered directly to the

Justice Department in Washington and were created to combat major white-collar crimes and the Mafia. This is not to mention the U.S. Attorney for New Jersey, right across the Hudson River, whose organized crime targets are often alleged to have been active in New York as well as New Jersey.

Locally there are the district attorneys for the five boroughs of New York City, each of whom sports his own Rackets Bureau. In addition, there is New York State's Organized Crime Task Force, an entity independent from the New York State attorney general's office, whose director at the time was the notoriously ambitious, smart, and publicity-conscious Ronald Goldstock.

These are only the prosecutors. There are also battalions, divisions, and legions of FBI agents assigned specifically to organized crime in the New York area, as well as the detectives assigned to the five district attorneys offices and the state's Organized Crime Task Force.

At the same time that Diane Giacalone in the Eastern District began thinking of using the relatively new RICO statute to obtain indictments (arising out of "tribute" allegedly paid to the Gambino family) against Neil Dellacroce and those whom she believed to work under him, the Organized Crime Strike Force, operating only a few floors away in the same building, was preparing its own RICO case. They were targeting the Gambino structure as they believed it to be and in particular the lower and middle echelons of the so-called family, including John Gotti. This created tremendous animosity in the competitive world of high-stakes prosecutions. Edward McDonald, the head of the strike force, was a likable and prominent prosecutor and an excellent trial lawyer who'd achieved some reputation on the basis of his successful cases brought against various politicians, including Senator Harrison Williams of New Jersey, in the ABSCAM prosecutions. It has been said in later accounts written on the topic that when McDonald first found out about Giacalone's investigation, he tolerated its coexistence, mainly because *only he* was receiving the fruits of electronic eavesdropping planted in the homes of Angelo Ruggiero, Paul Castellano, and Neil Dellacroce. And *only he* was to have the benefit of certain informants procured by Goldstock's New York State Organized Crime Task Force. Moreover, he had no faith that Giacalone had any chance of turning her allegations against Dellacroce and Gotti into an indictment or provable case.

In order to obtain convictions against John and his codefendants under the RICO statutes, Giacalone had to prove the existence of a "criminal enterprise," which is defined as a group of individuals or corporations "associated in fact" who committed crimes for the purposes of the group. She also had to prove that behavior by the defendants constituted a "pattern of racketeering." This requirement could be met by showing two or more violations of state or federal law by each defendant, even if the defendant had already been tried and convicted *and served a sentence* for those crimes. The idea here was that the RICO statute was seeking to punish the "enterprise" nature of the crime, as opposed to the underlying crime itself, for which the defendant may have already paid his debt to society. In John's case, as he had pleaded guilty to attempted manslaughter in the McBratney case in 1975 (for which he had been sentenced to four years in prison) and had a hijacking conviction in 1969 (for which he had also been sentenced), Giacalone would have no problem proving her two "predicate acts." Her problem would be demonstrating the existence of a "Gambino crime family" or a "Mafia," and that those acts were committed in furtherance of such. It became increasingly clear that the driven Giacalone was almost single-handedly going to proceed with her case, with the nominal backing of the then–Eastern District U.S. Attorney.

Too much lay on the table for these agencies to simply join together even to pursue what they would have called the "common good": There were careers to be made, names to go down in history, judgeships, mayoralties, maybe even presidencies. Who knew? Stranger things had happened. Teddy Roosevelt had been a crime-buster, and Thomas Dewey almost rode into the White House on his anticrime efforts. This ruthless and shameless competition among prosecutors to make a case against John would become a common theme in the history of John's days in court, and it makes perfect sense if you accept what I've been saying all along: John was vilified and built up in criminal importance by the government for the primary purpose of creating a vehicle to carry certain prosecutors to fame and fortune. The irony was that the more John was able to evade the government's efforts to jail him, the greater a prize he became.

The FBI had sided with McDonald in his war with Giacalone, partly because it had a strong, long-standing relationship with McDonald and partly because as usual the FBI had developed the strike force case from its inception. By contrast, Giacalone's case was essentially her own brainchild. Then one day Willie Boy Johnson allegedly turned over to his FBI handlers a document that he'd managed to get his hands on: a memorandum from Giacalone to the Justice Department outlining her whole case. According to at least one book on the subject, the FBI evidently traced the leak to Giacalone's office and allegedly decided to sever all contact with her office based on the perceived porosity of its security. Frankly, given the intensity of the internecine struggle for glory among New York's various prosecutorial agencies, I wouldn't be surprised to hear that a competing agency arranged for the document to fall into the wrong hands simply to provide the FBI with an excuse to take sides against Giacalone. Did the FBI ultimately desert Giacalone because she was ruthless and overbearing? I couldn't say. But she would go on without the help of the FBI. Despite McDonald's efforts and those of the FBI, the Justice Department in Washington would rule that Eastern District U.S. Attorney could make the final decision, and Giacalone managed to convince him that she should get the nod, and McDonald's case would be folded into hers and prosecuted by her.

So the stage was set for Greek tragedy. But contrary to what one's expectations might have been in the circumstances, the tragic hero was not to be John Gotti, rising to the heights and stricken down by fate, a victim of his own pride—not on *this* day, at least. This would be Giacalone's day, *her day* to bask in her victory, in that her case was chosen to go to indictment and trial, and *her day* to lose as well.

Willie Boy Johnson's alleged double-dealing finally caught up with him. Upon making a routine request to the Queens DA's office for any information that might assist her case, Giacalone was at once shocked and overjoyed to find that Willie Boy was a longtime informant for the FBI and the Queens DA—shocked because the FBI had never bothered to tell her this vital piece of information, and overjoyed because she envisioned Johnson testifying for her. Despite the FBI's bellows of rage, Giacalone decided that she would indict Johnson along with John Gotti and the others. The law would then require that she advise defense counsel that Johnson was an informant. In her fantasy, Giacalone might have believed that Johnson would seek the government's protection from his codefendants, that she would demand he testify in exchange for protection, and that he would agree. In short, Giacalone tried to squeeze Johnson. This was a common prosecutorial ploy. But as the FBI knew, Johnson was unsqueezable. Furthermore, they'd supposedly given him their word that he'd never be compelled to testify, and besides compromising what they claimed was their most important source of organized crime intelligence, they believed Giacalone was needlessly endangering Johnson's life. They pleaded with Giacalone to understand, but I'm told her attitude was "the FBI isn't helping me anyway, so to hell with 'em."

In the early-morning hours of March 28, 1985, as John Gotti and Willie Boy sat playing cards, the two of them, as well as John Carneglia, Tony Rampino, Nick Corozzo, and Gene Gotti were arrested at the Bergin. Charlie Carneglia (who was labeled a fugitive) and Lenny DiMaria (who was already in jail on other charges) were indicted as well. Later that day, in Judge Nickerson's courtroom, I first met John formally and represented him at his arraignment. Neil Dellacroce, who was indicted as well, was ill and was arraigned separately at his home at a later date. We all attended a so-called *Curcio* hearing, held to permit Neil to waive a possible conflict of interest in that Barry Slotnick (his attorney) and I (representing John Gotti, another defendant in the same case) were nominally partners. I recall John shaking his head at what he (and Neil) believed to be the artificiality of the proceeding. Neil was a tough guy's tough guy who knew the score legally; he knew whom he trusted and wanted as an attorney, and he didn't care a fig for conceptual legal structures such as that requiring a waiver of conflict of interest if two lawyers in the same case were affiliated in some way. Moreover, he and John both saw the rule as likely created to help the prosecution by giving them another way to disqualify counsel of choice. John and Neil waived the potential "conflict."

At the arraignment before Judge Nickerson, Giacalone acceded to bail for all the defendants except Willie Boy. The courtroom was a sea of blank faces staring at Giacalone, more than a few with mouths hung open. Chairs stopped scraping the floor, papers stopped shuffling, and all whispering ceased. And then as if on cue, all heads turned to Johnson. When asked by the court why Willie Boy warranted different treatment, at a side bar Giacalone claimed to the amazement of all that he'd been a government informant for over fifteen years.

The stunned silence was suddenly shattered by Willie Boy's shout that it wasn't true. I recall that John shook his head slightly, incredulity in his eyes, and then a knowing smile slowly appeared, a rueful comment on the depths to which the government would stoop. There was also in his expression a touch of pity: for a *goombara* who had allegedly failed to make the grade, meet the standard for friendship, for manhood. In the conference Judge Nickerson immediately convened in chambers, Johnson denied being an informant. Moreover, he said, his life was not in danger, and he did not fear any harm from John. Applying maximum pressure on Johnson, Giacalone moved to have him placed in segregation in prison to prevent him from being harmed by other prisoners. Of course, placing an inmate in segregation is an open advertisement to the prison population that he's an informant. The court acceded to Giacalone's request over the objections of Willie Boy and his attorneys.

Nine months later, after my meeting with John at the Ravenite and after the Piecyk trial, speaking to a few of us, John said, "How could that be true about Willie Boy going bad? You know, when I was on the lam on account of the McBratney thing, I came back into town specially to help him out with a beef he had goin'. And now they're sayin' that Willie Boy told 'em where to find me when I came in to help him? I come in here to save Willie Boy and he fingers me? I don't believe it. I hope they [the government] never make up something like this about you or one of your friends, Bruce." John never acknowledged to me, or anyone else that I know of, that he *believed* Willie Boy had betrayed him. My personal opinion is that the allegations about Willie Boy broke John's heart.

By March 1986 the Wilfred Johnson hearings were under way. They were proceeding at the same time as the Piecyk trial, so we were constantly shuttling between Brooklyn and Queens on the two cases. The purpose of the Johnson hearings was to determine the admissibility, at Giacalone's RICO trial, of Johnson's so-called "coconspirator" statements—the out-of-court statements Johnson made to or about the other defendants (John Gotti included). The legal issue was whether (1) Johnson was technically an agent of the government, which would mean he was *not* a coconspirator, in which case his out-of-court statements would *not* be admissible against the defendants; or (2) Johnson *was* a defendant and hence a coconspirator in the crimes in question, which would render his statements admissible. The issue was obviously of vital importance to the government's case, as it was clear that Johnson would be making no *in*-court statements. We argued that, as the government stated Johnson was working for them as a top-echelon confidential informant, whatever statements he made to some other witness about John Gotti or about anybody were not admissible but were hearsay because he was an agent (albeit secret) of the government.

These hearings were my first opportunity to stand up and actively represent John in a federal proceeding—cross-examining FBI agents and the like. Of course, as ever, the gallery was filled with John's closest friends, FBI agents and prosecutors, reporters, and an array of organized crime buffs. I was nervous at first, but once I actually stood on my feet, everything seemed to come naturally, and I felt strong and effective. The hearings gave me, and John, an opportunity to get comfortable in the milieu as well as with each other, and to witness each other's courtroom demeanor. These were my first glimpses from within the fishbowl that would be my home until a verdict would be rendered in March 1987, and beyond.

I was getting acclimated to it all: the courtroom, the judge, and in particular the stiff, meager young woman to whom I'd be effectively married for the next year of my life. Giacalone was dark haired and not unattractive, if schoolmarmish in appearance. She had the hard mouth and tough talk of a woman impersonating a cop; yet she was highly articulate, if somewhat strident, and above all compulsively well prepared. Her eyes were always on the move, slightly, almost vibrating with the tension of anticipating the next attack, the next affront. Her scholarly young cocounsel John Gleeson, who had been in the U.S. attorney's office for only a few weeks by then, remained in the background during the pretrial proceedings.

Those early days of the Giacalone trial had a placid quality, the stillness of a black and white photo one might find in a grandparent's album. Courtesy prevailed in the courtroom then, and amenities were observed, as if we were all whispering. The elderly and avuncular Judge Nickerson was still friendly and pleasant to me. Even my relations with the prosecution team were professionally cool but genial. And I was working with some of the best criminal lawyers around, including Gerry Shargel, Barry Slotnick, Jeff Hoffman, George Santangelo, Michael Santangelo, Richard Rehbock, Susan Kellman, and Dave DePetris. I became particularly close with George Santangelo, who developed into one of the leaders of the defense team and one of my best friends.

It came as no surprise when Judge Nickerson issued his decision ruling that the Johnson statements would be admitted in the government's case against the defendants. In the court's view, although Willie Boy was a government informant, he was also participating in criminal activity (albeit with the FBI's knowledge of and acquiescence to all his criminal activities, excluding those of violence); therefore his statements were made more in the course of his efforts on behalf of the "Gambino enterprise" than on behalf of the government.

But I was learning the way things worked in the conservative federal court, when it came to defendants whom the government had determined deserved to be imprisoned. And I shouldn't have expected things to be any different before U.S. District Court Judge Eugene H. Nickerson, whom I understood to be a wealthy WASP patrician who lived in Nassau County where he had formerly held the office of county executive. At best, every motion before him would be like a challenge to the champion in a boxing match: All close decisions would go to the champ or in this case the government. Even in ordinary circumstances the court would be highly reluctant to render a decision that critically harmed the government's drive to jail someone it deemed dangerous. In the case of someone like John, whose reputation for criminality had been bloated (purposefully) by the government, the court's reluctance would become nearly overwhelming. It was that proverbial railroad—one tie after another, one station stop after another, quietly in the subdued gentility of the federal court, but unswervingly moving toward the only possible ultimate destination: life in prison.

13

IT was during jury selection that ripples began to appear on the smooth surface waters of Giacalone's prosecution and storm clouds began to roll in. John's friend Frankie DeCicco, whom I'd met on that first visit to the Ravenite a few months before, was murdered by a car bomb. The government opined that rival gangsters were responsible, perhaps trying to kill John, and the media now had the immediate gory drama it most enjoyed selling the public. As a result of the media saturation of Mafia sensationalism and photos and film clips relating to John, of course provided by the government in an attempt to further vilify him and ensure a government verdict, it became impossible to empanel an impartial jury. In the course of our first efforts to find twelve suitable jurors and four alternates, the government never mentioned any need for an anonymous jury—everything was out in the open. Jurors were questioned as to what they knew about the defendants or the government's case, and we found that they all had preconceived notions born of the publicity with which they'd been bombarded.

Finally, on my thirty-eighth birthday, Judge Nickerson declared a mistrial, dismissing the first pool of potential jurors. Then he turned to Giacalone and spoke to her in a manner that clearly implied that she should move to have bail revoked. Giacalone so moved, seeking to have John, his brother Gene, and John Carneglia remanded into custody pending a verdict after trial. The government position was technically not based on the then-new Bail Reform Act, which would greatly facilitate the pretrial incarceration of defendants, but rather on the notion that the defendants had committed crimes while out on bail. John's "crime" was the alleged intimidation of a government witness who was to testify against him—not in Giacalone's RICO case but in the Piecyk case. It was argued that John or one of his friends had threatened Piecyk with physical injury if he were to testify in the Queens case against John or point him out in court as his assailant. This argument was belied by the witnesses in the Queens case, who'd testified that Piecyk had been intimidated not by John or his friends but rather by the media and the Queens DA squad detectives. More important, we submitted an affidavit from Piecyk himself swearing that he had been threatened by no one except the DA's office and that John had done nothing to prevent him from testifying in the assault case.

Nevertheless, in a decision that was as preordained as that in the Johnson hearings, on May 2, 1986, Judge Nickerson found it more likely than not that John or someone at his instance had threatened Piecyk and ordered that John's bail be revoked, granting him one week to settle his affairs and surrender into custody. Neither John Carneglia nor Gene Gotti was remanded, and in fact it was always our view that their inclusion in the government's motion was merely a sham, a pretense that John Gotti was not being singled out for remand. So only John Gotti (and Lenny DiMaria, who was already incarcerated on another matter, and of course Willie Boy Johnson, who was in "protective custody") would be imprisoned while awaiting and during the trial.

I left the courthouse after the decision was rendered and took a cab to Taormina in Little Italy, where John awaited the court's decision on the government's remand motion. He was sitting at a table with some friends, wearing a dark spring pullover shirt and gray slacks, and I approached with some trepidation. I was about to tell him that in a week he'd have to go to jail for perhaps the rest of his life, that he might never again walk about the earth freely, that he might never again see his family outside of a courtroom or a prison visiting room. Before I sat down, I had told him what my face had already conveyed. He asked me to sit down, wanting to know what we were going to do next—he wasn't dwelling on past losses but was already on to future action. I told him I could appeal the decision. He said, "Good," and then he said, "Don't worry about it. The fix was in [meaning the court had prejudged the motion in favor of the government before it heard any argument at all], and we're gonna win this thing because they don't have a case and you're gonna make the jury see that."

I did appeal Judge Nickerson's decision to the U.S. Court of Appeals for the Second Circuit (which encompasses the New York district courts), trying to obtain a stay of execution of the court's order, but it was no use. The appeals court is bound by law to give great deference to the trial court on issues of fact, and we had to demonstrate that the trial court's decision was "clearly erroneous," or that no view of the facts could possibly support the decision—an impossibly difficult burden to bear.

Much later, long after the jury had rendered its verdict in Giacalone's RICO case, friends told me that, as John sat in the Manhattan Correctional Center (MCC) from May 1986 to trial, lawyers from every big city in the country were visiting him there, on one or another contrived pretext or introduction, seeking to replace me and represent him in the case. The ethical constriction against an attorney soliciting an already-represented client proved no disincentive to them. But John wouldn't budge. Untried and raw as I was, I would be his voice.

John flourished in the MCC. The whole place, jailor and jailee alike, was aflutter with his arrival. It took him about half an hour to get acclimated and to begin holding court—in the cell block, with the other inmates, I imagine, and in the attorney conference rooms. Besides John's audiences with his inmate and unrelated lawyer courtiers, there were frequent meetings with his codefendants and counsel on the pending Eastern District case. In this forum procedures were different from those of the federal court in which he was to be tried, looser, more relaxed. But they had their own rules, John's rules, their own stringency. It was standard operating procedure among attorneys visiting clients in the MCC to "float" around, visiting *this* inmate client in one conference room and *that* in another, and visiting perhaps other past or potential clients. But not when they were visiting John. He promulgated an unofficial edict: When a lawyer was coming to the MCC to see him or another codefendant on his case, he would see no one else. John wanted the lawyers' full attention, as he would give his to someone who had come for a meeting with him. If an attorney were at the MCC to see another client or inmate, then he would not be seeing John on that day.

The first time I visited the MCC to see John after his remand, I was in a conference room chatting with a man whom the government would tout as the head of a different so-called family. John walked by the door, clearly displeased. At first I thought it was general ill humor in that he'd been incarcerated. But I was wrong. The MCC was to him a playpen—his playpen—and he was completely comfortable there. He was upset that none of the counsel in his case were in the conference room allotted to him, that instead they were chatting with another inmate, another *well-known* inmate. To John, his attorneys were acting in a cavalier manner he couldn't abide, and he let us know; initially he was cold and distant during our meeting, and finally in his inimitable way he reverted to the offense, asking rhetorically, "Did you fellas come to see me down here at 150 Park Row, or somebody else? 'Cause I've got two nice sunny-side-up eggs and a slice of bologna waitin' for me upstairs which I would like to attend to right now. I don't need anybody to hold my hand, and I don't need visits from lawyers here to see other people." It wasn't his intent to scold, to upbraid, or to make anyone feel bad—it was just a point of order, a subtle procedural matter in John's court. When he made his statement, I instantly understood. As in any other court, the maintenance and semblance of order were of importance: It wasn't about time, liberty, or money—proper order had its own significance.

On the other hand, John had no interest in micromanaging his defense team. Certainly many attorney-client conferences were conducted at the MCC prior to and during the trial, but John would often leave prematurely, throwing a cordial "See you fellas tomorrow" over his shoulder. This is not to say that important preparation would not sometimes take place in the course of these meetings, but most group lawyer-client jailhouse meetings are a colossal waste of time, rife with lawyer rhetoric and conceptual conversation along the lines of how many angels might dance on the head of a pin. John had no patience for that sort of thing. His attitude was largely laissez-faire when it came to defense strategy, telling the lawyers, "You guys decide what's best." John took enormous pleasure in the small enjoyments in life, like his eggs and bologna, or a hard physical workout in the gym. These things were available to him even in the confines of prison, and he wasn't about to forgo them in favor of some useless symbolic meetings with counsel. The minutiae of his legal defense simply didn't concern him.

What *did* concern him—immensely—was the propriety of his own conduct and demeanor and the face that his defense would bear in the eyes of the public, of potential jurors, and of those who actually wound up sitting in his jury box. John wasn't shy about expressing his opinion to me about this or that cross-examination; when he had no comment at all, I knew he thought I'd done a particularly good job. He had an uncanny sense of how a certain fact or attitude by the government or the defense would affect the public or a jury's opinion of him or the case against him. While he relied on counsel on matters of trial technique, his perception and sensitivity as to what might be well received by a juror, the press, or even the judge was as great as that of any trial lawyer I've ever known.

One common legal tactic he absolutely abhorred was the distancing of one codefendant from the others to gain some meager advantage for that defendant. For example, John was extremely vocal in his prohibition of severance applications on behalf of one codefendant made with language that would create the impression that one or more of the remaining defendants was an ogre. A legitimate motion by a codefendant under Rule 14 of the Federal Rules of Civil Procedure—the rule governing when the case against a defendant may be separated from those against his codefendants on the ground that forcing him to proceed with the others would unduly prejudice him—would have been acceptable to John if it were phrased in a manner sensitive to the defenses of his codefendants. Thus, a properly worded motion might state, "The government *claims* this or that about the codefendant, and *these claims* are so outrageous and inflammatory that my client can't get a fair trial." An *unacceptable* Rule 14 motion would be one whose language distanced or even *appeared* to distance—for *appearance* in the public or jurors' eyes was ninety percent of the battle to John—the defendant from the group by pointing the finger of guilt at another defendant. To John, this was an intolerable breaking of ranks that would ultimately give the enemy a tremendous advantage over *all* the defendants. It was also anathema to his credo.

So when an attorney made such a severance motion on behalf of a codefendant, which indelicately implied that going to trial with John would prejudice that codefendant, John was exceedingly offended, and the attorney was strongly upbraided. "If you make an application like that again, you'll be carried out of this fuckin' courthouse wrapped in that rug over there. You understand me? You see that elevator over there? You'll be taking it down without the elevator!" His language was colorful but not a real threat, and to the attorney's credit he recognized John's ire for what it was. These days, if a client spoke to a lawyer that way, the lawyer would likely go crying to the judge that he couldn't properly represent a client under such circumstances—in essence testifying against his client's character. But those were different days, and John was a different sort of client, one who brought out the best in his counsel and cocounsel.

In John's view, which was hard to dispute, the government *counted* on the defendants' climbing over one another to save their own necks—with severance motions, with conflict of interest motions, with cross-examinations designed to distance a particular defendant ("you didn't hear *my* client on that tape, did you?"), and with summations noting that a particular defendant didn't do what those other bad guys did. John saw it as his duty to protect the group from the fears and worst instincts of the individual codefendants and their self-serving or indifferent attorneys—to increase the odds of acquittal for all. Of course, the government view was that John was merely increasing the odds of getting *himself* off, at the expense of the individual rights of his codefendants. But John had no interest in jeopardizing anyone's chances at trial; rather, he was enforcing what he believed was the most effective way to fight the government: a unified defense. Even the government would likely agree that its strongest tactic was "divide and conquer."

The government would use John's leadership role in dealing with me and counsel for his codefendants to argue that it demonstrated him to be the boss of a so-called Mafia enterprise. Ultimately, on the basis of alleged technical conflicts of interest, the government was successful in

John's 1992 trial (which resulted in his conviction) in depriving John of his attorneys—removing me and others from the defense team. In John's view, the federal rules regarding such things as conflict of interest and purporting to protect the defendant were all a fraud, just another way to sink *all* the defendants in a case, to prevent defendants from influencing the outcome of the case through smart leadership (which the government would, of course, call *mob* leadership). Was he right? Obviously it depends on the facts of each situation. But I've certainly seen a lot of painstaking, excellent lawyering undermined and ultimately undone by a codefendant's attorney standing up to cross-examine and asking the witness, "You didn't see *my* client there, did you?" The clear message to the jury is that this lawyer is trying to distance his client from the others, a tacit admission that they are bad fellows, and *all* these lawyers are a bunch of carnival hucksters trying to pull the wool over our eyes as to their own particular defendant. To John, the offending lawyer and client alike were no better than informants. As far as John was concerned, a lawyer was just a lawyer; the onus was on the client to imbue the lawyer with his feelings, values, and ideals. So John would take the offending client *and* lawyer to task, and it wouldn't happen again in *that* case. I witnessed the process myself—it was impressive. And after upbraiding the lawyer, John would have lunch with him to show that there were no hard feelings going forward; having made his point, the team moved on.

In short, the prosecution is not interested in protecting the civil rights of criminal defendants they've tagged, targeted, or indicted; the prosecution is interested in obtaining convictions. John understood this and naturally didn't want himself or his friends to be convicted. The government would call his attitude dictatorial, or Mafia-type domination, but for me (still in my thirties) it was eye-opening and left an indelible impression. I was comfortable with John's view. For one thing, I found it refreshingly democratic—all for one, and one for all. I didn't see him taking advantage of his position to benefit himself; he used his influence to protect the whole.

John's whole life was about coming straight at a problem, directly, without circumambulating. Accordingly, he had little faith in technical defenses or motions in court. Rather than "get off on a technicality" or some slick maneuver, he'd prefer to remain in jail and maintain his credo of riding into battle head on—no posturing or taking "sneaky" ways out. And there was nothing wrong, in John's view or mine, with instructing or guiding the lawyers of John's friends. Would there be any question in a high-stakes corporate litigation that the lead party on one side or the other of the case, and its counsel, would control that side of the courtroom? Would anyone look askance at that?

In me John found the perfect vehicle for this sort of combat. Together we controlled the courtroom—or at least the defense table. Having John as a client was like discovering a new kind of freedom in the well. All I had to do was be myself, and the more I was myself, the more effective I was, and the more approbation I received from John, other lawyers, other clients, and spectators. My style, my direct approach, was just what he needed from an attorney. Other lawyers—perhaps more brilliant speakers, writers, thinkers—didn't have my peculiarly and aggressively straightforward style. Over the course of the seven-month trial, by March 1987, everyone on the defense team was in lockstep, and the trial had taken on a life of its own. I'd never been in a trial of such magnitude then—the multidefendant, endless RICO megatrial was new to me. I'd see many more thereafter, but I'd forever be spoiled by the quality of my client in that first case, defending John in U.S. District Court for the Eastern District of New York.

In August 1986 a second jury pool was convened from which we were to pick a jury. The jurors were to be "anonymous," a relatively new concept then that would become de rigueur thereafter in federal organized crime prosecutions. The practice is inherently prejudicial to the defense, for it conveys the impression to prospective jurors that the defendants are so dangerous that the jurors' very names need be withheld to protect them from violence. But we didn't experience the problems we'd had the first time around picking a jury, because John was in the MCC and was not the subject of massive media coverage every day when he arrived at the courthouse. Less publicity made it easier to find jurors untainted by governmental propaganda.

Another pleasant by-product of John's remand was the luncheon ritual that evolved. Lunch was delivered to us in the courtroom from the finest restaurants in New York: steaks, spinach, and potatoes from Peter Luger's in Brooklyn, sandwiches from the Second Avenue Deli, veal scallopini from SPQR, you name it. John's friends would carry the food in, and we would all partake. I can't recall who paid for these banquets, but I suppose it was one or another of John's friends. Seven grueling months on trial, with no time for the gym and lunches like these, were no doubt the beginning of the end of my figure. Under the Bail Reform Act, and given the extraordinary fact that John was incarcerated during the trial, the court had the right to permit this seeming extravagance. To his credit in the circumstances, Judge Nickerson allowed it. But it was seldom if ever permitted again in future cases. As denying bail became more prevalent, the courts have been loath to allow such lunch privileges, in that they might appear like a preference afforded the wealthy and denied the poor defendant. I disagree with the trend: Lunch in the courtroom permitted counsel to work with the client through the lunch hour, instead of scrambling through long lines in local eateries. Further, it added a civilized touch to the process, according a measure of dignity to the defendant, who has the most at stake in a trial and yet is usually treated as the most insignificant party in the courtroom.

14

Despite the animosity that arose in connection with John's remand, Giacalone and I had regained a certain cordiality by the time we made our opening statements to the jury in September 1986. She wore a blood-red dress for her opening. In the austere world of the law, a venue steeped in dull grays and browns, her color choice was highly dramatic. What was she thinking? I wonder. Was it just the next dress in her closet? Whatever the reason, I've no doubt she later wished she'd chosen a more subdued hue. Red was a poor choice in my opinion, associated as it is with carnality and venality; she appeared at the opening of the case embodying the government exactly as we sought to portray it: venal and bloodthirsty. From then on, the "Lady in Red" was how I'd refer to her when speaking to the jury.

In her opening Giacalone used a blackboard to describe the structure of the Cosa Nostra, drawing its alleged hierarchy in chalk, complete from *capo* through lieutenants to soldiers and so on down the chart. When she'd finished, I rose, strode directly to her blackboard, lifted an eraser, and erased everything she'd written there. It was an aggressive move, making the point that what was written on the board *was not evidence,* that it was only Giacalone's opinion until proven in that very courtroom to all the jurors' satisfaction. And I wanted the jury to know that I and my client had nothing but disdain for her opinion. Then I began my opening in earnest.

I told the jurors that the government was prosecuting John for his lifestyle—for gambling, for staying out late, for not having a nine-to-five job, for the friends he'd grown up with from childhood, for his loyalty to them, and most of all for his pride—and not for any crime they'd be able to prove he'd committed. Pride is no crime in this country. This was the first serious unveiling of the "lifestyle defense," which would be the cornerstone of John's legal posture thereafter. I also advised the jury that the indictment was essentially a rehash of old crimes for which John had already been prosecuted and for which he'd already served his time— "bad meat and bad potatoes" rendering a "rancid stew," I called it. By the time I began my opening, I knew everything about John—I was infused with his credo, his doctrine, his principles, and his theories as to the case and the government's intent. John's beliefs and my words were in perfect harmony and rhythm. My opening, a personal tribute to John and a vilification of the government's motivations, was swirling into a crescendo, peaking spontaneously—for I had long abandoned any script or notes I might have had—with a violent thrust of the indictment into a nearby trash basket. The act of slam-dunking the government document worked at that specific moment in time and space—and it would never work again in my opinion, although it has been tried by many others since, both in cinema and in the courtroom—because at that pivotal and frenzied moment the gesture was utterly real. It was how I felt, and it was a deeply cathartic moment for me. And in a strange way, although there would be several other high points—and low ones—over the next seven months, and although I remained in great doubt over the case's outcome until I heard the jury foreman's voice echo through the courtroom, I can say that the case fell into place for me at that moment. I was in a groove, like a baseball player at bat watching pitches coming toward him as though in slow motion, as though there were nothing he couldn't hit out of the park. This feeling, this realization was electrifying. Thereafter everything short of the verdict—the trial itself, and certainly the summations—was anticlimactic.

"You must really like your client," Giacalone whispered to me after my opening, leaning over in her chair at the prosecution table as I returned to my seat at the defense table.

"Yes, I do," I replied.

Giacalone's case, patched together without the help of the FBI, was indeed weak—nothing like the carefully prepared government case launched against John in his final trial. This case consisted mainly of old gambling tapes in which John might have been mentioned, rife with tough-guy talk but devoid of any admissions of criminal activity. Then there was the testimony of alleged coconspirators, which invariably backfired on the government; even people who had been present at the bar where McBratney was murdered years before were dragged in. No FBI Cosa Nostra expert (even then a fixture in organized crime cases to prove the existence and structure of a criminal enterprise) was proffered in the government's case. Other than the resurrected crimes for which John had already done time, the only "facts" alleged by the government concerned the notion that a portion of the proceeds of a certain IBI armored truck robbery had been paid to John and the Gambino family as tribute. The charge was based on the alleged statement of the brother of one of those accused of the IBI robbery, but even the so-called perpetrators themselves laughed at the suggestion that they'd paid anyone any part of their proceeds; John didn't know them, and they said that they'd never met John. In essence, the case was designed to have Willie Boy Johnson break down and testify as a government witness. Without Johnson the case was an engine without a driving wheel. Johnson's alleged police and FBI handlers had warned Giacalone that Johnson would never "break down" and that all she would achieve by asserting his informant status would be his murder. This is in fact what happened when the trial was over and Johnson was released.

The Giacalone team paddled desperately to keep the government case from foundering. For example, they presented the testimony of Dominic Lofaro, of Newburgh, New York, touting him as a "major Gambino defector," a "lieutenant." He was to be called as a surprise witness, they said, a "hitman's hitman." I went to the MCC to ask John about him. John laughed and said he'd never heard of him.

Lofaro's testimony revealed that he'd hung around a couple of small-time gamblers with whom John was acquainted and worn a wire against them. On one occasion he actually did catch John on tape when "Bones" and his son (the gamblers) visited John to complain about someone who'd opened a competing crap or card game in their neighborhood. The government was attempting to show that John had sufficient control to warrant the gamblers coming to him with their complaint. The "offending" gambler allegedly had ties to the Lucchese family, and John was heard to say, on the tape, something to the effect of "You tell him that I, me, John Gotti, I'll sever his motherfuckin' head if he's doing that." It made for a sexy soundbite but was probative of little else. John was also overheard to have said something like "I know everybody around here—I don't know anything about a card game. Who would give this guy permission? I'll go over there right now. I could use some exercise."

This sort of quote showed him to be the "Crazy Horse," the hands-on fellow that he was. Maybe there was some criminal aspect to his words, but they were not bullying words—they were protective, they were real, they were him. And from the jury's perspective they concerned friction (which evidently came to no violence) between two competing factions voluntarily engaged in the same petty criminal activity—unimpressive and remarkably hands-on for a Mafia don. In addition, a portion of the Lofaro tapes, recorded after the meeting with the Boneses and John, has the Boneses and Lofaro raving about John, saying something to the effect of "Boy, he's gonna be our man [boss] some day. What a tough-looking crew he's got, what a crew. What a class guy, a hoodlum's hoodlum." The government believed this sort of statement would bolster the assertion that John was an organized crime figure.

As it turned out, Lofaro originally took credit for seventeen murders, none of which he actually committed. At the end of the government's direct examination of Lofaro, Giacalone asked, as to each of the murders he'd initially told the government under oath that he'd committed, whether he'd in fact committed the murder. To the shock of all at the defense table, he admitted that he hadn't. Then the government asked why he'd lied to them, and he replied that he'd wanted to enhance his negotiating posture, offering his testimony in exchange for immunity with respect to certain crimes with which he was charged. A palpable, stunned silence fell over the courtroom. Massive reports of murders, complete with victims' names, dates, and places—they were all made up, or else they were committed by someone else and he was taking credit for them. He did in fact hurt a certain woman, and he was in fact charged with some narcotics offenses, but he was clever enough to use the government's Mafia hysteria to obtain a free ride for these offenses. By the time of trial the government knew Lofaro had lied prior to signing the cooperation agreement, which specifically stated that if he lied, he would forfeit all rights to immunity for the charges against him and all rights to be relocated after his testimony. But the government was so eager for even the crumbs of testimony that the liar Lofaro could afford, so desperate for the colorful "mob crew" talk about gambling, that it permitted him to testify, then allowed him to walk away from the crimes with which he was charged, and agreed to relocation, *even if it was compelled (as it was) by law to elicit testimony from him as to his previous lies.*

The thunder of my cross-examination of Lofaro was largely stolen by the government's direct. I merely took him through the murders he didn't commit once again, had him admit that he'd neither committed crimes nor shared any proceeds with John. This became the John Gotti liturgy for government witnesses in the case: "What crime did you commit with him? What proceeds did you split with him? When did you ever meet him? Under what circumstances?" It was highly effective, highlighting the paucity of Giacalone's evidence. Most importantly with Lofaro, I emphasized the lengths to which the Lady in Red was willing to go to obtain a conviction—even after Lofaro lied, even after he'd violated his cooperation agreement, even after he'd admitted that he'd lied, he was still not in jail, he had still never paid for any crimes he'd committed, he'd still never been sentenced. But the truth is that the government's direct was more harmful to the prosecution than my cross.

Another unfortunate witness choice for the government was Jimmy Cardinale. Cardinale was a one-time Bergin Club errand and sweep-up fellow who'd been to jail and had a heroin addiction problem. John was always picking up guys like Cardinale and trying to give them a new lease on life—a job, antidrug motivational lectures, what have you. Cardinale was another witness who had nothing bad to say about John. He testified that he'd personally helped Willie Boy Johnson strong-arm someone over a gambling debt, and I suppose that was largely why the government called him to testify. Prior to his testimony, though, Cardinale contacted us and made some terribly denigrating statements about the prosecution. When he did this, we immediately advised Judge Nickerson and thereafter taped Cardinale's statements and produced a transcript of the tape. Cardinale stated that the government was putting words in his mouth, threatening him and forcing him to say the things he'd say on the stand, and he personally attacked Giacalone, calling her a "slut" and a "whore." His statements were offensive and deplorable, arising from the government's failure to provide him this or that item of comfort in the jailhouse; when an informant-witness's expectations are disappointed, he invariably becomes vindictive. But the embarrassing statements of witnesses like him were also the government's just deserts. In their lust for convictions the prosecutors lie down with these dogs; when they wake up with fleas, they've no one to blame but themselves. Moreover, when the government doesn't get what it wants from these witnesses, it is no less vindictive, but it has far more powerful and destructive means of avenging itself, as, for example, Willie Boy's treatment by Giacalone and his resulting murder. Cardinale wound up picketing an FBI building somewhere out west with a placard complaining about the government's unfair treatment of witnesses. As to John, Cardinale's taped transcript said he was the finest man he'd ever known, noting that he'd helped free him of his drug addiction.

To be sure, the atmosphere in the courtroom had been somewhat strained since John's remand, and it grew incrementally more so through my opening statement, the Lofaro testimony, and Cardinale's statement. Outright hostility took over when, at the close of the government's case, Matthew Traynor appeared on the scene. Traynor called my office from jail, advising that he was supposed to be the government's lead witness but that he was no longer testifying for Giacalone. So the defense team went to the jailhouse to interview him. The good news was that Traynor gave us evidence of actual misconduct by the government in the case. The bad news was that he was clearly a lowlife with a screw or two loose to boot. When I told John about him and his testimony, John said he'd never heard of Traynor, and his immediate reaction (to which he held through verdict) was that Traynor's testimony was repulsive and could damage our image in the eyes of the jurors. He thought we should stay away from him. In fact, John thought the prosecution case was so thin that we should consider putting on no defense case at all, relying on the presumption of innocence in criminal matters for an acquittal.

Through Traynor we would be able to prove the government's control of the MCC for purposes of creating evidence against a defendant on trial. Jailhouse informants would be solicited with offers of virtually anything they might want to make their stay more pleasant in exchange for their singing the government tune the way the government wanted to hear it sung. When the purported informant didn't quite hit the notes required by the government, he'd be punished in one way or another.

Traynor was a codeine and Valium addict and an alcoholic as well. He was to have been the government's first witness until the MCC resident physician categorically refused to prescribe codeine and Valium for what he described as "shoulder soreness." So in order to circumvent the MCC physician, Giacalone pulled Traynor out of the MCC on a writ to see a private psychiatrist named Dr. Schwartz, at Beth Israel Hospital in Manhattan. He, of course, provided the codeine and Valium prescription for Traynor. Traynor said that he'd had a disagreement with Giacalone over what he would and wouldn't be able to say on the witness stand, and when he wouldn't agree to say exactly what she wanted to hear, he was sent to a federal prison in Sandstone, Minnesota; in the interim, as negotiations were in progress as to his upcoming testimony against John, he was being plied with drugs. We called Traynor to the stand to testify as to what he'd told us, and Judge Nickerson permitted us to do so.

But all hell broke loose in the courtroom when one of the defense team attorneys, the now-deceased Dave DePetris, sought to serve a subpoena on the wife of one of the prosecution team attorneys, John Gleeson, who was second-seating Giacalone. It appeared that Gleeson's wife, a nurse at Beth Israel Hospital, was the conduit between the prosecution team and Dr. Schwartz. This is not to say that she'd done anything wrong; she had merely provided the introduction, and I imagined it was not unreasonable for DePetris to want to explore for the defense what criteria were applied to determine that Dr. Schwartz would be the appropriate physician to see in obtaining a prescription for codeine and Valium for shoulder soreness. Judge Nickerson and the prosecution disagreed. In fact, they went wild.

The court dismissed the subpoena out of hand and severely chastised DePetris. Thereafter Judge Nickerson displayed nothing but overt disdain for the defense team. I suppose his position boiled down to, you can prove *some* government misconduct but not *too* much—not to the point where it might be embarrassing to a young prosecutor. I should point out that overall my only complaints as to Judge Nickerson's handling of the case were twofold: first, his remand of John pending trial; and second, his closing down the defense inquiry as to prosecutorial misconduct in soliciting the testimony of others. On the other hand, from Judge Nickerson's and the prosecution's point of view, the court afforded me far *too much* latitude in exploring the government's misconduct, and I believe that Judge Nickerson blamed himself for John's ultimate acquittal. In Gleeson, by the way, I was seeing the beginnings of a very good prosecutor and a fine attorney—one of the better trial prosecutors I've met, who did the rebuttal summation for Giacalone and who would later defeat me in the trial of Michael Coiro and would also successfully prosecute John in the 1992 Eastern District trial. After the attempt to subpoena his wife, Gleeson was hostile to the defense team and to me personally, making some highly disparaging comments about me in open court. I'm sure I directed a few choice words of my own to him as well.

Then on re–cross-examination of Traynor, the government asked him some open-ended question to the effect of "And what other favors do you say the government conferred on you?" Well, Traynor went to town, blurting out that Giacalone had given him a pair of her panties into which he might masturbate. Once again the courtroom went wild—the press, the gallery, the prosecution, and even the defense table. I should note that Traynor had conveyed this tidbit to us as well in the course of our debriefings with him, but we would never have introduced it, both because it was highly distasteful and because we didn't believe it to be true. John and I were both extremely embarrassed by the testimony, particularly given the media slant (and lie) on the event, which was that *I* had introduced the testimony. Legend has it that I tossed the panties in question on the prosecution table after my examination of Traynor. This is all a fabrication. I never did anything of the kind, and the panties never surfaced in the courtroom. Moreover, we never believed Traynor to be truthful in this regard or that Giacalone would be capable of such ridiculous and disgusting—and potentially damaging—behavior. John kept saying, "See, I told you we should never have called this crackpot to the stand."

By this point in the trial both sides had lost all restraint in their open animosity toward each other. I recall Giacalone standing and shouting at defense attorney George Santangelo, jabbing her finger in his direction, and George telling her to take her finger out of his face and stick it up her ass! We later heard that Judge Nickerson made findings of fact designed to hold the defense legal team in contempt if there were a conviction in the case. I suppose the thinking was that it would look like sour grapes to hold us in contempt in the event of an acquittal.

Then the government compounded the Traynor problem by calling some seventeen witnesses (mostly FBI agents, surfacing for the first time in the trial) in rebuttal, largely to respond to his testimony. All this did was reinforce Traynor's words in the jury's memory. One of the prosecution's rebuttal witnesses was Dr. Schwartz, the psychiatrist who had provided Traynor with prescriptions for codeine and Valium. In my cross-examination of Schwartz I really let him have it. Every question I asked was answered with a sanctimonious and haughty air that I couldn't stomach. Schwartz was determined to rationalize his essentially indefensible action—you can't rationalize codeine and Valium for a sore shoulder, particularly when the prison physician refused to issue such a prescription for the patient. In an excess of emotion at one point I slammed my pen down on the lectern, and the pen sprang up in the air. It was an unnecessary and silly action on my part, born of my frustration and anger over what the government had done for Traynor and over Schwartz's supercilious attitude. The pen incident was not good lawyering, but the cross-examination itself was. Giacalone complained to the court, in one of her many objections, that it wasn't a cross-examination but a mugging. So the judge chastised me, and we went on to summations.

As I've said before, the summations themselves were anticlimactic. I was already thoroughly imbued with John's credo, his feelings about life and the government's case. I had no need of instruction or direction in summation—as I stood before the jury, I was John Gotti.

The jury began deliberation on a Friday and ended on the following Friday, March 13, 1987. John and Willie Boy, as well as Lenny DiMaria, were in lockup downstairs, so our last few fancy lunches were in jail. I recall the day the jury came back with its verdict. We'd had a sumptuous lobster luncheon; Willie Boy was not permitted by the government to eat with us, allegedly for his own safety, so he'd had his lobster alone. At about ten of two the marshal advised us that there was a verdict. I remember being upset, because I'd been hoping for a hung jury—I didn't dare hope for a victory. I'd just wanted to get bail pending the retrial and walk John out of the courthouse. The last inquiry from the jury deliberation room had been to rehear *only* the direct testimony of Jimmy Cardinale. Everyone had been surprised. Why Cardinale, why only the direct, and why after five days of deliberation?

I should have known by then that one can gauge nothing about the direction jury deliberations are taking from the little hints along the way: this or that question, a look of disapproval or concern, or even a smile, from a juror, the number of hours or days the jury is out. Once that door shuts behind the jury, it's all up to them. Lawyers try to pretend it's not, just as they try to better their odds of prevailing by picking this fellow or excluding that lady in jury selection—some now would raise the latter to a science, hiring psychologists to assist in the selection process. But it's all largely folly from where I sit, just lawyers futilely attempting to control their clients' destiny, mimicking the vain struggle of us all, mired in the human condition, to affect our fate. Sit on a jury sometime, or talk to a juror, or read one of the many books out now recounting the experiences of jurors in their deliberations. You'll be astounded at the facts or issues that do—and *don't*—come into play in reaching a verdict. And you'll know once and for all how tenuous, how frightfully random, it all is when you're a criminal defendant. Even an innocent man might wind up spending a lifetime behind bars.

After the verdict was rendered, some jurors told news reporters that the initial polling on the first day was eleven to one for acquittal on the first ballot; one lady held out and thereby caused four more days of deliberation. John maintained a strong outward demeanor throughout, but I could see that he had been as unsure as anyone of the outcome. He'd sent word to all defendants and their attorneys that he wanted no reaction to the verdict—win or lose, one way or the other—just file in, file out. But when the verdict came in, his reaction was one of genuine relief: exhaling strongly and pulling in his fist in that way now common to athletes after victory in a sporting event. I don't believe John was ever complacent about his chances of prevailing. That's one reason I gave no credence to the government charges that a juror had been compromised by John or someone on his behalf. It should also be remembered that this was a very hard-fought trial. It was no "show."

For those of us at the defense table other than John, there was shock, euphoria, and relief. Only Jeff Hoffman, representing one of the defendants, had the presence of mind to effect the formality of asking the court to discharge the defendants. Judge Nickerson grudgingly granted the request, angrily thanked the jury, and abruptly stormed off the bench. In view of rumors I'd heard of another indictment being prepared by the government against John, I began to fear he'd be arrested before we could leave the courtroom.

The "Teflon Don," as the tabloids would dub him, was born. I never thought I'd walk John out of the courthouse a free man, but I did that day. A friend of his, now deceased, Tony Moscuzzio, picked John up outside on the courthouse steps and off he drove.

At around that time David Margolick of the *New York Times* wrote a column describing my Rambo style of defense, accompanied by a large cartoon of me in Sylvester Stallone's garb from the movie—you know, bandanna, cartridge belt. Similar articles surfaced at the same time, in the *New York Observer* and the *Manhattan Lawyer*, and *GQ* came out with an article about me called "Iron Bruce." That's who I was from then on in the public eye. And someone in the press coined the term *Brucified* to describe being cross-examined by me. Even the conservative *Wall Street Journal* would later, in an editorial, suggest that Congressman Dan Rostenkowski of Illinois could do with my legal assistance. So be it. I believed in my client, in what I was there in that courtroom to do, and I did it.

15

Before the verdict on March 13, 1987, I was certainly in the eye of John's tornado, but I was just an adjunct, his overcoat flailing in the wind around him. I was his attorney, bigger and louder than most, but I could have been anyone. Now I wasn't just anyone. Now I was one of the forces that *propelled* the wind—I'd *earned* my place in the storm.

I stepped from the federal courthouse in Brooklyn and embraced the bedlam.

Back at the office the forces that surrounded Barry Slotnick had already arranged a major press conference. This was in the simpler pre–O.J. Simpson days, before *Court TV,* before the age of massive media saturation. The mere press conference still prevailed. And no one knew better than Barry how to use the tool effectively. It was to be a live broadcast at seven P.M. on all stations. All the attorneys were there—George Santangelo, Mike Santangelo, Dave DePetris, Jeff Hoffman, Richie Rehbock, Susan Kellman, and of course Barry, snuggled perfectly, centrally—ever centrally—in their midst, the one and only microphone set before him, the one and only earpiece, that talisman of power, firmly in his ear.

I wasn't angry then, although I must admit some annoyance, but I still assumed the supremacy of Barry's public relations machinery, and I had a residual shyness in the spotlight outside the courtroom. I wasn't going to ask for center stage, so I stood by watching the preparations for a while, until the TV anchorwoman received a phone call from CBS. I understood later that one of the attorneys in the suite had called CBS himself and told them that they had the earphone on the wrong attorney, that Bruce Cutler and not Barry Slotnick had represented John Gotti. The anchorwoman turned from the phone, immediately asked Barry to remove the earphone, and motioned me over to the microphone. As an uncharacteristically nonplussed Barry rose and detached the earphone, I approached and took his seat in the center—Barry sat on the arm of my chair, smiling broadly, paternally, his arm around me— and we were on the air, and the anchorwoman directed her questions to me. I was nervous being on live television, and I remember telling myself to think of the camera as no more than a jury. I know I thanked all the other lawyers in the case and settled into what would become my life with John. And I thought, *Holy cow, I've arrived.*

By March 1987 I was no longer in the firm. As of August 1986, when the trial began, I was merely renting space in the suite.

When I began representing John, I also began to receive a fair amount of press coverage, and in deference to my rising star, my weekly draw was raised, as was my percentage of the proceeds from the business I would bring to the firm. It was getting harder and harder to tell who had brought in what client, since many of our clients knew, or were connected to, one another socially or in business; who had brought in which client had become a gray area. Even John had arguably been brought to the firm by Barry, who had represented Neil Dellacroce, John's friend. And the matter of whose client was whose would necessarily become more and more gray *after* my representation of John, as I would be receiving business arising directly from his notoriety. To this day, in this or that context or biography of John or Barry on television, I will occasionally hear Barry referred to as John's attorney. In short, if all the clients I obtained through publicity spawned in John's defense were to be allocated to Barry, I'd never have had any chance of achieving some real financial security.

I remember asking Barry one day in August 1986, after five years of working with him, on the eve of the Giacalone trial, to form a true partnership in which I would receive what I thought was a modest ten percent share of the firm's gross proceeds (knowing full well that given Barry's accounting methods, like those of many law firms, a percentage of the net was likely to prove worthless). I would have been happy to stay and grow with the firm. After all, there was still a great deal I could learn from Barry in terms of legal strategy and interacting with the media, and the troops around him were wonderful. But Barry didn't want to do it. So I went out on my own, and in June 1987 I moved upstairs to share a new office space with several attorneys, including Lionel Saporta (this book's coauthor), who had left the civil litigation firm with which he'd been affiliated and whom I knew from Poly Prep and the district attorney's office.

After the verdict in the Giacalone trial, prominent lawyers from all over the country—F. Lee Bailey, Gerry Spence, Howard Weitzman, Barry Tarlow, Geoffrey Fieger—were writing and calling to congratulate me on the victory. I was in the news almost every day between 1987 and 1989, when John would be indicted and go on trial again in Manhattan. And the very nature and texture of my relationship with John changed as well after the Eastern District verdict in March 1987. John seemed to enjoy having me around more and more. In fact, he did the unthinkable—he appeared in a lawyer's office, mine, not once but twice. On both occasions he caused a stir, arriving with an entourage of some ten friends and associates. He was always under investigation by one of the New York City District Attorneys' Offices, or the U.S. Attorney for the Eastern or Southern District, or some other local or federal agency, receiving this or that subpoena or demand, or was the subject of some major "exposé" in one periodical or another. So I started going to the Ravenite Club on Mulberry Street to confer with him about twice a month on average to discuss the government activity and his response thereto, if any.

At the same time I was doing what I do best: trying cases. One of the most important of these, during the period between the Giacalone prosecution and the start of Morgenthau's case in 1990, was the Joe Massino case in federal court in Manhattan (the Southern District of New York). Joe Massino was a then-reputed "captain" in the so-called Bonanno family and a close friend of John Gotti. I was representing Joe Massino's brother-in-law, Salvatore Vitale, and Sam Dawson was representing Massino. This case came on the tail end of a series of prosecutions that the government claimed arose from a bitter internecine struggle between rival factions of the Bonanno family; these prosecutions were, at least in part, the result of the unprecedented undercover work of FBI agent Joseph Pistone (alias Donnie Brasco), who allegedly infiltrated the Bonanno group and was later portrayed by Johnny Depp in the film *Donnie Brasco*.

Massino and Vitale were charged under the RICO law with many old predicate criminal acts involving hijacks of shrimp, tuna, and sneaker cargoes, and new predicate acts involving conspiracy to murder, and three murders of Bonanno rivals. Massino was already serving ten to twelve years on a previous Eastern District conviction in connection with some union offenses at the New York airports, and he was prepared to plead guilty in our case to a racketeering count of the indictment and agree to a twenty-year sentence (which would have meant serving only some twelve years under the law applicable at the time) to run concurrently with his earlier sentence. However, the government insisted that he plead guilty to one of the murder counts. Massino refused, so Sam Dawson and I went to trial against federal prosecutor Michael Cherdoff, who was fresh from his victory in the Commission case (and who would later become the U.S. Attorney for New Jersey and is now chief of the Justice Department's Criminal Division). The trial lasted six or seven weeks, beginning in May or June 1987, shortly after John Gotti's acquittal. Judge Robert W. Sweet, who is as erudite and gentlemanly a jurist as I've ever appeared before, presided over the matter. It was a fair and civilized trial in which both the prosecutors and the defense, as well as the bench, were courteous in every way.

The defense focused on the new crimes: the murders and conspiracies to commit murder. This was because, under the highly complex RICO law, per the interpretation prevalent at that time, the jury had to find a predicate act within five years of the indictment and within ten years of the previous predicate act. We believed that if we could create doubt in the jurors' minds as to the murders, the court would have to set aside the verdict as barred by the statute of limitations and acquit the defendants even if the jury found them to have committed the old hijacks. Sure enough, the jury found the defendants guilty under RICO, but only of acts that were some nineteen years old. In a major setback for the government, Judge Sweet set aside the verdict. Massino finished serving his time on the earlier conviction and has been free ever since. After the Massino trial I celebrated my fortieth birthday, having dinner at a New York steakhouse with Sal Vitale; my cocounsel, aide-de-camp, and close friend Bettina Schein; and Sam Dawson and his lovely wife, Joy.

This was my first case working with Sam. He was a mentor to me who would begin to look to me for advice as well. Sam was one of the lead counsel in the famous Commission case, a federal RICO prosecution brought by then–U.S. Attorney Rudolph Giuliani in the Southern District of New York against all the alleged bosses of the five New York Cosa Nostra families (Gambino, Bonanno, Genovese, Lucchese, and Colombo). It was charged in essence that the Mafia was a "criminal enterprise" and that the defendants had committed predicate criminal acts in furtherance of that enterprise. The verdict, convicting all the defendants in the Commission trial, was rendered while I was defending John Gotti in the prosecution brought by Giacalone in the Eastern District; all the Commission defendants received sentences of a hundred years' imprisonment. Presumably John was not named in the Commission case because Paul Castellano, who was assassinated before the case came to trial, was still alive and the alleged boss when the Commission indictment had been returned. Nevertheless, when the Commission sentences were meted out in January 1987, all of us struggling in the Eastern District case couldn't help but be depressed. John was upset as well and troubled that the defense in the Commission case had made the tactical decision to admit the existence of a Mafia or Cosa Nostra.

The Commission lawyers believed that the so-called Jaguar Tapes, whose source was a bug placed in the car of Salvatore Avellino, who at the time was the driver of an alleged Lucchese underboss, were damaging and that denying the existence of a Mafia didn't make sense to them as a trial strategy. Such an admission was anathema to John, partly because he believed it would be the kiss of death legally, but more importantly because his credo wouldn't permit making concessions of any kind to the government. He was rankled that these defendants would allow their attorneys to make this kind of admission even as a trial strategy. Sam Dawson, who was one of the lead attorneys in the case, acceded to the lawyers' (and the clients') consensus in admitting the existence of a Cosa

Nostra. As a result of the tumult created by this concession, Sam later asked me, as an attorney with the reputation building at the bar of one who never bows to government pressure, to see what I could do to remove any stain these admissions might have caused to the reputations of the lawyers in that case, with John and others. I was flattered that Sam had approached me. I did talk to John, who as I've said personally took issue with the clients' decision, although he had only the highest regard for those clients and their attorneys, especially the great Sam Dawson. (Sam was appreciative of my efforts, and I later worked with him on many cases successfully. I loved him and was devastated by his sudden death of leukemia at forty-nine years of age in early 1992.)

As a footnote on the matter of admitting the existence of a Mafia, even in 1989, during preparations for the Morgenthau trial, I and cocounsel obtained John's approval to seek assurances from the potential jurors in voir dire (the jury-selection process) that, if all the authorities are able to prove at trial is that John Gotti is part of some cartel or Cosa Nostra that the state may say exists, the jurors would not convict John on this basis alone. Perhaps it was a subtle distinction in presentation: Although there was no overt admission of the existence of the Mafia, there was a tacit acceptance of the possibility that the jury might find that it did exist. Thus we attempted to preserve our credibility with jurors whom we were going to be trying to convince of our and John's truthfulness, while at the same time preserving John's image as someone who would not cede an inch to government. Thereafter I no longer denied outright the existence of a Mafia—rather, I adopted attorney Jimmy LaRossa's line, first quoted after the trial of his client Anthony Scotto, to the effect that "I never saw any evidence of an Italian-American Mafia of the kind described by the prosecutors in this case." Later, I would refine that even further, saying more often, "Who cares about whether there is any so-called Mafia? What's important is preserving the rights of men under the Constitution." John would have denied, as a matter of honor and personal obligation, the existence of, or his membership in, any Cosa Nostra. At this late date I'd be seen as something of a crank if I were to suggest that the so-called Italian Mafia is for the most part little more than an association of men who have chosen a maverick and dangerous lifestyle, who cannot abide a nine-to-five job, who put more industry into earning a buck by getting around the system than it would take to earn one legally, who tend to gamble inveterately, and who largely cause injury almost exclusively to themselves—that is, to others who have chosen to live the same life. To suggest that the court time, tax dollars, manpower, newspaper space, and television air time expended on eradicating the Mafia could more effectively have been used elsewhere, would, I know, make me a proselytizer for my client. So I won't.

John Gotti was indicted again at the beginning of 1989, in a case that went to trial a year later in January 1990. I tried three other important cases in 1989. The first of these, the Peter Vario case, was tried before U.S. District Judge Jacob Mishler in Uniondale in the Eastern District of New York. Judge Mishler was gentlemanly and fair to me and my client throughout the trial, notwithstanding my reputation in that forum after the Giacalone trial before Judge Nickerson. Peter Vario, allegedly an associate of the Lucchese family and a Local 66 Carpenters' Union official, was the nephew of Paul Vario, whom the government believed was the Lucchese underboss.

My first and only encounter with Paul Vario, who allegedly was portrayed by Paul Sorvino in *GoodFellas,* was at the MCC in May 1986, during my first meeting with John Gotti, after his remand to await his 1987 trial. Paul Vario entered the conference room where John and I were talking. He respectfully, tentatively, asked John if he could speak to him for a moment. John rose and stepped to the side with Vario, and although I couldn't hear what was being said, their body language told the story, revealing that John was giving Vario an earful about something.

Paul's nephew Peter was being charged, in an indictment brought by the Eastern District Organized Crime Strike Force, with extorting money from contractors in exchange for looking the other way when the contractors were engaged in construction with nonunion labor—a pure labor racketeering and bribery case. Bruce Maffeo, an excellent attorney, was prosecuting the case for the government, and James LaRossa, Robert Katzberg, and Sam Dawson represented Peter's codefendants. For me the case was an early foray into the cadre of very well-prepared federal criminal defense practitioners and their methods of preparing for the government's "paper and tape" (documentary evidence) onslaughts. It was said that Jimmy LaRossa had "civilized" the practice of criminal defense law. His forte was responding to a formidable government documentary case the way a "white shoe" civil firm would handle a civil litigation: by dividing the duties among the various defense attorneys on the case and meticulously preparing a documentary defense, as opposed to relying in the main on a lapse in the government's case or emotionally swaying the jury. We met weekly in the Vario case to divide the duties and analyze the case. By contrast, the Gotti cases had required largely mental and physical preparation, focusing on the important themes until I would be living and breathing them, then just driving forward.

The only defendant alleged to have organized crime ties was Peter Vario, and at one point in our preparations my cocounsel and I joked about the possibility that all the other defendants might settle their cases with pleas of guilty, and I would be left to try the case alone for Peter. That's what happened; the codefendants all were offered and accepted nonracketeering pleas. On the eve of the trial Peter's daughter was tragically killed in a car accident. Judge Mishler adjourned the case for a time to permit Peter to recover mentally and emotionally, and then in the spring of 1989 I moved to the Marriott Hotel in Uniondale for two months to try the case. Thankfully for me, Sam Dawson remained in the case to assist me. *United States v. Vario* later became a landmark case, holding that the government may not obtain an anonymous jury simply by requesting one; Judge Mishler stated that something substantially more than the mere allegation of Mafia connections would be necessary. Allowing Vario to remain at liberty during his trial, Judge Mishler later remarked that in his long career on the bench, he'd never taken so much time in selecting a jury—nearly every prospective juror had been a relative of a police detective or an FBI agent. At the opening of the trial Peter and I were facing a tough-looking, conservative jury straight out of a defense attorney's worst nightmare. He was convicted. The only good thing about the case was that the court had been kind enough to expand Peter's bail conditions to permit Peter and his wife to travel abroad before his incarceration.

Three or four weeks after the verdict in the Vario case, I found myself in Manhattan Supreme Court before the prominent New York State Justice Leslie Crocker Snyder, representing the nephew of Carmine Persico, Teddy Persico Jr., on narcotics charges stemming from his alleged possession of eleven ounces of cocaine. I got along very well with the notoriously tough Justice Snyder, who treated me solicitously and, despite the jury's verdict of guilt, complimented me on my performance in the case. She was kind enough to permit Teddy to remain at liberty pending the verdict, despite her reputation for levying heavy-handed sentences. In fact, she did sentence my client unduly harshly to twenty years to life in prison, when in my opinion she could have handed down a more reasonable sentence of fifteen years to life.

Then, just before picking the jury in John's third trial, in November 1989 I defended Mike Coiro, the last remaining defendant to be tried in the 1983 federal RICO narcotics prosecution brought against Angelo Ruggiero, Gene Gotti, and others. It was a feather in my cap that I was asked to represent John's former attorney and good friend, and I did so pro bono, but it was truly a sad case to lose. The prosecutor was John Gleeson, who'd second-seated Giacalone in her prosecution of John Gotti. Mike was charged with aiding and abetting the main defendants' narcotics activities and with the predicate acts of facilitating those activities and obstructing justice. I tried a strong if desperate case, but I suppose Gleeson tried a better one, or at least had the better facts, and Mike was convicted.

The government spoke of Mike Coiro as more than just an attorney to Angelo Ruggiero and John, more even than just a friend; they spoke of him as a partner in crime and said he'd crossed the line. I don't think Mike crossed the line, although the government certainly made it look that way. Mike was no hoodlum or crook. His crime, if he committed any, was his too-strong willingness to help his clients. The evidence against him was principally a request by Ruggiero (caught on an electronic listening device planted in Ruggiero's home) that Mike hold certain funds for him that the government claimed were the proceeds of illicit activity, specifically narcotics sales. Mike made an innocuous reply that the government interpreted as assent. Gene Gotti added, in a moment of largesse, "Mike, you're not just our lawyer, you're one of us as far as we're concerned." Understandably, Mike responded, "I know it, Gene, and I feel that way, too."

Mike Coiro had come out of semiretirement in Florida to help Angelo Ruggiero settle the affairs of his brother Salvatore, who'd just died in an airplane crash. There was no evidence that Mike had actually held any funds for anyone. Mike also prepared a witness called into the grand jury to testify in connection with the investigation against Ruggiero, and the government alleged that he'd suggested that the witness color or slant his testimony. Mike Coiro, a lifelong lawyer and already well into his sixties, was given a fifteen-year sentence. I felt this was tremendously excessive, considering that at some point well before trial the government had offered to recommend that he receive no jail time if he pleaded guilty.

Mike was a victim of the Mafia hysteria, the governmental cause célèbre of the moment, the prosecutorial stairway to the stars. He was no different from so many lawyers who represent major corporations. The corporate attorneys, the David Boies types or the fellows who represent Microsoft, try to predict what the other side of the litigation (possibly the government) might say, and they will prepare their clients' testimony. Have they "gone too far" when they suggest to their clients the most beneficial slant their testimony should take, what specific words to use to best fit the applicable law? Have these attorneys "crossed the line"? Would anyone have taped their conversations with corporate executives with a view to incarcerating them for fifteen years? Does anyone tape the conversations of prosecutors preparing their agents or police officers to testify in court? Does anyone tape them browbeating their reprobate informants, trying to induce the "right" testimony, the "truth" from them, as necessitated by the case at hand? It isn't black or white; rather, these are all points on a very long spectrum with vast gradations of gray throughout the center. It's not what a lawyer does that matters so much as for whom he does it: The wrong client, the wrong corporation-of-the-day in the government's mind, and the lawyer's actions are criminal.

If the government wants to gain the edge in a high-end prosecution, or if it wants to intimidate a target's counsel or avenge itself for too stiff a defense in a prior litigation, it has so much power, so many means available to it to chill an attorney's representation of the client. The mere thought of being investigated by the government—the issuance of sub-poenas to friends, clients, other counsel, summoning them to a secret grand jury; visits by FBI agents; or demands for financial records—is enough to make many an attorney abandon a client entirely. Having the government actually *take* those steps usually means the ruination of a law practice, even if the government comes up with no indictable offense. Then if the government actually decides to *prosecute,* it's so easy to frame a charge of criminality in the vague terms of "conspiracy" or "aiding and abetting" or "facilitating." Even if a jury acquits, the attorney is forever tainted by the publicity surrounding the prosecution itself. The law affords no real recourse against a prosecutor; prosecutors have a quali-fied privilege against any lawsuits as long as they bring charges "in good faith," and proving "bad faith" is virtually impossible. Rare is the lawyer who won't buckle or flinch in the face of governmental pressure.

There are very definite things a lawyer can do that are unequivocally criminal. I don't believe Mike did them. His treatment was hypocritical, and his fate tragic.

I later helped, inadvertently and unwittingly, to compound the tragedy. Immediately after the conviction I asked Judge John McLaughlin for permission to take Mike to the Ravenite to see John for the last time. John had been like a brother to Mike, and he wanted to say good-bye before beginning his imprisonment. The judge kindly granted Mike Coiro's request. I escorted him directly to the Ravenite, where he met with John briefly in an upstairs apartment that I didn't even know existed. I awaited his return downstairs. Unfortunately, the apartment in question had been bugged, and he was caught in compromising conver-sations with John that in turn led to more perjury and obstruction of jus-tice charges against him. The same eavesdropping device would capture conversations that years later would be used by the government in its own way, along with other testimony, to convict John.

16

IN the early evening of January 23, 1989, the Gotti Wars continued. John Gotti was arrested as he left the Ravenite Club on Mulberry Street, by some twenty-three police officers, according to the authorities, although John's estimate is two or three times that. I met him at the Manhattan Criminal Court at 100 Centre Street and attended his arraignment on an indictment obtained by the office of New York County District Attorney Robert Morgenthau. Indicted with John were his old friend Angelo Ruggiero, who was battling cancer at the time, and Anthony "Tony Lee" Guerrieri. I argued that John should be released into my custody overnight until he could appear for a formal bail hearing the following morning. As John had never been a risk in terms of fleeing the jurisdiction, and since it seemed to me that the state had purposely arrested him in the evening in order that he would be forced to spend the night in jail, I thought it would not be unreasonable to let him go home to his family that night and appear voluntarily the next day. The assistant district attorney, Morgenthau's protégé Michael Cherkasky, argued that John should be treated like every other arrestee and jailed pending his hearing. I responded that, if John were like every other arrestee, the state wouldn't have sent a battalion of officers to peacefully arrest him on a three-year-old assault charge. Needless to say, John was not released that night. So I gave John the tuna fish sandwich I'd brought him and went home.

The next day I returned to a media frenzy at the criminal court. Reporters and newscasters from every organ and station in the country were there, together with curious law enforcement types, defense attorneys, and other interested onlookers. I argued for bail, before New York

State Criminal Justice George Roberts (the brother of outspoken former Bronx District Attorney and later New York State Supreme Court Justice Burton Roberts), on the basis that John had always returned to face any charges against him and had deep roots in the community. These were the principal criteria for bail in the courts of New York State, as distinct from the federal courts, where the danger posed by the defendant to the community may also be considered. Over the objections of Cherkasky, Justice Roberts granted John bail in the amount of $100,000 bond. I had come to court that morning armed with bail bondsman Irving Newman, who promptly posted the requisite bond, and I walked John out the front courthouse doors.

John's brother Peter and attorney Rich Rehbock picked us up in Peter's car, and we all took a ride to Queens, to Dominic's, a hot dog vending truck famous throughout Queens both for its great hot dogs and for the charm of its owner, Dominic. It seemed to be permanently parked on Woodhaven Boulevard, near the cemetery, and I was familiar with it myself from the early days of my marriage to Gladys. Dominic knew John, and anyone who mattered in Queens knew that the territory within a several-block radius of where Dominic's hot dog truck stood was his—exclusively. In that late morning, on January 24, 1989, I recall John had three hot dogs—one with mustard, one with mustard and onions, and one with mustard and sauerkraut—and he ate them slowly, deliberately, savoring every bite as he stood there exchanging jokes with Peter, Richie, and me. I ate three as well, the same way. We stopped off briefly at the Bergin, and then Peter drove me back to my office in Manhattan, where I began learning about the latest legal assault on John.

According to the authorities, in February 1986 John O'Connor, the vice president and business manager of Local 608 of the United Brotherhood of Carpenters and Joiners in New York City, visited the construction site of a new restaurant to be opened in Battery Park City in downtown Manhattan, to be called Bankers and Brokers. Allegedly noting that nonunion crews were working on the site, O'Connor demanded some sort of payoff. When he wasn't paid—or at least not enough—he supposedly sent a few thugs over to tear the place apart, causing some $30,000 in damages. Shortly thereafter, in May 1986, O'Connor was ambushed by four gunmen, who shot him in the buttocks and legs. O'Connor spent a month in the hospital recovering and then, later on, was himself arrested on labor racketeering charges.

An electronic eavesdropping device planted by the New York State Organized Crime Task Force at the Bergin intercepted a conversation in February 1986 between John and his friends and codefendants Ruggiero and Guerrieri, in the course of one of John's almost-daily haircuts, supposedly after John had been advised of O'Connor's activities at the Brokers and Bankers construction site. Out of a morass of largely unintelligible murmurings, the prosecution contended that the phrase "Bust 'im up" was uttered by John and further imagined that John was referring to retribution that O'Connor should suffer for his actions. The newspapers also quoted John as mentioning, "Put a rocket in his pocket," repeating the phrase daily for weeks, and intimating that this was what John wished would happen to O'Connor. Oddly enough, no such phrase ever surfaced even in the state's transcripts and fondest imaginings in court. But the phrase had legs and traveled far enough to become an integral part of John Gotti lore without relevance to its truth. To this day, if it matters to anyone, I remain bewildered as to the source of the phrase. Make it up, say it's so, print it in some tabloid—and it becomes so. Anyway, the prosecution cast John as England's King Henry II, who famously exclaimed "Will no one rid me of this meddlesome priest?" with reference to Thomas à Becket, as portrayed by O'Connor in the Morgenthau production. Of course, Henry's sub rosa directive was carried out by four knights who slew Saint Thomas at Canterbury. And in the prosecution's imagination John's four knights were four members of the notorious Westies, a Manhattan-based, largely Irish-American gang.

The state's belief in this regard was purportedly based on the statement of one James McElroy, an allegedly reformed Westie, a self-avowed killer, heroin addict, and indisputable lowlife, who had once killed a man for a hundred dollars in drug money, who was serving sixty years in federal prison on a conviction for his Westies activities, and who faced a state murder charge as well. While he was asking Morgenthau to send a letter to the federal judge who'd sentenced him urging a reduction in his sixty-year term, no jail time on his pending state charges, and the dismissal of the state drug charges pending against his girlfriend, McElroy suddenly recalled that he'd done "some things with Johnny Gotti." "I'd never rat out an Irish guy," he reportedly told one police officer, "but the Italians, who gives a shit?"

As it turned out, McElroy later admitted he'd never done anything whatever with or for John Gotti, but he *was* ready, willing, and able to testify that he was present when the then-reputed boss of the Westies, Jimmy Coonan, told another Westie, Mickey Featherstone (who later defected and became a government witness), that O'Connor had been shot up at the request of "the greaseballs." John's wishes in this regard were allegedly conveyed somehow from Ruggiero and Guerrieri to one Danny Marino (another friend of John Gotti) and from him to Coonan. This was enough for Morgenthau who, in his eagerness to win the race to lock up John Gotti, agreed to McElroy's terms.

Morgenthau's brethren in the federal system, the U.S. Attorneys for the Eastern and Southern Districts of New York, balked at bringing any case against John based on the Bergin tapes and the testimony of McElroy and, I understand, tried to convince Morgenthau that such evidence was too flimsy to convince a jury. But Morgenthau, who had himself been the U.S. Attorney for the Southern District from 1961 to 1972, had authority to prosecute for a crime committed in Manhattan. Another factor raised the stakes for his office as well as for John Gotti: New York State has a "persistent predicate felon" law, allowing the court to sentence a defendant convicted for a third felony to a term of twenty-five years to life in prison. John had had previous convictions for a truck hijack in his youth and attempted manslaughter (in connection with the McBratney killing). So Morgenthau's office, in conjunction with Goldstock's office, brought the case.

We were originally assigned Justice Jeffrey Atlas, who had a liberal reputation, but were later advised that Justice Atlas had recused himself and that the case had been reassigned to Justice Ed McLaughlin, a tough judge with a strongly conservative reputation. I spoke to John while he was away in Florida to convey the news. He and a few friends had gone there by train for a short stay; John looked forward to these periodic train trips on which he'd play cards through the night with his pals.

When I told him about the change in judges, he was quiet for a while, then asked me whether I thought Morgenthau had managed to "fix" the reassignment. I honestly answered that I didn't know. Then he abruptly said it didn't matter, that we'd win the case anyway.

Justice McLaughlin scheduled the trial for January 6, 1990. The prosecution's request for an anonymous jury was duly denied, as there was no provision for such under New York State law, but the court did order that the jury be sequestered (isolated from the moment the jurors were selected until the verdict was rendered). Codefendant Ruggiero's case was severed from John's on the basis that he was terminally ill and unable to stand trial. (Angelo passed away about a month before John's trial began.) John's only remaining codefendant was Guerrieri. I knew Tony Lee personally because, prior to the Giacalone case, Barry Slotnick had asked me to attend an arraignment in Queens Criminal Court for Tony Lee in a matter involving his alleged purchase of some stolen gold—and I managed to have bail granted for him. In this new matter Tony Lee would be represented by Gerry Shargel, and John Pollok would be assisting us with the necessary motion practice.

The trial began on schedule on January 6, and the jury-selection process took about a week. There was a camera in the courtroom every day, televising the case in its entirety. My opening statement focused on the abuse of process and the prosecutorial ambition that had brought John before a jury once again. I contrasted this waste of public funds with the enormous poverty in the city as evidenced by the large number of homeless and disenfranchised New Yorkers who had taken up residence in the alleyways literally adjacent to the courthouse steps. John and his friends would stop there every morning on the way to court and give a few dollars to the downtrodden. On the following morning *New York Newsday*'s Jimmy Breslin wrote the following article, headlined NO AID FOR HOMELESS, BUT MILLIONS FOR GOTTI, highlighting my opening statement:

> The measure of what government knows of the life outside its offices and official cars could best be seen yesterday when the trial against John Gotti opened at 111 Centre Street.

Here they were yesterday in the cold moist air, all by themselves, the more to inflict the scene upon you, refugees from a war, or from a corner of Calcutta, 15 homeless people stretched out under old dirty blankets, two of them inside cardboard delivery boxes. They were against the front wall of the court building. Those entering, the jury foremost, had to be stunned and shamed. Always, it is the first time when you walk by someone living on the street.

That they chose to begin the Gotti trial yesterday indicated that these state and city payroll people have no comprehension of the homeless of the city. For a couple of years now a group has sprawled out every night at this building, right at the "Small Claims" sign. They are invisible on weekdays, as they must clear out at 7 a.m. They stash their blankets and go wandering on the streets so as not to offend those coming to court. They return at 7 p.m. But they are there all weekend, and spend Saturday under dirty blankets, talking about food.

Because John Gotti is such a vital case, the opening day, which could have been Friday, was put off until yesterday. Why open such a case as this on Friday when Saturday will produce the larger Sunday readership and interest, which immediately affects television? Yet those maneuvering it for Saturday had no idea that these homeless, these lumps under blankets, would be there for the jury to pass, and right after them, a defense lawyer who can see.

Yesterday, Bruce Cutler, attorney for Gotti, carried the scene upstairs with him and got up and delivered an unusually effective opening statement.

"Why are the homeless out there?" Mr. Cutler asked the jurors. "The state says it can't care for them. But the state can use 100 police officers to arrest John Gotti.

"They can surveil John Gotti at the Bergin Hunt and Fish, they have the money for that. They can surveil John Gotti when he goes to Aniello Dellacroce [*sic*] house in 1984, they have the money for that. But they don't have the money to care for the homeless."

Gotti is charged with having somebody shoot a carpenter's union business agent named O'Connor. He lived to be indicted for his own troubles. The O'Connor shooting happened in May of 1986. Last year, throngs of police suddenly leaped out at Gotti as he walked onto Lafayette Street, around the corner from his Mulberry Street club-house, and arrested him on the charge.

Yesterday, Cutler put both hands on the railing and called loudly, "They came to shake up the neighborhood pretty good, shake up the neighborhood pretty good. They came with football helmets, with all these flak jackets and shotguns. It looked like they invaded Panama."

Noting that the state could have made such a massive effort against a gangster, and did not have money for the homeless was a red herring, an outrageous diversion, a total irrelevancy. Of course it is ludicrous to put the homeless and Gotti in the same speech. As Cutler wants to keep the gangster Gotti out of jail, nobility was hardly the motive.

And Cutler also happened to place his hand on the one most powerful emotion in the city. The daily sight of a person freezing on the sidewalk at their feet has been making everybody crazy for so long that now you can throw the homeless into an assault trial. As there are strict laws against burning public buildings down, the recourse is a vote. At 111 Centre St., the jury has the vote, and Bruce Cutler stood right up and asked them for it yesterday.

Where this takes Cutler and gangster Gotti through the rest of the trial is up to tapes and testimony and time and arguing. The case against Gotti can't be light, the publicity always is very bad for him and therefore he is in court with the chance to be imprisoned forever.

But yesterday, for one morning of one day of the Gotti trial, Bruce Cutler swarmed all over the courtroom and threw something out there that had a lot more meaning than the case in court. Their government does nothing and people feel that nearly every time they go for a walk, they wind up being marked lousy.

Outside in the afternoon, somebody had brought pizza to four of them wrapped in their blankets. They were asked what they had for breakfast. They each held up a slice of pizza.

"This is the only thing so far today," Gordon Taylor, 38, said. He wore a blue wool cap and had blankets pulled up to his chin.

"Not exactly. Somebody brought us coffee," Robby Healy, 28, who comes out of Arthur Avenue in the Bronx, said.

"Been here all winter?" they were asked.

"Been here for a year and a half," Glenwood Smith, 33, said. He had six blankets over him and one under him had the insignia of Ohio State University.

"Year," Taylor said. "We come here because it's better than the city shelter. They ain't safe."

"On and off for two years," Healy, from the Bronx, said.

"Do you stay here on the real tough nights?"

"Every night of the year," Gordon Taylor said. "The night before Christmas when it was so cold, we were right here. The only thing is when you have to get up to go to the bathroom. We walk right over there, go right in the park. Woo. That's cold."

Court was over and Bruce Cutler walked by, a brown hat pulled down. "Thanks for mentioning us," Healy, from Arthur Avenue said. "Hiya," Cutler said. He did not stop to cry. He was there to try to get a gangster off, not to express social outrage, although he did that better than anybody in a long time around here.

I started receiving calls from lawyers all over the country (Bobby Simone from Philadelphia, Oscar Goodman from Las Vegas, Barry Tarlow from Los Angeles), requesting a copy of the court reporter's minutes. Interestingly enough, shortly after the verdict was rendered in the case, the city erected a prisonlike fence across from 111 Centre Street, a spot south of the courthouse where the homeless would gather, prohibiting them from taking cover near the building. I presume the fence was built at least in part as a result of my comments. Would that they had engendered a more compassionate governmental response! Not likely. I remember John commenting simply that my opening was good but too long.

I enjoyed working with Gerry Shargel, and we made a good team. He was focused on the nuts and bolts of preparation for cross-examination, certain technical aspects of the electronic eavesdropping conducted by the prosecution and the tapes themselves, and I brought the passion to the case, concentrating on the broad themes of governmental over-reaching and vendetta. The courtroom atmosphere was powerfully positive for John, with the gallery filled every day with well-wishers and celebrities there to bask in his aura. I recall actors John Amos, Tony LoBianco, and Ray Sharkey appearing on separate occasions to root for John and to try to imitate his carriage for upcoming roles; two well-known prizefighters, heavyweight Renaldo Snipes and middleweight Buddy McGurk, also showed up to wish John well.

The prosecution case was flimsy at best. An Agent Wright (whom I jokingly dubbed Agent Wrong) testified as to the transcripts he'd produced from the Bergin tapes, but it often seemed as though the words in his transcript didn't match what was audible in the tapes. Moreover, the phrase "bust 'im up" or "bust 'em up" was inherently ambiguous without further amplification, and the prosecution had none. We argued cogently that John could well have been referring to the necessity for O'Connor's governing party in the union to be voted out of office and removed from control of the union. And Cherkasky's star witness, McElroy, was ineffective because of his nefarious background and because he had not been a part of the shooting and had nothing to offer the jury but rank hearsay. Furthermore, his credibility was bankrupt even on the things he said he'd heard from others.

Finally, the prosecution called Vincent "Fish" Cafaro to the stand. Cafaro was a close confidant of "Fat Tony" Salerno, boss of the Genovese family, the prosecution claimed—but only long enough to have him convicted of such in the Commission case and sent to prison for the rest of his life. Now, while prosecuting Vincent "Chin" Gigante, it was more convenient for the government to say that *he* had been the Genovese boss all along. Anyway, Morgenthau's office announced Cafaro's defection to the media as though it would ensure John's conviction, noting that Cafaro was the first "made member" of the Mafia to defect since Morgenthau turned Joe Valachi in 1962. In fact, all Cafaro testified to was that, as far as he knew (and under cross-examination he admitted having no particular basis to substantiate this belief), John was the boss of the Gambino family. Boss or no boss, if John hadn't done the thing he was charged with—ordering the shooting of O'Connor—the jury would have to acquit. At bottom the problem with the prosecution's case, other than the fact that it relied largely on the testimony of the reprobate McElroy, was that there was no evidence as to how John Gotti's supposed message was passed to the Westies. No one testified that Ruggiero or Guerrieri or even Marino contacted Coonan or any other Westie. Also unanswered went the obvious question of why—assuming, as Cherkasky suggested, that John was the boss of the Gambino family, with some three hundred soldiers of his own—John would need to reach outside the family to an Irish gang to "bust up" O'Connor.

One of the highlights of the defense case was calling O'Connor to the stand. O'Connor was represented by James LaRossa (the late Paul Castellano's attorney) in connection with the labor racketeering prosecution pending against him; he consented to be interviewed by Gerry Shargel and me in Jimmy's office. He told me that he didn't know who had shot him—violence is by no means unusual in the union world—and that he didn't believe John had been involved. I should note that O'Connor had no particular reason to be responsive to the district attorney's office. For one thing, Morgenthau was considering indicting him. And from a historical standpoint, there was no love lost between Irish union men and the authorities, who traditionally busted unions in aid of the monied establishment and more recently were often investigating union leadership for alleged corruption. More to the point, Morgenthau's office had often specifically investigated the carpenters' union and O'Connor himself. Also, it should be recalled that if the government's allegations were correct, O'Connor had broken the law when he wrecked Bankers and Brokers and was subject to prosecution for that as well. Finally, I think O'Connor enjoyed the attention he was receiving from the defense team. O'Connor was at the very least a likable maverick, a take-charge kind of person who was not afraid to act in aid of his membership.

We decided to have him testify for the defense. After all, he was the victim in the case, and the jurors had to be wondering why the prosecution hadn't called him to testify. Of course, Cherkasky hadn't called him because he was not cooperating with the prosecution's theory of the case, and because he didn't technically need O'Connor and could make do with the medical reports and the testimony of the police officers who responded to the crime scene. The jury had been left at the end of the prosecution's case with the impression that O'Connor was terribly injured. In fact, he had been badly hurt, but he had recovered fully and had no visible effects from the shooting. I wanted the jury to see him, first, to show that he was getting along all right, and second, to have him say that he had no belief that John had harmed him. Also, O'Connor had a certain charm, and his open friendliness to John in the courtroom was good for the jury to see. But O'Connor paid the price for bucking the district attorney—shortly after the verdict in the case against John, he was indicted for union corruption. In fact, Morgenthau's office came down hard on the carpenters' union thereafter, taking the view that it was a puppet of the Gambino family, and that O'Connor's actions at Bankers and Brokers were an aberration for which he'd been duly spanked by John. Was Morgenthau's office retaliating against the union and O'Connor for O'Connor's performance in the Gotti case? You tell me. But in one of the subsequent Morgenthau prosecutions involving the carpenters' union, in which O'Connor was a defendant, George Santangelo and I successfully represented two other union officials, Gene Hanley and Attilio Bittondo.

The summations in Morgenthau's prosecution of John were carried live on television and radio. I rose to the occasion and made a very powerful statement on John's behalf, focusing as ever on the overweening prosecutors and the unreliability of McElroy, a psychopathic killer and liar. I told the jurors that the prosecution had written a fictitious scenario to convict John: "They wrote the screenplay, but you ladies and gentlemen will write the final act—not they." After deliberating for three days, on February 9, 1990, the jury returned a verdict of not guilty. I understand that on the first vote six jurors were for acquittal, one was for conviction, and five were unsure. On the second vote there were six for acquittal and six for conviction. Then upon a rereading of our cross-examination of McElroy, there were eleven for acquittal and one for conviction. The final juror holding out for conviction was convinced to acquit on the fourth day of deliberations. When the verdict was announced by the foreman, an elated John punched a fist in the air, sighed with relief, and hugged me. I ushered John through the crowd to the street, where hundreds of onlookers were shouting congratulations to him. Even the homeless on the street were yelling out their approval of the verdict. John's friend Jackie D'Amico pulled his car up to the curb, John stepped in, and they drove off.

On television that afternoon I watched the jurors imitate my summation style for the reporters, peppering their renditions, as I had, with Irish and Italian phrases and with Shakespearean quotes. My style was identifiable—identified with *me*. I would later be on countless television programs, including CBS's *60 Minutes*, and in too many magazine articles to recall, discussing the case and John. It was the third trial and the third victory for John, and for me, in four years.

We didn't know then that even as John was being tried in Manhattan Criminal Court—and was being acquitted—the federal government had in its possession tapes obtained from the eavesdropping device planted in the little apartment upstairs from the Ravenite Club (the same bug that had captured Mike Coiro's conversations with John after Mike's conviction in November 1989) and that certain conversations on those tapes would be twisted and convoluted to become part of the evidence against John in his subsequent 1992 trial and conviction. Why didn't the federal government provide its tapes to Morgenthau's office? John being the menace to society that the government claimed, why allow him to run rampant through the streets of New York for another year before utilizing the evidence that was so damning? Because the government fervor was never about protecting society from John. Rather it was about *which* prosecutor would ascend to greater fame and stature on John's back. And there was no way the federal prosecutors would assist a state prosecutor in stealing their pedestal.

When the Piecyk circumstances arose, the Queens district attorney had proceeded without heeding the federal prosecutors' wishes. When the Eastern District U.S. Attorney's Office and Giacalone realized they had a so-called Mafia don within their grasp, they were going forward with or without the FBI's help—no one was going to steal their glory, even if it meant John might go free because the best possible governmental team didn't put on the best possible case. When the Manhattan district attorney came up with McElroy and the Bergin tapes, it didn't matter that the federal government said his case wasn't strong enough— he would do whatever most suited his own aggrandizement.

A case built around the federal government's Ravenite tapes wouldn't be brought until December 11, 1990, long after the tapes were garnered, and nearly a year after Morgenthau's debacle in the O'Connor case. What was the delay about? The Eastern and Southern Districts were fighting over which office would get to bring the case. It had no particular connection to Brooklyn or the Eastern District, as most of the predicate acts with which John would be charged had taken place in Manhattan. Still, Andrew Maloney's office in the Eastern District won the political struggle in Washington. The theme of Maloney's pitch was that his office was entitled to vengeance for its loss in the Giacalone case. The concern was not how long the ogre would stalk the street, nor which office was better equipped or had the stronger jurisdictional basis for criminal charges. The concern was personal ego and a pound of flesh. John had peed on Eastern District shoes, and the Eastern District would get even, by hook or by crook.

But until then the Teflon Don lived.

17

Those years, the mid- to late 1980s and early 1990s, were a kind of golden age of criminal, and in particular organized crime, litigation. There were so many massive Cosa Nostra cases and investigations going on at the same time, one could hardly keep track. And a mystique of invincibility was growing for me and those around me. I didn't have to think much about procuring business—the work came in easily, and the trials flowed one after another. I was often able to recommend to the codefendants in these cases that they retain the best counsel, counsel I could work with easily and who had specific fortes that complemented my own—a good legal writer, a good tactician and technician (like Sam Dawson)—thereby maximizing the camaraderie and the possibility of a positive outcome. And most of the cases I tried during this period, this brief hiatus in the Gotti Wars, had something in common: proud men with large egos who wanted to fight the government, who wanted to clear their names unequivocally, who wanted some sort of redemption. Could I, did I, ever give them that? I did the best I could, and in each instance the client was happy with my effort. Were the clients redeemed? I don't know. Who would redeem me—or the rest of us, for that matter?

In mid-1990 John's brother Peter Gotti put John back on the front page of the newspapers. Peter and the heads of the other "families" were indicted in the Eastern District of New York in an alleged bid-rigging scheme involving window replacement in low- and middle-income housing. The government claimed that organized crime was getting a portion of the proceeds from every window installed in a building in New York, and as hundreds of thousands of windows were installed in Manhattan, the government was talking about a lot of money. John asked me to appear for his brother, and I did. I remember Peter being surprised and flattered when I, John's lawyer, appeared at his bail hearing. I was flattered that he was so moved by my presence. I had him released on bail, and then later I won an acquittal for him at trial.

Two days before John's acquittal in the Morgenthau prosecution, there was a prominent article in the *New York Times* about a major "organized crime round-up" by the federal government in Chicago. Shortly afterward my office began to receive phone inquiries as to my availability to appear as counsel for the reputed underboss of the alleged Chicago family, on a team that included some of the best criminal lawyers in the country. I visited Chicago, met with lead and very highly regarded counsel Pat Tuite and the prospective client, Solly DiLaurentis, and agreed to represent him. I arrived in Chicago three days before trial was to begin, immediately after a jury had taken only thirty minutes to acquit Peter Gotti in the seven-month-long windows case in New York. The government's RICO case in Chicago, *United States v. Infelice and DiLaurentis*, was on the front page of every major paper in the country.

In 1989, before the windows case, I received a visit in my office in New York from Dr. Thomas Gionnis, a Newport Beach, California, doctor who also happened to be John Wayne's son-in-law. Gionnis was, at the age of twenty-one, then the youngest person ever to graduate medical school in the United States. He was under investigation for an assault on John Wayne's daughter in the course of a custody dispute in a hotly contested divorce, and he was ultimately indicted in Orange County. F. Lee Bailey was retained to represent Gionnis for bail purposes, and I was to step in for the trial, but because of other engagements in New York and Chicago, I couldn't appear. However, when his first trial resulted in a hung jury, I was able to appear in the Santa Ana courthouse in California for him, with my cocounsel Bettina Schein, in the spring of 1992 for a retrial. It was in the midst of the enormous turmoil of the earthquake, the Rodney King trial, and the subsequent riots. The Gionnis trial was one of the first-ever live transmissions on *Court TV*. Gionnis was convicted after a seven-week trial in "John Wayne County," but the conviction was ultimately reversed on appeal.

Then there was the Joseph Chili case. Once again I was able to assemble a team of solid criminal practitioners (Sam Dawson, Bob Katzberg, and George Santangelo) to fight a broad gambling and loan-sharking RICO indictment before Judge William Patterson in the Southern District of New York. I represented Joe Chili III, and Gerry Shargel had been independently retained to represent Joseph Chili Jr. We prepared the case for trial, but it never got there; it was settled with the clients accepting plea agreements. These clients were once again old-style fellows who were easy to deal with, demanding little but honesty from their lawyers and stand-up behavior in court.

Finally, I tried the Jerry Winters matter in federal court in Newark, New Jersey, before Judge Marion Trump Barry (Donald Trump's sister). I was lead counsel, and on my recommendation another defendant retained Sam Dawson in the case. I knew Sam wasn't himself when the case began, but as usual he more than carried his own. All the defendants in the case, which involved extortion and severe assaults in the context of a large candy-sales business, were acquitted except my client and two others (one of whose conviction was reversed on appeal). Although Winters was convicted, the success of his codefendants was generally ascribed to my performance in the case, and the client was pleased with the job I'd done. But Sam died a while after the Winters verdict, while I was in Chicago, and as I said earlier, I was very unsettled by his passing. I'd grown to admire him and considered him one of my closest friends. Perhaps I should have read his tragic youthful passing as a harbinger for me as well, forecasting a change in the winds that had borne my ship so favorably for so long.

In the meantime, as I was handling these cases between John's acquittal in the O'Connor assault trial and his next (and last) indictment, there were some important out-of-court developments in my own life. The first of these was my decision in May 1990 to move uptown to share a suite with Jimmy LaRossa's firm. I didn't become aware of Jimmy's prominence in the field of criminal law until he tried the case of Anthony Scotto (allegedly a *capo* in the Gambino family and now an eminent New York restaurateur), when he called Mayor John Lindsay and Governor Hugh Carey as character witnesses for Mr. Scotto. Jimmy has since become the dean of the New York criminal bar. Of all the well-known criminal lawyers in New York, my father was most impressed with him; he'd say, "You know that kid LaRossa? He's one of the top guys around." And my father wasn't prone to offering opinions, good or bad, about his fellow members of the criminal bar. When I moved into the beautiful suite at 41 Madison Avenue, I really felt as though I'd joined the New York Yankees.

Jimmy had also been Paul Castellano's attorney. He'd met with him just before Castellano went to Sparks on the night of his assassination and would later be called into the grand jury in connection with the federal government's investigation of Castellano's death. I'm often asked the ridiculous question whether Jimmy was afraid of John Gotti, in view of the allegation that John had murdered or conspired to murder his client. The answer is an emphatic no. After the Commission case Jimmy asked my permission to visit my client John in the MCC (where he was awaiting trial in the Giacalone matter) to discuss the posture that the defense team had taken in that case. Of course, I told him he was free to visit John any time he wished. You may recall that scores of other attorneys were visiting John then in the MCC without bothering to consult with me at all, so I was grateful for the respect Jimmy showed me. Jimmy always carries himself with class and dignity, as far as I am concerned.

Why would a top-notch criminal attorney come to see John or anyone else in the MCC to justify a position taken in court on a legal matter in which that person was not a client or even a defendant? It was for the same reason that Sam Dawson had had me intervene on his behalf with John. For better or for worse, John's voice carried a lot of weight with the knock-around street fellows. And if an attorney like Jimmy or Sam wanted to continue obtaining work from those fellows, if and when they'd be indicted in future federal megacases, it made good sense to see that John didn't have a negative opinion of them. Sure, John wasn't their client, and sure, they wouldn't prejudice their clients' rights to curry favor with a fellow who was not a client, but as long as they could politically do what was best for their clients *and* at the same time do so in a manner that was not offensive to those who were influential in that milieu or to their credos, there was no harm in being prudent—treading softly through yet another gray area between the white of the client's interests and the black of the lawyer's own.

At around the same time, June 1990, my friend Larry DeMann Jr. (who had attended Poly Prep with me and was my chiropractor) introduced me to Shonna Edan Valeska (née Wechsler), a successful New York free-lance photojournalist and onetime apprentice to the renowned Richard Avedon. Although I didn't know it when I first made her acquaintance, she would change my life forever. She was a walking contradiction: tough as nails and yet a former ballet dancer, a delicate princess who climbed mountains, fashionably sophisticated yet down to earth. She was every-thing I'd ever wanted in a woman. Her father was a law professor at American University Law School in Washington, D.C., so like Barbara, she had some understanding of what I did for a living. I, on the other hand, would only later develop any understanding of what *she* did for a living. This self-absorption on my part, and my inability to extend myself beyond my own world (the courtroom), would doom our relationship.

It didn't take us long to start seeing a lot of each other, and I met her family, but I don't think she was immediately enamored of me. She wasn't sure. I was self-centered and eccentric and disinclined to the social and cultural whirlwind that was her life. She later told me that in those early days of our relationship, when she'd expressed her reservations to her sister, saying that ours might not be the perfect match and that per-haps she should move on, her sister suggested that I was "a big, sexy guy" and that she should keep at it for a while. So Shonna kept at it. I always suspected that, were it not for the publicity I was getting and the glam-our of my representing John, and perhaps the photojournalistic oppor-tunities that might flow from such, she might have gone the other way. This troubled me at first, but then I decided, so what? If DiMaggio hadn't been DiMaggio, Marilyn probably wouldn't have been with him either. I began to court her in every way I knew how, doing all the laughable things that would-be lovers will do. I wanted her to like me.

I invited her to join me for an early July weekend in 1990, at the house in Sagaponack of one of my closest friends, Dave Goracy. She was reluc-tant at first, I think, to be trapped there with me for an entire weekend, but I assured her she would have her own bedroom, and she eventually agreed to come. We flew to East Hampton and drove to Dave's and spent the entire weekend just looking at each other and falling in love. I'm afraid we weren't very good guests that weekend, or maybe we were the *best* possible guests in that we kept to ourselves and hardly saw our hosts. I can still see Shonna in Dave's pool, framed by the turquoise water, one slender brown arm holding to the pool's slate edge in the blazing sun, requesting a drink from the house with a slight twisting motion of the other hand near her mouth. I like to imagine there's a defining moment in a relationship when you know you're in love—for me that moment, that hand motion was it. We became close that weekend and were inseparable for the rest of the summer. She even accompanied me on my trips to Chicago to meet with my clients there (stopping off in Gary, Indiana, where Shonna had grown up). I rented a house for us in Montauk, and we began to confide in each other, or I should say she began to confide in me. In whom can a criminal lawyer, or even just a man, like me confide?

That August 1990, while Shonna and I were in Montauk, John Gotti asked us to join him and his wife, Victoria, and a union leader friend of John's and his wife, at the Montauk Yacht Club. After hearing so much about Shonna, John had insisted on meeting her. Shonna's hair hung loose over her shoulders, and she wore an elegant black crepe dress over tight black capri pants. When we first arrived, I introduced Shonna to John and playfully mentioned that she'd been nervous about meeting him. Shonna was mortified, and John, chuckling, said, "*You* have nothing to be afraid of." She promptly retorted with a grin, "Were you afraid of meeting me?" And John burst out laughing. Then he leaned over to give her a kiss on the cheek, and when she too excitedly moved toward him, they bumped heads. This time they both burst out laughing, embarrassed. It was a sweet moment between two people for whom I cared a great deal.

As usual with John, we never looked at menus, and the food just kept coming. Waiters and busboys were lined up all night with platters upon platters of hors d'oeuvres, then second courses and third, and then a lobster course, followed by desserts. And the champagne never stopped flowing. Shonna was characteristically irreverent, her way of countering nervousness to find a comfortable place in company. She would later describe John as "sleek and suave, with beautiful salt and pepper hair, very flirtatious, spunky and charming, with a real sparkle in his eye, as though someone had sprinkled fairy dust on him." I don't know that John would agree about the "fairy dust." She found Victoria, John's slender, dark-haired, well-dressed wife, "beautiful but subdued, with a sad aura about her" that Shonna attributed to her son's accidental death. Although Victoria didn't drink down life with the same gusto as John, she jovially rose to the occasion, and Shonna loved her. In fact, she was overwhelmed to be in their company. Dinner lasted for hours, although the time seemed to flow by as quickly as the champagne, and when it was time to leave, John insisted that Shonna and I join him and Victoria, alone this time, for dinner the following night at Gurney's Inn, where they were staying. Dinner at Gurney's was more relaxed. Shonna was wearing an off-beat black and white Betsey Johnson outfit, a "young-gal-about-the-Village look," she called it, and I remember Victoria politely admiring it, although she herself was a far more conservative dresser. Once again we all had a great time.

John thought the world of Shonna, but I think he expected me to be with a showier woman. He was never anything but complimentary in the things he said, but he couldn't help projecting his preferences in women on me. Murray would say, if a woman is your cup of tea, she's your cup of tea, and Shonna was decidedly my cup of tea. John grew to care more and more for her, particularly when he was in jail and she sent him an extraordinary card she'd made herself; he was amazed at the work she'd put into it. Shonna once said that she only went out with me to get a picture of John. She was joking, but I think there was more than a grain of truth there.

I knew Shonna was special, and a part of me was hanging on for dear life, although my other side—the work-driven recluse—was in constant rebellion, angry at the changes she was causing in my pedantic, hermetically sealed social existence. Still, I would allow her to schedule the hell out of my out-of-court time, from jazz clubs to dance clubs, from the opera to the ballet, from symphonies to theaters, from small dinner engagements to sprawling parties. Then she'd still be up at six A.M. for a racquetball lesson, or a photography shoot, or a breakfast meeting. Her energy was astounding, and utterly exhausting. We whiled the hours away talking and eating at my place; of course, by evening that sort of inactivity had Shonna dragging me out to a show. But it was all right, I would tell myself. She was opening up new vistas, new universes for me. I was doing things I would never do before, and if I could survive the culture shock, I thought I'd be a better person for it.

But she never stopped. There were events, spectacles, shows, exhibitions of every sort, complete with extensive entourages, lined up infinitely into our future, all of them things I would never do on my own. I attended some, and others I made excuses for missing. Early in the relationship, toward the close of the summer, I took her to lunch at a place on Twenty-ninth Street that John used to like called Troubles (the irony in the name wasn't lost on me even then) for the express purpose of protesting our myriad excursions, our social whirl. I know it was early for this sort of a problem lunch, but the issue had evolved into the central theme of our disagreements. Shonna would bring out the very best in me, but what she brought out in the *other* me, the workaholic loner, was devastating.

The honeymoon was certainly winding down for us by December 1990, when I was trying the Jerry Winters candy case in Newark. I was distracted by the trial, and my efforts there precluded my exerting any effort in my private life. I had no energy left for Shonna's cultural events or even for the high-maintenance Shonna herself. When we did see each other, I was irritable, explosively so at times, and unpleasant to be around. She enjoyed coming to see me in action in the courtroom, but for the most part my work was killing us. We would stagger on in this way, through my heavy trial schedule, until September 1993, when we would break up for the first time.

This is not to say that there were not still some great times. One fond recollection is of the spring of 1991. John had been arrested again in December of the previous year, but his friend Cesar Gurino invited Shonna and me to a baseball game at Yankee Stadium. I was representing Peter Gotti at the time, and Cesar was a codefendant in that case. That evening at the game I wound up being named the Fan of the Night.

My face was up on the stadium's massive screen, and everyone cheered me and wished John and me the best of luck. We were seated in the VIP section of the stadium, right next to the on-deck circle in front of the Yankees' dugout, and I was presented with a bunch of awards and presents (videos, bats, hats), which I gave to Cesar's young son. Frank Howard (then the Yankees' batting coach, formerly a legendary slugger for the Washington Senators), Steve Howe (a Yankee pitcher and a big fan of John's), and a few other players came out to say hello to Shonna and me. I recall Don Mattingly nodding and smiling at us when he went into the on-deck circle; he had the biggest forearms and hands I'd ever seen in my life, like Popeye's. It was a terrific thrill for me, and Shonna was in heaven with all the attention. This was once again the world of John Gotti, where anything was possible, even though he was in prison. That euphoric DiMaggio-like feeling of invincibility was still in the air.

18

ON December 11, 1990, John Gotti was indicted, arrested, and remanded again.

Based on the government leaks to the press, we were expecting his arrest every day for two months before December 11, while I lived at the Gateway Hilton in Newark, on trial in the Winters case. In fact, I'd addressed a letter to every single judge in the Eastern and Southern Districts of New York, with a copy to the U.S. Attorneys for the Southern and Eastern Districts of New York, advising as John's longtime counsel that John was ready to surrender himself in a gentlemanly fashion upon notification of the return of an indictment, without need of an intrusive and bombastic public arrest. The letters were ignored.

I said to John several times, in my periodic phone calls from Newark, "You know, John, the British [his nickname for the government] are definitely coming, and you might as well stay up all night playing cards, so they don't wake you at five-thirty in the morning." He replied, "You're right—I'm gonna do that." He generally stayed up almost that late anyway. It was the government's standard operating procedure to make its arrests in the early-morning hours at the defendants' homes. But it didn't happen that way this time. This time was special. At about seven-thirty in the evening some thirty or forty FBI agents showed up in front of the Ravenite. The agent in charge, George Gabriel, entered and politely advised someone that he was there for John. Led to where John sat, he said, "John, I have to take you in. I have a warrant for you, Frank LoCascio, and Salvatore Gravano."

John looked up, smiled, and said, "All right." Then he turned to

Norman Dupont, the Ravenite's manager, and said, "Norman, I'll have that espresso now, please."

Gabriel and the other agents waited patiently while John finished his espresso. Then he stood up, and they took him into custody. It was his last moment of freedom.

I was waiting for Judge Marion Barry to charge the jury in the Winters case in Newark on the morning of December 12, but with Judge Barry's and Jerry Winters's permission (leaving cocounsel to cover for me), I returned to the scene of John's acquittal (some three years before) in the Eastern District federal courthouse in downtown Brooklyn, this time to attend his arraignment before Judge I. Leo Glasser. The prosecutor was John Gleeson, the man who'd second-seated Giacalone in her ill-fated stab at prosecutorial immortality, and it was most certainly payback time for the Eastern District U.S. Attorney's Office. That was the basis for the Eastern District winning the Justice Department's nod for the next shot at putting John in jail for life. This time John was indicted together with his alleged lieutenants Salvatore Gravano and Frank LoCascio. At the arraignment Gerry Shargel attended on behalf of his longtime client Gravano (for whom Shargel had previously obtained an acquittal in a tax evasion case before Judge Glasser), and LoCascio was represented by David Greenfield. I argued that the government was trying to have John do life in prison "on the installment plan," spending years in jail (cumulatively) awaiting trial and vindication each time the government brought one of its serial groundless, and fruitless, indictments. Of course, all three defendants were detained without bail on the ground that they were a "danger to the community."

For about a year thereafter, while the defense and the prosecution methodically prepared for trial, John and I were the subject of massive media exposure and press conferences, largely caused by the government's intermittent leaks of information about its case against John to various favored reporters, those who would most readily vilify John and cast a favorable light on the government. In this context, in March 1991, I received a warning from Judge Glasser to abide by Rule 7 of the Local Rules of the U.S. District Court for the Eastern District with respect to out-of-court comments to the media by counsel on a pending federal criminal case. The landmark case in point, *Gentile v. State Bar of Nevada,* was about to be decided by the U.S. Supreme Court, and we were supposed to be guided by the decision with respect to freedom of speech rights under the First Amendment of the Constitution in the context of attorneys talking to the media about pending federal criminal cases. In a five-to-four decision the Supreme Court held in *Gentile* that comment to the media was *not* protected under the First Amendment where there was a reasonable or substantial likelihood of affecting the opinions of prospective jurors, or where the comments were designed to do just that.

I was on trial at the time of Judge Glasser's order, defending Peter Gotti in the windows case. In addition, I was representing John in a *civil* (as opposed to criminal) RICO case brought by the federal government before Judge Leonard Sand in the Southern District of New York against John, many other alleged organized crime leaders, and the longshoremen's union, seeking enormous money damages, for partaking in an alleged union-controlling cartel. The case was ultimately of little importance because by the time it became ripe for trial, most of the defendants (including John) were already incarcerated. With the windows case and the civil RICO suit, as well as the government's incendiary allegations in the case before Judge Glasser, I was being queried every day by reporters on television and radio shows. The pressure to do what I could to shield my client from a government propaganda onslaught was enormous and intoxicating; it was exceedingly hard for me to remain silent. So I didn't.

In January 1991, wholly separate and apart from my comments to the media about John or the prosecution, the government filed a motion to disqualify Gerry Shargel and me as well as John Pollok (who'd been retained to assist in motion practice and appellate considerations) from further representing the defendants in the case. The motion surprised us in that we believed it had a low likelihood of success, and that the government risked being publicly perceived as fearing John's defense team and trying to treat him unfairly by depriving him of his counsel. We underestimated the government's resolve and its animus, as well as that of the judiciary. The government's argument for disqualification (based almost entirely on a few offhand and jocular comments by John, overheard by the Ravenite apartment eavesdropping device) had four alleged bases: (1) we were "house counsel" to the Gambino "organized crime family" in that we had been allegedly paid by John to represent other alleged Gambino family associates or members in past cases, and this tended to prove the existence of the Gambino family, a RICO enterprise; (2) we "might" be called by the government as witnesses against the defendants, our clients; (3) we had allegedly represented in past cases other "potential" government witnesses in this case and were therefore conflicted; and (4) as we were overheard in some innocuous tapes generated in the Ravenite while visiting John, we were rendered part of the evidence.

Concerning the allegation that counsel had been paid by the Gambino family to represent its associates, the government provided a list of some twenty such cases. Of these, I appeared in three: the Coiro case, the Peter Gotti windows case, and the Giuseppe "Joe" Gambino case (which had yet to be tried then). As I was admittedly never paid by Mike Coiro, the government (and the court) assumed that *John* had paid me. In fact, as I said, *nobody* had paid me: I represented Mike without a fee as a courtesy to an older and well-respected criminal practitioner, a fellow member of the bar, and yes, because being asked to represent him was an honor that could only enhance my reputation as a lawyer. This is no evidence of a criminal enterprise. In Peter Gotti's case any payments received by me were from Peter himself. In the Giuseppe Gambino matter I was paid a commensurate retainer by Giuseppe himself and *not* by John or anyone else. No comment by John on the government tapes ever stated that I was paid by him for any of these cases. John merely asserted in his inimitable style that he was disgusted with paying lawyers, and the government merely provided the court with a list of cases against alleged Gambino associates that were handled by Shargel, Pollok, or me. But that was all it would take.

As to our being potential witnesses against our clients or part of the government's evidence, there was never any chance in the world that the government would call me, or Gerry Shargel or John Pollok, to testify against John or his codefendants. And as far as the question of our names coming up on the government's tapes and the asserted possibility that we might become in the eyes of the jurors *unsworn* witnesses *for* our clients—that is, lead the jury to believe that, as we were mentioned or involved in the defendants' conversations, we somehow were more than just their lawyers but were also impliedly vouching for the legitimacy of their activities—this could easily have been remedied by redacting, or blacking out, our names where they appeared in the transcripts of the tapes. The government argued that this would be too difficult and that the jury would somehow guess the redactions were references to Shargel and me. This was absurd. The clear goal was to deprive the defendants of counsel with whom they'd successfully fought off the government before.

We decided that we would be better served by having other counsel represent us on the motion; you know, the old adage that a lawyer who represents himself has a fool for a client. You probably have also heard the competing adage to the effect that if you want something done right, do it yourself. I don't mean to cast aspersions on the attorneys who argued the motions on our behalf, for they did a great job. But with the benefit of twenty-twenty hindsight, we all wondered whether we mightn't naturally have brought greater passion to our defense. It's tough for successful, aggressive attorneys with oversize egos to sit by and watch others defend them in the courtroom. It feels unnatural, frustrating, and humbling—the kind of thing clients are compelled to endure all the time. Not that it would have mattered if we'd represented ourselves, for from the outset it seemed that no procedural issues that might have been even remotely outcome determinative would be resolved in John's favor. I was represented by Sam Dawson, Gerry Shargel by Herald Price Fahringer, and John Pollok by Victor Rocco—excellent counsel all. Norman Siegel, then of the New York Civil Liberties Union, filed an *amicus curiae* (a legal Latinate meaning "friend of the court") brief in support of our position; Judge Glasser rejected the brief, saying, "You're no friend of this court."

When the motion was argued, Judge Glasser was in one of his more caustic moods and receptive to none of the positions taken by our lawyers. It should have been no surprise to anyone when, in late July 1991, he issued his decision (in a published opinion dated August 1, 1991) disqualifying Shargel and me, while permitting Pollok to remain in the case as long as he (as a motion practice and appellate expert)

never appeared before the jury. Despite all the signs, his ultimate decision disqualifying Shargel and me was shocking to us, to our clients, and
to the bar generally, particularly as our clients were willing and eager to
waive any conflict of interest purporting to protect them. However, Judge
Glasser was technically within his discretion in disqualifying us, if only on
the basis of the appearance of impropriety, and if there were any rationale whatever for thwarting John's interests, those interests would be
thwarted. Parenthetically, buoyed by its success before Judge Glasser, the
government would a year or so later again try to disqualify me, this time
from representing Giuseppi Gambino at trial before Judge Peter K.
Leisure in the Southern District of New York; the court would deny the
government's motion there.

Judge Glasser's disqualification decision was a crippling blow to the
defense and a massive victory for the prosecution. Moreover, I cannot
help but believe that the Second Circuit Court of Appeals had somehow
been consulted and had conveyed its approval of such a move, prior to
the issuance of the trial court's decision, perhaps prior to the government's decision to make the motion at all. The government had decided
that this case was to be Gotti's Last Stand—nothing would be left to
chance. And in order for a ruling depriving John Gotti of his attorney
to stand, the court's written opinion would have to contain terribly
derogatory findings about me. Every allegation of the government, every
possible nefarious implication, carried the weight of fact in the court's
written opinion. Is it possible that, since Mike Coiro didn't pay me to try
his case, John had? Then John had. Is it possible that when John spoke
of how much he'd paid attorneys, he was speaking of me? Then he was.
The government's unsubstantiated allegation that I'd received payments
from John that I was not declaring to the IRS was thrown into the mix by
the court as well. The prosecutor John Gleeson even speculated, without
any substantiation whatever, in a letter submitted to the court on the
motion, that it was at John Gotti's insistence that I'd moved to my present offices at 41 Madison Avenue with Jimmy LaRossa. During the
O'Connor assault trial, when I first mentioned to John that I'd be moving to a suite with LaRossa and some others, he didn't care much one
way or the other; he said only, "Good for you. Those guys are great

lawyers, and it will be a good move for you professionally, if you're asking me what I think." There was no way the government was going to allow me to try Gotti's Last Stand, and if my character had to be impugned, if my career had to be derailed, sacrificed on the altar of governmental vengeance, then so be it; in fact, so much the sweeter for the government, for I'd been so vociferous in my client's defense, so irreverent of the government's authority. Now I'd be chastened. Now I'd feel the heavy hand of the government on my own head.

Like the attorneys in the case, John was somewhat taken aback by my disqualification. He'd always allowed for the possibility that Shargel would be lost to the case, because of the many witnesses and other alleged Gambino associates he'd previously represented, but he hadn't expected me to be disqualified as well. Then John's disbelief gave way to acceptance, accompanied by his comments as to how the government was now going to do whatever it wished with impunity, but he was not going to let it affect his attitude of defiance. Gravano's reaction was altogether different; he was devastated by the loss of Gerry Shargel and would never recover his composure thereafter. He remained disconsolate even after he'd obtained new counsel. Gravano would insist on looking up such things as the guidelines applicable to the plea agreements of government cooperating witnesses. This was the sort of thing that John viewed as implying weakness and would never do. Gravano was fading fast.

At the first scheduled court conference after Judge Glasser's decision had been issued, there was some confusion as to whether we would be permitted to appear on the clients' behalf prior to the actual trial of the case, so we (Shargel and I) did not show up in court. The defendants were produced, and Judge Glasser asked where their lawyers were. John replied, "I know it's your prerogative to do it, but you threw my lawyer off the case. Why'd you take away my counselor, Bruce, who's represented me for the past seven years?" And Gravano delivered his oft-quoted quip that "Jacoby and Meyers don't come to the MCC." The court advised John and Gravano that Gerry and I had not been disqualified for purposes of pretrial proceedings.

Then John said, pointing to prosecutor Gleeson, "I'm his only defendant. Why don't you disqualify him?" At that juncture John pointed to FBI Agent George Gabriel and said, "And Little Lord Fauntleroy, sitting next to him there . . ." The media garbled the entire statement, reporting John to have said, "Why'd you throw Bruce Cutler out? I've been his only defendant for seven years," and to have referred to Gleeson as Lord Fauntleroy. John was *not* referring to Gleeson as Fauntleroy, and he was saying that he had been *Gleeson's* only defendant for seven years, not *mine*. John was obviously aware that I'd had many low- and high-profile clients over the preceding seven years. But as usual, what made for a better story was taken for the truth, so the morning headlines read that John had directly affronted the prosecutor and indirectly his own counsel.

After duly exploring and ruling out all immediate possibilities of attacking Judge Glasser's decision, and recognizing that in the federal court system no appeal of a judge's ruling is permitted until a final verdict or judgment in the case, the defendants and we counsel tried to come to grips with the reality that John, as well as Gravano and LoCascio, would be going to trial without the attorneys who had previously been so successful for them. I should add that we had the option of pursuing a mandamus procedure, arguing to an appellate court that Judge Glasser had engaged in "conduct prejudicial to the effective and expeditious administration of the business of the courts" in disqualifying me, Shargel, and Pollok, but these motions are *never* won. We didn't fight battles we couldn't possibly win. New counsel had to be retained. For John, I would have suggested Jimmy LaRossa, but Jimmy would have been susceptible to another disqualification motion because of his representation of Paul Castellano, with whose murder John was being charged in the case. So I suggested Michael Kennedy, a New York lawyer whom I'd always admired very much: a really tough trial lawyer, with no fear of the government, who'd represented defendants in the Irish Republican Army cases as well as alleged organized crime figures. He had a combative style that would mesh well with John's, and yet his courtroom presentation was dignified in every respect. I also thought an Irish-American attorney, in the venerable New York tradition of eloquent Irish advocates, would be a novel look for John and would burnish his image.

John interviewed dozens of well-qualified attorneys, some of whom were eager to represent him and others reluctant, and in the end he selected sixty-nine-year-old Albert Kreiger, a friend of LaRossa's (from Florida) whose talents were legendary in criminal defense circles. Gravano selected Benjamin Brafman, and Dave Greenfield stayed on for Frank LoCascio. George Santangelo was supposed to come into the case to represent Frank at trial, but he was also a veteran of the government's Giacalone disaster, and he, too, would have to pay—the government would later make a motion to disqualify him as counsel, and the court would grant it. Tony Cardinale, an able Boston attorney, would come into the case to defend LoCascio at trial.

I stayed in the Gotti case through the futile pretrial motions; the trial was originally scheduled for September 1991. By September the new attorneys in the Gotti case had taken over, and the trial was adjourned. By mid-October I'd left New York for Chicago to begin the protracted Infelice-DiLaurentis trial there with cocounsel Pat Tuite, George Leighton, Kevin Milner, Terry Gillespie, Bobby Simone, and Tony Onesto.

At around that time the government put in a request that the defendants all be produced from the MCC to the prosecution offices for handwriting exemplars, supposedly relating to the tag-along tax fraud counts of the indictment. Such requests are granted routinely, so the defendants were produced without a fight. But the request appeared odd in that no dispute had arisen as to the authenticity of any defendant's signature, and the case was essentially a murder case, to which handwriting was irrelevant. The procedure was just a pretext for the government to meet secretly with Gravano to discuss the terms and manner of his defection. The defendants were all returned to the MCC, where Gravano was permitted to remain for another month. The longer the government was able to conceal his collaboration, the longer it would have a spy in the defense camp, hearing and partaking in further conversations about which he could later testify, and the less time the defense would have to prepare to rebut his testimony. In November Gravano was pulled out and placed in the Witness Relocation Program.

I returned to New York for the Veterans Day weekend, because the DiLaurentis case was adjourned from Friday through Tuesday. I was in my Madison Avenue office when I was notified of Gravano's defection to the government. Gravano's attorney, Ben Brafman, had been in the Bahamas on vacation when the government telefaxed him a letter from his client, discharging him. When I heard the news, I went to the MCC to see John. He'd already been told. But he wasn't crestfallen. Rather, he was his usual upbeat self and told me he'd seen it coming. Gravano had been acting aberrantly in jail, not shaving, not sleeping, frequently crying to his girlfriend on the telephone. His comportment had not been up to the street tough-guy standard.

It seemed to me that Gravano's defection didn't have the same impact on John as the news of Willie Boy Johnson's alleged informant status. The Willie Boy allegation had, as I've said, broken his heart, and he always displayed a certain incredulity with regard to it. He'd liked Willie Boy. In my opinion, he'd never liked Gravano. He'd known Willie Boy for most of his life; he hadn't met Gravano until the mid-1980s, although he'd heard of Gravano's *reputation* for toughness in Brooklyn. John was angry at Gravano's betrayal, not hurt. Gravano was to John a fellow who had claimed to be someone he wasn't—a tough guy—someone who'd never before been tested, never been in jail, and now that he was being tested, couldn't take it. After eleven months of bemoaning his fate on the telephone to his girlfriend, he was ready to roll over for the government. To John, Gravano was an "opportunist," a "coward," an "unprincipled creep who was only interested in making money and pushing weaker people around." He was a "punk" with no conscience or sense of the core meaning of life on the street.

The government's strategy was always to portray their turncoat informants as violent monsters (as distinct from *liars,* which is what I would try to demonstrate them to be), because the jury would then be led to infer that the defendants on trial must be *more* monstrous still. To this end the informants would be encouraged to exaggerate their résumés for violence. The more beastly the informant appeared, the warmer the government's embrace, for tactical reasons. This was why John and I would be upset by attorneys who fell into the government's trap by attacking an informant as a violent sociopath.

Perhaps what rankled John more than the lies Gravano told about him were the lies promulgated about Gravano himself. Gravano's account of his interactions with John, as described in his book *Underboss,* written with Peter Maas, gives the impression that they interacted as equals and that Gravano would behave in a confrontational manner around John. In my presence Gravano was always subservient, even obsequious with John. And Gravano was no enforcer, no man of heroic proportion, as *Underboss* and other books about John or Gravano would imply. Gravano's murders, except for *one* in which he admitted personally pulling the trigger, were all committed by others at his behest. He took *credit* for the rest, for the government's litigative purposes. John called Gravano an imposter, who lived the street life while the living was good but ran to the authorities when the going got tough. The point here is not to exalt the physical commission of violence over its mere orchestration, for murder is murder. Rather it's to let the truth be known as to what Gravano was all about in John's eyes.

I personally never knew Gravano well enough to explain his decision to testify for the government. I give no credence to the justification set forth among several others in his book, that John was going to physically hurt him or throw the blame on him in some way for their indictment or the criminal activity with which they were charged. That would not have been the John Gotti I knew. Gravano's words had the hollow ring of rationalization. The real proof of his character was in his actions after receiving the government's sweetheart deal and a new life in Arizona: He was arrested and convicted of peddling pills to high school kids and enlisting the aid of his own wife and children in his activities. Gravano, who admitted that he was responsible for some *nineteen* murders, served only eleven months in jail, then spent the rest of a five-year sentence on a minimum-security military farm, where he could come and go as he pleased, eat and drink what he wanted, and meet whomever he wished.

John's prosecution was ugly and getting uglier every day—and there was nothing I could do about it. I'd been neutralized, at least in the courtroom. I was still trying to do what I could for my client *outside* the courtroom, rebutting the leaked government propaganda that appeared daily in the media with respect to Gravano's upcoming testimony, and trying to shed light on the dark governmental processes at work to put

John behind bars at any cost. For my out-of-court efforts on John's behalf, the government would soon tear its pound of flesh from me. But in the meantime I was almost grateful to be in Chicago, doing battle for another in a different arena, my mind and body occupied, venting the frustration of standing on the platform as the government railroaded John.

19

I was treated royally by the clients, cocounsel, U.S. District Judge Ann Williams, and even by the prosecution in Chicago. While the Gotti Wars raged in New York, I was in self-imposed exile in Chicago, doing battle there from October 1991 through mid-March of 1992. The Chicago defendants—Infelice, DiLaurentis, and others—were not convicted of the murders and murder conspiracies charged in the indictment, but they were convicted of the gambling and extortion predicate acts. The clients were appreciative of the efforts and the result, and I was pleased for them, for they were model clients in every way.

The DiLaurentis trial was a refuge for me, a relief from uselessly standing by, watching the injustice of the Gotti trial day after day, eating myself up inside. I really liked Chicago, enough to open a pilot office there with my friend Kevin Milner. Chicago is an honest, true American city—I'd say the *most* American of cities. I sometimes fantasized about moving there, which probably expresses better than anything else how my treatment in the Eastern District of New York was making me feel. But Brooklyn dogged my steps, tracking me down even in Chicago. One evening after a hard day of trial, I went back to the hotel where I was staying to meet Bobby Simone, a first-rate Philadelphia lawyer who was my cocounsel in the case and would thereafter become a good friend. He'd brought me a copy of the most recent *New York Magazine,* and I was on the cover with John. The caption read something to the effect of THE MOB — WHERE WILL IT GO AFTER JOHN GOTTI. If it wasn't purporting to be John's death knell, certainly it was predicting he wouldn't be around much longer.

I would keep up with the Gotti trial through the New York and Chicago newspapers every day, and I would be back in New York every two or three weeks for a weekend. John was being tried for murder, extortion, racketeering, gambling, tax evasion, and obstruction of justice in the context of a RICO prosecution. The centerpiece murder was, of course, the Castellano assassination. Every tape proffered by the government was replete with John's denials of any hand in Castellano's death. The government claimed that these were all false exculpatory statements, yet it asked the jury to believe the other "admissions" in the very same conversations.

The prosecution's theory of the Castellano assassination was originally based on the testimony of Phil Leonetti, a widely acknowledged turncoat and nephew of Nicky Scarfo (the reputed boss of the alleged Philadelphia organized crime family in the late 1980s). Leonetti, who was reputedly the family's underboss, later defected, and he testified before a federal grand jury that John Gotti had met with Scarfo and told him (in the presence of Gravano) that John was responsible for the Castellano killing and that it had been done with the approval of the Commission. The government position was completely and cavalierly altered by the Gravano defection. Now the theory was that John had effected the killing with a small group of coconspirators, without the approval of the Commission, and that all those involved swore never to admit their involvement even privately among themselves. Of course, this change in the prosecution approach was mandated by the innumerable Gotti denials on the tapes.

But there was no common sense at all in the idea that John and his friends would believe themselves secure enough from eavesdroppers to admit to crimes that might open them to prosecution and multiple life imprisonment sentences, but not secure enough to admit to a particular internecine killing. If a juror were to believe that the "admissions" of criminal behavior were true, why would he disbelieve a denial? Why would he believe that a particular tidbit of criminal activity would be scrupulously kept under wraps? Are we to believe that John and his friends exercised two distinct levels of caution? (1) They *were not* cautious enough to refrain from admitting criminal activity that would expose them to life in prison. (2) They *were* cautious enough to avoid anyone knowing of one particular murder—that of the former alleged Gambino family boss? Isn't it easier to accept that they were either incautious, or exceedingly cautious, consistently across the board? Apparently not for a disingenuous prosecution; nor for the jury that was led astray. To this day I don't believe John had any hand in the Castellano killing. I do know that in his life he was always loyal, and in my opinion his personality didn't jibe with the Gravano version of having others do what he wanted done himself.

As to the other murders with which John was charged (excepting that of Tommy Bilotti, whose killing was, in the government's theory, a part of the Castellano assassination), only one person stood to gain, and did gain, from these deaths: Gravano, the man the government allowed to go free, the man the government used to blame John for the murders. In fact, even the government agreed that Gravano actually set up the murders of Louis Milito (a person whom Gravano called one of his oldest friends and his partner in a steel erection business), Louis DiBono (Gravano's partner in a drywall construction business that was developing embarrassing tax problems), and Robert DiBernardo. It's not even *alleged* that John had anything to gain from these deaths.

It's often said, by representatives of the government, that John convicted himself with his own words. Everyone who knew John, including government agents, knew that John had a dramatic, even operatic way of speaking. If John's grandiloquent statements were really admissions sufficient to convict him, why was Gravano so important to the government's case? Because everyone, including the government, knew that without Gravano's lies (bought and paid for with a token jail sentence and a new life in Arizona), John wouldn't and couldn't be convicted.

On my intermittent trips back to New York from Chicago, I'd be drawn to the courthouse in Brooklyn. John was my longtime client, a man I'd grown to care about, and yes, a man who'd "put me on the map" professionally. I would have shown up at the trial just as a matter of client relations and loyalty. But there was more. Something was dead or dying there in that courtroom in Brooklyn, and it wasn't John, for he'd never be broken. Our justice system lay stricken. Or perhaps it was my perception of the system that lay there broken—my innocence. I retained a certain naïveté, a belief that in our legal system, no matter who the defendant—or what his name or how advantageous it might be for the powers that be to convict him—the process would be the same. This innocence was ruptured, and I returned to that courtroom in Brooklyn whenever I could, a masochist probing the wound.

There he sat before his accusers, a man locked up before and during his trial, stripped of his attorney of choice, and faced with sequestered and anonymous jurors who'd been impressed with the dangers they would risk at the hands of the defendant or his confederates were they, the jurors, not cloaked in anonymity and guarded day and night by armed officers. It was a show trial whose outcome lay in no doubt, where the form of due process was observed without the substance. John's trial was no trial in the sense that we have come to expect in this country—it was an inquest, an abomination.

Government-skewed irregularities abounded throughout the process. In voir dire (jury selection), for example, unlike the usual time-consuming practice whereby the court questions the jury pool extensively to make sure the jurors ultimately selected are not biased, Judge Glasser curtailed the process, simply asking each juror if he (or she) could be fair. As reported in the press, when Frank LoCascio turned to John to protest, John just shook his head and said, "It's all mind over matter, Frankie—if you don't mind, it don't matter." He didn't mean that he didn't care. Rather, he was saying to the court, "You can't touch me. Sure, you can put me in solitary for life, but you can't touch *me*." It was the supreme expression of his defiance.

Another example of the special rules of procedure in effect for John Gotti: After Gravano already testified as to the contents of the taped conversations, a federal agent was called to interpret the same tapes again—a procedure ordinarily barred as "improper bolstering" of the witness's testimony. In John's trial it was permitted. Like the disqualification of his counsel, every government abrogation or vitiation of his statutory, judicial, and constitutional rights was rationalized on some legal theory or another, and it would hold up on appeal because John Gotti could not be allowed to flout the majesty of our prosecutorial and judicial system any longer.

When I returned (at the close of the DiLaurentis trial in Chicago in mid-March 1992) to see the last two weeks of John's trial, the courtroom atmosphere in Brooklyn was gloomy and suffocating. Judge Glasser was short-tempered, bearing a perennially sour expression, as though he were tasting something foul, and surely he was, for by reputation he was considered a fair and liberal judge, and this was not the way his courtroom was usually run. But for John Gotti, martial law would prevail. Although this was no multidefendant case, and John and Frank LoCascio were the only remaining defendants (now that Gravano had been embraced by the government), the courtroom was nevertheless ringed with armed federal marshals, lending the proceedings the air of some third world political trial—and not-so-subtly conveying to the jury again how dangerous John was. In the entire gallery there were perhaps two pews for the public; all other seating was reserved for prosecutors, law enforcement types, press, lawyers, and the defendants' family members. There would be no leisurely jovial lunches in *this* courtroom.

Gravano was off the stand, and the government was just mopping up. John was seated at the defense table next to Al Kreiger, flanked by Kreiger's wife and his legal assistant, just as he'd sat beside me at defense tables in the past. Although John's mein was as defiant as ever, the overall picture was very different now. For the first time John didn't seem larger than life in the well. A palpable absence of energy reigned. It was as if I'd suddenly gone deaf and were watching a trial in mime. Michael Kennedy once told me that, if all you did at counsel's table was shift paper from one place to another, you were playing the government's orderly game and you were bound to lose. John hadn't given up, and his defense team was doing a good job, so it was hard to point to one specific reason, a source of the lassitude that held sway in the courtroom. There were several perhaps.

The case was rushed to trial less than two months after the Gravano defection was announced; only later, after Gravano had already testified, did more and more information surface about him (like his narcotics trafficking and the other murders he had orchestrated that had not been disclosed to the defense or presumably to the prosecution) that would have been extremely helpful in cross-examining him. Postjudgment motions collaterally attacking the verdict were made on John's behalf on the basis of this information; these were of course summarily denied. Tony Cardinale, along with John Mitchell, had done an admirable job defending LoCascio, as did Al Kreiger for John. But of course, these attorneys were new to the case, and John was working with a lawyer who hadn't had much time to acclimate himself to John, his history, and the case against him. Al is a highly skilled attorney, but I never felt he was John Gotti's kind of lawyer. Although I'd spent a great deal of time with Al, trying to prepare him for the case, it is very difficult for an attorney to step in, as Kreiger did, with so little advance notice. And stylistically, I thought Al was a little too subdued for John. Lawyers are not robots; they are extensions of the client as seen by the outside world. Some lawyers can represent anyone, but some clients, John Gotti in particular, need a special kind of lawyer, one who can exude the client's persona in the well. For John, that lawyer was me and none other, because I knew him so well and because we'd developed that certain bond.

In the end I realized that the languor and lethargy that pervaded the Gotti trial was not sourced in the attitude of the clients or their counsel. Rather, it was the result of laboring for three months on trial in the certain knowledge, clearly conveyed by the court's interim rulings, that conviction was a foregone conclusion. If there could be no belief in the possibility of success, no hope, then there could be no drama, no suspense, no electricity, and ultimately no success. The government had put John on wheels, then firmly locked them onto tracks leading straight to the maximum security federal prison in Marion, Illinois.

I stayed in New York until March 31, 1992, through the summations. Then I traveled to Orange County, California, where for two months, beginning on April 5, I would be trying the Gionnis case. Taking no chances to the last, the Gotti prosecution moved to have a juror removed on the grounds that she was obstructing the deliberation process: Notes reflecting her thoughts and recollections of witnesses' testimony were found in her pocketbook, and she was removed. Two days after the summations (including the time it took to argue and decide the government's motion to remove the juror), the jury in John's case returned a verdict of guilt on all counts of the indictment. The famous civil liberties attorney William Kunstler (now deceased) filed a motion on John's behalf, first before Judge Glasser and then in the Second Circuit Court of Appeals, objecting to the removal of the note-taking juror, asserting that the government had purposefully interfered in the jury's deliberation process, and that it had tried to, and succeeded in, intimidating a jury, which had already been unduly intimidated by anonymity, sequestration, and the constant surveillance of federal marshals. The motion was denied. Nothing would be allowed to derail John's path to prison. Even though the verdict was no surprise, when I heard the news in California, I felt as though I'd lost a piece of myself.

I appeared in New York for John's sentencing on June 23, 1992. The proceeding was short—so short that even if you were in the courtroom and you'd turned your head for a moment, you might have missed it. The juxtaposition of this moment with the years that had preceded it was stunning. After almost a decade of mortal combat sprawled across five boroughs, television, newspapers, magazines, cinema, and four courtrooms in New York, the war ended so silently, you could hardly hear it. In the soundless, airless vacuum of that courtroom, I don't recall any more of the proceeding, but the newspapers reported that when Judge Glasser asked if John had anything to say before sentence was pronounced, John, in a dark double-breasted suit and a yellow tie and matching pocket square, calmly said no. The court then duly handed down the sentence that had already been written in stone long before his trial began: life in prison. *New York Newsday*'s Peter Bowles quoted a high-ranking FBI agent as having said, "He went out like a real boss, fair and square. That's the way he is supposed to do it, without a word."

When John's sentencing "hearing" was over, a riot erupted on Cadman Plaza, on the steps of the courthouse in Brooklyn. A large unruly crowd of about a thousand people, rankled by the perceived injustice of John's conviction, was venting its frustration, shouting and overturning cars. The U.S. Marshal's Office contacted me and asked me to do what I could to disperse the crowd. With a megaphone I was handed by a marshal, I entreated the rioters to calm down, saying that I shared their frustration but the last thing anyone wanted was for a violent demonstration to take place. I asked them to follow me over to Cadman Plaza Park, away from the courthouse steps, where they might peacefully protest if they wished to do so. They followed me, and the incident ended.

Accompanied by Gerry Shargel, John Mitchell, and Tony Cardinale, I visited John that night at the MCC. John said he thought he'd be shipped out to a maximum security facility that very night. I said that sounded impossible—it ordinarily took the Bureau of Prisons anywhere from six to twelve weeks to designate a facility. If a defendant is out on bail at the time of sentencing, he is often given a date by which to surrender directly to his designated facility; if he's already in custody at the time of sentencing, he might be transferred from the local facility (in New York, the MCC) in two months or more from the date on which his sentence was meted out. John shrugged and changed the subject. I'd told him that the reporter John Miller had asked me to appear on his television show that night. John asked me to do it, to let it be known that a destructive riot was not his way of dealing with any problems and that he wouldn't countenance that kind of action. I left the MCC and appeared on John Miller's show. Then I went home to my apartment to wallow in my own dark thoughts over John's fate—and my own.

As I lay down in my bed, John was being prepared for a journey. And at four-thirty on the morning of June 24, 1992, the day after his sentencing, he was transferred from the MCC in a massive police motorcade, complete with accompanying helicopters hovering above them all the way, to Teterboro Airport. He descended from the prison van into an airstrip flooded with security vehicles and personnel and lit so brightly that it might have been high noon, and he was immediately bundled onto a small jet with several federal marshals. He would later tell me that one marshal offered him cookies that the marshal's wife had prepared especially for him the night before. By seven-fifteen that morning, the door had slammed behind John in his cell in the maximum security lockdown Level 6 federal prison in southern Illinois.

When had it been decided that John Gotti was going to Marion? Was the Bureau of Prisons' cumbersome machinery moved to effect what ordinarily takes it two months in only a few hours, on that afternoon of June 23 after John's sentence was handed down and before he was whisked away in the early-morning hours of June 24? Or was the decision made before John's presentence report ever reached the court, or before John's verdict was ever returned by the jury, or even before he was brought to trial? And whose decision was it? Did the attorney general in Washington decide where John would be buried before the government had even placed a bug in the apartment over the Ravenite Social Club on Mulberry Street?—that secret decision known to all in the federal community right down to the kindly marshal's wife toiling away in her kitchen.

Within no more than a day or two after John was transported to Marion, as if on cue, his father, John Gotti Sr., passed away after a protracted battle with cancer. I, along with great numbers of John's friends and family, lawyers, well-wishers—and, of course, a battalion of FBI agents on surveillance—attended the wake.

PART THREE

The Aftermath

20

SHORTLY after I was disqualified from representing John Gotti in his 1992 trial, Judge Glasser appointed a special prosecutor to investigate and indict me for contempt on the ground that I'd disobeyed the court's order to refrain from commenting to the media about John or his pending case. At roughly the same time the government opened a grand jury investigation of me and Gerry Shargel, to look into our alleged receipt of unreported (to the tax authorities) legal fees received from the alleged Gambino family. Vengeance for the Giacalone case was the order of the day; John was to be imprisoned for life, and I was to be destroyed with him.

The basis for the grand jury investigation was certain operatic comments John made to Frank LoCascio and some other friends that were caught on the Ravenite tapes. In one of them John complained of attorneys who don't fight hard enough in court, suggesting that this was because they were afraid that they themselves would be investigated for "taking money under the table." As the government-acknowledged most aggressive courtroom fighter in quite some time, it is hard for me to imagine how John's comment could have been referring to me (if in fact it referred to anyone and was anything more than loose banter among friends). In the same conversation, John allegedly said, "If they really want to break Bruce Cutler's balls, what did he get paid off me? Three years ago I paid tax on thirty-six thousand." The government interpreted this to mean I'd received vast sums of money that I failed to report as income for tax purposes. It wasn't true. The investigation would hang over our heads for more than three years, and friends and clients would be questioned, subpoenaed to the grand jury, and generally harassed throughout that period.

As the government juggernaut rolled on, I too rolled on, doing what I'd always done—fighting cases in court as hard as I could. Notwithstanding John's conviction, I was at the zenith of my public approbation through 1992 and into 1993, when my father fell ill. The media pundits, who know little about anything, held me the biggest winner in the Gotti saga, suggesting that John had been convicted because I was unable to represent him.

After another trial in Chicago I began to prepare to try the Giuseppe Gambino case in the Southern District of New York (Manhattan), before Judge Leisure. Giuseppe is a distant cousin of Carlo Gambino, in whose honor the government christened the "crime family" of the same name. The indictment alleged a major transatlantic heroin narcotics conspiracy involving the so-called Sicilian faction of the Gambino family, with attendant murder and obstruction of justice charges. Several Sicilian witnesses were testifying against the client and his brother Giovanni Gambino, and Gravano was to testify as well about a murder allegedly committed on the Gambino brothers' behalf, with the "usual" John Gotti blessing.

The government, in its continuing effort to chill vigorous representation of its targets and to put me personally out of business, again tried to disqualify me from representing a client. Now the government, supposedly buttressed by Gravano's testimony, argued that in addition to being "house counsel" for the "Gambino crime family" per the prior disqualification ruling of Judge Glasser in the Eastern District Gotti case, I was also *myself* the target of a grand jury investigation and could conceivably compromise my client's interests to curry favor with the government in my own investigation. Ironically, the government was trying to remove me because I was too *vigorously* defending my clients, and it was doing so by arguing that I might lie down on the job to save myself. It was a ridiculous posture, once again using the conflict of interest rules, allegedly protective of defendants, to bludgeon them. Needless to say, Giuseppe Gambino was adamant in waiving any conflict. Moreover, I owe thanks to Judge Leisure, a true lawyer's judge, who gave all the issues a fresh, unbiased review and, despite the pressure of Judge Glasser's earlier decision and scathing opinion, denied the government's motion to disqualify. Nothing in Gravano's testimony alleged any wrongdoing by me, I was on no relevant tapes, and I wasn't alleged to be an unsworn witness. I was before a different judge in a different district, one with no agenda, and most important, John Gotti was not the defendant. Hence the different result. (I should note that, in contrast to the Eastern District case, the Southern District prosecutors, James Comey [now the United States Attorney for the Southern District of New York] and Pat Fitzgerald [now the United States Attorney for the Northern District of Illinois], tried the Giuseppi Gambino case honorably and professionally, refraining from ad hominem attacks on me.) The case was tried for five months, and on Memorial Day 1993 the jury advised the court that it was unable to reach a unanimous verdict. Later, before the case could be retried in January 1994, the client would take a plea and receive a fifteen-year sentence. And Judge Leisure would invite me to be one of the speakers at his fifteenth-anniversary celebration of his appointment to the federal bench, held at the Yale Club in 1999.

But I was far from out of the woods. The government's grand jury investigation continued, as did the special prosecutor's preparation of a contempt case against me stemming from my violation of Judge Glasser's quasi–gag order in the Gotti case. And there were other blows awaiting me.

After spending a weekend on Long Island, Anthony Barratta was introduced to me as a good friend of John Gotti and said, "I might need you one of these days." Sure enough, a few months later, while I was on trial in another matter in Chicago, I was contacted by Anthony's son on behalf of his father. Anthony Barratta, a well-spoken, debonair man-about-town and a contemporary of John's living in East Harlem, who was alleged to be a *capo* in the so-called Lucchese family, was indicted on the day of John Gotti's sentencing (June 23, 1992) in a federal multidefendant case before Judge Miriam Cedarbaum in the Southern District. I visited Anthony in the MCC, where he was remanded without bail pending trial, and he retained me, as many did, to present a vigorous defense. The prosecutor, David Kelly (now chief of the Southern District U.S. attorney's Criminal Division)—an astute attorney and someone I grew to like, who was also a weight trainer, had played football at William and Mary, and had once been a police officer in Amagansett—believed Anthony to be an evasive and highly sophisticated "street guy" and charged him in a major racketeering case involving allegations of serious violence. After one of Anthony's principal codefendants with a severe health problem pleaded guilty, and after a thorough study of the discovery and prospective testimony in the case, Anthony astutely decided he'd be better off pleading guilty and settling the case, without any embarrassing admissions of Cosa Nostra. He was facing a mandatory life sentence if convicted after trial, and with the assistance of my cocounsel and friend Jeffrey Stephens, we negotiated a palatable plea whereby he received an eleven-year sentence, one year of which he'd already served while awaiting trial. Defendants and counsel fighting these major federal racketeering indictments had come to recognize that the cards were increasingly stacked against them, and the trend became to settle these cases where possible rather than fight to the end.

It was one of the few times I'd had a client take a plea in a major racketeering case. Clients ordinarily retained me for a scorched-earth defense, and I simply didn't often have occasion to take a plea. Judge John Martin, who took Barratta's lengthy and carefully framed allocution (the wording by the defendant of his plea of guilty), was also struck by the anomaly and amiably quipped, "So you're rolling over now?" When I emerged from the courthouse on that day in early August 1993, I looked up gleefully into the falling rain and realized I had an unanticipated free month. So I arranged to rent a house on the beach in Napeague, Long Island, for Shonna and me for August. It was to have been a well-deserved vacation after years of fighting the Gotti Wars and other high-profile cases, and an opportunity to rekindle the dying embers of my romance with Shonna. It wouldn't turn out that way.

I don't usually believe in this sort of this thing, but I recall reacting with a sense of portent to the ringing of the telephone that day at the Napeague beach house. I let the phone ring until it stopped. When it rang again a few moments later, I reluctantly picked it up. It was my mother, and she had bad news: My father was at Long Island College

Hospital in Brooklyn. No, it wasn't his heart, she said. It was a stomach problem: My father had been diagnosed with cancer. Test results showed that the cancer had already reached his liver. As strong as he was, and as careful as he'd been with his diet, the genetic factor had prevailed—my grandmother had succumbed to the same illness. Had it not been for his diet, he might have been stricken earlier, and who knew how long the cancer had gone undiagnosed, for Murray was never sick and never went to doctors. He just didn't trust them.

Anger was my prevalent emotion—misplaced anger at Murray for being sick, for adding another burden to John's incarceration, the contempt inquiry, and the grand jury investigation. I couldn't acknowledge then the true source of my anger: a mounting series of adverse events beyond my control. There had been nothing I could do to save John from a guilty verdict and a life sentence. There was nothing I could do to stop the special prosecutor's contempt inquiry or the government's grand jury investigation. And most of all I could do nothing to cure my own father's cancer, to stave off the passage of time and approaching death—my parents', my own.

The more frightened and frustrated I was by the prospect of losing my father and by my own implied mortality, the more I sought refuge in anger and exercise. I visited my father at the hospital, and my younger brother, Richard, and older sister, Phyllis, both came to New York to see him. There was a steady stream of distinguished members of New York's bar and bench at his bedside. I particularly recall that federal Judge Henry Bramwell (before whom I had tried my first federal case, the Ferrugia matter) would walk over from the courthouse every day to visit my father. Judge Bramwell, a Republican whose impartiality made him unpredictable and hence equally unsettling to the defense and prosecution alike, had risen from poverty to become one of the earliest blacks appointed to the federal bench. Murray had known Judge Bramwell from Brooklyn Law School and from the Black Judges Caucus (of which Murray was a member), and they'd become close friends, although Murray's left-leaning politics were very different from Judge Bramwell's.

I don't remember a word of complaint from Murray about the pain or discomfort (which were doubtless considerable) while he was at the hospital, either before or after his first operation. But I recall picking him up at the hospital in a Minuteman car service to take him home shortly after that first surgery. My mother was waiting at the screen door, which she opened as I was helping my father up the few steps to our stoop. He stopped for a moment—to catch his breath, I thought. My gaze rose from his feet, which I'd been watching to make sure he didn't trip, to see Selma and Murray looking affectionately into each other's eyes, as if they were sharing a secret understanding, as though they'd arrived at a time and place they'd been working toward for a lifetime. She smiled and touched his arm gently and said simply, "Welcome home, Murray." I'd never seen my mother that way, so tender. It was a poignant moment that embodied everything that a man and a woman might mean to each other if they were lucky, a most private and tender moment that I felt almost embarrassed to be witnessing.

I said, "It's not bad being home, is it, Pop."

"Not bad? From hell to heaven," he replied, without taking his eyes from my mother's.

I can't recall Murray ever missing a single day of work before he got sick, and he returned to his office, and even court, a day or two after he was discharged from the hospital in late 1993. In December he and I attended our last Brooklyn Law School annual luncheon. I would attend the event (which used to be held at the Terrace on the Park in Queens but moved in later years to the Plaza Hotel or the Hotel Pierre in Manhattan) with him every year. We'd enjoy it and have a few drinks; I'd meet a few lawyers and judges. I liked it more in the early years, when it was still held in Queens. Later it became more of an annual obligation I'd undertake in deference to my father. That year, in 1993, the reunion held particular significance for me, a reminder of the days in my childhood when my father and I, the "men," would take our little trips together to the Riis Park beach or upstate for a "march."

Then Murray had a second colon operation, and another to remove his prostate. And I broke up with Shonna, taking out on her my anger at my father's plight and my own legal predicament. Almost immediately after we broke up, Shonna received a grand jury subpoena, calling her in to testify concerning my spending habits, which prompted my father to joke, with reference to the privilege prohibiting the prosecution from compelling a wife to testify against her husband, "You should have married her!"

That summer of 1993, one of the hottest on record, hardened into the winter of 1994, one of the coldest on record. In between, in or about October 1993, the Second Circuit Court of Appeals affirmed John Gotti's conviction, rejecting all arguments strongly proffered by the eminent, scholarly, and passionate Harvard professor (Anita Hill's attorney and my good friend) Charles Ogletree, who represented John on the appeal. My father continued to work until early spring 1994, when he was readmitted to the hospital. My own contempt trial was commencing, and by the beginning of June I had to get away from the demeaning and frustrating court proceeding, away from the daily sight of my father getting sicker. My brother, Richard, had returned from California to be at his bedside, so I leaped on my friend Dave Goracy's invitation to Sagaponack to revisit our athlete-men's retreats of the 1980s. The day after I arrived at Dave's house, on June 5, I received a phone call from my brother advising that our father was dead.

That whole time period is a blur to me now as I look back. I remember Shonna and both my ex-wives, Barbara and Gladys, calling to convey their condolences. I remember that I wouldn't go to the house in Flatbush because Richard was staying there. In his grief he'd made a remark I resented about my not having been at our father's bedside when he'd passed away. I suppose it's inevitable that I'd feel some guilt over not having been there at the final moment, but I was in fact relieved to have missed seeing my father die. And I hadn't faced the possibility of his dying when I left for the weekend.

I returned to Flatbush on the Monday after my father died. His funeral was on Tuesday. My eulogy tried to encapsulate the essence of what Murray had meant to me, what fatherhood had meant to me—that bulwark against the immense forces of the universe bearing down on a child, that great hand hoisting my little body from the roiling surf of Riis Park. Shonna attended the funeral and joined me in the limousine to the cemetery. Later, Shonna, Gladys, Barbara, and many other close friends all appeared at the Flatbush house, where my mother was sitting *shiva*.

My true farewell to Murray was a while after his death, when I tried a case he had pending in the federal court in Brooklyn on behalf of one of Murray's Dominican clients, Silvio Peña. The matter was before Judge Raymond Dearie, the former Eastern District U.S. attorney, Diane Giacalone's boss, before whom I'd also tried Peter Gotti's windows case. Both he and the federal prosecutor, an attractive woman named Dimitri Jones, treated me warmly and respectfully throughout the trial. In a fit of self-consciousness I recall telling the jurors that I was trying the case for my deceased father to avoid their assumption that the case was one of organized crime, given all the publicity surrounding the disqualification, the pending contempt case, and the grand jury investigation. Peña was acquitted of the substantive cocaine-trafficking counts against him but was convicted only of conspiracy, reducing his potential sentence. This, my parting gift to my father, was of the sort he would most have appreciated: a job well done.

I remember jokingly saying to Judge Dearie, "You know, judge, I'm a new man."

"I hope not," he replied.

21

SHORTLY before my father passed away, my trial on contempt charges began. The special prosecutor appointed by Judge Glasser was John Gallagher, his former law clerk, a former Eastern District assistant U.S. attorney, and a partner in the civil law firm of Corbin Silverman and Sanseverino. Gallagher's firm received roughly $300,000 from the government to prosecute me for speaking to the press in defense of John Gotti. Judge Glasser recused himself from hearing the case, as did Judge Nickerson, presumably on the grounds that they would not be (or would not be viewed as) impartial in their treatment of me. My attorneys moved to have the case removed from the Eastern District altogether and heard instead in the Southern District, across the river in Manhattan, but this would have resulted in *too* impartial a trial no doubt, so the motion was denied. The case was referred to Eastern District Chief Judge Thomas C. Platt, perhaps the most conservative judge in the district, whose courtroom was in Uniondale, Long Island, at least a full hour's drive from my apartment in New York. What did my "contemptuous" behavior have to do with Uniondale? Judge Platt was clearly a judge who believed before the trial began that I'd done wrong, and he was going to see that I was punished for it. John was paying for his sins, and now the day of reckoning for mine had come—most particularly for those sins I perpetrated in Judge Nickerson's courtroom in the Giacalone case. In fairness, I'd made it easy for them.

In the words of Second Circuit Court of Appeals Judge John McLaughlin, in the published opinion of that court:

The underworld exploits of John Gotti and the courtroom leg-
erdemain of his attorney, Bruce Cutler, are now the stuff of legend.
Cutler's last appearance on Gotti's behalf was in the United States
District Court for the Eastern District of New York (I. Leo Glasser,
Judge). Notwithstanding the court's pre-trial admonition and orders
to comply with Local Criminal Rule 7 of the Southern and Eastern
Districts of New York ("Local Rule 7"), Cutler spoke repeatedly and
heatedly to the media on the merits of the government's case against
his client.

Judge Glasser couldn't have made himself more clear, in his orders of
December 20, 1990, and January 9 and July 22, 1991, specifically prohibiting any deviation from Local Rule 7, which essentially restrained
attorneys from talking to the press about a pending criminal matter in
order to avoid affecting the opinions of prospective jurors. Of course,
before we'd ever appeared before the court in the case, the Eastern
District's U.S. attorney, Andrew Maloney, in a press conference convened
after John's arrest, had called John a "murderer, not a folk hero," and
boasted that "this time" the government's case, which included extensive
wiretap evidence, was stronger than the previous cases and would stick.
These comments were bolstered by frequent periodic leaks to the media
as to the nature of the evidence against John and the many "murders"
he'd committed. Poisoning the well of potential jurors in this way has
always been a common tool of the government. Certain reporters who
were patent shills for the government would get an exclusive in return
for printing the government slant on a case or a defendant; they'd "take
the king's shilling and do the king's bidding." As far as I am aware, no
prosecutor in history, let alone Mr. Maloney or any one of his subordi-
nates, was ever sanctioned or prosecuted for talking to the media in
derogation of Local Rule 7 or any gag order of a court.

Neither was any defense attorney, except for the attorney in *Gentile*,
where I understand the case was prosecuted without any severe result for
the attorney in question, for the express purpose of testing the constitutional limits of gag orders in view of our First Amendment rights of free
speech. How could John rebut the impressions on prospective jurors cre-

ated by the steady stream of government leaks both before and after his arrest, and before the court's admonition to refrain from speaking to the media? I would simply have to espouse John's point of view. Did this mean I was exposing myself to the court's wrath? No doubt, but I've never shrunk from taking that sort of risk on a client's behalf. And there was virtually no precedent that would lead me to believe that there was any great risk. I was only leveling the playing field. What I didn't understand at the time was that, where John was concerned, the government could no longer tolerate a level playing field. Such defense luminaries as William Kunstler, Herald Price Fahringer, and Ron Fischetti have often spoken to the press in the interest of the First Amendment and their particular clients. They were not held in contempt or prosecuted. But they weren't representing John Gotti, and their clients hadn't personally affronted or embarrassed any prosecutors. I was honestly naïve as to the existence of a different set of rules, or a different manner of enforcing the existing rules, that would apply to John and to those who sought to shield him from the government.

So between December 1990 and January 1992 I vigorously espoused John's point of view publicly, in New York's four major newspapers—the *Daily News, New York Newsday,* the *New York Post,* and the *New York Times;* in *Interview;* and on television, on *Prime Time Live, 60 Minutes, Thirteen Live,* and *9 Broadcast Plaza.* I went after the Eastern District of New York prosecutors, accusing them of bad faith and of manufacturing a case against John with the purchased assistance of self-avowed murderers; worse, I called them a "sick and demented" lot, "McCarthyites" who'd "orchestrated" a conspiracy to "get" John. I'd suggested to one reporter that it would be "John Gotti today—you [the public] tomorrow." In short, I'd told the truth. The government complained to Judge Glasser, whom at some point, angry over a ruling, I referred to as "His Majesty." He issued an order to show cause why I should not be held in contempt and appointed special prosecutor Gallagher to bring a case against me.

I was doomed from the outset, not just by the selection of Judge Platt to try the case but also by the very nature of the charge, which was a misdemeanor, carrying a maximum sentence of six months' imprisonment, which under New York law accorded me *no right to a trial by jury.* A jury would never have convicted me in the circumstances, but I was to be afforded no such escape.

Special prosecutor Gallagher tried a meticulously prepared case, putting into evidence all my statements to the media. The coup de grâce was his submission of my lecture at a Brooklyn Law School symposium on April 25, 1991, where I stated:

> I've really grown to appreciate and respect Anthony DeStefano from *New York Newsday*[,] Pete Bowles for *New York Newsday,* Lenny Buder for *The New York Times,* and Arnie Lubasch from *The New York Times* and some of the other reporters who I think do a conscientious job. Do I have selfish reasons? I have honest reasons that I don't want to alienate them, that I want the prospective veniremen out there to feel that I mean what I say and say what I mean, and if that can spill over and help my client, then I feel it's important for me to do that.

This self-evident bit of truth, obvious to lawyer and layman alike, would be held by the court to be a smoking gun, revealing my willfulness in violating the court's quasi–gag order and Local Rule 7.

My lawyer and good friend Sam Dawson had died by then, so I retained Fred Hafetz, a well-respected "white-shoe" criminal attorney (with whom I'd worked on the windows case), to represent me, with Robert Katzberg and the noted First Amendment lawyer Herald Price Fahringer acting as "of counsel," and they did a great job. Hafetz called prominent members of the New York criminal bar (such as Jimmy LaRossa and Jack Littman) to testify on my behalf that speaking to the media was the only means a defense attorney had of counterbalancing the government leaks to the press. He argued that although I'd violated Judge Glasser's unequivocal order, the orders and Local Rule 7 were unconstitutionally violative of First Amendment free speech rights. All these defense arguments were brushed aside, largely on the technical basis that, if I had had a problem with Judge Glasser's orders or the Local Rule, I should have appealed them (or sought some other extraordinary legal relief) when they were issued, rather than violating them first and attacking them afterward (the legal equivalent of shooting first and asking questions later). In the first place, the case law was far from clear that I had any right to any interlocutory appeal (an appeal before the Gotti trial was over and judgment entered), and moreover, as I said before, there was no precedent to warn me that I'd be treated differently from any other attorney who'd ever before violated a quasi–gag order.

Judge Platt convicted me, and on June 7, 1994, two days after my father died, I was sentenced. Judge Platt asked if I wanted an adjournment in view of my father's recent death. I said no thank you, and the court proceeded to sentencing. The sentence imposed was pure Platt—three years' probation, coupled with three conditions: (1) a ninety-day period of house arrest; (2) a 180-day concurrent period of suspension from practicing within the Eastern District of New York; and (3) six hundred hours of "non legal" community service. I was also ordered to pay a $5,000 fine. Later, recognizing that the $5,000 fine was illegal, the court vacated it and ordered me instead to pay "only" the cost of my "supervision" during the probationary period (amounting to about $7,500). Judge Platt's voice pronounced my sentence so quietly, I had to strain to hear. There are those who are more demonstrative and vocal, like John Gotti, like me. Then there are those who speak their softest when they are doing their worst—Judge Glasser, (now) Judge Gleeson . . . Judge Platt.

I recall sitting with Shonna (with whom I'd newly been reunited at my father's funeral) over breakfast at a diner near her apartment, perusing the newspapers—my sentencing was on the front page of each of New York's four dailies—trying to fully appreciate what had just happened to me. I know my attitude must seem ingenuous (or *dis*ingenuous) to the outside observer—I must (or *should*) have seen where my vehement representation of John could lead. But I didn't. I'd been doing what I was trained to do—represent a client as forcefully and effectively as I knew how, no matter how unpopular he might be with federal law enforcement and without thinking of the dangers to my own career. I'd done what I was supposed to, what so many great lawyers had done before me, and I'd done it successfully (at least until my disqualification). So how had I wound up convicted of such a charge? As I sat there in the diner with Shonna reading about myself, it was almost as if I were reading about another person, another lawyer.

Fred Hafetz was extremely eager to represent me on the appeal to the Second Circuit Court of Appeals, and I agreed to allow him to do so. But when the Second Circuit affirmed Judge Platt's decision and sentence, Fred lost his zeal for the fight and rather unceremoniously advised me that he didn't feel the likelihood of success warranted the investment of more of his time in seeking redress at the U.S. Supreme Court level (certiorari). I thought of the many clients who'd heard something similar from their attorneys, and I knew they'd felt as crestfallen as I had at the words. We filed no certiorari petition with the Supreme Court, and I simply faced my sentence. It is worth noting that the level of proof required to bring a case against an attorney for a violation of Local Rule 7 was later changed to counteract the decision in my case.

Judge Platt surely recognized that, had he given me a straight six-month term of imprisonment, he'd have martyred me. I don't want to say I'd have been happy spending time in jail, or even that I'd have preferred to do so, but Judge Platt's sentence was more onerous in many respects. My suspension from practice in the Eastern District and elsewhere was intermittent, drawn out over a two-year period, in view of the disciplinary proceedings in New York State (in which George Santangelo and Ed Panzer represented me pro bono), after which New York tagged on its own suspension period.

Meanwhile, the government's fever to avenge every aspect of John Gotti's Eastern District victory before Judge Nickerson still raged as strongly as ever. As its grand jury continued to investigate Gerry Shargel and me, the same prosecution team, led by John Gleeson, commenced yet another assault on the 1987 verdict. Once more, based on Gravano's testimony, the government alleged that one Bosko Radonjich, a native of the country formerly known as Yugoslavia and (oddly) the boss of the Irish Westies, had had some preexisting relationship with a juror on John's case named George Pape; it was alleged that Pape was paid some $60,000 by Radonjich for his vote. The juror, represented by Barbara Hartung, a highly competent attorney operating in the government cauldron of retribution, was convicted without any proof that he had ever received any monies. Radonjich was long gone, back to his native country. I don't believe that the 1987 case was fixed. I don't believe that John or any of the defendants believed the case was fixed. I was there.

That summer, in the shadow of the pending appeal of my contempt conviction, Shonna and I traveled to California, where the O. J. Simpson affair was in full swing. It was a media circus, in which, for a change, I was not gamboling about in the center ring. After Mafia madness had dominated the American legal stage for so long, a celebrity murder had taken the spotlight. Network stations across the country televised a caravan of police vehicles chasing a famous athlete and movie star's Jeep down California's coastal highway. Simpson was charged with the murder of his ex-wife Nicole Brown and her friend Ron Goldman, and seemingly ironclad physical and DNA evidence tied O.J. to the crime. Early on overtures had been made to my office as to whether I'd join the Simpson defense team, but I respectfully declined as Simpson was already represented by a supercompetent lawyer whom I knew and respected, Howard Weitzman (who'd twice represented John DeLorean in racketeering cases).

Shonna and I drove leisurely down Highway One, from San Francisco through Carmel and the extraordinarily beautiful Big Sur, to Los Angeles. It was an idyllic trip for us, almost convincing us that we might be able to make our relationship work. But by October 15, 1994, when I took Shonna with me to Columbus, Ohio, to receive an award and for a speaking engagement hosted by my close friend and colleague John Rion, we were again on bumpy ground. My rage at the passing of my father, and at John's conviction and my own, had made me self-destructive emotionally and simply impossible to live with. We broke up again and would stay apart for four years.

22

BEFORE too long I found myself once again in Judge Nickerson's courtroom in the Eastern District—my first return to the scene of John Gotti's victory over Diane Giacalone—in the William Cutolo case. I represented Frank Ianacci in that matter. Frank was one of my first criminal defense clients from my early days with Barry Slotnick; his father had been a close friend of Joe Colombo, who was assassinated in 1971. In this case, my role, and those of my cocounsel George Santangelo, Thomas Ashley, and others, were subordinate to lead counsel Jimmy LaRossa. Jimmy's client was charged with multiple murders, and our clients with conspiracy to commit various acts of mob warfare, so it was agreed by all that Jimmy would do most of the heavy lifting.

But Judge Nickerson's animosity toward me was palpable. I'd heard it said, and found it in my experience to be the case, that he was not the same judge he had been before John's 1987 acquittal; he no longer displayed any jocularity or patience whatever, particularly if the case before him had even a patina of organized crime involvement. He'd been embittered by the Gotti case, and unlike Judge Dearie and many other Eastern District judges, he treated me with undisguised hostility after 1987. In the Cutolo trial he upbraided me within hearing of the jury panel, for talking to my client during jury selection. I refrained from responding in kind, to avoid lending an acrimonious tone to what would be lengthy proceedings, but his ill treatment of me continued through the opening statements, the few cross-examinations I performed, and my summation. Some federal judges (such as, in New York, Judge Leisure, Judge Sweet, Judge Mishler, Judge Thomas P. Griesa, Judge Harold Baer,

Judge Arthur D. Spatt [at whose invitation I appeared to speak at the Inns of Court], and elsewhere, Judges John Corbett O'Meara of Detroit and Ann Williams of Chicago, Judge Herbert J. Hutton of Philadelphia, Judge Marion T. Barry of Newark, Chief Judge Walter H. Rice of Dayton, and others) allow themselves to enjoy watching an attorney at work in the theater of the courtroom, the show, the action. Others are boiling inside and uncomfortable if the attorney's flamboyance is unleashed and his efficacy is high. I think the latter judges are fearful that they might appear to have lost control of the courtroom. Judges who are in control don't fear appearances—they can afford to appreciate the joyous notes struck by the attorneys before them. Anyway, the Cutolo clients were acquitted in the end on December 20, 1994.

Shortly thereafter, in March 1995, I tried the Morgue Boys case. Ironically, it was heard in federal court before Judge Glasser in the Eastern District—the same judge who'd tried John Gotti's last case, disqualified me from representing John, and had me charged with contempt. As John had predicted when I mentioned I was before Judge Glasser again, he bent over backward to treat me courteously and fairly throughout the trial. The government's allegations were that a band of New York City police officers were accosting local drug dealers, assaulting them, stealing their money and drugs, making false arrests, and splitting the proceeds of their illicit activities at an abandoned morgue in a downtrodden neighborhood near the Seventy-third Precinct. The arrest of the officers had caused a media furor the year before, and at that time the mother of one of the defendants, Rick San Filipo, wrote me, and I agreed to take the case.

Representing a man in blue echoed simpler, less controversial times in my life and in my legal career: my father's days as a policeman, my assistant district attorney days riding homicide, and my many cases on behalf of policemen when I began my career as a defense attorney. I'd probably represented more policemen in my career than street fellows, and it usually felt unambiguous to me—no grayness, just black or white, and the men in blue were generally on the side of right. In the course of my career some of the most rewarding cases I've handled were in the defense of New York City policemen charged with misconduct or criminal activity (shakedowns, assaults, and the like). My father, the former policeman, viewed it as a debt to do his best for a copper in trouble—a debt that, I guess, I inherited. Some of the men in blue I've represented are among the finest people I've ever met.

Even when I was representing John Gotti, neither he nor I ever had any animosity toward the local police. We took on the alphabetized *federales* (FBI, CIA, ATF, IRS, INS), the robotic, cool, detached suits; we never had a problem with the local policeman doing his job. I can't say John was ever in love with the police, even the local man on the beat, but he understood their necessarily hands-on, day-in, day-out local community involvement. And when I took on the Morgue Boys case, John gibed, "Maybe they'll root for 'Papa Schultz' now, too!" ("Papa Schultz" and "Johnny O'Fritz" were two jailhouse nicknames of uncertain origin picked up by John in his younger days in the state penitentiary. He detested nicknames, because government informants would invariably have colorful ones, so I never heard anyone use the nicknames to refer to him in his presence, but he would sometimes use them to refer to himself—"They're gonna fuck with Johnny O'Fritz? Let's see!" People would call him John or Johnny, Chief, or Grandpa—because by the age of forty-two he'd had his first grandchild—and some others, like his brother Peter, might call him Junior, as he'd been named after his father, John.) I think the cops *did* root for John, for the most part. They respected his toughness. Most ordinary street policemen have nothing against knock-around tough guys who don't bother the local police, and the local police don't bother the street fellows except when they're caught doing something illegal. There's a certain grudging respect for those who carry themselves a certain way, and the local cops had that for John Gotti. Even Detective Kenneth McCabe, who testified against John as an organized crime expert with the Kings County District Attorney's Office Squad (called by the federal government to testify that John Gotti had become the new boss of the alleged Gambino crime family, during the 1986 hearings to jail him pending his trial before Judge Nickerson), testified that John had always garnered this kind of respect even prior to his alleged ascendancy.

But if I had any expectation that the Morgue Boys case (which incidentally arose in the Brownsville–East New York area, in which John Gotti grew up) would be some unambiguous oasis in the vast desert of moral complexity that had become my practice, I would be sorely disappointed. Two attorneys with whom I was good friends were representing Keith Goodman and Frank Mistrella, other policemen-defendants in the case: Steven Worth, who had been in the Homicide Bureau with me in the district attorney's office, and Edward Jenks of Mineola, both excellent trial men. We shared the load in the case, although as I represented the lead defendant, I led the cross-examination attack on the police informant government witnesses (who were themselves police officers). The prosecutors (Chuck Gerber and John Caruso) relied principally on the testimony of these police informants, who admitted to having taken part in the crimes. The case also involved electronic eavesdropping testimony; these tapes turned out to be damaging and opened my eyes to just how rotten some who wore the uniform could be. The crudity and meanness of the police informants (who were awarded sweetheart deals by the prosecutors in exchange for their testimony), as reflected by the prosecution's own tapes, were utterly repugnant. I've always believed there is nothing worse than a bully in uniform—an imposter, a phony tough guy, hiding behind the privileges of the shield. The thug–police informants admitted to crimes of great brutality, evincing an absolute disdain for the badge and uniform they wore. Ironically, one of these police informants (who admitted on cross-examination to having his wife steal things and file false insurance claims for him) was a distant cousin of Salvatore Gravano.

Another police informant testified on direct examination that he'd been a drug dealer prior to becoming a police officer, when he was employed by Amtrak. I asked the witness, "You were an Amtrak employee, a top union job, making a good salary, feeding your family. Why would you sell drugs?" I figured, what could he say to hurt me?

"I don't know," he answered.

"Didn't you have any moral compunctions?"

"What's that mean?"

"Weren't you concerned that you were doing something wrong, evil?"

"Yes."

"Then why did you do it?"

"For the money."

"Money for what? How much did you make? What did you do with the money?"

"To buy things."

The witness left Amtrak, became a cop, and continued selling drugs. The badge and uniform were sold so cheaply—not out of some perceived necessity but out of an utter amorality. I believe the jurors' disgust with the government's witnesses carried the day for the defense, as so often happens.

I was soured by the case, particularly by the police informants. I was personally offended by what I heard on the tapes, on which the government witnesses were mostly speaking. I'd largely been away from representing policemen for most of the Gotti era, and the conversations I heard bespoke a new, more intolerant, and nastier state of mind in the squad car. There were racial overtones to the allegations, in that the victims (although drug dealers are not generally sympathetic characters themselves) were largely black. They were highly offensive and just plain hard to listen to. I think back to my father and the detectives with whom I'd worked in the district attorney's office—they were not scholarly types, but they had a nice way about them, and they behaved like gentlemen. I couldn't imagine them speaking this way about other human beings, no matter what color or religion, no matter how poor. The racial divide between those who controlled the government and the most economically deprived segment of the population had widened dramatically during my career. The Morgue Boys affair was not a racial case per se, but it was to me a precursor to the horrific Abner Louima case (where a black man was unspeakably tortured, or allowed to be tortured, in the precinct house by a cowardly lowlife in a blue uniform), which would arise a few years later. In light of what I'd heard in the Morgue Boys case, the facts of the Louima matter appalled but did not shock me. Obviously this behavior isn't commonplace with policemen—I'd represented many in departmental hearings at around the same time who were totally honest and forthright, harboring no racial malice. But the Morgue Boys case, where the oppressors were relatively privileged and wore the colors of authority, and where the victims were society's disenfranchised, was highly demoralizing to me. I am not referring here to our clients in the case, whom I liked—we did a terrific job for them, and the trial resulted in a hung jury and an acquittal.

On December 7, 1995, I began the house arrest portion of my sentence for the crime of lauding John to the media. This would last until March 8, 1996. During that time I was permitted to go to my office during the business day (although I was prohibited from practicing law) but required to be home by a specific time. On the weekends I was allowed two hours of liberty on each of Saturday and Sunday (increased to three and four hours in my second and third months of house arrest, respectively), because I was single and needed to shop for groceries and other necessities. I wore a monitoring bracelet twenty-four hours a day, and I had to pay for the privilege, forwarding monthly money orders to a V. I. Monitoring Corp. somewhere in the Midwest.

During my house arrest I recall receiving a funny greeting card from John Gotti in Marion—I don't know where he managed to obtain it. A man on the front of the card was relaxing in a chaise longue, with sunglasses and a cigar, reading a newspaper, a cocktail complete with tiny umbrella in hand. Inside the card John had written, "Enjoy house arrest." When I first read the card, I was glad to hear from John, and it made me laugh. Then wonder and admiration crept in, that he would even think of me and my paltry sentence as he sat for the rest of his life in a hellhole like Marion.

When the house arrest was completed, there still remained the endless probation—three years of monthly reporting to a probation officer, frequent urinalysis, filing financial reports—and of course, the six hundred hours of community service. I enjoyed performing the community service portion of my sentence. I made great friends providing assistance to the servicemen and their relatives who would visit the Douglas MacArthur USO Center at the Soldier's, Seamen's, Airmen's Club on Lexington Avenue, and I will always remember its kindly British executive manager, Hazel Cathers, who lives to help others. She became a good friend, and I grew to love her dearly. Yet in what way could my "contemptuous" behavior be construed to have taken something from the "community," to have warranted my "services"? How did that portion of my sentence make any sense? It was Judge Platt's way of humbling me. No doubt he was projecting how he might have felt had *he* received such a sentence. Community service doesn't humble me—I'm at home helping others, especially those in the armed forces.

My mother had grown thinner and had not been her robust self for several years prior to my father's death, but she seemed to embrace the challenge of life without him. She decided to sell their Flatbush home and move into a lovely apartment with a terrace in the Georgetown Building on Ocean Avenue and Avenue S in Brooklyn. My mother had always loved the idea of apartment living, but it was anathema to my father. Now she'd be able to indulge herself. I was trying a three-month racketeering case in Newark federal court, but my sister came in from Syracuse to help my mother with the move, and a client of mine who had a demolition company sent one of his trucks to assist as well.

Within a day of moving into her new apartment, my mother fell ill—she was unable to walk or drive her car. Tests conducted at New York University Hospital seemed to reveal that she had a cancer of some sort, although it was never clear to me what kind or where it was located. Her physicians said it was colon cancer, but she had none of my father's symptoms and nobody was suggesting surgery. She wasn't in the hospital for any length of time, and for a three-month period I'd visit her every Sunday in her apartment (after five days a week on trial in Newark and every Saturday at the gym). Over that time she grew more and more ill, until she passed away on August 9, 1995, fourteen months after my father. Her death was even harder for me to accept than my father's, compounding the impact of his passing. I also think I was closer to her somehow. She was my protector—from the outside world and even from my father's tough, quasi-military sense of order. I had respect and love for my father, but I had an affinity for my mother, sharing her mercurial, sometimes saturnine nature. Shonna came to my mother's funeral, but unlike our reunion at my father's passing, our relationship was not rekindled this time, and I had to face the aftermath of my mother's death alone.

As when my father died, my brother, Richard, was at my mother's bedside. Phyllis had already returned to Syracuse. Neither my brother nor my sister had had as close a relationship with our mother as I did. Unlike me, they didn't share her temperamental personality. They are brilliant, cooler people—more emotionally detached, like our father. In fairness to Richard, he had it tougher than I growing up, for my mother was already thirty-six years old when he was born; my grandparents were no longer of an age to help her rear a child, so she relied instead on the assistance of Phyllis and a nurse. Both Phyllis and Richard were more my father's children than hers. But she was always there for me, her firstborn son.

All in all, compared with the sentences meted out to my clients, Judge Platt's sentence was not all that harsh. In some ways it was a welcome relief at the absolute nadir of my life. I'd lost my father, my mother, Shonna, and much of my law practice. I needed the time to exercise and meditate, to reflect on where I'd come since my days as a newly married young law student and assistant district attorney. I was still under sporadic suspensions in various jurisdictions, stemming from my contempt conviction, and I was still on probation. Moreover, the government's grand jury investigation into me was still open.

Gerry Shargel and I expected some sort of indictment to be manufactured by April 15, 1995. An article in the *Daily News* had called it "imminent." We knew we'd done nothing wrong, but we also knew that the government could get a grand jury to indict anyone, and our very indictment (even assuming ultimate acquittal) on tax evasion or even racketeering charges would be devastating to Gerry's career and to what was left of mine. We met at J.G. Melon's for cheeseburgers one day to discuss how we would deal with the indictment. But it gladly came to naught. Just after my mother's death, well before April 1995, the government announced that, doubtless to its great disappointment, it was closing its grand jury investigation without indictment. There is no question that, had the government been able even remotely to prosecute us—me in particular—it would have done so. But I was (and remain) scrupulously honest in my financial dealings, as in all other aspects of my life.

The government's announcement was the beginning of my rebound. I still had interim suspensions pending here and there and a period of probation remaining, but I was trying to put my parents' death behind me as best I could, and in June 1997 I walked into the stately Second Department Appellate Division building on Monroe Place in Brooklyn, where I'd first been admitted to the bar of the State of New York in 1975 and where I'd married my second wife, Barbara. I received a warm reception there from the clerical and administrative staff, and I obtained five copies of my newly unblemished admission certificate at five dollars apiece.

23

My conviction more than woke me up as to what can happen when a criminal attorney is too confrontational in a client's behalf. I don't wonder at my feelings for John. He'd attained a direct link between himself and his actions—every minute of every day of his life, at any given moment, he was exactly what he wanted to be, *in* that moment. Most people live at the other end of the spectrum; they spend almost no time at all being who they want to be. A friend of John's once said, "John would rather spend a day as a lion than a lifetime as a lamb." I'd spent a lifetime *approximating* John's presence, living a culturally permitted facsimile thereof, donning sports uniforms, standing on sheaves of paper, extending my reach with this diploma, that certificate. Is it really so hard to understand *my* appreciation for the true lion striding through the jungle?

I suppose society can't tolerate a lion in the parlor, or even walking down Broadway. But it's easy to admire the magnificent and so efficient piece of existential machinery that *is* the lion—the real thing, not the facsimile, the diluted, vague, societally acceptable version. John *is* the real thing, the one whom an actor or director would seek to imitate. Today's so-called gangster might look to those very actors and directors, might watch *Mean Streets, GoodFellas,* or the *Godfather* trilogy to learn how to speak or act—third-generation reproductions wearing *Sopranos* T-shirts, consciously embracing their status as reflections of reflections. There was no *Godfather* when John first met Neil Dellacroce. Sure, there were *Scarface, Public Enemy,* and *White Heat,* but even if John had seen them, they were anachronistic by the 1950s, and John didn't need a movie to show him how to act. In a world that's learned to settle for the Madison Avenue or Hollywood-created counterfeit of the original, John was decidedly the thing itself. The government recognized him as the real deal, a true leader whom they had to decapitate. This is a point on which even the government and I could agree.

I'd insulated myself from the many deaths that had occurred in John's life while I'd been representing him. They were somehow part of my professional life, as though they'd transpired in a novel I was reading or in sheaves of transcripts I was wading through. They were remote from the life of that boy who grew up in a nice home with Murray and Selma in Flatbush, and attended Poly Prep and Hamilton, and wrestled and played football, and married his high-school sweetheart fresh out of college. Those deaths were part of a world that I didn't understand and couldn't try to understand better, perhaps for fear that I'd be impeded in doing my job.

In the end, despite my press clippings, I'm not John or even his semblance, and I couldn't be. His world is not mine, and I will never understand that world any more than I will ever truly understand the predatory brutality of survival on the Serengeti Plain. Perhaps I've been a lion of the courtroom, roaring to get my point across. But the feature that most distinguished John from the rest of us was his unyielding, immutable, unswerving presence—as he was then, so he is now, and so shall he ever be. That which does not change, does not adapt, is destined to perish, at least physically. John never changed who he was, notwithstanding incarceration, isolation, insulation, disease, physical incapacitation, and whatever else his enemies on the street, the government, or life threw at him. I, on the other hand, have tried to adapt in order to survive as a lawyer.

A woman's quirky little mannerisms at twenty are not as cute or sexy at fifty—they become idiosyncrasies, overwhelming her personality. A fastball pitcher at twenty-two won't be able to survive in the big leagues at thirty-seven, relying solely on a heater—the fastball is still there when he needs it, but he must develop other tools, a curve, a slider, a knuckleball. Nor could I have the same courtroom persona at fifty-three as I did at thirty-seven. That which is effective or attractive in youth must mutate for the sake of seemliness, if only to avoid becoming a parody of itself. I recall reporters saying that I had a "swashbuckling way" and a "grandiloquent physicality" in court. In describing my friend Harry Bachelder (a former top military officer and U.S. attorney), who was on trial representing a notorious drug dealer in federal court, the *New York Post*'s Mike McAlary (now deceased) wrote, "Next to the defendant, he was the second most formidable looking defense lawyer in New York City, the first being Bruce Cutler."

At fifty-three my physicality is somewhat tempered. I can still wield a broadsword when necessary, but I'm also willing to try diplomacy, to use a kind word, a tack that I now find often beneficial to the client. Not every battle requires a scorched-earth policy; not every prosecution is a vendetta. I avoid the ad hominem attacks, unless provoked. I try to reserve my fight for the courtroom, on the record, as opposed to the courthouse steps. What worked for John Gotti, for me, in the 1980s and 1990s may not necessarily be the most advantageous for me and my new client base. As I grow older, the naïveté of my younger trials is gone, and I muster indignation and rage against the system only when it is needed. I have a sense that life and its injustices will go on after the trial is over, no matter what the outcome. With maturity things have become less cut and dried for me, less black and white. Gradations of gray are more prevalent.

In letters to Judge Platt in connection with my sentencing on the contempt conviction, Harry Bachelder thanked me for "teaching [him] courage in the courtroom," and former federal Judge George Leighton of Chicago described me as a "warrior," which I still am. But in order to survive professionally, my weapons have had to be diversified. It was a matter of survival. I'd survived the disqualification in John's last case; I'd survived the contempt conviction and the house arrest and suspensions; and I'd survived the loss of both parents in the space of a year. After their deaths I strongly considered leaving the practice of law to teach law or rhetoric at the college level, or at Poly Prep, but I realized that I'd have soon been bored by such a lifestyle. The court is my battlefield, my home.

I'd never learned to establish continuity in my New York business through reciprocal relationships with other attorneys, and certainly not through maintaining warm relationships with the government, which can refer enormous amounts of business to the lawyer who knows how to play ball. I'd never learned how to play ball. So when times were tough, I had to learn to play alone. And mostly I played out of town.

I was being at least consulted on nearly every big criminal case in the country, and I was also receiving numerous requests to appear for speaking engagements, many of which I accepted. Whenever I needed a financial or emotional booster, an interesting or remarkable client would appear out of the blue, requesting my services abroad, keeping me afloat. It was on the road that I resurrected myself and rediscovered my sense of rhythm in the courtroom, my self-confidence.

In June 1997, with the ink still wet on my freshly laundered New York State admissions certificate, I proffered it to Chief Judge William G. Cambridge, sitting in the U.S. district court in the Omaha District of Nebraska, to permit me to represent Harold Maradon, an expatriate living with his wife in Costa Rica. Maradon had been indicted in a fraud case that encompassed the whole country but centered in Nebraska. I stayed at a huge, sumptuous, and empty Omaha hotel, reminiscent of the one in Stanley Kubrick's *The Shining*, complete with eerie staff haunting its cavernous hallways. In the end Maradon decided that the stakes were too high, given the venue, the judge, and the prospective jurors, and that he'd best enter into a plea agreement with the government. To this end Bettina Schein and I negotiated a solid agreement with the prosecutor, whereby the client would spend seven and a half months in a federal medical facility and pay $1,050,000 in restitution to the victims and an additional $50,000 fine to the government. Maradon agreed to return to the United States from Costa Rica and take his medicine, and the plea agreement was executed by every U.S. attorney in the country to make sure that all possible charges were covered. Harold was happy, and the government was happy.

The only fly in the ointment was Judge Cambridge, who had a cold vindictiveness. The court had the option to give Maradon probation rather than jail time, especially in light of the enormous restitution involved. But Judge Cambridge was unhappy even with the agreed-upon incarceration period, believing, I imagine, that hanging would be too good for Maradon. On sentencing Harold was remanded right then and there, a highly unusual procedure on a first offense in a white-collar criminal case; it is more common to permit the defendant to self-surrender to a particular facility on a mutually agreed date. The court's action was especially harsh given that Maradon had voluntarily submitted to its jurisdiction in the first place, having returned under no duress from Costa Rica, a country with which the United States has no extradition treaty. Judge Cambridge stormed off the bench even as I was requesting reconsideration on the immediate remand, and he seized Maradon's bail and instantly applied it to the levied fine. My experience in his courtroom stands out in my memory as my most unpleasant out of New York, and I felt as though the judge's animosity was directed in part against me personally.

Another call came from Michel Guerin, who was interned at the federal prison in Otisville, New York. I traveled with my close friend and cocounsel George Santangelo to Quebec to meet with friends of Guerin's, fellow members of the Canadian chapter of the Hell's Angels Motorcycle Club. As I was still on probation then for my contempt conviction, I was obliged to notify my probation officer prior to leaving the country. He advised me that I would have to apply and pay for a special permit of free passage from the Canadian consul, as I was a "convicted criminal." So I obtained my special permit.

The Canadian Hell's Angels' image is totally different from the beer-guzzling, biking image of the American clubs. I never saw a motorcycle throughout my stay in Quebec (at the Loewe's Le Vogue in Montreal). The fellows I met there were always impeccably dressed in European fashion, well groomed, clean shaven, and French speaking, of course. These were serious men, exceedingly courteous to me, and the U.S. government viewed them as highly sophisticated and dangerous criminals. I understood that the HA members attended conventions all over the world except in the United States, where they would so often be arrested. The HA credo was "no burn," which is to say, their word was their bond— if they said they'd do something, be somewhere, hand over something, they would. I took the case partly because I identified with the rebellious, disciplined, antiestablishment attitude of the HA, and partly because I liked Michel.

The prosecution alleged that Guerin and two others had sold drugs in Plattsburgh, New York, to two Haitians engaged in a sting operation for the government. Two men were arrested on the spot, and it was claimed that the third perpetrator, who leaped through the motel window and escaped, was Guerin. He was a fugitive for a year and a half and subsequently spent four years in a Canadian prison, battling extradition to the United States to face the charges pending in a Syracuse court. He was then extradited and imprisoned at Otisville to await trial. Eventually he decided to plead guilty and take a ten-year sentence, with credit for the time he'd already served in Canada.

An added benefit of representing Guerin was that I had an opportunity to return to French Canada, where I'd traveled by train at the age of eleven with my father. We had stayed at the Château Frontenac in Quebec City and went ice skating together. It was a significant trip in my young life then, and returning to Quebec brought my father back to me, even if only for a few hours here and there.

At the end of 1997 I was in Detroit, the Southern District (Eastern Division) of Michigan, before Judge John Corbett O'Meara, one of nicest judges before whom I've ever had the pleasure of appearing, where I represented Tony Giacalone. Tony (whose name was spelled identically to that of the prosecutor in John Gotti's second trial but was pronounced differently), a high-level figure in the so-called Detroit family, was an old-time stand-up tough guy, much like John (although much older) in that he was concerned with how he carried himself, his choice of words, his home, and his family. He was always a pleasure to represent, particularly considerate and solicitous with me. The press liked to allege that he'd been involved in the notorious disappearance of labor leader Jimmy Hoffa, and in the film *Hoffa* the figure portrayed by Armand Assante was allegedly patterned after him. Of course, Tony was never charged with harming Hoffa, and he always asserted that he knew nothing about his disappearance.

He and the entire hierarchy in Detroit were indicted in 1996 for activities reaching back thirty years. In December 1997 I moved to the

Townsend Hotel in Birmingham (a beautiful Detroit suburb) to begin what was projected to be a three-month trial. After preparing for six weeks for trial against prosecutor Keith Corbett (an excellent attorney and ex–New Yorker with whom we got along well), my friend and cocounsel Dominic Sorise and I unexpectedly won a motion to sever Tony's trial from that of his codefendants. The basis for the decision was that Tony was too ill from a renal failure to stand trial at that time. Tony's family was obviously relieved that he wouldn't have to sit through grueling proceedings while undergoing dialysis three times a week. (The eighty-three-year-old Tony Giacalone recently passed away of his illness.) Judge O'Meara was enormously gracious, inviting me into chambers to welcome me, to initiate me into the history of the State of Michigan, and to admit me to the Eastern District of Michigan, not *pro hac vice* (for this case only) but as a regular member of the Michigan bar, permitting me to practice there whenever I wished.

On Christmas Eve 1997 a child named Michael Cutler—my son—was born. My close friend and football teammate from Poly Prep, Michael DiRaimondo, was named one of two godfathers to Michael (the other being his mother's brother-in-law).

My friend, the prominent New York attorney Edward Hayes, had always commended the comfort and spiritual contentment of a life of bourgeois domesticity. I'd met Ed in 1981 at the wake of the son of John McNally, a private investigator, former policeman, and friend of Barry Slotnick and Frank King; John McNally's son was tragically killed at eighteen, and I continue to contribute to this day to a scholarship fund established in his name at the Tottenville High School in Staten Island. Eddie and I became soul mates, and I was named godfather to one of his two children, his daughter Avery (the other being his son, John). Eddie, who'd been a Bronx homicide assistant district attorney when I was doing the same in Brooklyn, went on to become attorney to the stars, including Robert De Niro, Harvey Keitel, and the Andy Warhol estate; Tom Wolfe dedicated his novel *The Bonfire of the Vanities* to Ed, who played Virgil to Wolfe's Dante in his voyage through New York's sometimes infernal legal processes.

I'd never sought out fatherhood—my work was always paramount. Recognizing how much attention I'd required from my parents, perhaps I feared I'd be unable to give so much to anyone. But as I gazed down at my beautiful son for the first time, I saw Murray and Selma and maybe even a hint of all those who came before, and I saw myself and all those who might come after, and I was overcome—with love, with pride, with confusion, and with some sadness. It was as though Michael's birth had to happen as it did, if only to teach me the folly of a life in mad, ego-centric pursuit of career success.

The child was christened Michael, an M for Murray, and I want to be in his life as much as possible. I find myself thinking of him at the odd-est times, in the moments when my mind has disengaged from one task and has yet to engage another: as I await takeoff in the airport of some new city, or as I race down the FDR Drive to court in a car service, or as I walk back to my seat after a cross-examination, or in the hollow, cav-ernous silence, echoing the last syllable of a jury foreman's verdict, as I rise with my client—and I surrender to the joy of knowing Michael exists. I dream of living long enough to see him grow up straight and strong, bright, tough when he needs to be, yet sensitive. I hope to see him play football at Poly Prep and maybe even at an Ivy League college. I want to toss a baseball around with him and advise him on a career path. I want to see him happily married with children of his own—my grandchildren. In short, I wallow shamelessly in the most foolishly sentimental waters of fatherhood at times. I wish Murray and Selma were around to see it.

I told John about Michael on a visit to Marion after his birth. The cir-cumstance of having a child and not being married to his mother is not ideal. In that light, John cautioned me, wagging a finger, "Remember, Bruce, if you don't love your son, no one will." When I reassured him, he broke out in a big grin and said it was about time I had a son. Then he asked jokingly where his cigar was. I said I wished I could give him one. He'd had three sons and two daughters of his own, and he'd always told me I'd been missing out on one of the most important parts of life. He was right.

24

UPON my return from Detroit after the Tony Giacalone matter, in March 1998, a couple of months after my son was born, John asked me to help *his* "boy."

He'd been charged with racketeering in the construction industry, gambling, and extorting money from a club called Score's. The prosecution again moved to disqualify me. It was the government's view that young John was running the alleged Gambino family for his father, and the prosecutors sought to introduce the same old tapes from the Ravenite Club and some new ones from Marion, arguing once again that I was "house counsel," conflicted, and an unsworn witness for my client. This time I argued against the motion myself (with papers prepared by Bettina Schein's husband, Alan Futerfass, who appeared on the cover of the *New York Law Journal* in connection with the motion) before Judge Barrington D. Parker, a fair-minded, warm, and brilliant judge sitting in federal court in White Plains, New York. Judge Parker's father was a close friend of George Leighton, an attorney with whom I'd worked in Chicago and a former federal judge who'd written to Judge Platt on my behalf in connection with my contempt sentencing. Judge Parker denied the government motion, and I remained on young John's case as cocounsel to Gerry Shargel, who'd been retained as well.

I never believed young John was running anything for his father. In fact, when he was arrested, a police Mafia "expert," Joseph Coffey, spoke on television news, saying that young John was not even a "made man" because membership was open only to full-blooded Italians, and John Gotti's wife is half-Jewish. (In fact, she is half–Russian Orthodox and *not*

Jewish.) Leaving aside the issues about who can be made and who can't, taking money from a strip club would have been anathema for the John Gotti I knew and therefore for his son, too. I'm not trying to beatify John, but he is a traditionalist: Narcotics and pornography had nothing to do with his life.

After spending ten months in jail prior to trial, young John was granted bail and was able to post a ten-million-dollar fully secured bond, agreeing to house arrest and to wear an ankle bracelet. Unlike John, his son had the business opportunities to garner substantial assets from successful and legitimate businesses. These assets and his many friends and neighbors helped him to secure his bond. He wanted to spend time with his young children and to be free to prepare his defense in this, the trial of his life, a privilege of which his father was deprived in both of his federal trials.

Young John's case was never tried because he opted instead to plead guilty and accept a sentence of seventy-seven months (ten of which he'd already served) in prison. It was said in the media that John was opposed to his son's pleading guilty and that young John hesitated to take the offer for a long time because of John's disapproval. To be sure, John was not in favor of pleas, particularly where, as here, he believed his son innocent of the charges. And although young John highly valued his father's opinion, he wanted to minimize the risk of successive prosecutions that might result in life imprisonment. His decision was a reasonable one, given young John's age and perceived duty and his wish to be at home to help raise his four small children. His father's life, and the choices he believed necessary, were borne of the poverty and difficulty of his childhood environment. His son understandably didn't feel the need to sacrifice his life, as he saw his father had done. Young John was his own man, as evidenced by his taking the plea, a decision that was his alone to make. He pleaded guilty to racketeering in connection with a business fraud and gambling predicate act—*not* the Score's allegations.

In late 1998 or early 1999 I received a script and a request to interview for a role as an attorney in an upcoming John Herzfeld film called *15 Minutes*. It was to star Robert De Niro, Edward Burns, and Kelsey

Grammer. I showed up at the Plaza Athénée, auditioned, and was awarded the part. The timing was perfect for me, as I had two or three months off between pending trials when the film was to be shot. It was the first time I could remember having a job where there was nothing at stake. I could have fun being my characteristically expressive self, which everyone appeared to enjoy. And working with the talented John Herzfeld in the billion-dollar movie business reminded me of working with John Gotti in that anything was possible. If you wanted a B-52 sitting there in an hour on the lot—poof! it was there. Whatever it took to get everything just right would be provided, miraculously, effortlessly.

That feeling of team camaraderie that has always been so important for me, whether in scholastic sports or in the courtroom in a sprawling RICO prosecution, was there on the movie set during production in Los Angeles. Kelsey Grammer took me under his wing, helping me to acclimate myself, and I worked very closely with Karel Roden, a terrific Czech actor who plays my client in the film. I met Robert De Niro, who personally thanked me for working on the project. He has played so many characters in film of the sort I've represented in real life (Jimmy Conway in *GoodFellas,* young Vito Corleone in *Godfather II,* Johnny Boy Chavello in *Mean Streets,* Al Capone in *The Untouchables*) that I felt almost as though I were meeting a potential client. I put a lot of time and effort into the film and learned a lot, and then it just ended. It was over, and everyone scattered to the four winds, and I was left with a feeling of emptiness and disappointment. When I'd finish a big trial, another indictment and trial would await my immediate attention, but now there was no next film for me. Not yet.

There's no question that there's a close connection between acting and litigating, especially for me. My style is to adopt the role of my client before the jury. Most lawyers don't need to do this. I do. In order to get motivated, to enter the zone of indignation, sorrow, or elation I seek, I need to feel that I'm portraying someone other than who I really am. Call me a method lawyer. My courtroom style is a performance, acting. I may be a lot of things, but I'm not my client, any more than De Niro is Al Capone. I am, after all, a real lawyer defending a real client accused of real crimes.

15 Minutes (whose title is a reference to Andy Warhol's famously ironic comment on the amount of time in the public eye we can each expect to garner in our lives) explores the notion that in our society the publicity surrounding a person, idea, or event is more important than the idea, person, or event itself; the image created in the media is more real to the public than the original. This should have been no news to me, a veteran of the Barry Slotnick school of law and the struggle with the government on John Gotti's behalf. The script called for me to play a criminal lawyer named Arnold Getman. I didn't care for the name, so Herzfeld told me to pick any name I wished. When I asked why I couldn't use my own name, I was told that I could. In fact, using my own name fit in perfectly with the film's mingling of true New York media personalities with the script-created ones; news correspondent Peter Arnett and actress Roseanne Barr both played themselves in the movie. Using my own name also played into the film's conceptual need to blur the borders between an objective reality and that created by the televised (and presumably cinematic) media.

Unfortunately, I didn't fully appreciate the transformation that my character could undergo in the rewriting and editing of the film. In the final version my character wound up decidedly unscrupulous; my choice of names appeared misguided, in that it might bolster the already widely held public misperception that attorneys who represent alleged mobsters are only in it for the money and are inherently dishonest, or worse yet, it might imply that I myself am dishonest. I tried to tell myself that people would understand I was an actor playing the scripted role of someone who happened to have my name in the movie and that I was not playing *me*. But this film was purposefully inhabiting the gray landscape between reality and fiction, and I wandered, albeit naïvely, into that territory. So when the dust settled on the film in the can and on the celluloid strips on the cutting-room floor, I was less than pleased.

It's easy to see why the filmmakers loved me playing "Bruce Cutler," underscoring the film's central theme. But why had *I* been susceptible to such a blunder? Was this a confession of sorts—working out my own ambivalent feelings about my profession, about the difficulty in reconciling the role I play in an undeniably violent world? I don't think so, because the principal manner in which I insulate myself from the world that I often find myself vigorously defending is scrupulous personal honesty. But for that inner compass, as Murray had taught me when I was young, surely I would lose my bearings. Or did I lunge at an opportunity to hear my own name shouted out loud on the silver screen? In this respect did I perhaps mirror Bruce Cutler of the courts, accepting the representation of one viewed by the government as the devil incarnate?

At some point in the history of the relationship between the underworld and those who would chronicle their behavior, the moral nature of an ostensibly criminal act came to be weighed in its social context. Those bold enough could grab the headlines and achieve heroic status by committing immoral acts against an even *more* immoral society, where the hunger of some and the enormous wealth of others made morality seem less black and white and more a relative concept (Robin Hood, Rob Roy, Ned Kelly, Pancho Villa, William Wallace, and Jesse James, to name a few, without digressing completely into the realm of violent political or revolutionary activists). There is a traditional symbiosis between the public's romantic affinity for the outlaw, free and untamed, and the chronicler's natural need to pander to public affinities.

But the venerable relationship between criminal actor and reporter took a more troubling turn in this country, perhaps during the Depression era—at least if we're to believe Hollywood in the 1960s, which brought us sympathetic criminal characters willing to sacrifice all for stardom, such as those portrayed by Warren Beatty and Faye Dunaway in Arthur Penn's *Bonnie and Clyde,* and Martin Sheen in Terence Malick's *Badlands.* With these characters, the primary goal was not to avoid starvation or to strike a social blow but rather specifically to achieve media exposure. The morality of an act was less important than its potential for public notoriety. In today's world of complete media immersion, the role of morality has been diluted still further: Any act, no matter how immoral—in fact the more immoral the better—may be redeemed by public fascination. In *15 Minutes* the criminal characters are not at all sympathetic; rather, the focus is on the media's, and almost everyone else's, automatic acceptance (accompanied by the occasional hypocritical denial) of the notion that publicity trumps morality every time, that a fleeting moment of fame trumps *every*thing.

In deciding to appear in *15 Minutes*, I never expected to be facing questions relating to who I really was professionally, and perhaps if I hadn't made the mistake of playing a character dangerously similar to me with my own name, I *wouldn't* be facing such questions. But I did. I'd been enticed by fame, by public image rather than substance. But ultimately, although I undeniably find the sound of my own name seductive, I still believe I practice law largely for two reasons: to make a living, and to prevent injustice where I can.

Before *15 Minutes* had even been shown in the theaters, I found myself representing a member of the Bukharian community (Sephardic Jews who settled in Uzbekistan after Queen Isabella's expulsion of the Jews from Spain in the fifteenth century), rather than a Czechoslovakian as in the film, and my Bukharian client was innocent. Aaron Bangiew, a jeweler on Forty-seventh Street in Manhattan, who had immigrated to this country with his family, was charged with money-laundering and stolen property offenses and was facing incarceration and, worse yet for him and his family, deportation. His trial, which was heavily attended by Bukharians, including several Soviet World War II heros, ended in a hung jury on some charges and acquittal on others. I became very close with the Bukharian-American community during the case, and Shonna and I attended a wonderful victory party at Rasputin's in Brighton Beach, Brooklyn. Aaron and his rabbis who attended the party referred to me as "a messenger from God," and I believe they truly meant it. Aaron's son, conceived during his home detention, was named after me. I hope it's because of clients like Aaron that I continue to practice law.

25

AFTER the film I reprised my role of roving paladin, responding to pleas for assistance from upstate New York, Los Angeles, Philadelphia, and Ohio, the most interesting of which was the case of one Darron Lamont Bennett. Bennett had called my office from jail when I was on trial in Goshen and was insistent that he would have no one but me represent him, so Ed Panzer, a close friend and experienced federal practitioner who'd worked with me in many past cases, discussed the case with Bennett's Los Angeles counsel, Roger Rosen.

Bennett and his siblings were born and bred in the war zone of South Central Los Angeles. He'd survived a half-dozen near-death experiences on the streets where he grew up, to open successful real estate and jewelry businesses. Bennett counted among his close friends, virtually every black person in the Los Angeles *Who's Who,* from basketball star Magic Johnson to celebrity lawyer Johnnie Cochran. He'd recently been married to an exquisitely beautiful woman, Jill Johnson, in an elaborate function held at Beverly Hills' Four Seasons Hotel. One of his brothers, Ron Finley, is a premier West Coast clothier who counts many stars of the professional athletic and entertainment worlds among his clientele. Darron's older brother, Brian Bennett, was reputed to be one of the country's major drug traffickers and, according to the government, was responsible for the first marriage between the Colombian drug cartels and black dealers. While I was representing John Gotti in the late 1980s, Roger Rosen had represented Brian in a highly publicized fourteen-month trial in Los Angeles federal court. Brian was convicted and is presently serving a heavy sentence in federal prison for various narcotics offenses.

Meeting Bennett for the first time was a treat: He was articulate, handsome and bright, and very respectful toward me. As with most of those whom I have represented, I liked him on the spot. Bennett's case, which was to be tried in Manhattan because the government had alleged that certain cocaine shipments by him were destined for New York, was assigned to senior Judge Thomas P. Griesa, who presided in the majestic penthouse courtroom of the new federal courthouse at 500 Pearl Street. The lead prosecutor was the extremely effective Mark Mukasey (son of Southern District Judge Michael B. Mukasey). I think the government was surprised the day we arrived to pick a jury. The prosecutors, who appeared to honestly believe Bennett was a major narcotics kingpin like his brother, expected to the last that Bennett would plead guilty and roll over on his drug connections. We had a difficult time selecting a jury, as everyone on the panel seemed to know me and have a predisposition one way or another as to my previous clients, but I was pleased with the panel eventually chosen. The case didn't garner a lot of publicity, presumably because Bennett was a West Coast rather than a New York personality, so the courtroom was never packed, but Bennett's family was there every day.

The government theory was that Bennett had taken over his brother's narcotics business and shipped hundreds of kilograms of cocaine by air, truck, and rail to Chicago, Detroit, New York, Newark, and Washington, D.C. The evidence against him was the testimony of two reprobate defectors from the Colombian cartel and a black man in jail in Detroit who would testify that he'd once met with Darron Bennett to discuss drugs. And the trucking company used to ship two of the packages (one to Chicago and one to Tenth Avenue in Manhattan) belonged to a friend of Darron Bennett who coincidentally bore the same name as his brother, Brian Bennett. The hole in the government's case was the absence of any connection between Bennett and the cocaine shipments in question—he had had no contact with the cocaine or the documentation attendant to the shipping process.

The Colombian informants were recidivist narcotics traffickers anxious to avoid their own lengthy prison sentences. In cross-examination I made it clear that they'd lied to the government about much of what they'd done, that the government had later caught them in their lies, and that the government nevertheless kept their sweetheart deals in place in order to present their testimony against Bennett to the jury. It emerged that these particularly loathsome witnesses, using innumerable aliases, had done terrible things to people, such as tricking them into involvements in narcotics traffic. Most important, however, their narcotic nexus to Bennett was never established. The government relied on a ledger that had supposedly been kept by the chief informant and supplied through his attorney. Ed Panzer and I argued that the ledger was a fraud and had been created after the fact by the shrewd informant to bolster his own testimony, which he knew would be a hard sell to the prosecution (or a jury) standing alone.

Then the government stretched and presented a third witness from Detroit, who headed a crack-distribution business there. Cross-examination revealed that he, too, had received a sweetheart deal from the government relative to his prison sentence and that he, too, had lied to the government about his past; in fact, he had lied about his future as well, continuing to deal drugs in jail while awaiting sentence pursuant to his plea agreement. As with the Colombian witnesses, the government kept his cooperation agreement in place so that he would testify against Bennett. Three witnesses—three miscreants whom the government had to convince a jury had become honest, despite lies to the government before and after cooperation agreements by which they would each avoid substantial jail time. Three frauds.

But the greatest damage to the government's case came from within. A flippant young FBI case agent testified for the government and appeared evasive and haughty on the stand; she was one of those agents of boundless ambition who'd seen *The Silence of the Lambs* and wanted to be Agent Clarice Starling. Her attitude was: Your client is black and wealthy with a brother serving life for a narcotics violation, so he must be guilty. At one point I asked her, "Do you have any personal stake in the outcome of the case?" It's a throw-away kind of question, usually parried by an experienced agent with "No, I'd just like to see the right thing done." I had nothing to lose by asking it, and something told me she was arrogant enough to take the bait. Sure enough, there was a long and pregnant pause before she responded, which did as much to damage the government's cause in the jury's eyes as anything else in the case. Race was not an overt issue in the trial, but it was subliminal, and the agent's attitude brought it boiling to the surface. As Ed Panzer brought out in the presentation of the defense case, Bennett was a true, legitimately successful black entrepreneur who had a declared income of well over two million dollars a year from real estate and the jewelry business. Unlike a drug dealer he had an honest income to support his lifestyle. He could easily have been perceived by the several blacks on the jury as a victim of the bias and envy of the FBI.

I felt a concentrated energy in the Bennett case that I'd not fully experienced in some time. My power in the well—to persuade, to fight as a partisan—was back. In the aftermath of the disqualification, the contempt conviction, the house arrest, and the suspensions, I had been somewhat insecure and uncertain. Bit by bit I was reestablishing myself in my own mind. I fully recovered my old self in the Bennett case and delivered an especially effective summation that capped a beautifully fought case.

Despite the weaknesses in the government case and what I believed were strong performances by me in cross-examination and summation, I can't say I felt secure about the trial's outcome. After all, the defendant was a stranger (from Los Angeles), a black man charged with major narcotics distribution in New York, arrested while carrying a handgun, with a prior (albeit youthful) narcotics conviction for which he'd served time in prison, and a brother reputed to be one of the biggest drug dealers in the country's history. In short, I was still negotiating for a plea even while the jury was out deliberating, but Darron would have none of it.

It was the first time in a long while that I'd walked a client out of confinement, from the courthouse to freedom. I've done it only three times in my career: once early on with John Gurino, then in John Gotti's first case in the Eastern District before Judge Nickerson (he'd been out on bail during our successful defenses in the Piecyk and O'Connor cases), and finally with Bennett. The euphoria of counsel and client is inexpressible. The sense of the power of deliverance is overwhelming. I'd never seen Darron Bennett as a free man before the jury's verdict of acquittal was read. During the trial his comportment was so hopeful, his belief in me so complete, that our walk from the courtroom to Pearl Street was as great a joy as I've ever experienced.

In August 1999 a TNT television documentary called *Family Values*, an exposé of Hollywood's love affair with Cosa Nostra, was produced by Joseph and Sandra Consentino. The Consentinos interviewed me (and Jimmy LaRossa, at my suggestion) at length for the documentary, and I developed a good relationship with them. It was through the Consentinos that I agreed to host an upcoming A&E/History Channel production documenting the most famous trials of the twentieth century.

The preview of *Family Values* in Astoria, Queens, was attended by a strange mix of the real and the surreal: actual street fellows and their families as well as those who portrayed them in film; lawyers and those who played lawyers; people who'd spent time in jail and those whose time was spent behind the camera; and of course, the full cast of *The Sopranos*, the HBO television series. It was here that I met George Borgesi's mother, as well as George's brother Anthony. Over a year later, in the spring of 2001, a few months after I'd finished trying the Bennett case, I checked into the hotel to prepare to represent George Borgesi in the Merlino case in federal court in Philadelphia. While there I renewed my relationship with Bobby Simone, a Philadelphia criminal lawyer and friend from the heady days of the DiLaurentis case in Chicago, who spent four years in prison for "crossing the line" in his defense of the so-called Philadelphia family and who has just published a book about his experiences; he has been reinstated to practice law in the federal courts.

The government charged Borgesi and the lead defendant Joe Merlino in the usual sprawling RICO prosecution with various predicate acts of murder and conspiracy, and charged them with other less serious acts of criminality as well. The reputed boss of the Philadelphia Cosa Nostra, Ralph Natale, had become a government witness and was implicating all his so-called underlings in his confessed criminal activities. The government had considered recommending to the U.S. attorney general that the relatively new federal death penalty be applied. The statute in question, 18 U.S. Code, Section 1959(a)(1), was enacted in 1994 to permit the death penalty to be meted out to those convicted of murder in aid of racketeering. I recall sitting down with Michael Stiles, Philadelphia's U.S. attorney, to argue against applying the death penalty to the Merlino-Borgesi case. I got along well with Stiles (who was not a Republican and was replaced when George W. Bush was elected president and now is an executive with the Philadelphia Phillies baseball franchise). The Justice Department ultimately decided not to seek the death penalty. It felt odd to be sitting down before trial with a U.S. attorney to plead that the death penalty not be applied, as if the case had already been lost.

The fair and patient Judge Herbert J. Hutton presided over what would be a four-month trial in a cramped courtroom (without a lectern or court reporter), involving eight defense lawyers and four highly experienced and professional prosecutors. Ed Jacobs from Atlantic City, who represented Merlino, the lead defendant, ran the case and did an excellent job. But it was hard for me and the other lawyers to be role players; to do my best, I need to play the full nine innings. Despite my role-player status, the clients had great expectations of me, as a designated hitter from New York (the city to which, according to the government, Philadelphia looks for everything from culture to mob families), and I felt tremendous pressure to save the lives of these young clients. I did my best with the help of Louis M. Natali Jr., on the faculty of the Temple University School of Law, and attorneys Steven Patrizio, Jack McMann, and Emmet Fitzpatrick—we made a great team.

The government tapes entered in evidence against the defendants revealed a culture of young Philadelphians following the Hollywood road map for street life behavior. Our clients carried themselves with dignity, as men, at all times. But many of the lines on the tapes, often uttered by the government witness-informants, came straight from Scorsese and Coppola films; for example, the alleged boss was heard to tell someone on the phone to wait while he watched a part of *Godfather II*, and a certain bookmaker in the case was called Don Fanucci, after a greedy character from the same film who would always want to "dip his beak." It was amazing to me, for I'd heard serious tapes (as well as jocular ones, interpreted by the government to be serious) in other cases in New York, Chicago, and Detroit, words uttered by serious men, not parroting a movie. I stood agape, watching the final stages in the erosion of the credo, the character of the street life. Like so much else in the American culture, it seemed that even the so-called underworld had become homogenized.

Yet in the patterns formed by the detritus of a dead way of life, I discerned the theme of my defense. I cross-examined the government informants using Sicilian underworld slang terms, words like *babania* (narcotics), *petitis* (traitor), *infamità* (a terrible sin), and *disgraziata* (disgrace), to establish my expertise in Cosa Nostra for the jury and for the witnesses. The informants were taken aback (and the clients were impressed) that a lawyer who was not Italian knew so much about the language and mores. My goal was to show that the informants knew nothing about this Cosa Nostra. The main government informant, Natale, deluded himself and tried to delude our young clients into thinking he could find legitimate business employment for them. The government position was that the defendants and the Philadelphia family were an extension of, and responded to, the New York families. It was my argument, hammered home in a dramatic and effective summation, that the government had won—there was no more Mafia, if there'd ever been one, and anyone of underworld note was in jail, dead, or a turncoat. The government's case was not Cosa Nostra—it was Cosa Natale, the grand delusion of the informant, to which the government cynically subscribed.

The result, which the local newspapers attributed in substantial measure to our joint lawyering efforts, was successful: The defendants were acquitted of all the serious charges (RICO murder and conspiracy). The convictions in the case were for the lesser charges of stolen property, gambling, and extortion. We did a great job, but I couldn't help sensing something funereal about the case, as if I'd been sweeping up in the wake of a burial retinue.

EPILOGUE

On a normal weekday night at the Ravenite Club there'd be some forty fellows sitting at the various tables or milling around talking, playing cards or having coffee or espresso or a club soda. (I never saw alcohol being consumed there.) John Gotti would usually show up at about six forty-five or seven in the evening. When the door would open for him, everything stopped—no words were spoken, no cups or glasses tinkled, no cards slapped the table. When he entered, you could hear your own heartbeat and the fellow's next to you. Iggy Allogna, John's friend since John was fourteen years old, would lift the coat from John's shoulders and hang it on the coat rack to the right of the entrance, and everyone would rise. Then John would greet each person in the room individually, with a handshake or a hug or a kiss and a personal remark: "How are the kids?" "What happened with your wife's operation?" "Did you ever get your car fixed?" "Glad to hear it!" The process would take no less than fifteen or twenty minutes, and nobody rushed it or intruded on him. There were many small square, pizza-parlor-type tables throughout the room, and only one large round one in the back. His arrival formalities complete, John would stride to the round table—his table, where only certain people would sit with him. I would sit there at times when I was at the club. This was the ritual of John's entrance to his domain. I'd witnessed it on only one or two occasions, because if I went there in the evening at all, it was usually after John had already arrived. It was a startling sight.

I remember a Christmas event at the Ravenite one year in the late 1980s. Enormous crowds of very well-dressed well-wishers of all stripes arrived from all over the country in a demonstration of complete fealty and appreciation, to partake in John's greeting ceremony. It took more than two hours. The government would say it was just a "mob hierarchical event," John's insistence on which would greatly aid the FBI in gathering information and would lead to the ruination of John and his friends. For John, the formality, the observance of ritual, the ceremony were more important than what the government might or might not do. And he wouldn't alter one syllable of those rites in deference to the authorities.

Ordinary Friday evenings were my favorite times to visit the Ravenite back then. Most of the fellows wouldn't be there on that day; it was John's quiet time. Perhaps his brother Peter would be there, and maybe his friend Bobby Boriello, who was a giant of a man, physically reminiscent of Luca Brasi of *The Godfather*. John would order a predinner "snack." Often I'd decline the sandwiches, on the theory that it would spoil my appetite for dinner later, but sometimes John would cajole me into partaking. "C'mon, counselor! You need to keep up your strength. You're lookin' a little peaked there." One time I stayed for the raft of sandwiches John ordered in from the Broome Street Café, a few blocks away— cheeseburgers, tuna, various cold cuts on pita bread. They were great, worth spoiling your appetite. And John would hold forth on this subject or that. I remember one evening when we'd heard that a certain ballplayer had received a contract extension for several million a year.

"That's good! He had *that* comin' to him," John said. He liked the phrase—for an athlete who'd been acknowledged at contract time, a lawyer who'd garnered a well-earned bonus, a friend who'd been acknowledged in the curbside life . . . and sometimes for those who'd received more unpleasant but (in John's estimation) just deserts. John had his own sense of the cosmic order, of what was fair and what wasn't. "Remember," he would say, "to the other guy, *you're* the other guy." It was a simple statement of complex advice. On those Friday evenings John would expound his rules for an orderly life, or we'd talk about some upcoming legal motion, or maybe we'd just talk about sports. He had a keen sense of humor, could tell a good tale and had lots to tell. He enjoyed talking and I enjoyed listening.

John's not down at the Ravenite on Friday evenings anymore, or any other time. Neither am I. Until his recent passing I had to go a long way to talk to him.

But John Gotti was still in charge, as when I'd met him in Judge Nickerson's courtroom. He'd been in federal custody then, some sixteen years ago, even as he had been at the end, although he was only awaiting bail and trial at that time and more recently was serving a life sentence in isolation. But he was still in command of himself and his immediate environment. Maybe this is the only kind of command that matters anyway. Maybe the illusion that one may control vast armies, corporate structures, even one's own family unit is just that—a grand illusion. Maybe the best we can strive for is to command ourselves, and if we succeed, we will have achieved a great deal and furthered the best interests of the world around us. John remained in charge of John, against all odds. Defiantly, to the very end.

On June 24, 1992, John arrived at the federal prison in Marion, Illinois. Marion was built in 1963, and in 1983 it was converted into a maximum-security lockdown facility—what was then (and at the time of John's internment) the only federal prison with a Level 6 designation, a *super* maximum-security facility. (The legendary federal maximum-security prisons, such as Leavenworth, Lewisburg, and Lompoc, are only Level 5.) The portion of Marion that is designated Level 6, where John was incarcerated, is not to be confused with "Camp" Marion, a minimum-security facility nearby, where former Cincinnati Reds baseball star Pete Rose served his six-month sentence for tax evasion. At that time the Marion Level 6 facility had an inmate population of about 240 and roughly from three to five guards per inmate. Inmates are in their individual cells for roughly twenty-two hours and fifty minutes a day. They are permitted to spend an hour and twenty minutes per *week* outdoors in the fresh air.

John's regimen in his cell was, like him, unchanging day in and day out. He'd organize his meals around what he could salvage of the prison fare that was edible, supplemented by the fruits, nuts, or raisins he could obtain from the commissary; he particularly enjoyed his instant oatmeal breakfasts. He'd exercise vigorously—push-ups, sit-ups, and step-ups from the floor onto the concrete slab that underlay the thin mattress he'd roll up and push to one end of the slab. He'd lean against the rolled-up mattress to read, and he read voraciously—fiction, nonfiction, anything of interest—until lights out.

Marion was created for the federal inmate who was alleged to be uncontrollable in the ordinary penitentiary milieu. It came to be used as well as a way station for the high-profile "gangster." Allegedly important Mafia figures (such as Carmine Persico, Nicky Scarfo, and Gaetano Badalamente) would be interned there for thirty-six months, presumably to be softened up with a jolt of extreme deprivation—isolation and insulation—and then transferred elsewhere. John was there for almost *ten years*. Isolation and insulation. And retribution. The government believed that John was a leader of such personal power and magnetism that he would hold sway within the walls of any ordinary prison, and perhaps without those walls as well, if he were not isolated and insulated from the prison population and the outside world. Maybe the government also believed that his leadership role in the largest criminal organization in the country warranted one kind of interment in Marion for as long as he breathed and another only thereafter. The government certainly believed that John owed it a debt for the chase, for the affront to its power, for the public ridicule.

At first I'd visit John about once a month at Marion. Later I'd visit about once every two months, then about four times a year, except during the two years of my intermittent suspensions (between 1995 and 1997), during which I was prohibited from visiting John. The visiting area there was no more than a booth in a lockdown unit. I wasn't even allowed to bring in a soda; John would get perhaps a container of milk and a piece of fruit during a visit. I wasn't permitted to eat anything, and he'd usually leave the fruit, maybe drink the milk. I understood the family visits to have sometimes been difficult in that they were closely monitored, and John had to see his family members through a glass partition across a heavy desk and speak to them via phones located on either side of the partition. To my anger—and I'm sure John's—tapes of his family visits were mysteriously obtained by the media, then televised.

After about three years at Marion John developed a severe infection in his gums, which was belatedly treated in prison with surgery (in the course of which he received some 154 stitches). A few years later he developed a severely painful lump in his neck. He was diagnosed with head and neck cancer, which is more often than not terminal. John was given a twenty percent chance of surviving surgery by the attending physician. His response to the medical staff was pure John, expressed with equanimity, "Better me than some five-year-old child. How 'bout a nice plate of lasagna before the operation?" He didn't get the lasagna, but he did get the operation.

So he was moved several hundred miles from Marion to the Medical Center Federal Prison at Springfield, in southwestern Missouri, for surgery and then thirty-six radiation treatments. And one day later he was transferred right back to the isolation and insulation of Marion. When I saw John shortly after surgical procedures of the most invasive sort imaginable and radiation, he'd been able to reestablish his persona and his physique, to relearn how to make himself understood verbally and how to swallow. Then, as predicted, the cancer returned. And he was transferred to Springfield once again for extensive chemotherapy.

Despite the horror of John's illness, he was quick to point out that his transfer to Springfield had some benefits for those visiting him. The visiting rooms were larger and had some amenities (like vending machines), and the attorney conference room was a spacious area adjacent to the family visiting room. I recall being pleasantly startled to feel John slap my knee on one visit—I had resigned myself to never getting that close to him again in this life. And best of all for John, he could share a meal with someone, something he hadn't done in eight years. This treat was short-lived, of course, as after his operation in January 2000 he was no longer able to eat solid food.

On another visit John told me, somewhat jokingly, "You know, Bruce, a lot of fellows in prison, they dream about girls—you know, the smells, the touch of soft skin. Me, I dream of a nice big bowl of rigatoni." He was kidding a bit, but I think at bottom he really meant it. He'd always loved food, whether from New York's finest restaurants or from a humble hot dog vendor, as long as it was the best of its kind. He was never gluttonous but had a genuine love for the process, and he would always eat slowly, deliberately, as if trying to retain the memory of each bite, as if it might be his last. He'd drink throughout his meals, and yet I never recall seeing any noticeable effect of the alcohol he consumed. He would ask me to describe meals I'd recently eaten; at first I would balk, fearing it would be torture for him, but then I realized he derived a vicarious pleasure from my descriptions. He recently told me, "Bruce, I'm just trying to get healthy again, so I can have one last meal, and then I'll go." When I last saw him, John was down to about 155 pounds from the 200 pounds he'd carried when he was in the limelight. He knew what I knew—that solid food was impossible for him to eat because of the surgery and the radiation treatments. Yet he wouldn't give in.

After his operation, nearly three years before he died, the tabloids monthly predicted (presumably based on the usual government leaks) his imminent death. But John continued, asking no quarter and receiving none. When I would come to visit, he would rise from the wheelchair (a required means of transport under Bureau of Prisons regulations for the medical unit), lighter of course, given his inability to eat as he once did, but it was definitely John, his dignity intact. When anything of importance arose requiring that his family be contacted, he insisted on doing it himself, as a matter of pride, as an outright refusal to capitulate to the forces that assailed him whether from without or within. Anyone else would have succumbed long before, but John emerged without a scintilla of self-pity. He took pleasure where he found it. His will to live, to overcome, was astounding.

John Gotti became a local celebrity in Marion and in Springfield as well. The townpeople loved him and wrote to him all the time. When I was in town, John would ask me to visit this or that restaurant from which he'd heard, and I was astounded by the reception I'd receive, especially in Springfield (where everyone sounds like Mickey Mantle, not far from Joplin where Mantle played minor league ball). I was amazed at the thousands of letters John received, from well-wishers, people down on their luck seeking advice, people who just wanted to reach out and touch him. And he sincerely tried to respond to those that warranted a response— honestly, in the spirit in which the letters were sent, in his own candid manner, remarkably open to the relatively banal concerns of those who wrote. He received a letter from my close friend and confidant Victor Juliano, an ex-detective who'd spent twenty-two years in the New York City Police Department and who'd assisted me in defending John from the beginning. Vic requested an autographed photo of John for his seventieth birthday. As infirm as he was, John sent the photo to Vic, whom he hadn't seen or spoken to in eleven years, having inscribed it, "To my good friend Victor, a man's man and a true friend, rare in this day and age, your friend always, John Gotti." John felt, correctly, that the photo was important to Vic, so he took the time to write a personal note and send it.

When John would see me on a visit, he would still remember to ask about this or that client, recalling clearly a comment I might have made on a prior visit months before, or he'd ask whether I got hurt financially in this or that case. He continued to instill in me and those around me—including his doctors and nurses—a sense that we were in it together, partisans all, not criminally but as men. Those who came in contact with him in his prison environment enjoyed his company; some of this was no doubt a function of his celebrity, but it was also the force of his personality.

At the end I couldn't help but contrast him with Gravano. After receiving a slap on the wrist from the FBI and praise from Judge Glasser and prosecutor Gleeson (who received a lifetime appointment to the federal bench soon after John's sentencing), Gravano wound up involving his own family in major narcotics trafficking in Arizona (where the government had placed him in the Witness Protection Program). FBI agents and their families socialized with this mercenary Tasmanian devil, yet had the temerity to criticize me and others for becoming friendly with John. On the one hand there is Gravano, a liar, a coward without any moral code or fidelity to anyone including his own wife and children; and on the other hand there was John, who in the words of Albert Kreiger didn't have a hypocritical or insincere bone in his body. The government chose to let Gravano go free (until his recent criminal endeavors in Arizona) and to kill Crazy Horse, burying John alive in Marion, poisoning him and then letting him fester until dead.

On my last visit to John in Springfield in January 2002, I arrived at nine A.M. John had already been awake since six. When he'd been free, he'd seldom open his eyes before eleven, because he'd seldom go home to bed before five A.M. I entered the attorney conference room, and John rose with some effort and strode toward me, to shake my hand. He still looked the same in a thinner frame, his walk still unmistakable. The *New York Daily News*'s Michael Daly once told me he'd never seen a walk quite like John's before—not cocky, but forceful. It was still there, undiminished despite all the diminishment. We sat and we talked for some seven hours over a two-day period.

Honestly, it's hard to recall specifically what we spoke of. We reminisced about cases and investigations, old and new. We talked of various clients and lawyers, past and present, and of people we knew, some dead, some in jail. We discussed college football and professional baseball. We spoke of Theresa Brewer, a torch singer of the 1950s whose tunes John recalled fondly, and of "Red" Levine, an old-time Jewish gangster who grew up in John's old neighborhood and whom John admired for his stand-up reputation. We did *not* discuss substantive legal strategy as applied to his internment, for he wouldn't permit any petitions to get him out of prison or even out of solitary confinement in Springfield, and although medical experts had advised him that there was a marked correlation between untreated dental infection and incidence of head and neck cancer, he refused to entertain any thoughts of a malpractice suit or filing a grievance with prison authorities.

It was all so terribly hard for me, reminding me of watching my parents die, and I felt ashamed of how hard I was taking it, for I could see how much harder it was for him, how much stronger he was than I. I didn't have the mettle to watch him struggle, to watch the doctors and nurses and guards hovering. I didn't have the courage to watch him die, to watch a piece of myself die.

Those trips to remote Springfield and Marion were long and punishing for me. John surely recognized this, although I would never have mentioned it to him, for if I had, he'd have refused all visits. Perhaps the John Gotti of a decade ago wouldn't have permitted the visits at all. But I'd heard from others who'd visited him that our meetings were particularly enjoyable to him because they were motivated neither by obligation nor by gain; I wasn't a family member, and I wasn't a lawyer looking for help for another client who'd once been an enemy or a friend. I visited John, despite the pain of seeing him in his discomfort, years after I'd ceased fighting cases with him, because I cared about him. And if the pleasure John derived from my visits may be perceived as a weakness, then it's of the sort that makes us human, and therefore binds John, and you and me, even more tightly: John the convicted outlaw and badman, you the "good" people, and me patrolling the gray purgatory in between.

I saw John, who turned sixty-one on October 27, 2001, sitting across the table from me, and suddenly I was awestruck by the realization that he was still John Gotti, the man I'd met some sixteen years before, when he was forty-four and I thirty-six. So much of who he was, the manner in which he influenced those around him in his heyday, was perceived to be and was a function of his appearance and physical presence: his mien, his posture, his gestures, the force of his voice, his million-dollar smile. Some of that was gone as I sat before him, destroyed as he was by the disease and its treatment. Yet it *was* the same face, untouched despite the surgery and treatment. His unaffected eyes, nose, and ears, his full head of silver hair—John Gotti hadn't changed one iota.

As long as he was around, I'd continue to sign my letters to him, "*Forza* [strength], Love Bruce." And if there are those in my profession or elsewhere who find that offensive, then fuck them.

John Gotti passed away on June 10, 2002.

As John became a lightning rod for every prosecutor seeking glory, reputation, or promotion in the world of law, I, too, became a lightning rod for controversy. Certain potential clients steered clear of me for fear that my presence would increase the prosecution's wrath against them. It was as though John's ship, in sinking, would create a whirlpool, sucking in all around. Would this be my fate? The government certainly tried its best, but no.

Representing John brought me notoriety at a relatively young age for the law, a profession whose practitioners more often than not seek the anonymity that permits them to quietly maximize their financial gain. I've never been good at maximizing my potential as a salable commodity, and maybe I never will be. Money has never been my idol. Rather, my motto has always been "Look for a fight, and if it's a good fight, get in it." This is what keeps me excited about the law, what keeps me going. I've taken risks for my clients, by fighting my way in court, and I've paid the price. Although I've done things in my personal life that I would change if I could, professionally I wouldn't change a thing. Looking back on my past twenty-five years in the law, I'm proud of what I see. But at fifty-three I'm too young to spend the rest of my life retrospectively, nostalgically.

So I suppose this is technically a *half-*autobiography, part one of an autobiography on the installment plan, for although I cannot say what the future holds, I do know that you will hear more of me, from me. My primary focus will, of course, always remain the fight. I'm still excited by the life-and-death stakes of the criminal matter, and I find that my experiences at the hands of the government while representing John have heightened my awareness of the importance of our constitutional guarantees inside and outside the purview of the criminal case. Whether in the criminal or civil context, I envision myself as a constitutional shield for those under the hammer of government. In the words of the Reverend Dr. Martin Luther King Jr., "Injustice found anywhere is a threat to justice everywhere," even if that injustice is being perpetrated today on someone whose name happens to be John Gotti.

As Mike Coiro said to John the first time I represented him in court, "This is Bruce Cutler. . . . He'll stand up for you today." That's what I do. That's what I'll continue to do.

INDEX

Lionel René Saporta, a graduate cum laude of Yale College and Boston University School of Law, is presently a writer of fiction and nonfiction living in Brooklyn and East Hampton, New York, with his wife, daughter, and stepson. Also an attorney, he grew up in Brooklyn, attended Poly Prep with Bruce Cutler, spent three years with Cutler in the Kings County District Attorney's Office, and later shared offices with him in private practice for three years.